Alternatives in Refrigeration and Air Conditioning

Alternatives in Refrigeration and Air Conditioning

S.C. Kaushik
Professor
Centre for Energy Studies
Indian Institute of Technology, New Delhi

A. Arora
Assistant Professor
Department of Mechanical Engineering
Delhi Technological University, Delhi

P.S. Bilga
Professor
Department of Mechanical Engineering
Guru Nanak Dev Engineering College
Ludhiana

I.K. International Publishing House Pvt. Ltd.
NEW DELHI

Published by
I.K. International Publishing House Pvt. Ltd.
S-25, Green Park Extension
Uphaar Cinema Market
New Delhi–110 016 (India)
E-mail: info@ikinternational.com
Website: www.ikbooks.com

ISBN 978-93-84588-38-0

© 2016 I.K. International Publishing House Pvt. Ltd.

All rights reserved. No part of this publication may be reproduced, stored in a retrieval system, or transmitted in any form or any means: electronic, mechanical, photocopying, recording, or otherwise, without the prior written permission from the publisher.

Published by Krishan Makhijani for I.K. International Publishing House Pvt. Ltd., S-25, Green Park Extension, Uphaar Cinema Market, New Delhi–110 016 and Printed by Rekha Printers Pvt. Ltd., Okhla Industrial Area, Phase II, New Delhi–110 020.

"In sweet memory of late Dr. Norman R. Sheridan
Queensland University, Brisbane, (Australia) for introducing the subject
to the authors and inspiring for writing this book.

and

Dedicated to
All young Researchers, Teachers/Students
and Practising Engineers in the field of
Refrigeration and Air Conditioning

Foreword

Air conditioning and refrigeration are integral parts of modern life. At home and in the industry, it is difficult to imagine life without these technologies. Whether it is personal comfort or the need to store food at home, or large-scale refrigeration or climate control in food industry and factories, cooling technologies have become central to modern living. Unfortunately, these technologies come with a very high energy cost, besides having a detrimental and degrading influence on the environment. There are strong reasons, therefore, to examine alternative options in refrigeration and air conditioning, and this has been a major area of research activity for the last few decades. Naturally, some of the advances provide very useful and energy efficient alternatives, which are also simultaneously environment friendly.

This book makes an effort to bring these advances to the reader in a simple and lucid manner. Starting from basic principles, the book presents a comprehensive and up-to-date treatment of the subject, with a clear focus on alternative refrigerants, alternative cooling cycle analyses, and alternative technologies. The exposition of the exergy concepts for energy efficient refrigeration systems is another very important feature of the book.

The authors have a long experience of teaching and research in these fields, and the result is a book, which captures their deep understanding of the subject. I have no doubt that the readers – students, teachers and professionals — will find the exposition of the subject to be accessible and useful. I congratulate Professor Kaushik and his associates for a commendable piece of work, which will go a long way in meeting a long-felt need.

Dr. Surendra Prasad
Former Director, IIT Delhi
Professor of Electrical Engineering
Indian Institute of Technology Delhi

Preface

The refrigeration and air conditioning (RAC) technology has been passing through a changing phase. The problems of ozone depletion and global warming have created a void in the refrigeration and air conditioning industry, which have led the refrigeration community to search for alternative refrigerants and alternative refrigeration technologies. The ozone depletion caused by CFCs and HCFCs demands alternative refrigerants, which are non-ozone-depleting. The issue of global warming has come into picture at a later date but its importance can't be discarded. The alternative refrigerants are present-day's necessity for vapour compression technology which has gone into turmoil due to the above-mentioned problems. However, alternative refrigerants are not the final solution and hence alternative refrigeration technologies are also been looked upon as solution to problems of ozone depletion and global warming. This has led to enhancement in research in the unconventional refrigeration technologies, viz., absorption and adsorption refrigeration, desiccant cooling, thermoelectric/thermoacoustic refrigeration, gas cycle refrigeration, ejector cooling, metal hydride refrigeration, vortex tube refrigeration systems, etc. The frequent and recent advances in this field have prompted and inspired the authors to share the know-how of upcoming refrigeration technologies and alternative refrigerants with the RAC community by writing a book on alternatives in refrigeration and air conditioning which can cater to the present-day need of postgraduate engineering students, teachers, researchers and practising engineers. The book contains the following ten chapters:

Chapter 1 : introduces concepts and scope required for further understanding of the text.

Chapter 2 : focuses on alternative refrigerants in vapour compression systems and alternate cycles and their thermodynamic analyses.

Chapter 3 : highlights conventional and advanced vapour absorption systems. It stresses on single and multi-effect vapour absorption cooling systems.

Chapter 4 : provides analysis of the combined compression and absorption systems.

Chapter 5 : focuses on ejector cooling systems and integrated options.

Chapter 6 : emphasizes on vortex tube refrigeration system and its integrated options.

Chapter 7 : deals with thermoelectric refrigeration. Due to use of newer materials, the energy efficiency and design of such systems have improved and hence it is desirable to discuss the same in the book.

Chapter 8 : describes a novel refrigeration option based on thermoacoustic phenomenon using sound waves for refrigeration which is an upcoming research based technology.

Chapter 9 : deals with vapour adsorption systems covering both physisorption and chemisorption systems which also include metal hydride and chemical heat pump systems.

Chapter 10 : discusses conventional evaporative cooling suitable for hot and dry climates and then desiccant based cooling systems suitable for hot and humid climates.

The novel approach of system analysis using exergy concepts has also been incorporated, wherever possible, keeping in view the energy conservation and energy efficiency considerations.

S.C. Kaushik
A. Arora
P.S. Bilga

Acknowledgements

We gratefully acknowledge the cooperation and help of the followings:
- Director/Deputy Director of Indian Institute of Technology Delhi and Head, Centre for Energy Studies for providing necessary facilities and financial assistance from the institute.
- Vice Chancellor of Delhi Technological University, Delhi and Principal, Guru Nanak Dev Engineering College, Ludhiana for everlasting inspiration and necessary help.
- Professor M.S. Sodha, former Vice Chancellor of DAVV University, Indore; Lucknow University, Lucknow and Bhopal University, Bhopal for his guidance and everlasting inspiration to the authors in research and daily life.
- Dr. Surendra Prasad, former Director, IIT Delhi and Emeritus Professor, Electrical Engineering Department, IIT Delhi for writing the Foreword of this book.
- Dr. R.K. Pachauri, Director General, the Energy and Resource Institute, Delhi for his guidance and support from time to time.
- Late Dr. Norman R. Sheridan, Queensland University, Brisbane, Australia for introducing the authors and giving valuable guidance in the field of refrigeration and air conditioning.
- Professor P.K. Kaw, former Director, Institute for Plasma Research, Gandhinagar, Gujarat, for his support and inspiration.
- Energy experts of scientific community and organizations whose works have been extensively consulted, referred and used in this book.
- Special thanks to Professors S.P. Sukhatme, C.P. Arora, R.S. Agarwal, P.L. Dhar and Sanjeev Jain for various stimulating discussions from time to time.
- Several colleagues specially Professors S.C. Mullick, T.C. Kandpal, V. Dutta and L.M. Das at CES, IIT Delhi for their cooperation and friendly advice during the preparation of this book.
- Faculty and research associates: Drs. K.A. Subramanian, Vamsi Krishna, Dibakar Rakshit, A. Mahesh, V. Baiju and Research students specially Manoj Dixit, S. Manikandan, Rahul and Ravita for their association and help in the preparation of the book.
- Last but not the least, we wish to place on record our appreciation for their family members (wife and children) for their endless patience and giving enough time for completing this book.
- And above all, the almighty God for giving power and strength to write this book with intellect and wisdom.

S.C. Kaushik
A. Arora
P.S. Bilga

Contents

Foreword — vii
Preface — ix
Acknowledgements — xi
About the Authors — xvii

1. Introduction — 1
  1.1 Historical Developments — 1
  1.2 Energy Scenario and Cooling Demand — 2
  1.3 Environmental Considerations and Alternative Cooling Technologies — 5
  1.4 Thermodynamic Basis of Energy and Exergy Analysis — 7
  1.5 Conclusion — 12

2. Alternative Refrigerants and Cycles for Compression Refrigeration Systems — 14
  2.1 Introduction — 14
  2.2 Alternative Refrigerants — 16
  2.3 Alternate Vapour Compression Refrigeration Cycles — 23
  2.4 Thermodynamic Analysis of Vapour Compression Refrigeration Cycle with Alternative Refrigerants — 31
  2.5 Conclusion — 85

3. Vapour Absorption Cooling and Advanced Absorption Cycles — 89
  3.1 Introduction — 89
  3.2 Single Effect and Series Flow Double Effect Generation Water-Lithium Bromide Vapour Absorption Refrigeration (VAR) Systems — 92
  3.3 Parallel Flow Double Effect Generation 'VAR' System — 113
  3.4 Triple Effect Generation Water-Lithium Bromide Vapour Absorption Cooling Systems — 121
  3.5 Half Effect Generation Water-Lithium Bromide Vapour Absorption Cooling System — 134
  3.6 Resorption System — 146
  3.7 Dual Loop Flow Type Double Effect Vapour Absorption Refrigeration System — 146
  3.8 Conclusion — 148

4. Compression-Absorption Combined Cooling Systems — 153
  4.1 Introduction — 153
  4.2 Absorption Recompression Refrigeration (ARR) System — 155
  4.3 Compression Absorption Refrigeration (CAR) System — 161

4.4 Compression-Absorption Cascade Refrigeration (CACR) System	170
4.5 Conclusion	182

5. Use of Ejector in Refrigeration and Air Conditioning — **185**

5.1 Introduction	185
5.2 Historical Background of Ejector and Its Applications	186
5.3 Ejector	196
5.4 Ejector Refrigeration System	201
5.5 Performance Improvement of Ejector Refrigeration System	202
5.6 Ejector Integrated Vapour Compression Refrigeration System	206
5.7 Transcritical CO_2 Compression Refrigeration System with Ejector-Expansion Device (TCCRSEJT)	214
5.8 Ejector Integrated Absorption Refrigeration Systems	216
5.9 Conclusion	218

6. Vortex Tube Refrigeration Systems — **225**

6.1 Introduction	225
6.2 Historical Background	226
6.3 Theory of Ranque-Hilsch or Vortex Effect	226
6.4 Various Phenomena in a Vortex Tube	232
6.5 Modeling of Temperature Separation in a Vortex Tube	236
6.6 Exergy Analysis of the Vortex Tube	252
6.7 Vortex Tube Integrated Refrigeration and Air Conditioning Systems	259
6.8 Conclusion	290

7. Thermoelectric Refrigeration — **296**

7.1 Introduction	296
7.2 Thermoelectric and Thermomagnetic Effects	297
7.3 Analysis of Thermoelectric Refrigerator	300
7.4 Multistage Thermoelectric Refrigeration Systems	304
7.5 Actual Thermoelectric Refrigeration System	310
7.6 Design Aspects of Solar Thermoelectric Refrigerator	315
7.7 Choice of the Thermoelectric Materials	318
7.8 Improvement in Thermoelectric Figure of Merit	321
7.9 Conclusion	324

8. Thermoacoustic Refrigeration — **327**

8.1 Introduction	327
8.2 Working Principle of Thermoacoustic Cycle	328
8.3 Types of Thermoacoustic Refrigerators	331
8.4 Components of Thermoacoustic Refrigerator	332

8.5 Thermodynamics of Thermoacoustic Refrigeration	334
8.6 Design Parameters of Thermoacoustic Refrigeration	335
8.7 Applications and Merits of Thermoacoustic Refrigeration	337
8.8 Conclusion	338
9. Vapour Adsorption Systems: Physisorption and Chemisorption Systems	**340**
9.1 Introduction	340
9.2 Description of Physisorption System: Vapour Adsorption Cooling Cycle	341
9.3 Developments in Solar Adsorption Cooling	345
9.4 Chemisorption Systems	352
9.5 Metal Hydride Systems	358
9.6 Conclusion	360
10. Evaporative and Desiccant Based Air Conditioning Systems	**365**
10.1 Introduction	365
10.2 Air Psychrometrics	366
10.3 Evaporative Cooling	368
10.4 Desiccant Materials	371
10.5 Developments in Solar Desiccant Cooling Systems	373
10.6 Conclusion	380
Index	**383**

About the Authors

S.C. Kaushik is Professor and former Head, Centre for Energy Studies, Indian Institute of Technology (IIT), Delhi. He obtained his Ph.D. in Plasma Science from IIT Delhi after his distinguished First Position in Master's degree in Science (Electronics) from Meerut University. His research fields of activities include Thermal Science and Engineering; Energy Conservation & Heat Recovery, Solar Refrigeration & Air Conditioning, Solar Architecture, and Thermal Storage & Power Generation. He has made significant contributions in these fields as evident by his above 400 research publications in journals of repute at national and international levels. He is a pioneer researcher on Exergy Analysis and Finite Time Thermodynamics of Energy Systems at national and international level and a leading expert on alternative refrigeration and air conditioning technologies. He has co-authored/edited several books on energy related topics. He has visited several countries as visiting scientist/fellow and invited speaker in the USA, Canada, France, Germany, Australia, the Netherlands, Mexico, Saudi Arabia and Ethopia. Prof. S.C. Kaushik has been coveted Top Academic Performer on all India level in the subject area of Energy for the last five years as declared by International Comparative Performance of India's Research Base in a recent DST Report.

Akhilesh Arora is Assistant Professor, Department of Mechanical Engineering, Delhi Technological University (erstwhile Delhi College of Engineering) Delhi. He has also served at College of Military Engineering, Pune in the capacity of Lecturer for two years. He was Associate Professor and Head of Mechanical and Automation Engineering Department at Indira Gandhi Delhi Technical University of Women, Delhi from 2012 to 2014. He obtained his Ph.D. and Master's Degree from the Indian Institute of Technology, Delhi in the years 2010 and 1997 respectively. His research areas are Refrigeration and Air Conditioning, Thermal Science and Engineering; Energy Conservation and Waste Heat Recovery. He has published number of research papers in international and national journals. Dr. Arora is an active member of ASHRAE and ISHRAE.

Paramjit Singh Bilga is Professor, Department of Mechanical Engineering, Guru Nanak Dev Engineering College, Ludhiana. He completed his B.E. (Mechanical) from Panjab University, Chandigarh in 1989, M.Tech (Energy Studies) and Ph.D. from Centre for Energy Studies, Indian Institute of Technology, Delhi in 1994 and 2009 respectively. He has twenty-five years of teaching and research experience in the field of design, analysis, modeling, simulation and optimization of thermal energy, environment and safety engineering systems. He has done the project and research work in the area of refrigeration and air conditioning.

CHAPTER 1

Introduction

1.1 HISTORICAL DEVELOPMENTS

The role of refrigeration and air conditioning (RAC) is vital for the overall development of the whole world. One even cannot think of progress in various fields without the use of RAC. It has applications almost in every sector of life. The RAC technology, which the world is enjoying today, is the outcome of the efforts of scientists, technologists, engineers and technicians, etc., of this field, and developments occurred in the other fields such as manufacturing technology, material science, automation and computer sciences, etc.

Refrigeration goes back to ancient times using first stored ice, vaporization of water, and other evaporative processes for cooling applications. Numerous investigators in different countries studied phase change physics in the 1600s and 1700s; their fundamental findings set the foundation for "artificial" (man-made) refrigeration. Oliver Evans first proposed the use of a volatile fluid in a closed cycle to freeze water into ice (Evans, 1805). He described a system that produced refrigeration by evaporating ether under vacuum, and then pumped the vapour to a water-cooled heat exchanger to condense for reuse. While there is no record that he built a working machine, his ideas probably influenced both Jacob Perkins and Richard Trevithick. The latter proposed an air-cycle system for refrigeration in 1828, but again did not build one. Perkins, however, did so with his invention of the vapour compression machine in the 1830s, and thus introduced actual refrigerants as we know them. His 1834 patent describes a cycle using a "volatile fluid for the purpose of producing the cooling and freezing and yet at the same time condensing such volatile fluids, and bringing them into operation without waste" (Perkins, 1834). Many refrigeration experts recognize his landmark contribution with identification of this mechanical vapour compression approach as the Perkins Cycle.

The most efficient technology used today throughout the world is vapour compression refrigeration (VCR). This consumes mainly high-grade energy. Other popular technologies, which are commercially used and running on low-grade energy are, Vapour Absorption Refrigeration (VAR) and Ejector Refrigeration (ER). The performance of systems working on these technologies, VAR and ER, is quite poor. The VAR and ER systems generally use the waste steam/heat available in the plants to produce cooling effect.

The promising futuristic and ecofriendly technologies which are still under development stages are Compression Absorption Refrigeration (CAR), Thermoelectric Refrigeration, and Ejector and Vortex Tube Integrated Refrigeration Systems. The CAR has been developed by combining features of the VCR and VAR. The literature shows that CAR systems offer better energy efficiency in comparison to VCR and VAR systems. The ejector or vortex tube integrated compression refrigeration systems exhibit enhanced performance. The thermoelectric refrigeration is going to become the most desirable option as it does not use any kind of refrigerant fluid. There are other RAC options also such as thermoaccoustic refrigeration, vapour adsorption cooling, metal hydride cooling and desiccant cooling, etc. Some of these emerging technologies are discussed in brief in the proceeding chapters of the present book.

1.2 ENERGY SCENARIO AND COOLING DEMAND

The industrial and economic growth of the most countries during the past few decades has resulted in significant growth of energy demand. Figure 1.1 shows the global energy mix and the contributions of energy carriers presently and future predictions in the exemplary path. The present scenario of energy mix is mostly based on oil, coal and gas. The futuristic projections tilt toward usage of solar energy. The estimate for the year 2100 highlights the great importance of solar energy in this scenario by using Photovoltaics (PV) and Concentrated Solar Power (CSP).

Not only, the way by which energy demand is met by a certain energy carrier mix is crucial to the exemplary path, but also the greater energy productivity enhancement. Such an increase can be

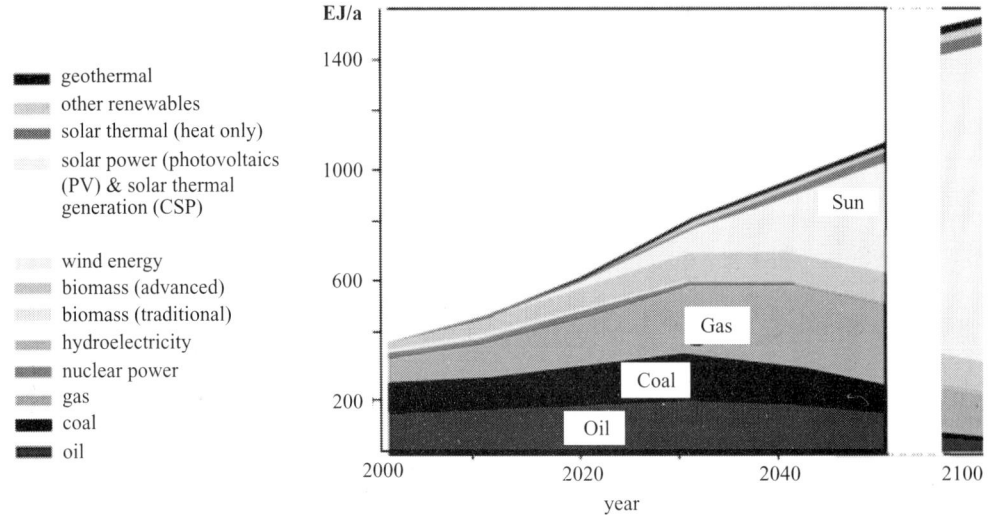

Fig. 1.1 Global Energy Mix: The Exemplary Path
(*Source:* WBSU – World in Transition – Towards Sustainable Energy Systems, German Advisory Council on Global Change, Berlin, 2003)

achieved in many ways, for instance through price-induced reduction of energy demand, which leads to efficiency improvements in both energy conversion and final energy use, as well as through sectoral structural change and altered settlement and transportation structures, or changed consumer behaviour. In brief, this can prevent so much energy use that energy productivity enhancement becomes one of the main pillars of the exemplary path. Figure 1.2 represents the energy enhancement in the exemplary path by assuming 1.6% annual increase in energy productivity from 2040 onwards as compared to historic figure of 1% annually.

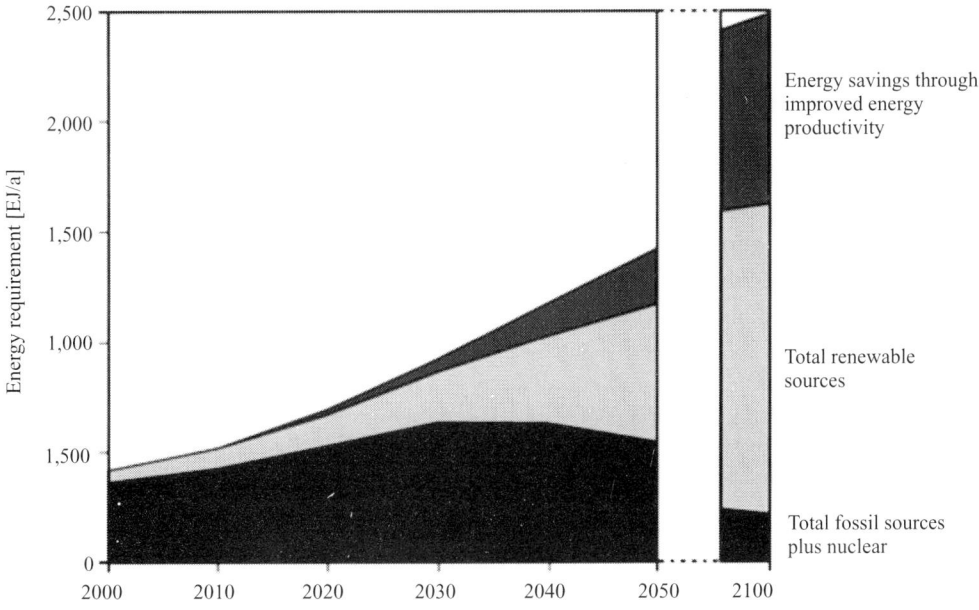

Fig. 1.2 Energy Efficiency Enhancement in the Exemplary Path
(Courtesy: WBSU – World in Transition – Towards Sustainable Energy Systems,
German Advisory Council on Global Change, Berlin, 2003)

The exemplary global transformation of energy systems embraces four key components:
(1) Major reduction in the use of fossil energy resources
(2) Phase out of the use of nuclear energy
(3) Substantial development and expansion of new renewable sources, notably solar energy
(4) Improvement of energy productivity far beyond historical rates

This is, in principle, technologically and economically feasible over the next 100 years for the stabilization of CO_2 concentrations in the atmosphere at a maximum of 450 ppm.

Figure 1.3 shows the presently installed capacity of power plants in India. The total installed capacity of power plants is 168945 MWe. The major contributors to power production are the thermal power plants. The contributions of renewable energy based power plants is only 11.04%.

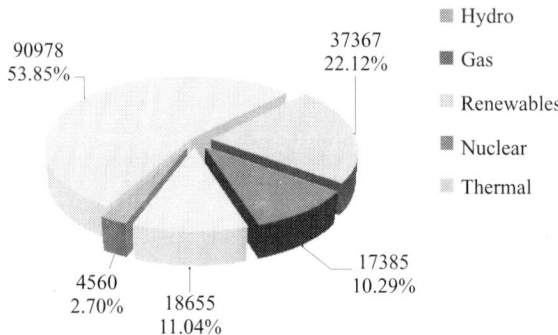

Fig. 1.3 Fuel-wise Installed Power Plant Capacity Breakup (in MW and %)
(*Source:* Strategic Plan for New and Renewable Energy Sector for the Period 2011 – 2017, MNRE, GOI, Feb., 2011)

During the past few decades, the economic growth of the most countries has resulted in significant growth of refrigeration and air conditioning industry. The advancement in refrigeration and air-conditioning sector has a major impact on energy demand which approximates to 15% of total energy consumption in the world (http://jp1.estis.net/includes/file.asp?site=ecanetwork&file=B8C63FD5-4B18-4E8F-A1C3-63B2DF9A5F0D). The annual energy consumption in RAC applications in residential and commercial sectors in the United States of America is 18.30% and 14% respectively, and that for India is 20% in residential buildings and 34% in commercial buildings (refer Fig. 1.4).

The current installed cooling capacity in India demands 35,000 MWe which is 20.7% of the total installed power plant capacity. As India is progressing, the demand for installation of new refrigeration and air conditioning systems is also increasing. Table 1.1 shows the additional energy required annually in India due to additional demand of cooling applications. There is a requirement of additional energy of 1450 MWe annually to meet additional cooling needs.

Table 1.1 Annual Additional Energy Required for Additional Cooling Applications

Application	Capacity Range (kW$_c$)	Capacity Addition Annually (MW$_c$)	Annual Additional Energy (MW$_e^*$)*
Domestic Refrigeration	0.3 – 7.0	40	13.33
Comfort Cooling	1.75 – 3500	2240	746.67
Cold Storage – Fruits and Vegetables	15 – 500	80	26.67
Cold Storage – Deep Freezing	15 – 100	85	28.33
Space Cooling	100 – 1000	300	100
Industrial Refrigeration	15 – 500	120	40
Industrial Process Cooling	100 – 7000	1500	500

* Assuming COP = 3

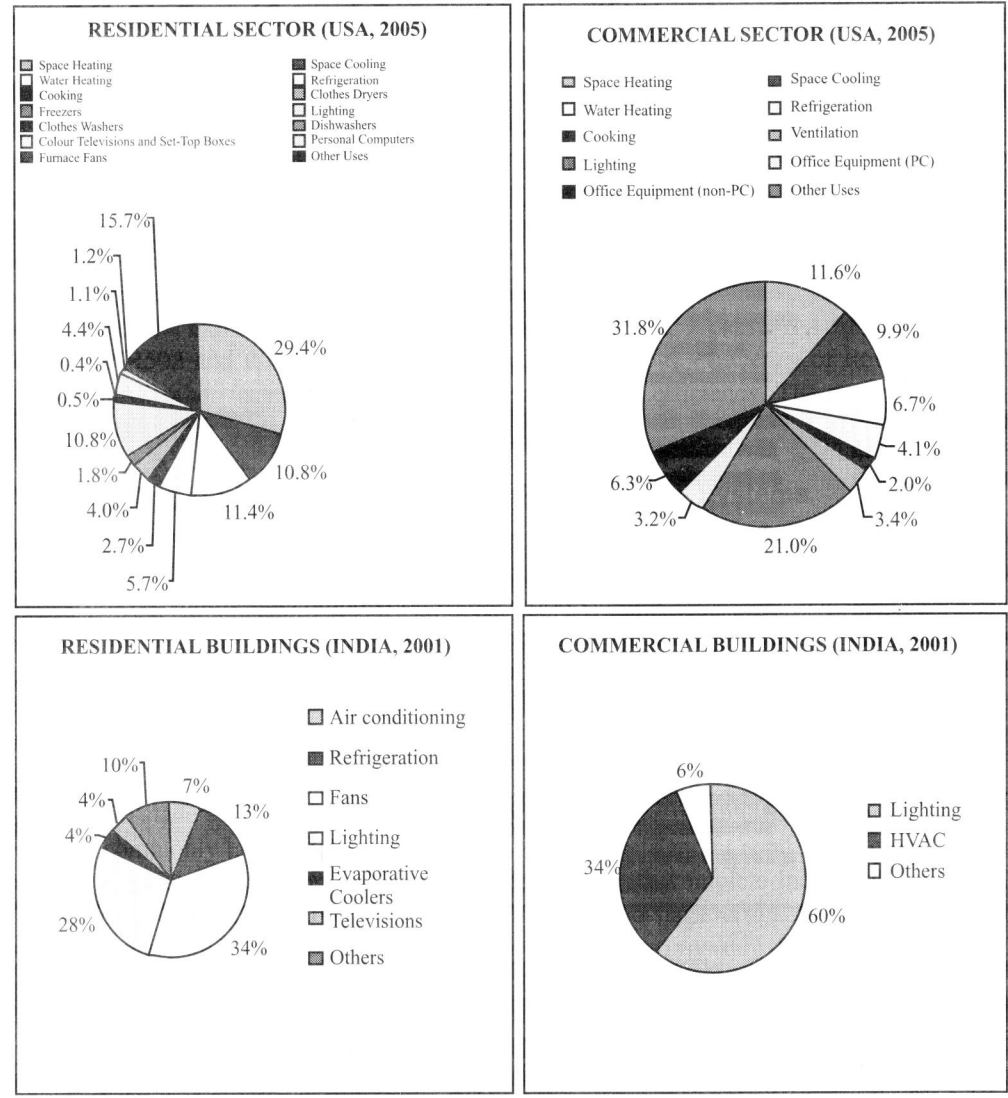

Fig. 1.4 Energy Consumption Pattern (Singh, 2009)

1.3 ENVIRONMENTAL CONSIDERATIONS AND ALTERNATIVE COOLING TECHNOLOGIES

Presently, the major demand of refrigeration, air conditioning and heat pumping load is being met out by the vapour compression systems. Conventionally these systems were operated on halogenated compounds such as chlorofluorocarbons and hydrochlorofluorocarbons because of their favourable thermophysical and thermodynamic properties. However, these compounds contain chlorine and were found to cause ozone layer depletion (Molina and Rowland, 1974) and hence the Montreal

Protocol was established in 1987 to phase out consumption and production of ozone depleting substances (ODSs). Further, Hydrofluorocarbons (HFCs) were brought in use as short and mid-term replacements to CFCs and HCFCs, however, HFCs are not permanent solution due to their high Global Warming Potential (GWP) (Agarwal, 2003). The GWP of HFC refrigerants is 500-3000 times greater than CO_2 (Hanaoka et al., 2002). HFCs contribute to direct global warming when released in the atmosphere on their leakage from the refrigeration system. Another factor is their low energy efficiency which accounts for indirect global warming. The low energy efficiency results in increase in consumption of electrical energy and eventually the amount of CO_2 released in to environment from the power plant is also increased. Thus, HFCs also contribute indirectly to global warming. The issue of global warming led to the signing, by the international community, of the Rio Convention in 1992 and then the Kyoto Protocol (KP) in 1997. The KP seeks to put HFCs and other Greenhouse Gases (GHGs) in one basket of controlled substances. It aims at the reduction and control of GHG emissions which cause climate change. Although CFCs and HCFCs also conribute to global warming, the KP does not address these substances, since these are already controlled under the Montreal Protocol with specific phase out regimes. After the Kyoto Protocol, United Nations framework on climate chage conventions are held regularly at various places in the world. In one of the recent conventions at Copenhagen in 2009, under the umbrella of United Nation framework on climate change, (http://unfccc.int/meetings/copenhagen_dec_2009/meeting/6295/php/view/reports.php) several key elements on which there was strong convergence of the views of governments, included the long-term goal of limiting the maximum global average temperature increase to no more than 2 degrees Celsius above pre-industrial levels, subject to a review in 2015. There was, however, no agreement on how to do this in practical terms. It also included a reference to consider limiting the temperature increase to below 1.5 degrees—a key demand made by vulnerable developing countries.

Thus, the RAC systems are major consumers of natural fuels on one hand and environment degraders on the other hand. Further, with the growth of economy of the country, the affordability of a common man to invest on RAC systems is increasing day by day. The cooling energy demand is thus on rise all over the world.

The challenge for refrigeration scientists and industry is that of increasing the energy efficiency of refrigeration equipment so that consumption of energy is reduced and available energy can be conserved. The conservation of available energy will help in reducing the GHGs and climate change. In the past two decades many research and development studies have established that replacement of restricted ODSs by alternative refrigerants involves substantial changes in the design of various components besides low energy efficiency of the system. Thus, there is utmost need to invent eco-friendly and more efficient alternative technologies for RAC applications. This will not only save natural fuel reservoirs but also has minimum environmental degradation impact and has high energy efficiency. Efficient alternative RAC technologies will thus save the planet from degradation. Better energy efficiency translates into lower operating costs for users and less environmental impact.

The problems of ozone layer depletion and global warming can also be completely eliminated by the use of eco-friendly alternative refrigeration technologies such as VAR and ejector refrigeration systems. These RAC systems can be operated using waste heat.

VAR systems use non-ozone depleting substances such as ammonia-water and water-lithium bromide pairs of refrigerant and absorbent. The climate change can thus be avoided by using eco-friendly VAR technology. But it presents only half a solution to the problem because the energy efficiency of VAR and ejector refrigeration systems is low (Aphornratana and Eames, 1995).

Another eco-friendly option is CAR technology which has been developed by combining features of the VCR and VAR. The literature shows that CAR systems offer better energy efficiency in comparison to VCR and VAR systems. However, it is still at developmental stage and their study is a subject of investigation (Herold *et al.*, 1991; Pratihar, 2001).

The vortex tube or Ranque-Hilsch tube or Ranque tube or Hilsch tube or Maxwell's Demon or vortex refrigerator is a simple device which produces hot and cold gas streams simultaneously from a same source of compressed gas. This device has no moving parts and is very simple in construction and operation. George Joseph Ranque, a French metallurgist associated with a steel company, invented the vortex tube in around 1930 (Ranque, 1933) and obtained the French patent in 1932 and U.S. patent in 1934 (Ranque, 1934). If it is used individually then its performance is very poor. But if it is integrated in other RAC systems, then the performance of the systems enhances (Singh, 2009) which leads to reduction in the energy demand and environmental degradation.

Thermoelectric refrigeration is also an option as the systems based on this option do not use any kind of refrigerant fluid which may damage the environment. But at present these systems also have poor performance due to poor figure of merit of thermoelectric materials.

Thermoacoustic refrigeration is another futuristic option emerging from the concepts of acoustics, cooling and thermodynamics. In addition vapour absorption based chemical reaction and metal hybrid based heat pumps are also emerging as new technologies using dry absorption/adsorption processes in the heating ventilation air conditioning and refrigeration sector. For air conditioning in hot and humid climates, desiccants based dehumidification and cooling is another simple and appropriate cooling option for supermarkets.

1.4 THERMODYNAMIC BASIS OF ENERGY AND EXERGY ANALYSIS

Any thermodynamic system/control volume, in general, has the following three types of energy interactions with its surroundings:

1. Work transfer interactions
2. Heat transfer interactions
3. Mass transfer interactions

Thus, the energy analysis of a thermodynamic system involves the balancing of the influx and efflux of energy associated with the above-mentioned interactions of the system with its surroundings. The difference between these influx and efflux energies is always zero as the energy is always conserved. The performance or the energy efficiency (which is the ratio of desired effect to the energy input) is based on the first law of thermodynamics or energy conservation. This first law efficiency provides information only about the energy losses occurring in the system and does not measure the approach of system towards ideality. For example, any system can have 100% first law efficiency if no energy losses are present, but this does not imply that the system is an ideal one. Thus, to determine more meaningful efficiencies analysis involving the second law of thermodynamics must be considered. The second law of thermodynamics states that ideal performance is achieved for a system in which all reversible processes are occurring. The actual performance is thus ascertained by the second law of efficiency defined as the ratio of the desired output effect to the maximum possible output effect. For understanding this, the concept of available energy or exergy is introduced here.

The exergy of a system is defined as the maximum theoretical useful work (shaft work or electrical work) that could be done by the composite of the system and a specified reference environment that is assumed to be infinite, in equilibrium, and ultimately to enclose all other systems. Alternatively, exergy is the minimum theoretical useful work required to form a quantity of matter from substances present in the environment (Bejan et al., 1996). Typically, the environment is specified by stating its temperature, pressure and chemical composition. Exergy is not simply a thermodynamic property, but rather is a co-property of a system and the reference environment. It is therefore, a measure of the available energy state of the system with respect to that of the environment.

Energy is conserved whereas exergy is always destroyed and generally is not conserved. A limiting case is when exergy would be completely destroyed, as would occur if a system were to come in equilibrium with the environment spontaneously with no provision to obtain work.

It is also necessary to define surroundings, environment and dead state before explaining the significance of energy and exergy analysis.

1.4.1 Surroundings, Environment and Dead States

The term surroundings refers to everything that is not included in the system. The term environment applies to some portion of immediate surroundings coupled with the system, the intensive properties of each phase of which are uniform and do not change significantly as a result of any process under consideration. The system coupled environment cannot be regarded as free of irreversibilities. All significant irreversibilities are located within the system and its immediate surroundings. Internal irreversibilities are located within the system. External irreversibilities reside in the immediate surroundings. Environment is modelled as a simple compressible system, large in extent, and uniform in temperature T_0 and pressure p_0. Where T_0 and p_0 may be taken as the average ambient temperature and pressure, respectively, for the location at which the system under consideration operates.

When the pressure, temperature, composition, velocity, or elevation of a system is different from the environment, there is an opportunity to develop or produce work. As the system changes state toward that of environment, the opportunity diminishes, ceasing to exist when the two, at rest relative to one another, are in equilibrium. This state is called the dead state. At the dead state, the conditions of thermodynamic equilibrium, i.e., mechanical, thermal and chemical equilibrium between the system and surroundings are satisfied. Another type of equilibrium between the system and environment is a restricted form of equilibrium where only the conditions of mechanical and thermal equilibrium must be satisfied. This state is called restricted dead state.

1.4.2 Exergy Components

The different types of exergy transfer that correspond to the various energy transfer forms are as follows:

1. Exergy associated with a work transfer
2. Exergy associated with a heat transfer
3. Exergy associated with a steady stream of matter or mass flow

The exergy associated with a work transfer is equivalent to the work itself as work is a high grade energy. The exergy of a heat interaction is determined from the maximum work potential that could be obtained from it using the environment as reservoir of zero grade thermal energy. This exergy associated with heat transfer is known as thermal exergy. On the other hand, the exergy associated with a steady stream of matter or mass flow is equal to the maximum amount of work obtainable when the stream is brought from initial state to the dead state by processes during which the stream may interact only with the environment. This exergy is further divided into kinetic exergy, potential exergy, physical exergy and chemical exergy, etc. Thus, in the absence of nuclear, magnetic, electrical and surface tension effects, the total exergy of a stream E can be divided into four components: physical exergy (E^{ph}), kinetic exergy (E^{ke}), potential exergy (E^{pe}), and chemical exergy (E^{ch}).

Thus,

$$E = E^{ph} + E^{ke} + E^{pe} + E^{ch} \tag{1.1}$$

The total specific exergy on a unit mass basis is given by:

$$e = e^{ph} + \frac{1}{2}V^2 + gz + e^{ch} \tag{1.2}$$

The sum of kinetic, potential and physical exergies is also referred as the thermomechanical exergy. The kinetic exergy and potential exergies of stream of matter are ordered forms of energy and thus fully convertible to work.

Physical exergy of a stream of matter is expressed using the following equation:

$$e^{ph} = (h - h_0) - T_0(s - s_0) \tag{1.3}$$

where h_0 and s_0 are specific enthalpy and specific entropy of the stream of matter at dead state.

The chemical exergy is expressed as,

$$e^{ch} = \mu - \mu_0 \tag{1.4}$$

where μ is the chemical potential of the substance at a given state temperature and pressure and μ_0 is the chemical potential of the substance at environmental conditions T_0 and p_0. The physical and chemical exergies are disordered forms of energy and can be determined with respect to the interaction of stream and the environment. The physical exergy is equal to the maximum amount of work obtainable when stream of matter is brought from initial state to the environment state (defined by pressure and temperature) by physical processes involving only thermal interactions with the environment. Further, this physical exergy of a stream of matter has two components resulting from temperature and pressure difference between stream and environment and are known as thermal component and pressure component of exergy.

1.4.3 Exergy Balance

According to Bejan et al. (1996), combining the first and second law of thermodynamics the exergy rate balance applied to a fixed control volume is given by:

$$\sum \dot{m}_i e_i - \sum \dot{m}_e e_e + \sum_i \dot{Q}_i \left(1 - \frac{T_0}{T}\right) - \dot{W} - \dot{ED} = 0 \qquad (1.5)$$

where ED is the exergy destruction due to entropy generation and is also known as irreversibilities.

The first two terms are exergy input and output rates of the flow, respectively. The third term is the exergy associated with heat transfer $\dot{Q}$, which is positive if it is entering into the system. It can also be regarded as work obtained by Carnot engine operating between T and T_0, and is therefore equal to the maximum reversible work that can be obtained from heat energy Q. W is the mechanical work transfer to or from the control volume. This equation states that the rate at which exergy is transferred into the control volume must exceed the rate at which exergy is transferred out; the difference is the rate at which exergy is destroyed within the control volume due to internal irreversibilities. Exergy destroyed is also known as exergy consumption and it is also proportional to entropy creation or generation specified by:

$$I = T_0 S_{gen} \qquad (1.6)$$

The above relation is also known as Gouy-Stodola relation.

1.4.4 Energy and Exergy Analyses

The energy analysis is an outcome of the first law of thermodynamics (i.e., the law of conservation of energy) and it gives no information on how, where, and how much the system performance is degraded. Energy analysis method is adopted for computing the energy efficiency of the thermal systems. Moreover, energy efficiency can either be the ratio of same forms of energy (i.e., low grade energy like heat energy) or of two different forms of energy, i.e., high grade energy and low grade energy. Thus, energy efficiency does not present the true performance of the thermal system. It is the measure of absolute performance and defined as the ratio of desired output energy to the input energy supplied as given by:

$$\eta = \text{(Desired Output Energy)/(Input Energy Supplied)} \qquad (1.7)$$

It does not provide any information about the irreversibilities occurring in various components of the thermal system and is also silent on the issue of maximum efficiency that can be achieved. Thus, it cannot be ascertained that which component of the energy system requires design improvement.

The exergy analysis of a thermodynamic system or control volume uses the combined first and second laws of thermodynamics or the exergy balance method applied to the system or control volume. However, the exergy is not conserved like the energy. There is always a positive difference between the influx and efflux of exergy associated with the real processes or interactions of the system or control volume with its surroundings due to the degradation of energy because of entropy generation within the system or control volume. This positive difference is termed the irreversibility due to occurrence of real processes. For ideal case of reversible processes, the irreversibility is zero.

Exergy analysis based on the second law of thermodynamics is a powerful tool in the design, optimization, and performance evaluation of energy systems. Exergy analysis of a complex system can be performed by analyzing the components of the system separately for exergy destruction. The evaluation of exergy destruction contributes towards identification of the sites of exergy destruction, i.e., system component with maximum exergy destruction. Identifying the main sites of exergy destruction shows the direction for potential improvements (Dincer and Cengel, 2001). Thus, exergy analysis takes into account the irreversibility losses (exergy destruction) appearing in the energy system, which could help in measuring the true performance of the system.

1.4.5 Exergetic Efficiency

Similar to first law efficiency (energy efficiency), using exergy analysis method, the performance of the energy system can be computed using exergetic efficiency. There are various ways of defining exergetic efficiency are given below:

(i) The exergetic efficiency (ε) is defined as the ratio of exergy output from the boundary of system/control volume to the exergy entering into it.

$$\varepsilon = \frac{(\text{Exergy})_{out}}{(\text{Exergy})_{in}} \qquad (1.8)$$

The exergy balance applied to the control volume/ system gives,

$$(\text{Exergy})_{in} = (\text{Exergy})_{out} + (\text{Exergy Destruction})_{total}$$

From the above equations, exergetic efficiency can be given by the following equation:

$$\varepsilon = 1 - \frac{(\text{Exergy Destruction})_{total}}{(\text{Exergy})_{in}} \qquad (1.9)$$

(ii) Exergetic efficiency can also be defined as the ratio of exergy rate of the product to the exergy rate of fuel (i.e., exergy rate of input energy). Exergy rate of product is the difference between exergy rate of fuel and irreversibilities (i.e., exergy destruction or losses) occurring in the system. Alternatively, it can be said that exergetic efficiency is an index which compares the actual system performance (energy efficiency) to the maximum performance (i.e., Carnot efficiency) which can be achieved. Thus, its value is always less than 1. Thus, it indicates that there is scope for improvement of system performance.

(iii) **Exergetic efficiency using the concept of efficiency defect**

Efficiency defect is given by:

$$\delta_i = \frac{\dot{ED_i}}{E_{input}} \qquad (1.10)$$

where 'i' stands for the particular component of an energy system, ED_i is the exergy destruction in that particular component and E_{input} is the exergy input to the system. (system here is considered to include a number of components.)

The total of efficiency defects is linked to the exergetic efficiency of the system by means of the relation given below:

$$\eta_{e_x} = (1 - \sum_i \delta_i) \qquad (1.11)$$

As shown in the above equation, the exergetic efficiency is a function of the sum of efficiency defects in system components. The efficiency defects in system components give an indication of the level of irreversibility. It is an important information because it helps in identifying the component in which maximum irreversibility is occurring. The design of the component in which exergy destruction or efficiency defect is the largest, may be improved to reduce the exergy destruction and increase the exergetic efficiency. The computation of exergetic efficiency also assists in identifying system operating parameters corresponding to which the exergy destruction (irreversibility) is least and exergetic efficiency is highest. This information cannot be obtained by even computing maximum coefficient of performance (COP).

It is important to mention here that COP is the first law efficiency and exergetic efficiency is the second law efficiency (which is always less than 100%). Sometimes under certain conditions the system exhibits positive values both for COP and exergetic efficiency, but there may be negative values of exergy in or out in some components of the system. The negative values of exergy destruction are not possible as it disobeys the second law of thermodynamics and thus, the working of the component is not possible. So, it is always essential to apply exergy analysis to each component of a system because conclusions based on only COP and energy efficiency may be misleading.

Lior and Zhang (2007) analysed the magnitude of differences of energy and exergy analysis as well as the first and second law efficiencies and showed the errors that can be made if the equations and systems are not defined carefully.

1.5 CONCLUSION

It can be summarized that the refrigeration sector is undergoing upheavals of ozone layer depletion and global warming of the environment. The result is climate change. To overcome these issues, alternative refrigerants and alternative refrigeration technologies are the need of the hour. However, the low energy efficiency of alternative refrigerants and technologies is a matter of great concern. Improving the energy efficiency of refrigeration systems is a must. Hence, one must identify the areas for improvement of a refrigeration system. The first and second laws of thermodynamics can facilitate the above objective. The combined treatment of the first law of thermodynamics is law of conservation of energy hence it is unlikely to give any information about the irreversibilities. The exergy analysis based on second law of thermodynamics is to be preferred to first law analysis as it facilitates the identification of components which require design improvement to minimize irreversibilities. Thus, it presents an opportunity to improve the system performance and efficiency. This can also ensure in conservation of available energy or exergy. Thus, the true meaning of energy conservation is exergy conservation and true system performance efficiencies is the second law efficiency.

REFERENCES

Agarwal, R.S., 2003. Alternatives to CFC and HCFCs for Refrigeration and Air-conditioning: Emerging Trends. *Proceedings of Workshop on Alternative Refrigerants and Cycles,* IIT Delhi, 1-8.

Aphornratana, S., Eames, I.W., 1995. Thermodynamic Analysis of Absorption Refrigeration Cycles using the Second Law of Thermodynamics Method. *International Journal of Refrigeration* 18 (4), 244–252.

Bejan, A., Tsatsaronis, G., Moran, M., 1996. Thermal Design and Optimization. John Wiley and Sons, USA, 143–156.

Dincer, I. and Cengel, Y.A., 2001. Energy, Entropy and Exergy Concepts and Their Roles in Thermal Engineering. *Entropy* 3, 116-149.

Evans O., 1805. The Abortion of a Young Steam Engineer's Guide, Philadelphia, PA, USA.

Hanaoka, T., Ishitani, H., Matsuhashi R., Yoshida, Y., 2002. Recovery of Fluorocarbons in Japan as a Measure for Abating Global Warming. *Applied Energy* 72, 705–721.

Herold, K.E., Howe L.A., and Radermacher R., 1991. Analysis of a Hybrid Compression-Absorption Cycle using Lithium Bromide and Water as the Working Fluid, *International Journal of Refrigeration*, 14, 264–272.

http://jp1.estis.net/includes/file.asp?site=ecanetwork&file=B8C63FD5-4B18-4E8F-A1C3-63B2DF9A5F0D

http://unfccc.int/meetings/copenhagen_dec_2009/meeting/6295/php/view/reports.php

James M Calm. The next generation of Refrigerants- A historical review, *Ecolibrium*, p.24–33, November 2008.

Lior N., Zhang N., 2007. Energy, Exergy, and Second Law Performance Criteria. *Energy,* 32, 281–296.

Molina, M. J., and Rowland F. S., 1974. Stratospheric Sink for Chlorofluoromethanes: Chlorine Atom Catalyzed Destruction of Ozone. *Nature* 249, 810.

Perkins J., 1834. .Apparatus for Producing Ice and Cooling Fluids, patent 6662, UK.

Pratihar, A.K., Kaushik S.C. and Agarwal R.S., 2001. Thermodynamic Modeling and Feasibility Analysis of Compression-Absorption Refrigeration System. *Proceedings of the International Conference on Emerging Technologies in Air-conditioning and Refrigeration*, Delhi, India, 207–215.

Ranque G.J., 1933. Experiments on Expansion in a Vortex Tube with Simultaneous Exhaust of Hot Air and Cold Air. *Le de Physique et le Radium*, 4(7), 112–114.

Ranque G.J., 1934. Method and Apparatus for Obtaining From a Fluid under Pressure Two Currents of Fluids at Different Temperatures. U.S. Patent No. 1952281.

Singh P., 2009. Study of Vortex Tube Integrated Alternative Refrigeration & Air Conditioning Options for Energy Conservation. Ph. D. Thesis, CES, IIT, Delhi.

Strategic Plan for New and Renewable Energy Sector for the Period 2011–2017, MNRE, GOI, Feb., 2011.

WBSU – World in Transition – Towards Sustainable Energy Systems, German Advisory Council on Global Change, Berlin, 2003.

CHAPTER 2

Alternative Refrigerants and Cycles for Compression Refrigeration Systems

2.1 INTRODUCTION

Refrigerant is the substance which is used as working fluid in a thermodynamic cycle, undergoes a phase change from liquid to vapour and produces cooling. These are used in refrigeration, air conditioning, and heat pumping systems. They absorb heat from one area, such as an air conditioned space, and reject it into another, such as outdoors, usually through evaporation and condensation, respectively. These phase changes occur both in absorption and mechanical vapour compression refrigeration systems, but they do not occur in systems operating on a gas cycle using a fluid such as air (ASHRAE Handbook, 1997, Fundamentals, Chapter 18).

The chronological evolution of refrigerants has been shown in Table 2.1 (Radermacher and Hwang, 2005). The first refrigerant used in a continuous refrigeration system by William Cullen in 1755 was water. But the credit for building the first vapour compression refrigeration system goes to Jakob Perkins who used sulphuric (ethyl) ether obtained from India rubber as refrigerant. In the beginning, the goal was to produce refrigeration only; whatever substance gave the desired results was used as refrigerant. All refrigerants were either flammable or toxic at that point in time. In 1930's, Thomas Midgley introduced R-12 (CF_2Cl_2, i.e., Dichlorodifluoromethane) and R-11 ($CFCl_3$, i.e., Trichloromonofluoromethane) as nontoxic and nonflammable refrigerants. This led to developments of a series of CFCs and HCFCs which deemed to be stable, nontoxic, nonflammable, and having a desired boiling point substances. During that period the main objective was safety and durability. The CFCs were not only being used as refrigerant but also as solvent, foam blowing agent, aerosol and in fire extinguishers. Later in 1980s, it was discovered that halogens and CFCs and other related substances react with ozone layer in atmosphere and thin down the ozone layer because of their higher atmospheric lifetime. Halogen, which results from the breakdown of CFCs in atmosphere combines with greenhouse gases and enhances the global warming threat. International consensus was made to stop production and use of halogenated refrigerants for which Montreal Protocol came into existence in 1987. In 1997, Kyoto Protocol was ratified to limit the green house gases causing global warming. Recent attention towards depletion of stratospheric ozone layer and global warming put a question mark to the present use of refrigerant in refrigeration and air conditioning (RAC) industry.

Table 2.1 Evolution of Refrigerants

Year	Refrigerant	Chemical Makeup, Formula	First Developer/User
1755	Water	H_2O	William Cullen
1834	Caoutchoucine sulfuric (Ethyl) Ether	Distillate of India Rubber; $CH_3\text{-}CH_2\text{-}O\text{-}CH_2\text{-}CH_3$	Jacob Perkins
1840s	Methyl Ether (R-E170)	$CH_3\text{-}O\text{-}CH_3$	Charles Tellier
1850	Water/Sulphuric Acid	H_2O / H_2SO_4	
1856	Ethyl Alcohol	$CH_3\text{-}CH_2\text{-}OH$	
1859	Ammonia/Water	NH_3 / H_2O	
1866	Chymogene Carbon Dioxide	Petrol Ether And Naphtha (Hydrocarbons) CO_2	T.S.C. Lowe
1860s	Ammonia (R-717) Methyl Amine (R-630) Ethyl Amine (R631)	NH_3 $CH_3(NH_2)$ $CH_3\text{-}CH_2(NH2)$	David Boyle/Carl Von Linde
1870	Methyl Formate (R611)	$HCOOCH_3$	
1875	Sulphur Dioxide (R764)	SO_2	Raoul Pictet
1878	Methyl Chloride (R-40)	CH_3Cl	C. Vincent
1870s	Ethyl Chloride (R-160)	$CH_3\text{-}CH_2Cl$	
1891	Blends of Sulphuric Acid with Hydrocarbons	$H_2SO_4, C_4H_{10}, C_5H_{12},$ $(CH_3)_2CH\text{-}CH_3$	
1900s	Ethyl Bromide (R160B1)	$CH_3\text{-}CH_2Br$	
1912	Carbon Tetrachloride Water Vapour (R718)	CCl_4 H_2O	
1920s	Isobutane (R-600a) Propane (R290)	$(CH_3)_2CH\text{-}CH_3$ $CH_3\text{-}CH_2\text{-}CH_3$	
1922	Dielene (R-1130)	$CHCl = CHCl$	
1923	Gasoline	Hydrocarbons	
1925	Trielene (R-1120)	$CHCl = CCl_2$	Carrier and Waterfill
1926	Methylene Chloride (R-30)	CH_2Cl_2	
1931	R12	CF_2Cl_2	Midgley
1932	R11	$CFCl_3$	Midgley
1934	R22	CF_2ClH	

Contd...

Year	Refrigerant	Chemical Makeup, Formula	First Developer/User
1980s	R123	CF_3CCl_2H	
1980s	R124	CF_3CFClH	
1980s	R125	CF_3CF_2H	
1990s	R134A	CF_3CFH_2	
1990s	R407C	R32/R125/R134a (23/25/52 wt%)	
1990s	R410A	R32/R125 (50/50 wt%)	
1990s	R404A	R125/R143a/R134a (44/52/4 wt%)	
2000s	R417A	R134a/R125/R600 (50/46.6/3.4 wt%)	
	R422A	R134a/R125/R600a (11.5/85.8/3.4 wt%)	
	R423A	R134a/R27ea (52.5/47.5 wt.%)	
	R432A	R1270/ RE170 (80/ 20 wt%)	
	R433A	R1270/R290 (30/70 wt%)	

These protocols initiated the search of alternative refrigerants and cycles, which are not only environment-friendly but also, consist of all desired properties of an ideal refrigerant and cycle. The HCFCs can be utilized as interim solution because of its low Ozone Depletion Potential (ODP) and Global Warming Potential (GWP) compared to CFCs. HCFC-123, HFC-134a can be used in low pressure and medium pressure systems. As no single substance can pass under ODP, GWP, toxicity, flammability, cost and efficiency criteria simultaneously, the search has already been started to find suitable azeotropic and zeotropic blends. These synthesized substances are expected as potential refrigerants in future which can fulfil desired requirements.

2.2 ALTERNATIVE REFRIGERANTS

Before the discovery of ozone hole in early 1970s, the refrigeration and air-conditioning industry was relying heavily on CFCs, HCFCs, Halons (BFCs) and their azeotropics. CFCs contain only chlorine, fluorine and carbon atoms but they cause ozone depletion (ODP of CFCs varies between 0.3 and 1) and have very long atmospheric life time (a few centuries). CFCs which have been in extensive use are R-11, R-12, R113, R114, R115, etc. Halons or BFCs contain bromine, fluorine and carbon atoms. For example, R-13B1 and R12B1 are BFCs. Their ozone depletion potential is very high, for example, ODP of R-13B1 is 10 and it was in use since 1995 for very low temperature vapour compression refrigeration systems. However, after the discovery of ozone hole the era for alternative refrigerants has started.

Alternative Refrigerants and Cycles for Compression Refrigeration Systems 17

The word alternative refrigerants was coined in the late 1980s. These are those working fluids which are used as a transitional/service and drop-in-replacements/substitutes in vapour compression systems/cycles with or without system design modifications or retrofitting in order to counteract the environmental and ecological issues. A drop in alternative refrigerant is one which can be used directly in the existing refrigeration system without any modifications and keeping the existing lubricating oil. The drop-in-replacements/substitutes (alternative refrigerants) have the same thermo-physical properties as the refrigerant it replaces and can accomplish the same task and does not require equipment redesign. First, especially for old equipment with remaining life, a drop-in process is recommended. The various types of alternative refrigerants have been shown in Fig. 2.1.

A retrofit alternative refrigerant is one which when used in the existing refrigeration plant would also entail the replacement of the lubricant oil, the expansion valves and certain other elements of the system. The retrofit alternative refrigerants are primarily intended as service refrigerants for older plants with view on phasing out of the previously used refrigerants in them with certain retrofitting to meet the legal regulations (Bitzer Refrigerant Report 16).

Long term alternative refrigerants such as hydrocarbons, ammonia or carbon dioxide require major design modifications in the existing plants or altogether a new plant design.

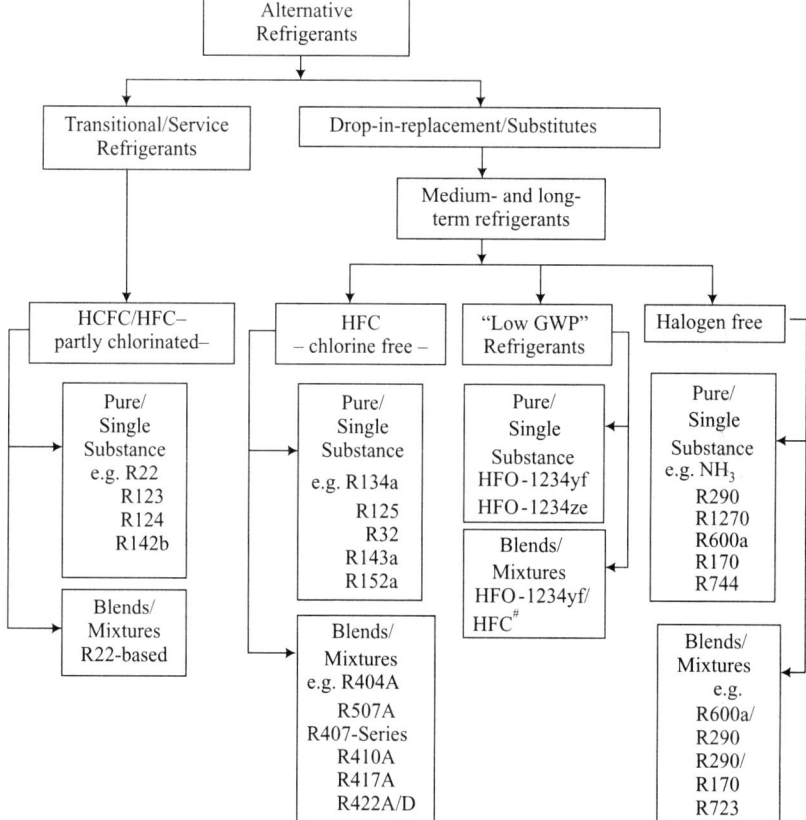

#In development and test phase

Fig. 2.1 Types of Alternative Refrigerants

Further, the alternative refrigerants are classified as:
1. Pure refrigerants (i.e., single component refrigerants)
   (a) Natural refrigerants, i.e., inorganic and organic compounds
   (b) Hydrofluorocarbons (HFCs)
2. Refrigerant mixtures/blends
   (a) Based on the number of pure components, i.e., binary / ternary /quaternary
   (b) Azeotropic /near azeotropic / zeotropic (non-azeotropic)

2.2.1 Pure Refrigerants (Single Component Refrigerants)

Natural refrigerants are those substances which exist in our biosphere, for example, air, water, ammonia (which are inorganic compounds) and hydrocarbons. Natural refrigerants are non-ozone depleting and also have negligible global warming potential. On the other hand, HFCs contain only hydrogen, fluorine, and carbon atoms and cause no ozone depletion but have high global warming potential. HFCs group include R-134a, R-32, R-125, and R-245ca.

2.2.2 Refrigerant Mixtures/Blends (Multi-Component Refrigerants)

(a) Binary/Ternary/Quaternary Refrigerant Mixtures/Blends

Binary mixtures, as the name suggests, consist of two pure refrigerant components mixed in required proportions. For example, R-410A is a binary mixture of R125 and R32 in equal proportions by weight. Similarly, ternary and quaternary mixtures contain three and four pure refrigerants respectively. R-407C is a ternary mixture comprising R-32/ R-134a/ R-125 (23%/52%/25% by weight).

(b) Azeotropic Mixtures/Blends

An azeotropic is a mixture of multiple components (refrigerants) of volatilities that evaporate and condense as a single substance and do not change in volumetric composition or saturation temperature when they evaporate or condense at constant pressure. HFCs azeotropics are blends of refrigerant with HFCs. ASHRAE assigned numbers between 500 and 599 for azeotropes. HFCs azeotrope R-507, a blend of R-125/R-143, is a commonly used refrigerant for low-temperature vapour compression refrigeration systems.

An azeotropic mixture has a temperature-pressure-concentration diagram where the saturated vapour and saturated liquid lines coincide at a range of concentrations, as shown in Fig. 2.2. At the point or region where the saturated vapour and liquid lines merge, the mixture of the two substances behaves with the properties of a single substance, having properties different from either of its constituents.

Even combinations that are azeotropic at certain concentrations at one pressure may not be perfectly azeotropic at another pressure, as shown in Fig. 2.3.

The azeotropic concentration changes as the pressure increases. Usually, the azeotropic region shifts toward the high concentration of the low- temperature boiling refrigerant (material A in this case). Even the azeotrope R-502, which has been used successfully for many years, was subject to fractional distillation at certain operating pressures.

(c) Near Azeotropic and Zeotropic (Non-Azeotropic) Refrigerant Mixture/Blends

A near azeotropic is a mixture of refrigerants whose characteristics are near those of an azeotropic. It is thus named near azeotropic because the change in volumetric composition or saturation

Alternative Refrigerants and Cycles for Compression Refrigeration Systems 19

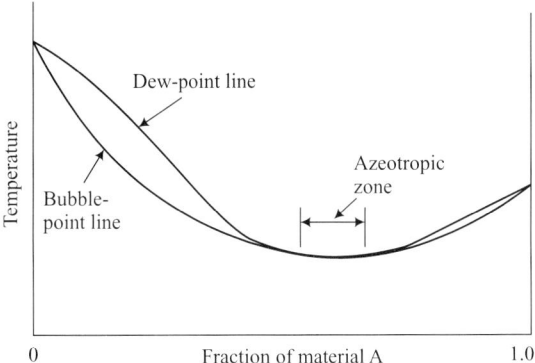

Fig. 2.2 An Azeotropic Mixture of Materials *A* and *B* in the Range of 50-to-60% of *A*

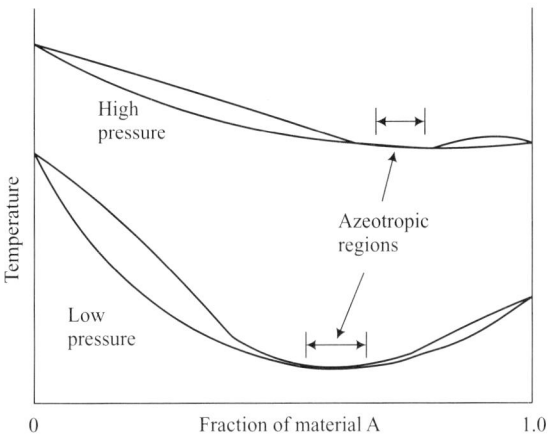

Fig. 2.3 A Shift in the Azeotropic Region as the Pressure Changes

temperature is rather small, such as, 0.5°C to 1.1°C. ASHRAE assigned numbers between 400 and 499 for near azeotropic/zeotropic refrigerant mixtures. R-404A (R-125/R-134a/R-143a) and R-407B (R-32/R-125/R 134a) are HFCs near azeotropic refrigerant mixture (NARM). R-32 is flammable; therefore, its composition is usually less than 30% in the mixture. HFCs near azeotropic refrigerant blends are widely used for vapour compression refrigeration systems.

The diagram of the temperature-pressure-concentration relationship of an ideal zeotropic mixture might appear as shown in Fig. 2.4. In a zeotropic mixture, the concentration of the two substances in the vapour is different from that in the liquid at a given pressure and temperature. There are certain applications where the properties of a zeotrope are advantageous, such as in the autocascade system for ultra low temperatures. For conventional industrial refrigeration systems, however, the zeotrope of Fig. 2.4 has the drawback that if a leakage from a system occurs that is in vapour form, for example, the composition of the lost vapour will be different from the original charge. Replacing the lost refrigerant thus requires some analysis of the concentration of the refrigerant mixture remaining in the system. A shift in composition causes the change in evaporating and condensing temperature/pressure. Another characteristic of the zeotrope when boiling or condensing is that its temperature changes, in contrast to a single refrigerant whose temperature remains constant during a constant-pressure evaporation or condensation. This characteristic could be an advantage on some systems which are designed for this behaviour, but it is not the standard expectation in an industrial refrigeration system. The difference in dew point and bubble point during evaporation and condensation is called glide, expressed in ΔT. Near azeotropic has a smaller glide than zeotropic. The midpoint between the dew point and bubble point is often taken as the evaporating and condensing temperature for refrigerant blends.

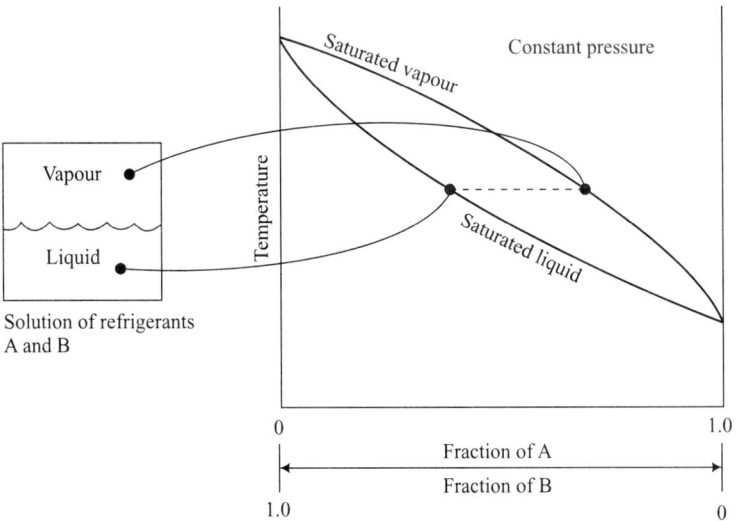

Fig. 2.4 An Ideal Zeotropic Mixture of Substances A and B

HCFCs near azeotropic and HCFCs zeotropic are blends of HCFCs with HFCs. They are transitional or interim or service refrigerants.

2.2.3 Factors/Indices Related to Environmental and Ecological Concerns

(a) Ozone Depletion Potential (ODP): The ozone depletion potential is defined as an index that indicates the ability of refrigerants and other chemicals to destroy stratospheric ozone molecules based on a value of 1.0 for R11.

(b) Global Warming Potential (GWP): The Global Warming Potential (GWP) is an index relative to the global warming impact of CO_2 is used to compare various greenhouse gases. The impact of the emission of the various greenhouse gases on the climate varies according to their atmospheric life. CO_2 is one of the main greenhouse gases in the atmosphere so, the GWP of CO_2 is fixed as 'one' and the global warming impact of the various gases is compared with that of CO_2. Since the atmospheric lifetime of the greenhouse gases varies therefore GWP is also time dependent. The time dependency spreads from 20 to 100 years and written as GWP_{20} (for 20 years), GWP_{100} (100 years lifetime), etc. Emission of one kg of R-134a is roughly equivalent to emission of 1300 kg of CO_2 in 100 years, so GWP_{100} of R-134a is 1300. Thus, the GWP of a greenhouse gas is an index relative to that of CO_2 to trap heat radiated from earth to space.

(c) Total Equivalent Warming Impact (TEWI): It is the factor to evaluate the environmental effect of GHGs in an appliance. TEWI provides the measure of the environmental impact of GHGs from manufacture, operation, service and end of life disposal of the equipment. It takes account of both the emissions of refrigerants and indirect emissions due to energy consumption and fossil fuels used. TEWI combines the effects of direct emissions of refrigerants (and also the foam insulation blowing agents) from appliance during its lifetime with the indirect emission of CO_2 from the combustion of fossil fuels and generation of electricity use by the appliance or the system.

TEWI is defined by the following equation:

$$TEWI = m_{ref} \cdot GWP_{ref} \cdot Z + m_{ba} \cdot GWP_{ba} + t.E.f$$

where   TEWI : Total Equivalent Warming Impact in kg CO_2
m_{ref} = Mass of refrigerant in kg
GWP_{ref} = Global Warming Potential of refrigerant in kg CO_2
Z = Number of charges of refrigerant during service life
m_{ba} = Mass of blowing agent in kg
GWP_{ba} = Global Warming Potential of blowing agent in kg CO_2
t = Service life of appliance in year
E = Annual energy consumption of appliance in kWh/yr
f = CO_2 factor of energy conversion in kg CO_2/kWh_{el}

The first two terms on the right hand side of the TEWI equation are for direct contribution of refrigerants during servicing and blowing agents to global warming and the last term is for the contribution to global warming due to the energy consumed during the lifetime of the appliance. There can be many additional factors to consider, for example, emission during the initial charge and recovery of refrigerants at the end of life of the appliance.

The TEWI can also be expressed as follows (Bitzer Refrigerant Report 16):

$$TEWI = (GWP \times L \times n) + (GWP \times m [1 - \alpha_{recovery}]) + (n \times E_{annual} \times \beta)$$

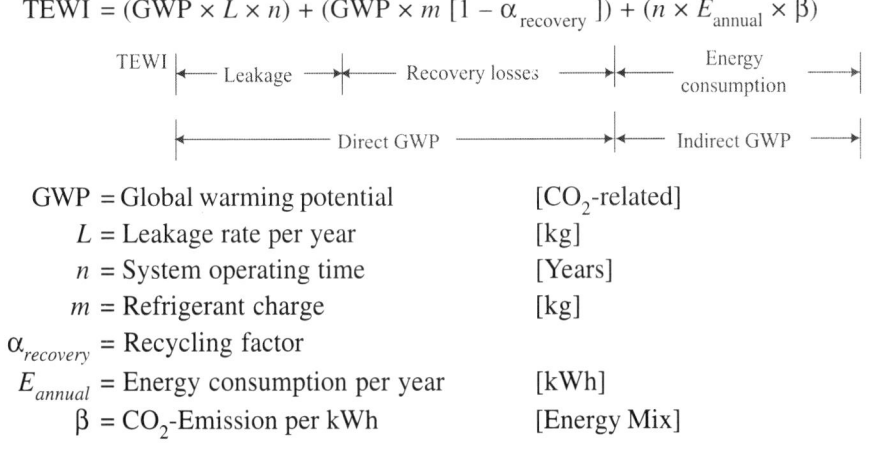

GWP = Global warming potential   [CO_2-related]
L = Leakage rate per year   [kg]
n = System operating time   [Years]
m = Refrigerant charge   [kg]
$\alpha_{recovery}$ = Recycling factor
E_{annual} = Energy consumption per year   [kWh]
β = CO_2-Emission per kWh   [Energy Mix]

22 *Alternatives in Refrigeration and Air Conditioning*

In addition, loss in energy efficiency due to inappropriate service practices also indirectly affects the environment as it leads to more emission of CO_2 for the additional energy required. Therefore, it is essential for the service sector to know the most efficient methods of charging the system and the energy efficiency degrades with successive charging of blends over a period of time. They must also be aware of the effect of good service practices on the energy efficiency of the refrigerating systems and appliances.

(d) Life Cycle Climate Performance (LCCP): In order to comprehensively analyze global warming impacts, the concept of Life Cycle Climate Performance (LCCP), has been evolved from the earlier concept of TEWI.

LCCP calculates the cradle-to-grave climate impact of direct and indirect greenhouse gas emissions including inadvertent emissions from chemical manufacture, energy embodied in components, operating energy, and emissions at the time of disposal or recycle. The calculated LCCP also accounts for location-specific electrical generation efficiency and power mix and is sensitive to assumptions of the system lifetime, emission losses, and the integration time interval used in the calculation of the GWP of GHGs. Energy efficiency is often revealed as the most important strategy for reducing primary energy demand and its emissions.

(e) *Eco-efficiency* (Bitzer Refrigerant Report 16): Eco-efficiency is used for the purpose of a more objective assessment of refrigeration plants. It is based on the relationship between added value (a product's economic value) and the resulting environmental impact.

With this evaluation approach, the entire life cycle of a system is taken into account in terms of:
- Ecological performance in accordance with the concept of life cycle assessment as per ISO 14040.
- Economic performance by means of a life cycle cost analysis.

This means that the overall environmental impact (including direct and indirect emissions), as well as the investment costs, operating and disposal costs, and capital costs are taken into account.

The studies confirm that an increase of eco-efficiency can be achieved by investing in optimized plant equipment (minimized operating costs). Hereby, the choice of refrigerant and the associated system technology plays an important role.

2.2.4 Alternative Refrigerants as Substitutes of Conventional Refrigerants for Different Applications

Table 2.2 shows the various alternative refrigerants for different applications. In domestic refrigeration systems R134a is a long-term substitute for R12. However, there are some problems with the use of lubricant with R134a. The traditional mineral and synthetic oils are not miscible (soluble) with R134a and are therefore only insufficiently transported around the refrigeration circuit. Immiscible oil can settle out in the heat exchangers and prevent heat transfer to such an extent that the plant can no longer be operated. The new lubricants based on Polyol Ester (POE) are more suitable for this application with R134a. Another problem with R134a is its high global warming potential which is 1300 times in comparison to CO_2. The mixtures of hydrocarbons have low global warming potential, such as R290, R600, R600a, etc., are also being looked as long-term alternatives. The transcritical CO_2 refrigeration technology is another option for replacing long-term alternative refrigerants used in vapour compression refrigeration system.

Similar to above, R432A and R433A (Park *et al.*, 2008a, *b*) are near azeotropic mixtures and are good long-term 'drop-in' environment-friendly alternative refrigerant to replace HCFC22 in residential air conditioners and heat pumps due to their excellent thermodynamic and environmental properties. It is also compatible with traditional mineral oils.

The common adjustments to retrofit R22 and R502 refrigeration systems with alternative refrigerants are as follows:

R-404A and R-507: These blends can be used to retrofit R-22 and R-502 systems. However, thermostatic expansion valves need to be changed to the appropriate R-404A model. In some cases, the superheat setting is required to be adjusted. Discharge pressures would also increase, although discharge temperatures would come down. An oil change to POE is also required.

R-422B and R-422D: These blends will have lower capacity than R-22 in the same system, and in many cases there will be an increase in pressure drop that may require changing thermostatic expansion valves or distributors. They contain hydrocarbons that will help circulate mineral oil in smaller systems. Larger systems, especially ones with receivers, will need addition of POE to help keep the mineral oil from being stranded.

R-407A and R-407C: Both products have the closest capacity and run-time property match to R-22. All products will have lower discharge temperatures than R-22. R-407A is a closer match at lower application temperatures, such as in commercial refrigeration applications. R-407C will work better in medium temperature and air conditioning applications. Replacement of mineral oil with POE is recommended.

R-422C: This blend can be used to retrofit low temperature R-22 systems. The performance characteristics (pressure/temperature) will look much like R-404A, but with a drop in capacity of up to 10%. The hydrocarbon additive will help circulate mineral oil around the system. In larger systems, however, some oil holdup may occur in the receiver. Addition of POE will solve this problem.

R410A: It is to be used in new equipment for air conditioning and heat pumping applications. Higher pressure and capacity exclude R-410A from being used as a retrofit blend for R-22.

2.3 ALTERNATE VAPOUR COMPRESSION REFRIGERATION CYCLES

The conventional VCR cycle with pure refrigerant is well understood by most of the students, researchers and practitioners and needs no explanation. In this section Lorenz Cycle and Transcritical Carbon Dioxide Compression Refrigeration Cycles are discussed.

Table 2.2 Alternative Refrigerants Based on the Area of Application (Bitzer Report 16, National Refrigerant Reference Guide, 2011)

Refrigerant Type		Designation	Pure/Blend (Composition by weight %)	ODP, GWP	Substitute for	Lubricant	Area of Application
Transitional/ Service	HCFC	R22	Pure	0.055, 1700	R502, R12⊖	Mineral oil or Alkyl-benzene	Frozen food cabinets, Quick freezing, Supermarket refrigeration, ice machines
		R124	Pure	0.022, 620	R114⊖, R12B1	Mineral oil or Alkyl-benzene	High ambient air conditioning
		R142b	Pure	0.065, 2400			
	HFC	R401A	R22/152a/124 (53.0/13.0/34.0)	0.037, 1130	R12, R500	Combination of mineral oil and Alkyl-benzene or polyolester lubricant	Medium* and low** temperature commercial and industrial direct expansion refrigeration systems
		R401B	R22/152a/124 (61.0/11.0/28.0)	0.04, 1220			Low** temperature commercial and industrial direct expansion refrigeration systems, R12 air conditioning
		R409A	R22/124/142b (60/25/15)	0.048, 1540		Compatible with mineral oil, Alkyl-benzene and poly-olester lubricant	Medium* and low** temperature commercial and industrial direct expansion refrigeration systems and non-centrifugal air conditioning
		R402A	R125/290/22 (60.0/2.0/38.0)	0.021, 2690	R502	Compatible with mineral oil and Alkyl-benzene	Low** temperature commercial and industrial direct expansion refrigeration systems

Category	Type	Refrigerant	Composition	ODP, GWP	Replaces	Lubricant	Applications
		R402B	R125/290/22 (38.0/2.0/60.0)	0.033, 2310			Ice machines
		R403B	R-290/22/218 (5.0/56.0/39.0)	0.031, 4310			Very low temperature single stage refrigeration, it is also a replacement for R13B1
		R408A	R-125/143a/22 (7.0/46.0/47.0)	0.026, 3020		Compatible with mineral oil, Alkylbenzene and polyolester lubricant	Medium* and low** temperature commercial and industrial direct expansion refrigeration systems
Substitutes (Long-term Alternatives)	HFC	R134a	Pure	0, 1300	R12, R22$^\ominus$	Compatible with polyolester lubricant for stationary equipment and polyalkaline glycol for automotive air conditioning systems	Household appliances, refrigeration (commercial and self-contained equipment), centrifugal chillers and automotive air conditioning
		R152a	Pure	0, 120	Used as part components of blends	N.A.	N.A.
		R125	Pure	0, 3400			
		R143a	Pure	0, 4300			
		R32	Pure	0, 550			
		R227ea (CF$_3$-CHF-CF$_3$)	Pure	0, 3500	R12B1, R114$^\ominus$	Polyolester oil	Suitable for air conditioning devices functioning in high temperature environments, high temperature heat pumps, and thermal collectors.

Contd...

Refrigerant Type	Designation	Pure/Blend (Composition by weight %)	ODP, GWP	Substitute for	Lubricant	Area of Application
	R236fa (CF_3-CH_2-CF_3)	Pure	0, 9400	R114		Low-pressure centrifugal chillers and industrial process refrigeration systems. Existing centrifugal chillers and industrial process refrigeration systems
	R23	Pure	0, 12000	R13, R503	Polyolester oil	Very low temperature refrigeration, below –40°C to –73°C, medical freezers and environmental chambers
	R404A	R143a/125/134a (52/44/4.0)	0, 3780	R22, R502		Medium and low temperature commercial and industrial direct expansion refrigeration and ice machines
	R507A	R143a/125 (50/50)	0, 3850			Medium and low temperature commercial refrigeration and industrial refrigeration
	R407A	R134a/125/32 (40/40/20)	0,1990			Medium and low temperature commercial and industrial direct expansion refrigeration
	R407F	R134a/125/32 40/30/30	0, 1705			Medium and low temperature refrigeration, suitable for R22, R502, R404A and R407A replacement
	R422A	R125/134a/600a (85.1/11.5/3.4)	0.3040		Compatible with mineral and alkylbenzene oil	Medium and low temperature commercial and industrial direct expansion refrigeration

R437A	R143a/125/134a	0.1680	R12, R500	Compatible with traditional and new lubricants; in most cases no change of lubricant type during retrofit is required.	Automotive air conditioning systems designed for R12 stationary air conditioning systems. Medium temperature stationary refrigeration systems designed for R12, such as supermarket display cases, food storage/processing. It is a service refrigerant and replacement for HCFC blends such as, R401A and R409A.
R407C	R134a/125/32 (52/25/23)	0.1650	R22	Polyolester oil	Medium temperature commercial and industrial direct expansion refrigeration and A/C
R417A	R600/134a/125 (3.4/50.0/46.6)	0.2240		Mineral, alkyl benzene or fully synthetic lubricants	Commercial refrigeration display cabinets
R417B	R600/134a/125 (2.75/18.25/79)	0.2920		Mineral oil	
R422D	R125/134a/600a (65.1/13.5/3.4)	0.2620		Compatible with mineral and alkyl-benzene Oil	Medium and low temperature commercial and industrial direct expansion refrigeration
R427A	R32/125/143a/134a (15/25/10/50)	0.2010		Polyolester oil	To replace R-22 in existing equipment for a wide range of temperatures. To retrofit low temperature refrigeration units as well as air conditioning installations. Many industrial installations in commercial refrigeration (supermarkets,etc., in industrial refrigeration

Contd...

Refrigerant Type	Designation	Pure/Blend (Composition by weight %)	ODP, GWP	Substitute for	Lubricant	Area of Application
	R438A	R32/134a/125/600/601a(8.5/44.2/45/1.7/0.6)	0, 2150		Compatible with mineral oil, alkyl-benzene oil	(food industry, pharmaceutical industry, etc.). Many water chillers have been successfully converted to 427A. Existing direct expansion refrigeration and air conditioning systems.
	R410A	R125/32(50/50)	0, 1980	R22$^\Theta$, R13B1$^\Delta$	Polyolester oil	Air conditioning equipment and heat pumps. Only for newly manufactured equipment, not retrofitting R-22 systems.
	ISCEON MO89	R125/FC-218/290 (86/9/8) (FC-218 Perfluoropropane)	0, N/A	R13B1$^\Delta$	Compatible with traditional and new lubricants mineral, alkyl-benzene or Polyolester oil	R13B1 in very low temperature direct refrigeration systems, expansion, very low temperature refrigeration (below-40°F to 100°F)/(below 40°C to-70°C), including: freeze driers, medical freezers, environmental chambers
	R508A	R116/23 (61.0/39.0)	0, 11940	R503	Polyolester lubricant with hydrocarbon additives	Very low temperature refrigeration (low stage of a cascade system).
	R508B	R116/23 (54.0/46.0)	0, 11950			Biomedical freezers
Low GWP	R432A	R1270/ RE170 (80/20)	0, less than 5	R22	Compatible with the conventional mineral oil	R432A is a good long-term 'drop-in' environmentally friendly alternative to replace

Alternative Refrigerants and Cycles for Compression Refrigeration Systems 29

	Refrigerant	Composition	ODP, GWP	Replaces	Lubricant	Application
	R433A	R1270/R290 (30/70)	0, less than 5	R22	Compatibility with the conventional mineral oil	HCFC22 in residential air conditioners and heat pumps
Halogen free	R717	Pure	0, 0			Residential air conditioners and heat pumps due to its excellent thermodynamic and environmental properties with minor adjustments.
						Industrial plants, supermarkets
	R723	R717/R-E170 (DME) (60/40)	0, 8 (502)	R22 (R502)	As lubricant mineral oils or (preferred) poly-alpha olefin	Commmercial refrigeration plants
	R600a	Pure	0, 3	R114, R12B1	Works with the existing lubricants mineral, alkyl benzene	Domestic refrigerators and freezer
	R290		0, 3	R22 (R502)	Mineral oil	Residential A/C units and heat pumps
	R1270		0, 3	R22 (R502)		Medium and low temperature systems, e.g., liquid chillers for supermarkets
	R170		0, 3	R13, R503		Low temperature refrigeration
	R744		0, 1	Diverse		Trans-critical cycle.

*****Medium Temperature Refrigeration**–Refrigeration applications that normally run evaporator temperatures between –18°C and 5°C.
******Low Temperature Refrigeration**–Refrigeration applications that normally run evaporator temperatures between –40°C and –18°C.
High Temperature Refrigeration–Refrigeration applications where the evaporator temperature normally runs higher than –1°C.
θ Alternative refrigerant has larger deviation in refrigerating capacity and pressure.
Δ Alternative refrigerant has larger deviation below –60°C evaporating temperature.

2.3.1 Lorenz Cycle

The main assumption of Carnot cycle (single, two or mixed phase) is that the cycle accepts and rejects heat at a constant temperature level throughout the heat absorption and heat rejection process. However, in most applications, heat is supplied by or rejected to a fluid (air, water, brine etc.) whose temperature changes during the heat exchange processes (Fig. 2.5). This leads to the presence of so called *pinch point*, and thus, at points 1 and 4 and at points 6 and 8 the temperature differences between both fluid streams become very small, consequently, the heat transfer rate becomes less. The entropy generation or exergy destruction is also small due to the fact that heat is transferred over a small temperature difference. In contrast, at points 2 and 3 and also for points 5 and 7, the temperature differences are large, the heat transfer rate is more as well and so is the entropy production or exergy destruction.

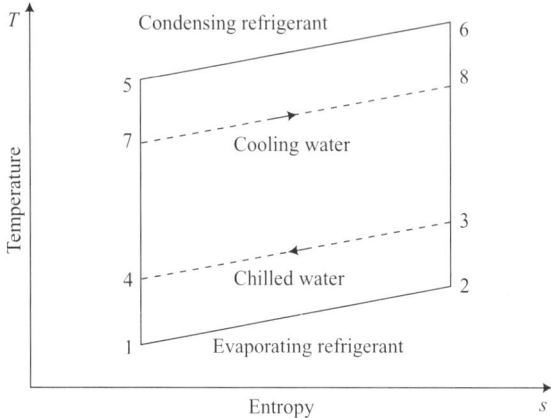

Fig. 2.5 Heat Exchange in Refrigerant Mixture/ Blend and Heat Absorbing and Rejecting Medium

The Lorenz Cycle (Radermacher and Hwang, 2005) addresses this issue. This cycle is for a working fluid that changes its temperature (i.e., it has a temperature glide) during the course of its phase change (evaporation and condensing) (Fig. 2.5). In the ideal case, the change in temperature throughout during both phase change processes matches that of the fluid (heat absorbing or rejecting medium), thus the overall entropy production or exergy destruction reduces significantly and the exergetic efficiency of the cycle improves to the largest possible extent, too. The refrigerant mixtures/ blends have the potential to approach the requirements of the Lorenz cycle depending on the degree to which they match the application glide in both the evaporator and the condenser. It should be noted that even for the same heat exchange area, a Lorenz cycle for the blends that better matches the source and sink glides gives higher cycle efficiency because it operates at an improved mean temperature than the corresponding pure refrigerant.

2.3.2 Transcritical Carbon Dioxide Compression Refrigeration Cycle (Singh, 2009)

In a transcritical carbon dioxide compression refrigeration cycle, the supercritical CO_2 is expanded to a subcritical state. The throttling loss is very large as compared to conventional refrigeration systems due to the higher pressure change during the expansion. Thus, basic transcritical carbon dioxide refrigeration cycle offers lower efficiencies when compared to HCFCs and HFCs.

In the basic transcritical CO_2 compression refrigeration cycle (Fig. 2.6), the refrigerant vapours coming out of the evaporator (EVA) is compressed using compressor (COM) above the critical conditions (P_{cr} = 73.77 bar and T_{cr} = 31.10°C) and then cooled in a gas cooler (GCO). Then, this cooled gas is throttled in an expansion valve (EXP) and low temperature and low pressure liquid CO_2 enters the evaporator.

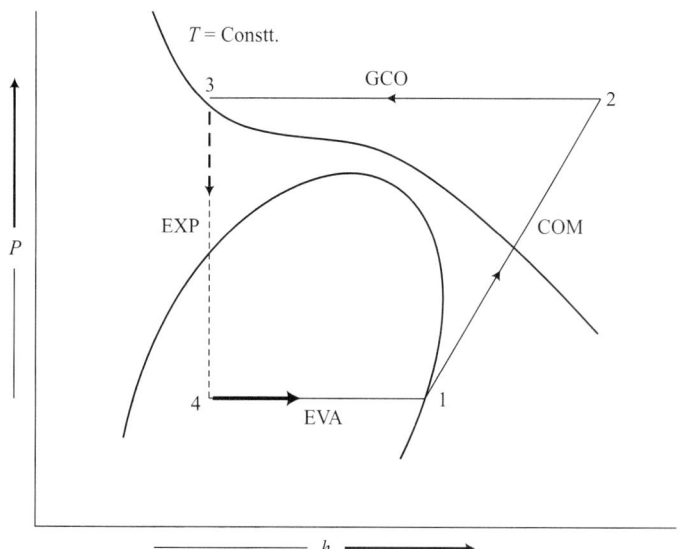

Fig. 2.6 Transcritical Carbon Dioxide Refrigeration Cycle

2.4 THERMODYNAMIC ANALYSIS OF VAPOUR COMPRESSION REFRIGERATION CYCLE WITH ALTERNATIVE REFRIGERANTS

2.4.1 Vapour Compression Refrigeration (VCR) System

Aprea *et al.* (1996) reported that VCR systems are normally used for cold storage and supermarket refrigeration. These systems operate between condensing temperature of 35°C and evaporating ones in the range of (–)40°C to 0°C. The suitable working fluid for these applications is the refrigerant R502 which is an azeotropic mixture of refrigerants HCFC22 and R115. Both of these refrigerants are harmful to ozone layer. The ozone depletion potential for HCFC22 and R115 is 0.055 and 0.4 respectively (Calm and Hourahan, 2001). Aprea *et al.* (1996) experimentally evaluated general characteristics and system performances of substitutes for R502 in a refrigeration plant. They examined R402A, R402B, R403B, R408A, R404A, R407A and FX40. All substitutes showed performances very close to those of R502 except R403B whose COP was found to be about 8% lower than that of R502.

Döring et al. (1997) carried out an experimental study of R507 to measure thermodynamic data viz., vapour pressures, liquid densities as well as the volumetric behaviour of the gaseous phase and presented the data in the form of mathematical correlations. Their theoretical results revealed that the compressor discharge temperature for R507 was approximately 8 K below in comparison to R502. The pressure ratios, refrigerating capacity and coefficient of performance of R507 were 4-5% higher than that for R502. Camporese et al. (1997) selected refrigerant mixtures as potential short- and mid-term substitutes for CFC12 and R502 and experimentally investigated their influence on the solubility of various lubricant oils by measuring critical solubility temperatures. The experiments were conducted to compare refrigerating capacity, COP, discharge temperatures and mass flow rates. Two low-boiling hydrocarbons (HC290 and HCC270) were used as the components for new mixtures due to their lower GWP and better availability than HFC32.The mixtures selected for new units were: HC290/ HFC134a, HFC125/HC290/HFC134a, FC125/HFC143a/HC290, HFC125/ HFC143a/HCC270 and HFC32/HFC125/HFC143a. The mixture of HFC143a/HC290/HCFC22 (28/2/70 by mass %) showed the best performance and its COP and cooling capacity were found to be higher in comparison to R502.

Göktun (1998) presented an outline of refrigerant blends, with and without HCFC, as alternative refrigerants to R502 and compared these blends on the basis of ecological and thermo-physical properties. R404A was found to be the best alternative refrigerant on the basis of COP, volumetric cooling capacity and pressure ratio among all other refrigerant blends. Sami and Desjardins (2000a) presented the test results of performance evaluation of R407B, R507, R408A and R404A as substitutes to R502. Their results revealed that R408A blend has a superior performance than R502 but it is characterized by high discharge pressure compared to R502.

Rakhesh et al. (2003) carried out experimental study on a heat pump with refrigerants HCFC22, R407C and R407A. R407C heat pump-chiller systems offered higher exergy efficiency than those operating with HCFC22. In the case of R407A systems; the exergy efficiency was higher than that of HCFC22 at condensing temperatures less than 50°C. Aprea et al. (2004) compared the performance of HCFC22 and its substitute R417A (R125/R134a/R600 46.6/50/3.4% by mass). R417A does not require a change of lubricant type and it is compatible with mineral, alkyl benzene and ester oils. The experimental results revealed that the COP and the exergetic efficiency of the plant was higher when HCFC22 was used as working fluid in comparison to R417, both when the plant operates as a heat pump and as a water chiller.

Aprea and Renno (2004) experimentally investigated the energetic and exergetic performance of a VCR plant for cold storage application using both HCFC22 and its substitute R417A. The evaporation temperature selected for the operation of the plant was (–)5°C, 0°C and 5°C and air temperature on condenser side was 32°C. The results showed that COP for HCFC22 was 15% greater than R417 whereas exergy destruction for R417 was 14% greater than HCFC22. Arcaklioglu et al. (2005) numerically calculated the rational efficiency and component based irreversibility ratios of a cooling system based on the second law of thermodynamics using HFC and HC based pure refrigerants, such as, HFC32, HFC125, HFC134a, HFC143a, HFC 152a, HC290, HC600a and their binary and ternary mixtures, along with CFC12, HCFC22 and R502. The effect of temperature glide, occurring at the condenser and evaporator, on the rational efficiency of the cooling system was evaluated. Their calculations revealed that rational efficiency increases parallel to increasing temperature glide values of both the condenser and the evaporator. The irreversibility in condenser was found to vary between 40 and 55% of the total irreversibility.

Stegou-Sagia and Paignigiannis (2005) carried out the irreversibility analysis in a single stage vapour compression cycle with refrigerant mixtures R401B, R401C, R402A, R404A, R406A, R408A, R409A, R410A, R410B and R507. Various parameters of the cycles were changed within a suitable range, and the results obtained were plotted in graphs of exergy efficiency factors or presented in Grassmann diagrams and tables. It was recognized that there will likely not be any major universal substitutes. Some fluids may be better suited for certain applications than others.

Xuan and Chen (2005) carried out an experimental study of a ternary near-azeotropic mixture of HFC161/HFC125/HFC143a (10/45/45 wt%) as a drop in replacement for R502. Without any modification to system components, experimental tests were performed on a VCR plant with a reciprocating compressor, which was originally designed to use R404A, a major substitute for R502. The experimental results under two different rated working conditions indicated that the pressure ratios of this new refrigerant were nearly equal to those of R404A. Under lower evaporative temperature, its COP was almost equal to that of R404A and its discharge temperature was found to be slightly higher than that of R404A, while under higher evaporative temperature, its COP was found to be greater than that of R404A and its discharge temperature was lower than that of the latter. This new refrigerant achieved a high level of COP and hence was considered a promising retrofit refrigerant to R502.

Park and Jung (2007a) studied the performance of two pure hydrocarbons and seven mixtures composed of propylene (R1270), propane (R290), HFC152a, and dimethylether (RE170, DME) in an attempt to substitute HCFC22 in residential air conditioners. Their test results revealed that the COP of these mixtures was up to 5.7% higher than that of HCFC22. Most of the fluids showed similar capacity to that of HCFC22 except that of propane. The compressor discharge temperatures were reduced by 11-17 °C with these fluids. No problem was found with mineral oil since the mixtures were mainly composed of hydrocarbons. The amount of charge was reduced up to 55% as compared to HCFC22.

Park and Jung (2007b) experimentally tested two pure hydrocarbon refrigerants, R1270 (propylene) and R290 (propane), and three binary mixtures composed of R1270, R290 and R152a in a refrigerating bench tester with a scroll compressor in an attempt to substitute R502. The test results showed that all refrigerants tested had 9.6-18.7% higher capacity and 17.1-27.3% higher COP than R502. The compressor discharge temperature of R1270 was similar to that of R502, while those of all the other refrigerants were 23.7-27.9°C lower than that of R502. For all alternative refrigerants, the charge was reduced up to 60% as compared to R502. There was no problem with mineral oil, since the mixtures were mainly composed of hydrocarbons. These alternative refrigerants offer better system performance and reliability than R502 and can be used as long term substitutes for R502 due to their excellent environmental properties.

Park *et al.* (2008a, b) experimentally tested the thermodynamic performance of R433A, R432A and HCFC22 in a heat pump bench tester under air conditioning and heat pumping conditions. Both R432A and R433A offer similar vapour pressure to HCFC22 for possible 'drop-in' replacement. The test results showed that the COP of R433A was 4.9-7.6% higher than that of HCFC22 while the capacity of R433A was found to be 1.0-5.5% lower than that of HCFC22 under both test conditions. The COP of R432A was found to be 8.5-8.7% higher than HCFC22 and its refrigerating capacity was 1.9-6.4% higher than that of HCFC22 under both test conditions. The compressor discharge temperatures of R432A and R433A were lower than that of HCFC22. The amount of charge required for both these refrigerants was 50-57% lower than that of HCFC22 due to their low density. Overall, both these refrigerants are good long-term environment-friendly alternatives to replace HCFC22 in residential air conditioners and heat pumps due to their excellent thermodynamic and environmental properties with minor adjustments.

Chen (2008) carried out the performance analysis, using simulation software, of R410A as a long-term alternative refrigerant with zero ODP (ozone-depleting-potential) for replacing HCFC22 in a split-type residential air conditioner. It was deduced that the adoption of R410A could be helpful for air conditioners to decrease their heat exchanger size or improve their operation efficiency for power saving. Moreover, compared to HCFC22, R410A could, in fact, help alleviate its overall impact on global warming through significantly reducing the indirect global warming impact.

Arora *et al.* (2007) carried out the exergy analysis of a vapour compression refrigeration system with HCFC22, R407C and R410A. The results were computed for actual vapour compression cycle without liquid vapour solution heat exchanger. It was concluded that R410A is a better alternative as compared to R407C, with high coefficient of performance and low exergy destruction ratio when considering for refrigeration applications. For air-conditioning applications, R407C is a better alternative than R410A.

Lorentzen (1995), Calm (2008), Wang and Li (2007) and Riffat *et al.* (1997) had advocated the use of natural refrigerants such as ammonia, propane, CO_2, water and air. The fourteenth refrigerant report released by Bitzer International specifies R717 (ammonia), R723 (60% ammonia + 40% dimethyl ether), HC290 (propane) and HC1270 (propylene) as long-term halogen free alternative refrigerants to HCFC22 and R502 (Bitzer International: Refrigerant Report no. 14-Edition A-501-14, 2007). Palm (2008) reported that vapour pressure curves of the propane and propene are quite similar to those of HCFC22 and ammonia, indicating that the application areas would be same. Recently, air conditioning provided by ammonia refrigeration systems has found applications on college campuses and office parks, small-scale buildings such as convenience stores, and larger office buildings. These applications have been achieved by using water chillers, ice thermal storage units, and district cooling systems (http://www.iiar.org//aaranswers_history.cfm?). Pearson (2008) reported that the benefits of using ammonia for water chilling applications have been reported by many authors. Moreover, ammonia is widely used in industrial systems for food refrigeration, cold storage, distribution warehousing and process cooling. It has more recently been proposed for use in applications such as water chilling for air conditioning systems.

Siller *et al.* (2006) have reported that ammonia is not a contributor to ozone depletion, greenhouse effect or global warming. Thus, it is an "environment-friendly" refrigerant. Ammonia has no cumulative effects on the environment and a very limited atmospheric lifetime. Applications for ammonia based refrigeration systems include thermal storage systems, heating ventilating and air-conditioning chillers, process cooling and air conditioning, district cooling systems, supermarkets, convenience stores and air conditioning.

Arora (2009) carried out the detailed theoretical analysis of a vapour compression system for the selection of alternative refrigerant for R502. R502, used for low temperature applications (like supermarket applications) down to – 40°C, is an azeotropic mixture of R22 and R115 and both these refrigerants are harmful to ozone layer. The ozone depletion potentials for R22 and R115 are 0.055 and 0.4, respectively (Calm and Hourahan, 2001). Most promising alternatives, viz., HCFC22, R507A, R404A and R717 were considered for performance comparison with R502. The effects of various parameters, viz., sub-cooling, superheating, effectiveness of liquid vapour heat exchanger, isentropic efficiency of compressor and pressure drops in evaporator and condenser on the COP and exergetic efficiency of the VCR system are evaluated. The results are then used to identify the best refrigerant for R502. The effects of these parameters on efficiency defects in system components are also evaluated. Table 2.3 illustrates the thermo-physical and environmental characteristics of R502, R404A, R507A, R22 and R717 refrigerants.

Table 2.3 Thermo-physical and Environmental Characteristics of R502, R404A, R507A (Xuan and Chen, 2005), R22 and R717 (http://www.iifiir.org/en/web files.php?rub=2)

	R502	R404A	R507A	R22	R717
Molecular weight (g/mol)	111.63	97.60	98.86	86.5	17
Critical temperature (°C)	80.7	72.1	70.9	96.2	135.3
Critical pressure (MPa)	4.02	3.74	3.79	4.99	11.33
Bubble point[a] (°C)	– 45.4	– 46.5	– 46.7	– 40.8	– 33.3
Dew point[a] (°C)	– 45.4	– 46	– 46.7	– 40.8	– 33.3
Temperature glide (°C)	0	0.5	0	0	0
ODP	0.221	0	0	0.055	0
GWP	4500	3800	3900	1500	0
Safety group	A1	A1	A1	A1	B2

[a] Bubble point and dew point are saturation temperature under standard atmospheric pressure, 101.325 kPa.

2.4.1.1 Description of VCR System and Cycle

A schematic diagram of actual vapour compression refrigeration system with liquid vapour heat exchanger (LVHE) is shown in Fig. 2.7. The various departures from the ideal cycle are shown on p-h (Fig. 2.8) diagram, viz. superheating of suction vapour, sub-cooling of liquid refrigerant and

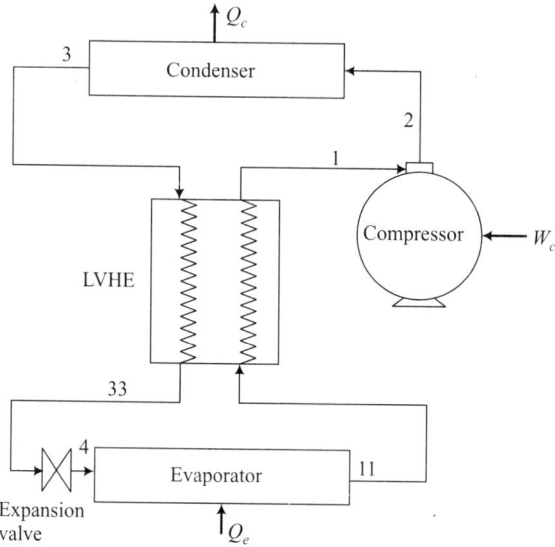

Fig. 2.7 Schematic Diagram of a Vapour Compression Refrigeration System with Liquid Vapour Heat Exchanger (lvhe) (Dincer, 2003)

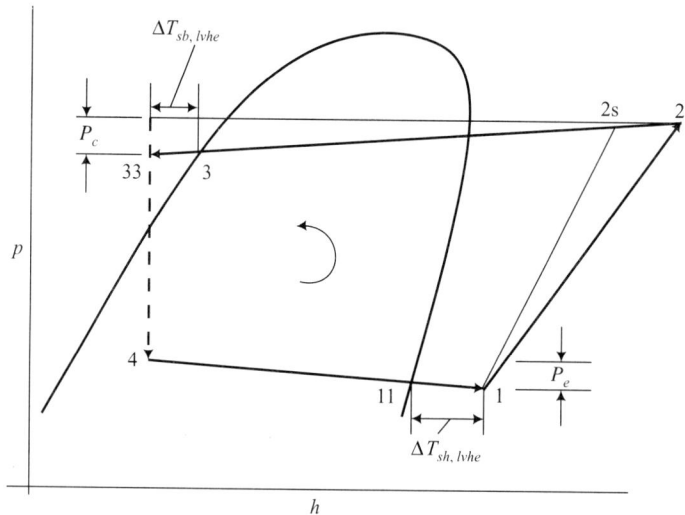

Fig. 2.8 P-h Diagram of VCR System Shown in 2.8(a)

pressure losses in evaporator and condenser. The main components of a vapour compression refrigeration (VCR) system are evaporator, compressor, condenser and a throttling device (expansion valve). A liquid vapour heat exchanger may be incorporated between the liquid line and suction line to transfer the heat from hot liquid refrigerant leaving the condenser to the cold suction vapour entering the compressor. It improves the overall system performance in some cases.

The various processes occurring in the cycle are heat abstraction from cold room (processes 4-11), actual compression process in compressor (processes 1-2) whereas isentropic compression is represented by process '1-2s', heat rejection in condenser (processes 2-3), throttling in expansion device (processes 33-34) and processes taking place in 'lvhe' are heat transfer from high pressure and high temperature liquid refrigerant leaving the condenser to low pressure and low temperature saturated vapour refrigerant leaving the evaporator (i.e., processes '3-33' causing sub-cooling of liquid refrigerant and processes '11-1' causing superheating of suction vapour in 'lvhe' respectively).

Alternatively, the processes of sub-cooling '3-33' can also occur in condenser and processes '11-1' may take place in evaporator if there is no 'lvhe'. The processes '3-33' illustrate the sub-cooling of the liquid refrigerant ($\Delta T_{sb, lvhe}$) and processes '11-1' represent the superheating of suction vapour ($\Delta T_{sb, lvhe}$). The pressure drops in evaporator and condenser are represented by δP_e and δP_c respectively. The pressure drops at inlet and discharge valves of the compressor are neglected.

2.4.1.2 Thermodynamic Analysis

For the thermodynamic analysis of the VCR system, the principles of mass conservation, energy balance (first law of thermodynamics) and exergy balance (second law of thermodynamics) are applied to each component of the system. Each component can be treated as a control volume with inlet and outlet streams, heat transfer and work interactions. The governing equations of mass conservation for a steady state and steady flow system are:

$$\Sigma \dot{m}_i - \Sigma \dot{m}_e = 0 \tag{2.1}$$

The first law of thermodynamics yields the energy balance of each component of the VCR system as follows (Şencan et al., 2005):

$$\Sigma(\dot{m}h)_i - \Sigma(\dot{m}h)_e + \left[\Sigma \dot{Q}_i - \Sigma \dot{Q}_e\right] + \dot{W} = 0 \tag{2.2}$$

The cooling COP of the VCR system is defined as the heat load of the evaporator per unit power input to the compressor and is expressed as:

$$\text{COP} = \frac{\dot{Q}_e}{\dot{W}_{comp}} = \frac{\dot{m}(h_{11} - h_4)}{\dot{m}(h_2 - h_1)} = \frac{h_{11} - h_4}{h_2 - h_1} \tag{2.3}$$

where $\dot{m}$ is the mass flow rate of the refrigerant and h is the enthalpy of working fluid at corresponding state points.

The second law of thermodynamics derives the concept of exergy, which always decreases due to thermodynamic irreversibility. Exergy is defined as the measure of usefulness, quality or potential of a stream to cause change and an effective measure of the potential of a substance to impact the environment (Dincer, 2003). When the kinetic and potential energies are neglected, specific exergy of a fluid stream can be expressed as (Arora, 2009):

$$e = (h - h_o) - T_o(s - s_o) \tag{2.4}$$

where e is the specific exergy of the fluid at temperature T. The terms h and s are the enthalpy and entropy of the fluid, whereas, h_o and s_o are the enthalpy and entropy of the fluid at environmental temperature (or dead state temperature) T_o (K). According to Bejan et al. (1996), the exergy balance applied to a fixed control volume is given by the equation (2.5).

$$\Sigma \dot{m}_i e_i - \Sigma \dot{m}_e e_e + \dot{Q}\left(1 - \frac{T_0}{T}\right) - \dot{W} - \dot{E}D = 0$$

or

$$\Sigma \dot{E}_i - \Sigma \dot{E}_e + \dot{Q}\left(1 - \frac{T_0}{T}\right) - \dot{W} - \dot{E}D = 0 \tag{2.5}$$

The first two terms are exergy input and output rates of the flow, respectively. The third term is the exergy associated with heat transfer $\dot{Q}$, which is positive if entering into the system. It can also be regarded as work obtained by Carnot engine operating between T and T_0, and is therefore equal to maximum reversible work that can be obtained from heat energy $\dot{Q}$. $\dot{W}$ is the mechanical work transfer to or from the system, and the last term ($\dot{E}D$) is exergy destroyed due to the internal irreversibilities. The principle exergy destruction factors in a process are friction, heat transfer under temperature difference and unrestricted expansion.

Exergy Balance and Exergetic Efficiency

For each individual component of VCR system incorporating 'lvhe', the exergy destruction rate equation for each component is written as follows:

Evaporator

$$\dot{E}D_e = \dot{E}_4 + \dot{Q}_e\left(1 - \frac{T_0}{T_r}\right) - \dot{E}_{11}$$

$$= \dot{m}(h_4 - T_0 s_4) + \dot{Q}_e\left(1 - \frac{T_0}{T_r}\right) - \dot{m}(h_{11} - T_0 s_{11}) \tag{2.6}$$

where T_r is the temperature of cold room.

Compressor

$$\dot{E}D_{comp} = \dot{E}_1 + \dot{W}_{comp} - \dot{E}_2 = \dot{m}T_0(s_2 - s_1) \tag{2.6a}$$

Condenser

$$\dot{E}D_c = \dot{E}_2 - \dot{E}_3 = \dot{m}(h_2 - T_0 s_2) - \dot{m}(h_3 - T_0 s_3) \tag{2.7}$$

Throttle valve

$$\dot{E}D_t = \dot{E}_{33} - \dot{E}_4 = \dot{m}(h_{33} - T_0 s_3) - \dot{m}(h_4 - T_0 s_4) = \dot{m}T_0(s_4 - s_{33}) \tag{2.8}$$

Liquid vapour heat exchanger

$$\dot{E}D_{lvhe} = \dot{E}_3 - \dot{E}_{33} + \dot{E}_{11} - \dot{E}_1$$

$$= \dot{m}((h_3 - h_{33} + h_{11} - h_1) - T_0(s_3 - s_{33} + s_1 - s_{11})) \tag{2.9}$$

The total exergy destruction rate in the VCR system is the sum of the exergy destruction rate in each component and it is expressed using equation (2.10).

$$\dot{ED}_{total} = \dot{ED}_e + \dot{ED}_{comp} + \dot{ED}_c + \dot{ED}_t + \dot{ED}_{lvhe} \qquad (2.10)$$

The exergy rate balance for a thermodynamic system is expressed by equation (2.11).

$$\dot{EF} = \dot{EP} + \dot{ED}_{total} \qquad (2.11)$$

where $\dot{EF}$ and $\dot{EP}$ are the exergy rates of the fuel and the product respectively. The performance of a system based on the second law of thermodynamics is expressed using exergetic efficiency (Bejan et al., 1996). Exergetic efficiency is the ratio of exergy rate of product ($\dot{EP}$) to the exergy rate of fuel ($\dot{EF}$). It is given by equation (2.12).

$$\text{Exergetic efficiency} = \eta_{ex} = \frac{\text{Exergy in product}}{\text{Exergy of fuel}} = \frac{\dot{EP}}{\dot{EF}} = 1 - \frac{\dot{ED}_{total}}{\dot{EF}} \qquad (2.12)$$

In a vapour compression refrigeration system, product exergy ($\dot{EP}$) is the exergy of the cooling load of the evaporator, i.e.:

$$\dot{EP} = \dot{Q}_e \left| \left(1 - \frac{T_0}{T_r}\right) \right| \qquad (2.13)$$

and exergy of fuel is the compressor power input, $\dot{W}_{comp}$. Hence, exergetic efficiency is given by equation (2.14).

$$\eta_{ex} = \frac{\dot{Q}_e \left| \left(1 - \frac{T_0}{T_r}\right) \right|}{\dot{W}_{comp}} = \frac{COP_{vcr}}{COP_{rr}} \qquad (2.14)$$

where COP_{vcr} is the coefficient of performance of VCR cycle and COP_{rr} is the coefficient of performance of reversible refrigerator operating between T_o and T_r.

Non-Dimensional Exergy Destruction

The ratio of the exergy destruction rate in a component to the total exergy destruction rate of the system is defined as non-dimensional exergy destruction of the component. The non-dimensional exergy destruction (Ψ_i) is given by equation (2.15) (Kilic and Kaynakli, 2004).

$$\Psi_i = \dot{ED}_i / \dot{ED}_{total} \qquad (2.15)$$

where i represents the particular component under consideration. By using equation (2.15), the significance of each component's contribution into the total exergy destruction rate of the system can be found.

Efficiency Defect

Likewise non-dimensional exergy destruction, efficiency defect (δ) is another approach of quantifying the non-dimensional contribution of each component towards irreversibility. It defined as the ratio between the exergy destroyed in each component to the exergy flow required to sustain the process (Kotas, 1985). The exergy flow required to sustain a process in the VCR system is the electrical power input to the compressor. Accordingly efficiency defect is given by equation (2.16).

$$\delta_i = \dot{ED}_i / \dot{W}_{comp} \qquad (2.16)$$

where i stands for a particular component. The efficiency defects of the components are linked to the exergetic efficiency of the whole plant by means of the relation (2.17).

$$\eta_{ex} = (1 - \Sigma \delta_i) \qquad (2.17)$$

Exergy Destruction Ratio (EDR)

EDR is the ratio of total exergy destruction rate in the system to the exergy in product (Said and Ismail, 1994) and it is given by equation (2.18). EDR is related to the exergetic efficiency by equation (2.19).

$$\text{EDR} = \frac{\dot{ED}_{total}}{\dot{EP}} = \frac{COP_{rr}}{COP_{vcr}} - 1 = \frac{1}{\eta_{ex}} - 1 \qquad (2.18)$$

$$\eta_{ex} = \frac{1}{1 + \text{EDR}} \qquad (2.19)$$

Simulation Model

A computer code is developed using Engineering Equation Solver (EES) software (Klein and Alvarado, 2005) and used to accomplish the energy and exergy analyses of the VCR system shown in Fig. 2.7 and corresponding P-h diagram shown in Fig. 2.8. The computer code was based on mass, energy and exergy balance equations and the available subroutines of the state equations for the thermodynamic properties of various refrigerants in EES. The input parameters are entered into the parametric table. With the given parameters, the program calculates at all state points of the cycle the values of temperature, enthalpy, entropy, mass flow rate, heat transfer rates and exergy destruction in each component of the VCR system. The COP, exergetic efficiency, non-dimensional exergy destruction, efficiency defects and exergy destruction ratios are also computed by the code.

Influence of Operational Parameters on the System Performance

The influence of various performance parameters on COP and exergetic efficiency is discussed in this section. The data assumed for computation of results is specified below.

1. Mass flow rate of refrigerant (m) : 1 kg s^{-1}
2. Dead state temperature and pressure (T_0 and P_0) : 25°C and 1.01325 bar
3. Isentropic compression efficiency (η_{comp}) : 0.75
4. Compressor motor's efficiency (η_{elect}) : 1
5. Pressure drop in evaporator (δP_e) : 20 kPa.
6. Pressure drop in condenser (δP_c) : 20 kPa.

7. The effects of sub-cooling in condenser and superheating in evaporator are neglected.
8. Difference between evaporator and cold room temperature, $(T_r - T_e)$ is taken as 2°C.

Two groups of variations in COP, EDR and exergetic efficiency are plotted. In the first group condensing temperature is varied between 30 and 60°C for a constant evaporating temperature equal to − 40°C. In the second case condenser temperature is taken as −55°C whereas evaporator temperature is varied. The data assumed for plotting Figs. 3.5 to 3.10 is identical as mentioned above except the evaporator and condenser temperatures are (−) 25°C and 55°C respectively.

2.4.1.3 Results and Discussion

Effect of Condenser Temperature

It is observed from Fig. 2.9 (a) and (b) that with increase in condenser temperature, COP of the VCR system reduces. This happens because with the increase in condenser temperature, the dryness fraction of the liquid refrigerant at the exit from expansion device increases and consequently the cooling capacity goes down. Simultaneously, the power consumed by the compressor also increases because of increase in pressure ratio across the compressor and both these factors cause COP to reduce. The COP of R507A is slightly better than R404A. R502 exhibits better COP than both R507A and R404A. The COP for R502 is 3.7% to 25% and 5.7% to 23 % higher in comparison to R507A and R404A respectively. Ammonia and HCFC22 outperform the other three refrigerants. The HCFC22 performs better in comparison to ammonia up to 40°C condensation temperature whereas above 40°C, ammonia gives better COP.

The exergetic efficiency is expressed using equation (2.14). For constant evaporation temperature, only numerator (i.e., COP_{vcr}) changes and denominator (COP_{rr}) remains constant therefore exergetic efficiency is directly proportional to COP of the VCR cycle. Thus, it also reduces, likewise COP, with increase in condenser temperature. The trends of variation in exergetic efficiency are similar to trends of COP curve for these refrigerants. Even the percentage difference in exergetic efficiency of R502 and R507A and R404A are also analogous. Exergy destruction ratio is inversely proportional to exergetic efficiency as expressed in equation (2.19) and hence EDR curves are approximate mirror images of exergetic efficiency curves as depicted in Fig. 2.9 (b).

Effect of Variation in Evaporation Temperature

In the second group (Refer Fig. 2.10 (a), (b) and (c)), the variation of COP, EDR and exergetic efficiency, with evaporating temperature (− 40°C to 0°C) for a constant condensing temperature equal to 55°C, have been plotted. The COP increases with increase in evaporator temperature because of reduction in compressor work and increase in cooling capacity. The COPs of R507A and R404A are lowest among the refrigerants considered. In the ascending order of COPs, the sequence of refrigerants is R404A followed by R507A, R502, R22 and R717. The COP curves of R507A and R404A overlap each other at 55°C condenser temperature. This overlapping of COP curves may not exist at condenser temperatures above or below 55°C because the values of COP obtained at 55°C were identical for − 40°C evaporation temperature also (refer Fig 2.8 (a)). The COP of R502 is approximately 17% higher at −40°C and 10% higher at 0°C in comparison to R507A and R404A.

42 *Alternatives in Refrigeration and Air Conditioning*

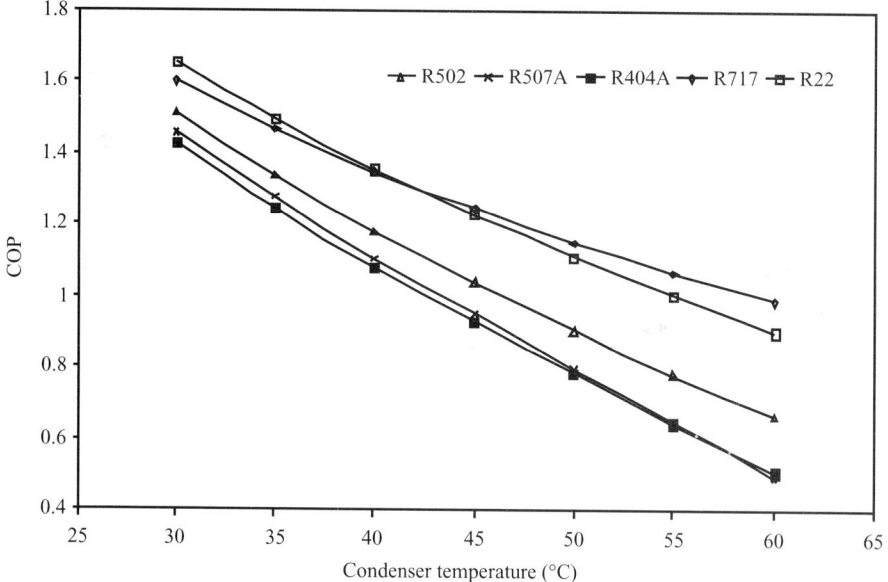

Fig. 2.9(a)   Variation of COP versus Condenser Temperature

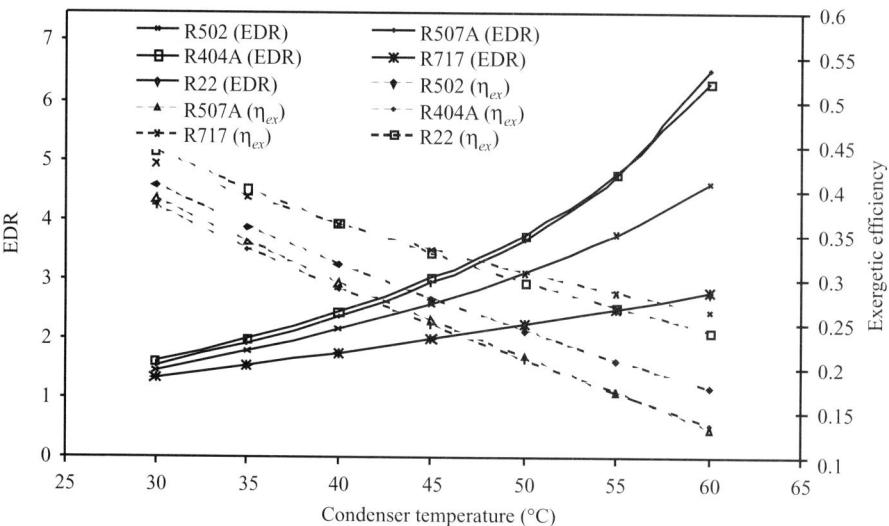

Fig. 2.9(b)   Exergetic Efficiency and EDR versus Condenser Temperature

Figure 2.10 (b) and (c) illustrates the variation of EDR and exergetic efficiency with evaporation temperature. The trend of EDR curves is approximate mirror image of exergetic efficiency curves. The reason for such a behaviour has already been explained. Figure 2.10 (c) illustrates the variation of exergetic efficiency with evaporator temperature. The significant feature of Fig. 2.10 (c) is the rise and fall of the exergetic efficiency with increase in evaporation temperature. Such behaviour of

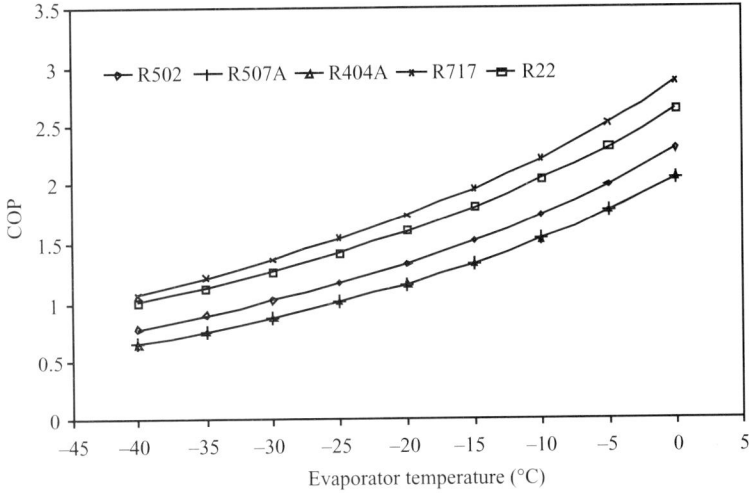

Fig. 2.10(a) Variation of COP versus Evaporator Temperature

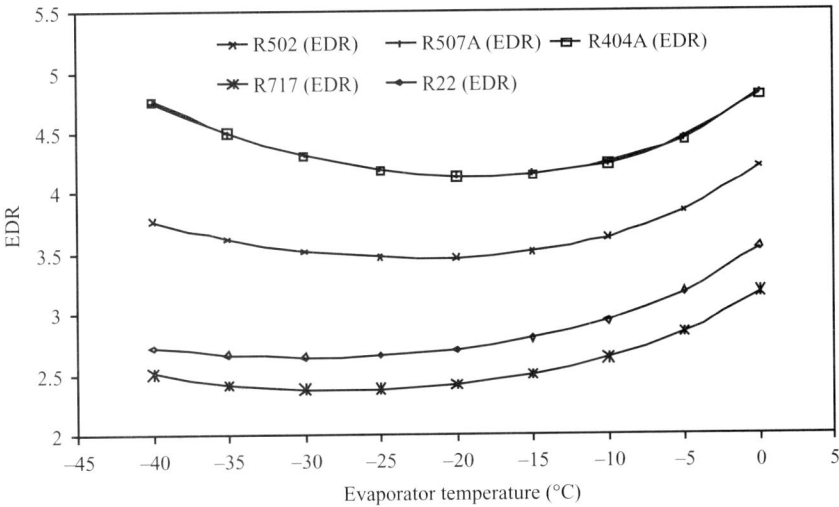

Fig. 2.10(b) EDR versus Evaporator Temperature

the exergetic efficiency can be explained on the basis of the definition of exergetic efficiency given by equation (2.14). The numerator of this equation, i.e., exergy of cooling effect expressed by $\dot{Q}_e |1 - T_0/T_r|$ is the product of $\dot{Q}_e$ and $|1 - T_0/T_r|$. With increase in evaporator temperature, cooling capacity ($\dot{Q}_e$) increases however the magnitude of term $|1 - T_0/T_r|$ reduces since T_r approaches T_0. The power consumed by the compressor decreases with increase in evaporator temperature. Thus, both $\dot{Q}_e$ and $\dot{W}_c$ have positive effect on increase of exergetic efficiency whereas the decreasing value of term $|1 - T_0/T_r|$ has a negative effect on increase of exergetic efficiency. The combined

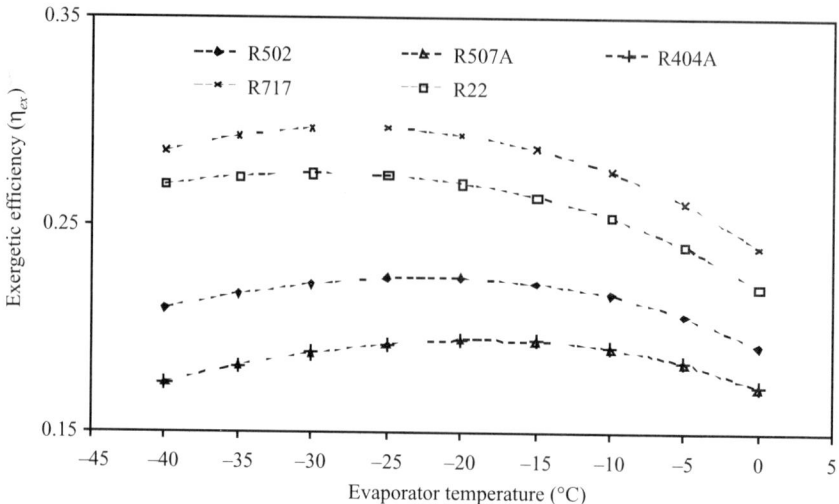

Fig. 2.10 (c) Exergetic Efficiency versus Evaporator Temperature

effect of these factors is to increase the exergetic efficiency till the optimum evaporation temperature (corresponding to maximum exergetic efficiency) is achieved. Beyond optimum evaporation temperature, the overall effect of these factors is to reduce the exergetic efficiency. The exergetic efficiency of ammonia is highest followed by HCFC22, R502, R507A and R404A. The curves of exergetic efficiency of R507a and R404A also coincide.

Efficiency Defects in System Components

The efficiency defects in components of the VCR system are represented in Fig. 2.11 (a) and (b). It can be observed that efficiency defect in condenser is maximum for ammonia among the refrigerants considered in this study. Ammonia is followed by HCFC22, R502, R404A and R507A. The throttle valve is the component in which the second largest efficiency defect occur for refrigerants R507A, R404A and R502. These are followed by HCFC22 and R717. The efficiency defect in evaporator is lowest for all the five refrigerants. The efficiency defects in components of a VCR system depend upon the refrigerant being used. Thus, the improvement in the exergetic efficiency of a VCR system depends upon the improvement in the design of the component in which efficiency defect is maximum. The maximum or minimum value of efficiency defect depend on the refrigerant being used in the VCR system along with the application for which the system is being used.

Effect of Sub-cooling in Condenser

The effect of sub-cooling of liquid refrigerant in condenser on COP and exergetic efficiency is shown in Fig. 2.12(a) and (b). It can be observed that both COP and exergetic efficiency increase with increase in degree of sub-cooling. The sub-cooling reduces the flashing of the liquid refrigerant at exit of the throttle valve. This causes the cooling capacity to increase for the constant mass flow rate of the refrigerant whereas compressor power input remains constant, thus COP increases. The exergetic efficiency is directly proportional to COP of the VCR cycle (Refer equation 2.14) hence it

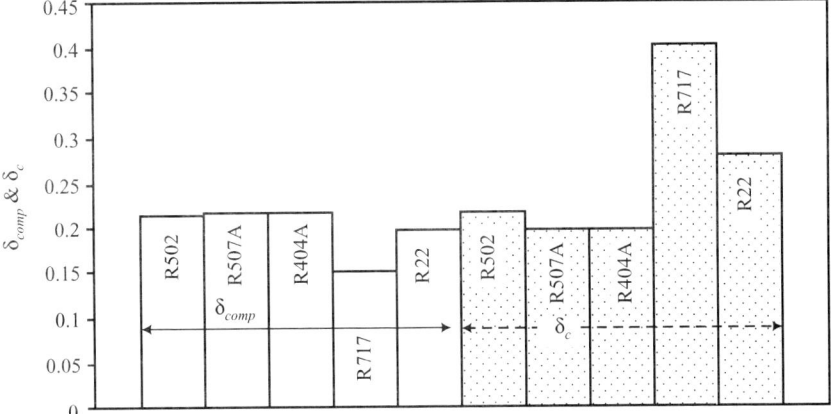

Fig. 2.11(a) Efficiency Defects in Compressor and Condenser for Various Refrigerants

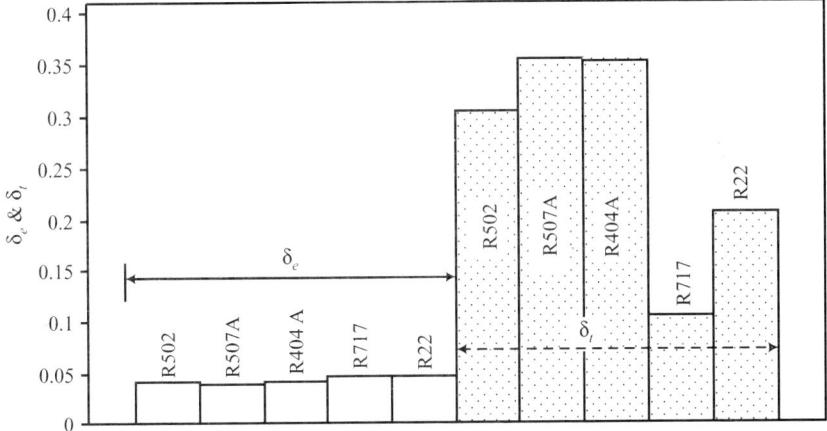

Fig. 2.11(b) Efficiency Defects in Evaporator and Throttle Valve for Various Refrigerants

also increases with increase in degree of sub-cooling. The increase in COP and exergetic efficiency is maximum for R507A among all the refrigerants and lowest for ammonia.

Effect of Superheating in Evaporator

Figure 2.13(a) and (b) illustrates the effect of superheating in evaporator on COP and exergetic efficiency. It is observed that superheating does not have much effect on COP and exergetic efficiency as shown in these figures. In case of R717 there is slight drop in COP and exergetic efficiency. For other refrigerants there is slight increase in COP and exergetic efficiency.

Effect of Effectiveness of Liquid Vapour Heat Exchanger

The effect of variation in effectiveness of liquid vapour heat exchanger on COP and exergetic efficiency is shown in Fig. 2.14 (a) and (b). It is understood that COP and exergetic efficiency increase with increase in effectiveness of 'lvhe'.

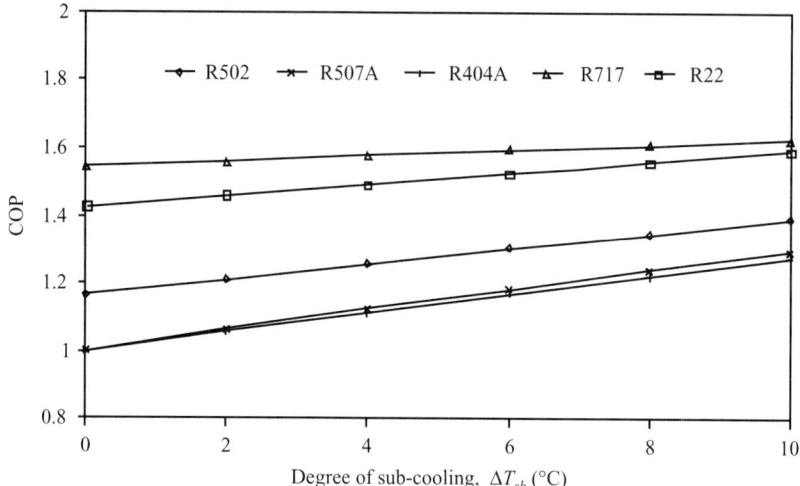

Fig. 2.12(a) Effect of Sub-cooling in Condenser on COP

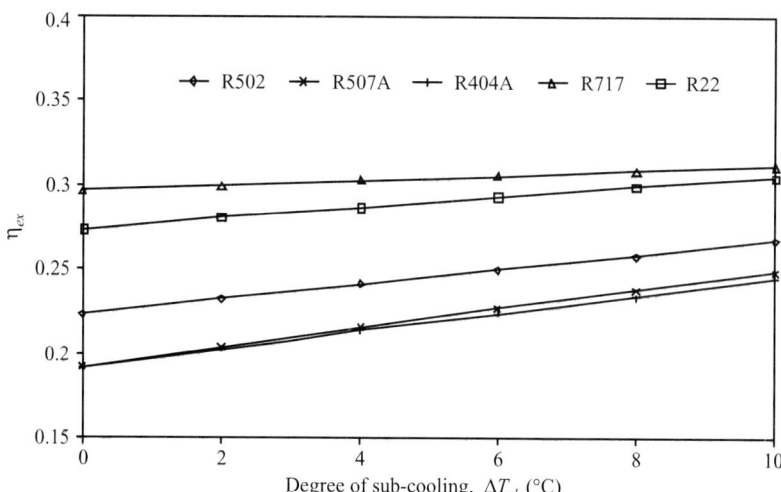

Fig. 2.12(b) Effect of Sub-cooling in Condenser on Exergetic Efficiency

An increase in effectiveness of 'lvhe' from 0 to 1 causes sub-cooling of saturated liquid in 'lvhe' which accounts for increase in the cooling effect. Simultaneously superheating of the suction vapour causes isentropic compression to take place along the isentropic lines having reduced slope thereby increasing the compressor power input. The combined effect of both these factors accounts for increase in the COP of all refrigerants under consideration except ammonia for which COP reduces. The exergetic efficiency of R502, R507A, R404A and HCFC 22 increases whereas it reduces for R717.

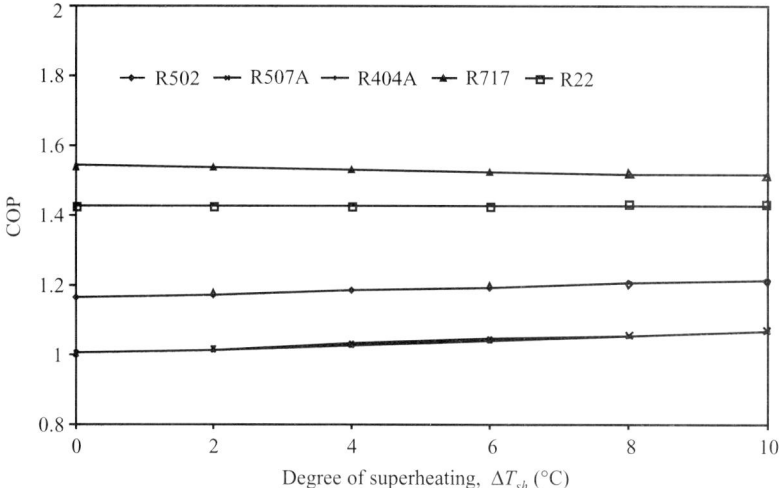

Fig. 2.13(a) Effect of Superheating in Evaporator on COP

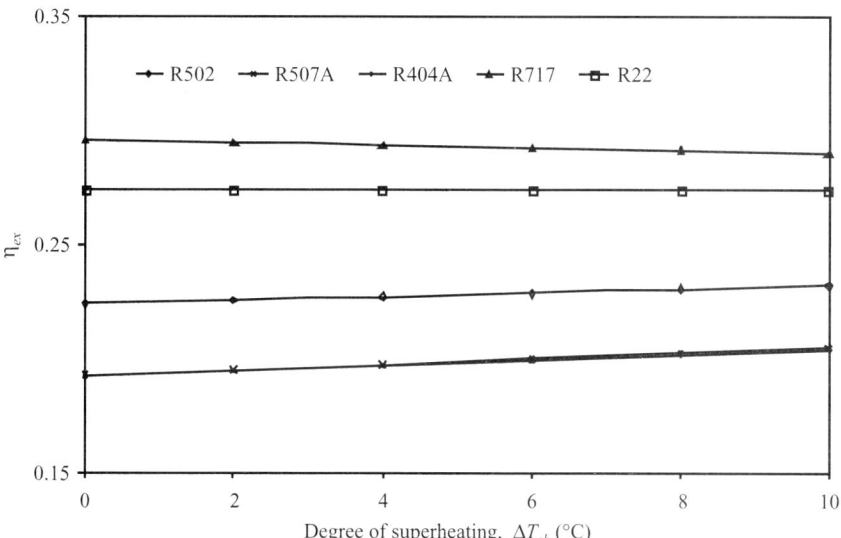

Fig. 2.13(b) Effect of Superheating in Evaporator on Exergetic Efficiency

Effect of Variation in Isentropic Efficiency of Compressor

Figure 2.15(a) and (b) show the effect of variation in compressor efficiency on COP and exergetic efficiency. With increase in isentropic efficiency of compressor, the power required to compress the refrigerant vapour reduces whereas the refrigerating capacity remains constant and the COP increases. The exergetic efficiency being directly proportional to COP (refer equation 2.14) also

48 *Alternatives in Refrigeration and Air Conditioning*

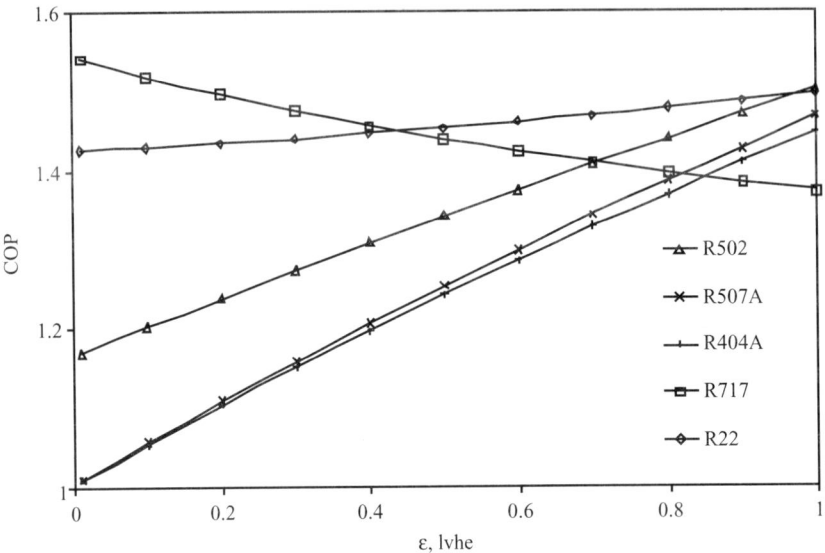

Fig. 2.14(a)   Effect of Effectiveness of 'lvhe' on COP

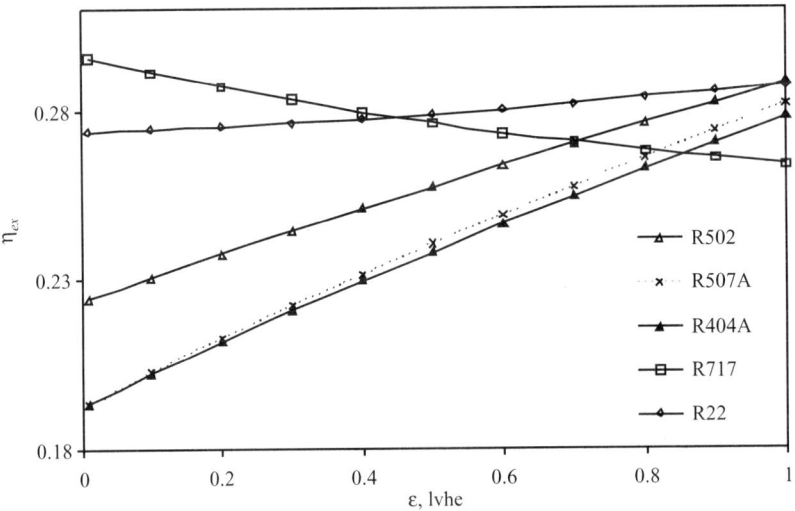

Fig. 2.14(b)   Effect of Effectiveness of 'lvhe' on Exergetic Efficiency

increases. The exergy destruction ratio is inversely proportional to the exergetic efficiency, therefore it decreases. The increase in COP and exergetic efficiency with increase in isentropic efficiency is highest for ammonia.

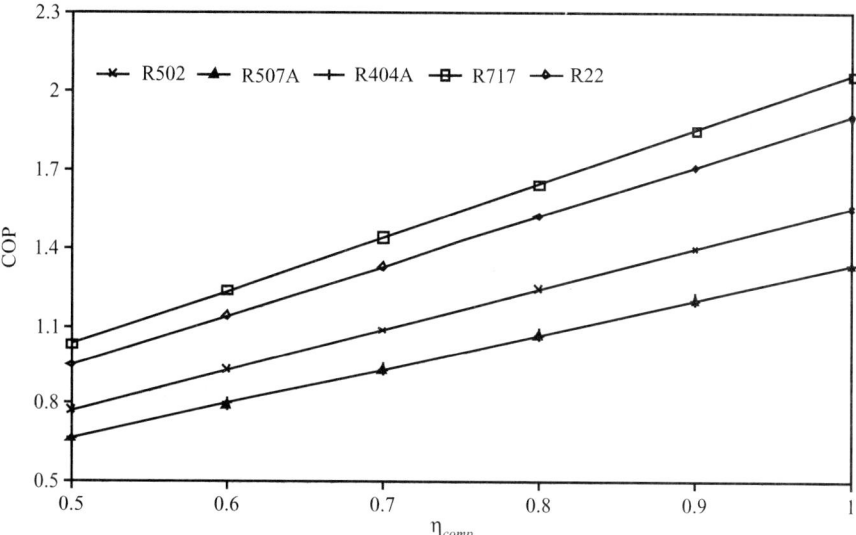

Fig. 2.15(a) Effect of Compressor Efficiency on COP

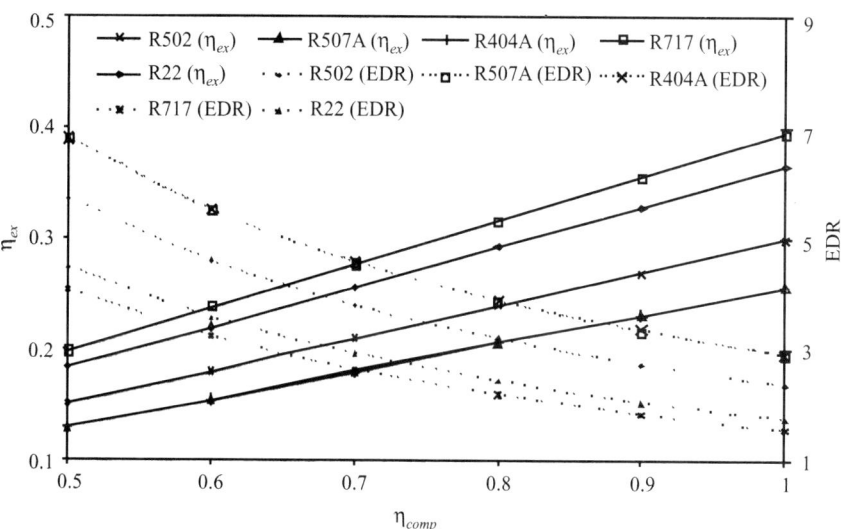

Fig. 2.15(b) Effect of Compressor Efficiency on Exergetic Efficiency and EDR

Effect of Pressure Drops in Evaporator and Condenser

The effects of pressure drops in evaporator and condenser on COP and exergetic efficiency are shown in Fig. 2.16 (a) and 2.16 (b). It is observed that the COP reduces with increase in pressure drop in evaporator. As the pressure drop in evaporator enhances, there is decrease in cooling capacity due to reduction in specific refrigerating effect and pressure ratio across compressor increases

50 *Alternatives in Refrigeration and Air Conditioning*

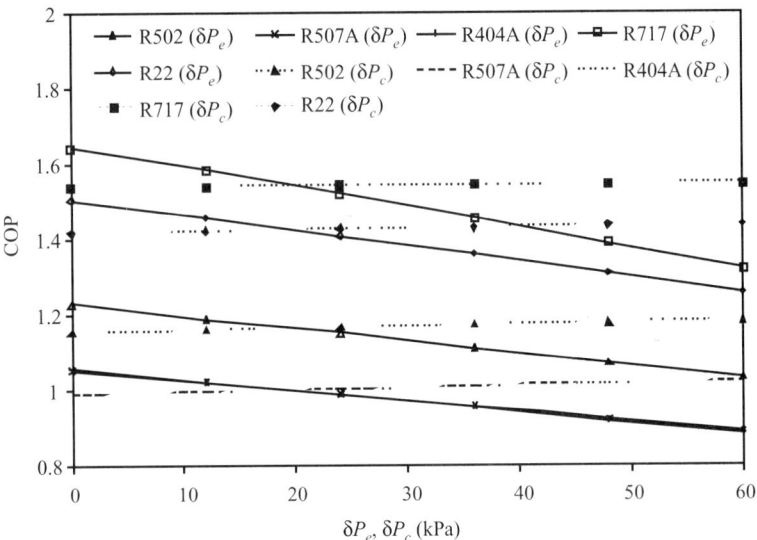

Fig. 2.16(a) Effect of Evaporator and Condenser Pressure Drops on COP

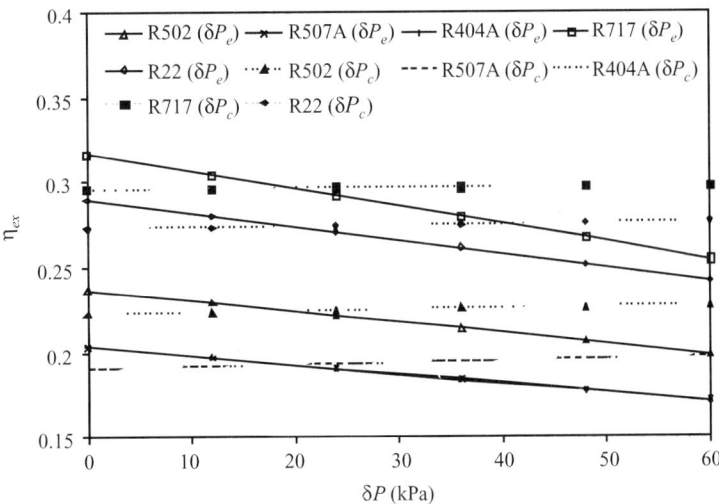

Fig. 2.16(b) Effect of Evaporator and Condenser Pressure Drops on Exergetic Efficiency

causing compressor power to increase. Hence, both COP and exergetic efficiency reduce. The pressure drop in condenser is beneficial since the exit state from expansion device is now nearer to the liquid saturation curve and hence specific refrigerating effect increases for the same specific compressor work. Thus COP and exergetic efficiency increase. The increase in the values of COP and exergetic efficiency is small.

Effect of Performance Parameters on Efficiency Defects in VCR System Components

Table 2.4 shows the effect of sub-cooling, superheating, effectiveness of 'lvhe', isentropic efficiency of compressor and pressure drops in evaporator and condenser on efficiency defects occurring in

Table 2.4 Effect of Various Parameters on Efficiency Defects in Various Components
(Base Cycle Parameters: $T_e = -25°C$, $T_c = 55°C$, $m = 1$ kg/s, $\eta_{comp} = 0.75$, $(T_r - T_e) = 2°C$)

	Efficiency defect (δ) and exergetic efficiency (η_{ex})	Base cycle	Sub-cooling by 10 K in condenser ($\Delta T_{sh,c}$)	Superheating by 10 K in evaporator (ΔT_{sh})	Effectiveness of 'lvhe' (ε, $lvhe = 1$)	Isentropic η_{comp} increases from 75% to 80% (i.e. $\eta_{comp} = 80\%$)	Pressure drop (δP_e) in evaporator increases to 60 kPa	Pressure drop (δP_c) in condenser increases to 60 kPa
R 502	δ_{comp}	0.2139	0.2139	0.208	0.1718	0.172	0.2126	0.2138
	δ_c	0.2162	0.2336	0.2328	0.4000	0.220	0.2058	0.2186
	δ_e	0.0423	0.0445	0.0396	0.03616	0.0452	0.1042	0.04252
	δ_t	0.3037	0.241	0.2871	0.05744	0.3239	0.279	0.2974
	δ_{lvhe}	–	–	–	0.04642	–	–	–
	η_{ex}	0.2239	0.267 (19.25%)	0.223 (−0.41%)	0.2882 (28.71%)	0.239 (6.74%)	0.1984 (−11.39%)	0.2277 (1.70%)
R507A	δ_{comp}	0.2177	0.2177	0.2121	0.1762	0.1748	0.2168	0.2177
	δ_c	0.19547	0.21831	0.21286	0.38948	0.19921	0.18634	0.19823
	δ_e	0.0386	0.04141	0.03568	0.03385	0.04117	0.09628	0.03882
	δ_t	0.356	0.2741	0.3342	0.06314	0.3796	0.3296	0.3485
	δ_{lvhe}	–	–	–	0.05554	–	–	–
	η_{ex}	0.1923	0.24853 (29.25%)	0.20512 (6.67%)	0.2818 (46.54%)	0.20522 (6.72%)	0.17095 (−11.10%)	0.19677 (2.32%)
R404A	δ_{comp}	0.2165	0.2165	0.211	0.1755	0.1739	0.2156	0.2165
	δ_c	0.19673	0.21921	0.2152	0.38921	0.2013	0.18825	0.20033
	δ_e	0.04098	0.04427	0.03801	0.03806	0.0437	0.1011	0.04122
	δ_t	0.3534	0.2757	0.3314	0.06272	0.376	0.3252	0.3457
	δ_{lvhe}	–	–	–	0.05636	–	–	–
	η_{ex}	0.1924	0.24437 (27.01%)	0.20435 (6.21%)	0.2781 (44.54%)	0.20512 (6.61%)	0.1699 (−11.69%)	0.19633 (2.04%)
R717	δ_{comp}	0.1502	0.1501	0.1449	0.1181	0.1218	0.1424	0.1501
	δ_c	0.40196	0.40842	0.41974	0.53661	0.40067	0.40158	0.40285

Contd...

52 *Alternatives in Refrigeration and Air Conditioning*

	Efficiency defect (δ) and exergetic efficiency (η_{ex})	Base cycle	Sub-cooling by 10 K in condenser ($\Delta T_{sb,c}$)	Superheating by 10 K in evaporator (ΔT_{sh})	Effectiveness of 'lvhe' (ε, lvhe = 1)	Isentropic η_{comp} increases from 75% to 80% (i.e. η_{comp} = 80%)	Pressure drop (δPe) in evaporator increases to 60 kPa	Pressure drop (δPc) in condenser increases to 60 kPa
	δ_e	0.04681	0.04757	0.0448	0.03706	0.04991	0.1114	0.04685
	δ_t	0.1045	0.0819	0.09997	0.02596	0.1114	0.09057	0.1027
	δ_{lvhe}	–	–	–	0.01865	–	–	–
	η_{ex}	0.2965	0.31198 (5.22%)	0.29058 (–1.99%)	0.2636 (–11.1%)	0.31619 (6.64%)	0.25412 (–14.29%)	0.29753 (0.35%)
HCFC 22	δ_{comp}	0.1955	0.1955	0.1897	0.1566	0.1576	0.1924	0.1955
	δ_c	0.27765	0.29006	0.29476	0.43976	0.28042	0.26806	0.2793
	δ_e	0.04626	0.04782	0.04394	0.03803	0.04935	0.1113	0.0464
	δ_t	0.2067	0.1616	0.1969	0.04465	0.2205	0.1865	0.2023
	δ_{lvhe}	–	–	–	0.0339	–	–	–
	δ_{ex}	0.2738	0.30502 (11.4%)	0.27465 (0.31%)	0.2871 (4.86%)	0.29212 (6.69%)	0.24179 (–11.69)	0.27651 (0.99%)

the systems components for the refrigerants under consideration. It is observed that providing a liquid vapour heat exchanger is the best option of increasing the exergetic efficiency for R502, R404A and R507A. The increase in the value of exergetic efficiency for HCFC22 is little whereas in case of R717, the exergetic efficiency reduces nearly by 11%. Sub-cooling in the condenser is the second best option to improve the exergetic efficiency of R502, R404A and R507A. The best option to increase the exergetic efficiency of R717 is to improve the isentropic efficiency of compressor. Improving the isentropic efficiency of the compression process by 5% also improves the exergetic efficiency of all other refrigerants by approximately 6.6%. The pressure drop in condenser also account for meager increase in exergetic efficiency by about 1.7%. This is because of reduction in efficiency defect in throttle valve. The superheating in evaporator causes the exergetic efficiency to reduce for R502 and R717 whereas it enhances the exergetic efficiency of R507A, R404A and HCFC22. All other factors lead to reduction in exergetic efficiency. The percentage change in exergetic efficiency is also specified in braces in last row.

2.4.2 Multistage Vapour Compression Refrigeration Systems

Multistage refrigeration systems are usually used for applications which demand very low temperatures which cannot be obtained economically through a single-stage system. According to Dincer (2003), multistage compression refrigeration systems are needed where large temperature and pressure differences exist between the evaporator and the condenser. For VCR systems operating at temperatures below (−)30°C and pressure ratios higher than 6, a multistage VCR system is employed. Some of the disadvantages of the high pressure ratio across compressor are (i) decrease in volumetric efficiency leading to decrease in system capacity and COP (ii) high discharge temperature.

For two-stage VCR cycles, the inter-stage pressure (corresponding to the minimum compressor work) is commonly taken as the geometric mean of the refrigerant condensation and evaporation pressures, which is only applicable for a perfect gas with complete inter-cooling between the stages. However, Threlkeld (1966) demonstrated that for two-stage refrigeration cycles, the optimum inter-stage pressure is much different than the geometric mean pressure. Baumann and Blass (1961) found that optimum occurred at a point quite close to equal pressure ratios in the stages when compressor and motor efficiencies are considered. Goseny (1967) studied the two-stage refrigeration system for various situations; however, he has not given any general expressions for the optimum inter-stage pressure. Keshwani and Rastogi (1968) determined the optimum inter-stage pressure in a two-stage VCR system for refrigerant CFC12. Their research was based on minimization of overall compressor work. Arora and Dhar (1973) used the discrete maximum principle, discussed by Katz (1962), to solve the problem of optimum inter-stage pressure allocation in multistage compression systems for CFC12, with and without inter-cooling between the stages. The results showed that the optimum inter-stage pressure approximately equals the geometric mean of the condensation and evaporation pressures. However, when the flash inter-cooler was incorporated, they found a considerable difference between the geometric mean and the optimal pressure values. Prasad (1981) determined the optimum inter-stage pressure in a two-stage VCR system for the refrigerant CFC12. Their study was based on the maximization of COP. The results revealed that the inter-stage temperature of a two-stage VCR system is given by the geometric mean of the condensation and evaporation temperatures. Zubair and Khan (1995) showed that the optimum inter-stage pressure for a two-stage refrigeration system can be approximated by the saturation pressure corresponding to the arithmetic mean of the condensation and evaporation temperatures. Zubair *et al.* (1996) found that optimum inter-stage pressure for refrigerant HFC134a for maximum COP of the system was close to the saturation pressure corresponding to the arithmetic mean of the refrigerant condensation and evaporation temperatures. The analysis was performed for evaporator temperature (–)30°C, condenser temperatures ranging between 40 and 70°C, compressor efficiency 65%, sub-cooling equal to 3°C and pressure drops in condenser, evaporator and suction line equal to 5%. It was also shown that the system irreversible losses are lowest at an intermediate saturation temperature near to arithmetic mean of the condensation and evaporation temperatures. Nikoladis and Probert (1998) used the exergy analysis to examine the behaviour of two-stage compound compression-cycle, with flash inter-cooling, using refrigerant HCFC22. The condensation and the evaporation saturation-temperatures were varied from 298 to 308K and 238 to 228K respectively. They concluded that greater the temperature difference between condenser and the environment or the evaporator and the cold room, the higher is the irreversibility rate. However, they did not compute the optimum inter-stage temperature (pressure) based on exergy analysis method. Ratts and Brown (2000) computed the optimum reduced intermediate temperature for HFC134a in a two-stage VCR system using entropy generation minimization method. It was revealed that this method gives better results than geometric mean method for evaluation of inter-stage temperature. Ouadha *et al.* (2005) investigated a two-stage VCR system using propane and ammonia and optimized inter-stage pressure for maximization of exergetic efficiency. It was shown that the optimum inter-stage pressure was very close to the saturation pressure corresponding to the arithmetical mean of the refrigerant condensation and evaporation temperatures. Their results were valid for the evaporator temperature of –30°C, condenser temperatures ranging between 30 and 60°C and compressor efficiency equal to 80%. However, they did not consider the effects of sub-cooling, superheating and pressure drops in system components for computing the optimum inter-stage temperature (pressure).

2.4.2.1 Description of Two-stage Vapour Compression Refrigeration (VCR) System

Figure 2.17(a) and (b) shows the schematic and ln *P-h* diagram for a two-stage VCR cycle. The main components of the cycle include low-pressure (LP) compressor, flash inter-cooler, high-pressure (HP) compressor, condenser, expansion valves and evaporator. The refrigerant from the suction line at state 1 is compressed by the LP compressor to the flash inter-cooler, which is operating at an inter-stage pressure. The separated vapour, due to throttling (during the processes 5-6) and de-superheating of the compressed vapour at state 2, yields the increased mass of vapour entering the HP compressor at state 3. The high-pressure superheated vapour leaves the HP compressor at state 4 and enters the condenser. The vapour is condensed in the condenser at state 5 before being throttled to the flash inter-cooler pressure at state 6. The saturated liquid after the flash inter-cooler at state 7 enters the expansion valve, where it expands to the evaporator pressure. At state 8 it leaves the expansion valve and enters the evaporator. In the evaporator, it receives heat from the refrigerated space.

2.4.2.2 Thermodynamic Analysis of the Two-stage VCR Cycle

The thermodynamic analysis of the two-stage VCR system involves the application of principles of mass conservation, energy conservation and exergy balance.

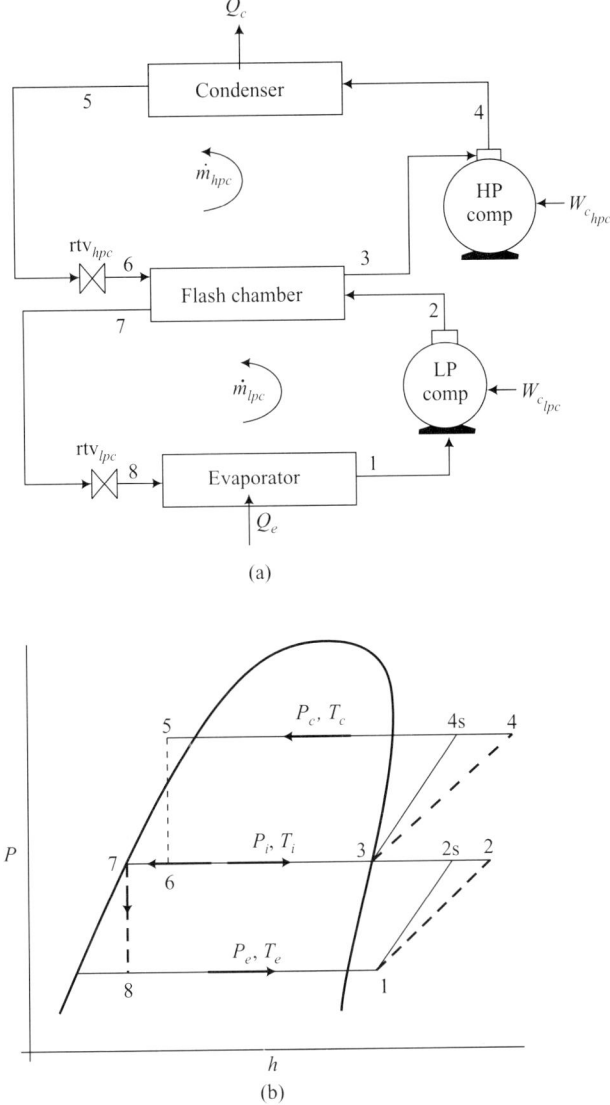

Fig. 2.17(a) Schematic Diagram of a Two-stage VCR System
(b) ln P-h Diagram of Two-stage System Represented in Fig. 2.17 (a)

Mass Balance

The mass flow rates through the LP and HP stages are $\dot{m}_{lpc}$ are $\dot{m}_{hpc}$ respectively.

Energy Balance of two-stage system

The energy balance of a two-stage VCR cycle allows the computation of mass flow rates through LP and HP stages and heat and work transfer rates through various components.

Energy balance across evaporator

The energy balance across evaporator allows determining of mass flow rate in LP stage. The refrigerant mass flow rate (in kg s^{-1}) through the evaporator for $\dot{Q}_e$ kW of refrigeration capacity is given by

$$\dot{m}_{lpc} = \dot{Q}_e/(h_1 - h_8) \qquad (2.20)$$

Energy balance across flash chamber

The energy balance across flash chamber allows computing the mass flow rate in the HP stage. The energy balance across flash chamber is specified through equation (2.21)

$$\dot{m}_{lpc}(h_2 - h_7) = \dot{m}_{hpc}(h_3 - h_6) \qquad (2.21)$$

COP of two-stage VCR system is given by equation (2.22)

$$COP_{ts} = \frac{\dot{Q}_e}{\dot{W}_{comp_{lpc}} + \dot{W}_{comp_{hpc}}} = \frac{\dot{m}_{lpc}(h_1 - h_8)}{\dot{m}_{lpc}(h_2 - h_1) + \dot{m}_{hpc}(h_4 - h_3)} \qquad (2.22)$$

The equations pertaining to exergy balance of two-stage VCR system are not presented and the same can be easily reproduced with the background presented earlier in this chapter. The exergetic efficiency is given by the equation (2.23).

$$\eta_{e_x} = \frac{\dot{Q}_e\left|\left(1 - \frac{T_0}{T_r}\right)\right|}{\dot{W}_{comp_{lpc}} + \dot{W}_{comp_{hpc}}} = COP_{ts}\left|\left(1 - \frac{T_0}{T_r}\right)\right| = \frac{COP_{ts}}{COP_{rr}} \qquad (2.23)$$

2.4.2.3 Results and Discussion

Figure 2.18 shows the comparison of results of Zubair et al. (1996), Zubair and Shaw (1986) and the theoretical model discussed in section 2.4.2.2. The values obtained from theoretical model are within ±5% of the theoretical and experimental values specified in above works. The minor difference between the theoretical and experimental results occurs due to the assumption that isentropic efficiency of compressor is invariable. In a real system, the isentropic efficiency of compressor is a function of evaporator and condenser pressures.

In order to show the enhancement of the two-stage compression cycle performance with respect to single-stage cycle, the following normalized forms of COP, irreversibility and exergetic efficiency are introduced: $COP_n = COP_{ts}/COP_{ss}$ where n represents normalized, ts two-stage and ss single stage. Similarly, normalized irreversibility is given by $\dot{I}_n = \dot{I}_{ts}/\dot{I}_{ss}$ and normalized exergetic efficiency by $\eta_{ex_n} = \eta_{ex_{ts}}/\eta_{ex_{ss}}$.

The data specified below for the computation of the base results is referred from Ouadha et al. (2005).

1. Cooling capacity (Q_e) : 3.5167 kW
2. Isentropic efficiency of compressors, (η_{comp}) : 80 %
3. Evaporator temperature, (T_e) : – 30°C
4. Condenser temperature, (T_c) : 40°C, 55 °C
5. Refrigerants : HCFC 22, ammonia and R410A

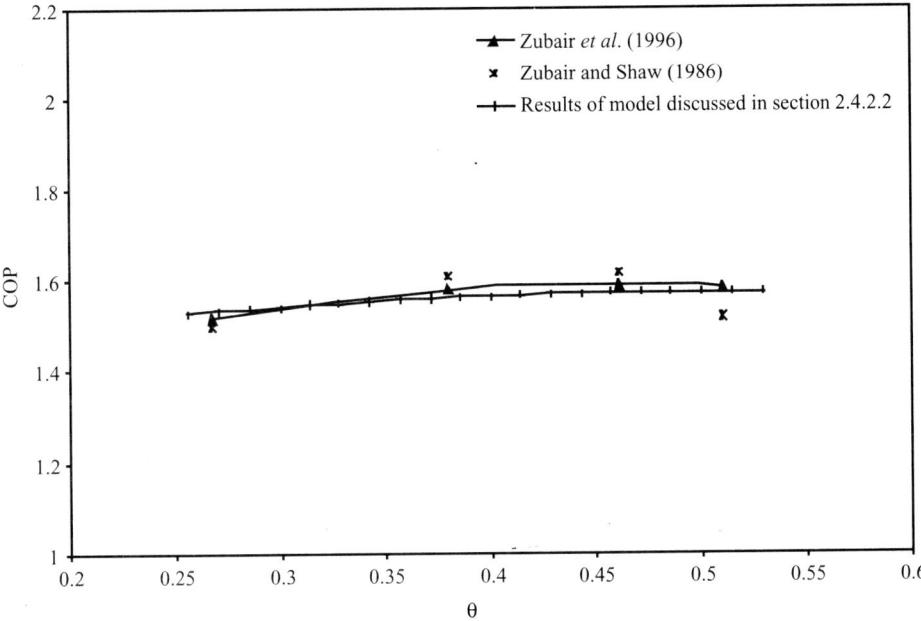

Fig. 2.18 Validation of Calculated Values (Present Work) with Calculated Values of Zubair *et al.* (1996) and Experimental Values of Zubair and Shaw (1986) ($T_e = -28.89$ °C, $T_c = 48.89$ °C, Refrigerant = HCFC22, Sub-cooling = 2.8°C, Return Gas Temperature (i.e., Temperature of the Refrigerant Entering the Compressor = 18.3 °C, $\eta_{comp} = 65\%$)

Ouadha *et al.* (2005) computed the optimum inter-stage saturation temperature (pressure) for ammonia and propane. Their results were valid for two-stage system operating between a constant evaporating temperature of –30°C and condensation temperatures of 30, 40, 50 and 60°C. The effects of sub-cooling of refrigerant, superheating of suction vapour and variation in isentropic efficiency of compressors were not included in their study. The present study incorporates the effect of above parameters. The COP, exergetic efficiency and exergy destruction (irreversibility) are plotted as a function of the reduced inter-stage saturation temperature given by the following expression:

$$\theta = \frac{(T_i - T_e)}{(T_c - T_e)} \tag{2.24}$$

where T_i, T_e and T_c are intermediate, evaporation and condensation temperatures, respectively. This epresentation form is chosen to simplify the illustration of the considerable influence of the intermediate temperature (pressure) on the COP, exergy destruction (irreversibility) and exergetic efficiency of the system.

Effect of Variation in Reduced Inter-Stage Saturation Temperature (θ)

Figure 2.19 (a) and (b) are plots of normalized COP versus dimensionless reduced inter-stage saturation temperature for the refrigerants HCFC22, R410A and R717. It is observed that the normalized COP of the system first increases and then decreases with the increase in reduced inter-stage saturation temperature. The maximum value of normailsed COP is observed to occur at an intermediate saturation temperature nearly halfway between the condensation and evaporation temperatures. The normalised COP of R410 is highest among the three refrigerants followed by HCFC22 and ammonia.

The performance of the two-stage VCR system (as compared to the single stage VCR system) increases as the difference between the condensation and evaporation temperatures increases as depicted in Fig. 2.19(b). Further, it is observed that the optimum reduced inter-stage temperature (θ_{opt}) also increases with increase in condensation temperature from 40°C to 55°C. For HCFC 22, θ_{opt} changes from 0.5143 to 0.5294, for R410A it changes from 0.5429 to 0.5647 and for R717 it increases to 0.4824 from 0.4714. The optimum inter-stage saturation temperature corresponds to the condition of minimum compressor work in both the stages. If 'θ' remains unchanged and the condenser temperature is increased, it causes the compressor work to increase in HT stage though the compressor work in LT stage remains same as previous and the total compressor work increases. As the value of 'θ' increases, it is observed that the compressor work in LT stage increases whereas in HT stage it reduces and total compressor work is also observed to reduce and it becomes minimum at 'θ_{opt}'. Moreover, the optimum inter-stage saturation temperature (θ_{opt}) is near to arithmetic mean of evaporator and condenser temperatures (Zubair and Khan, 1995; Zubair et al. 1996), hence it is clear that 'θ_{opt}' increases with increase in condenser temperature for a constant value of evaporator temperature.

Figure 2.20 (a) and (b) represents the effect of reduced inter-stage saturation temperature on exergetic efficiency. The trends of exergetic efficiency curves are exactly same as that of normalized COP versus reduced inter-stage saturation temperature. Another important point to note here is that maximum exergetic efficiency and maximum COP occur for identical value of θ_{opt} as shown in Fig. 2.21. This happens because both normalized COP and normalized exergetic efficiency have same formulae as proved below.

$$COP_n = \frac{COP_{ts}}{COP_{ss}} = \frac{\dot{Q}_e / \dot{W}_{comp}}{\dot{Q}_e / (\dot{W}_{comp_{lpc}} + \dot{W}_{comp_{hpc}})} \qquad (2.25)$$

$$\eta_{exn} = \frac{\eta_{ex_{ts}}}{\eta_{ex_{ss}}} = \frac{\dot{Q}_e \left|1 - T_0/T_r\right| / \dot{W}_{comp}}{\dot{Q}_e \left|1 - T_0/T_r\right| / (\dot{W}_{comp_{lpc}} + \dot{W}_{comp_{hpc}})} = COP_n \qquad (2.26)$$

Figure 2.22 is a plot showing the normalized irreversibility (exergy destruction) versus reduced inter-stage saturation temperature. This graph is an exact mirror image of the graphs of the normalized COP and normalized irreversibility versus reduced inter-stage saturation temperature. The effect of variation in condensation temperature, though not shown in this graph, can be easily predicted, i.e., with increase in condensation temperature the dimensionless irreversibility reduces since dimensionless exergetic efficiency is increasing.

Effect of Sub-cooling, Superheating and Compressor Efficiency on Optimum Inter-stage Saturation Temperature

Figure 2.23 (a), (b) and (c) shows the effect of sub-cooling, superheating and compressor efficiency on the reduced inter-stage saturation temperature for refrigerants HCFC22, R410A and R717

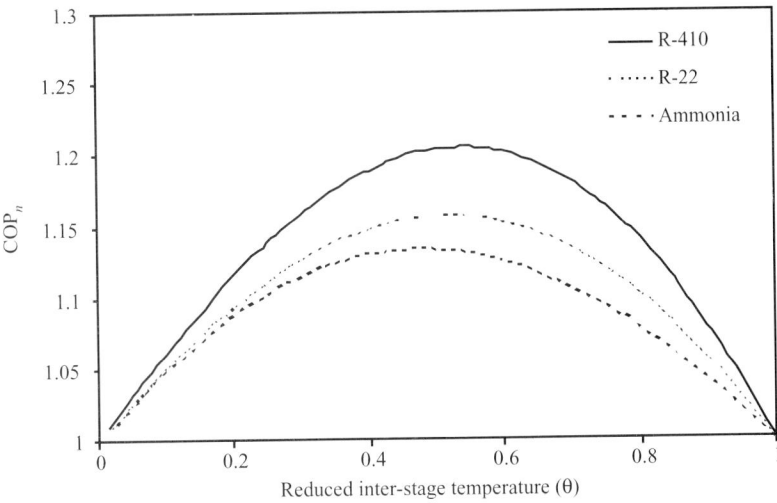

Fig. 2.19(a) Variation of Normalized COP versus Reduced Inter-stage Saturation Temperature ($T_c = 40°C$)

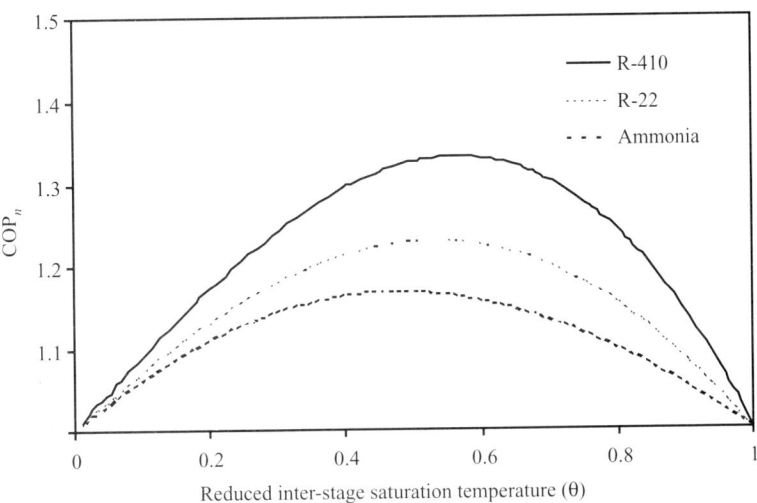

Fig. 2.19(b) Variation of Normalized COP versus Reduced Inter-stage Temperature ($T_c = 55\ °C$)

respectively. It is observed that optimum inter-stage saturation temperature (θ_{opt}) increases with increase in condensation temperatures for HCFC22 and R410A. Reducing the evaporation temperature does have minute influence on θ_{opt} for HCFC22 and R410A in comparison to R717. In case of R717, the θ_{opt} reduces by about 1.6%. Reducing the compressor efficiency by 10% increases the

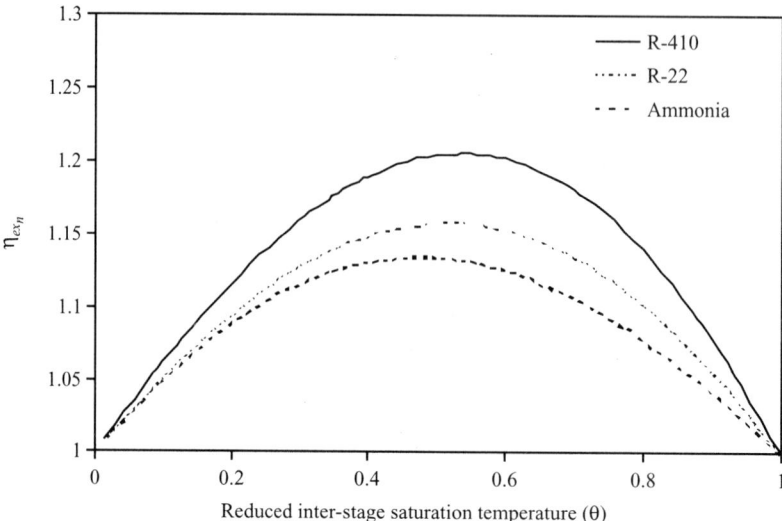

Fig. 2.20(a) Normalised η_{ex} versus Reduced Inter-stage Saturation Temperature ($T_c = 40°C$)

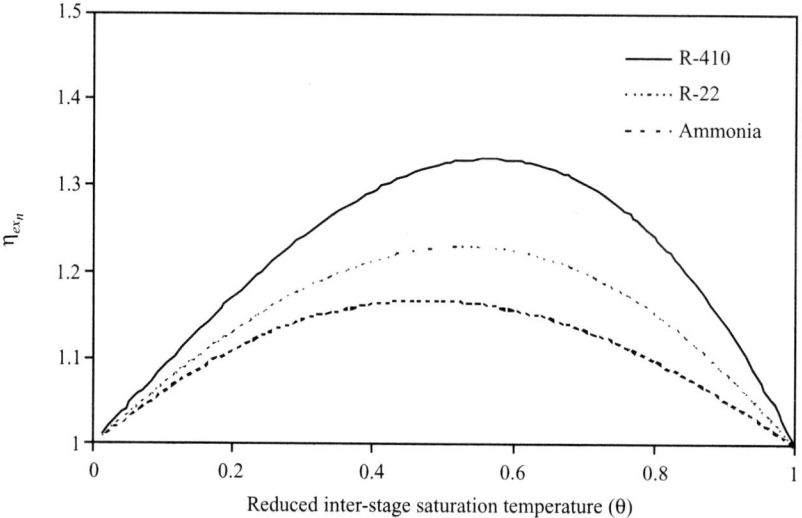

Fig. 2.20(b) Normalised η_{ex} versus Reduced Inter-stage Saturation Temperature ($T_c = 55°C$)

θ_{opt} approximately by 2% for HCFC22 and R410A whereas for R717, the value of θ_{opt} increases by about 0.2 to 0.3%. Superheating the suction vapour by 10°C brings down the θ_{opt} by about 2% for HCFC22 and R410A at lower values of condensation temperatures between 30°C and 50°C. However, at condensation temperatures higher than 50°C, θ_{opt} increases in comparison to θ_{opt} when there is no superheating. This increase is however insignificant. The influence of superheating of suction vapour for R717 is significant. It is observed that θ_{opt} reduces by about 6.5% at 30°C condensation

Alternative Refrigerants and Cycles for Compression Refrigeration Systems 61

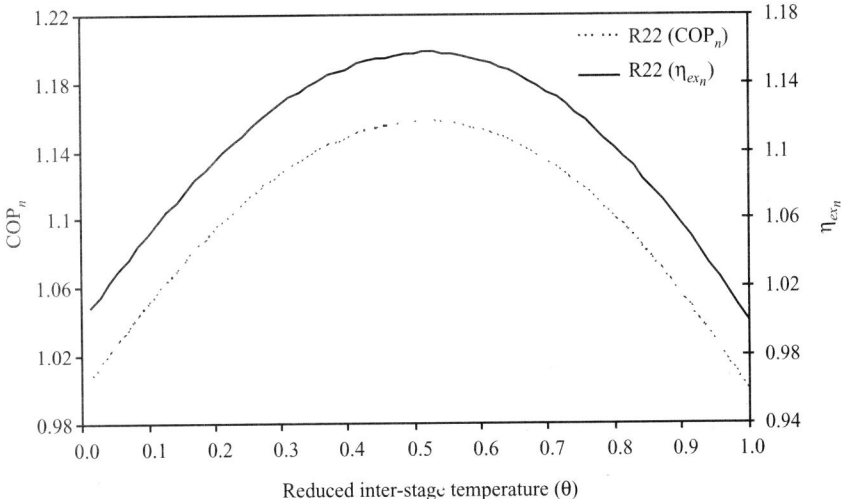

Fig. 2.21 Normalized COP and Normalized η_{ex} for HCFC22 ($T_c = 40°C$)

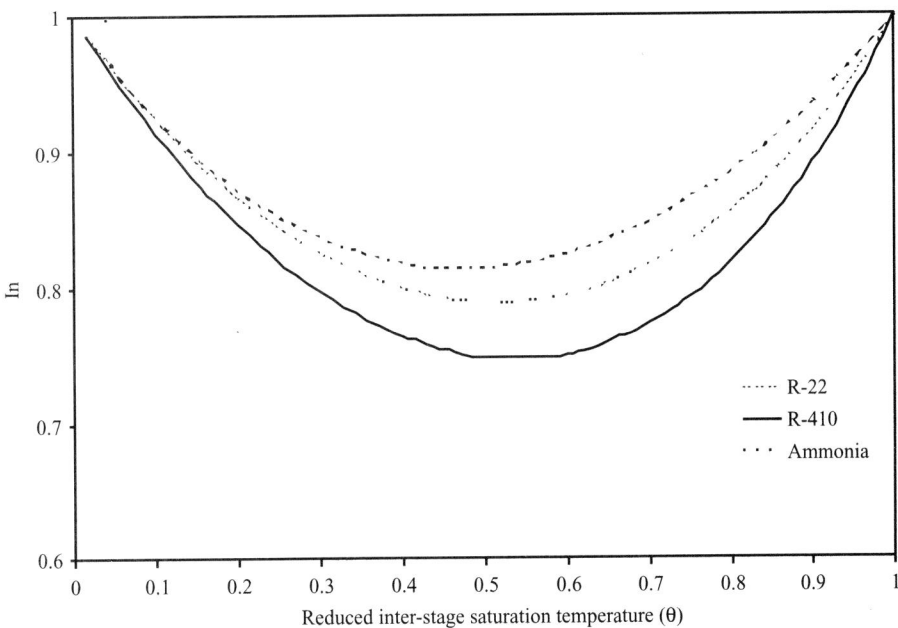

Fig. 2.22 Normalised Irreversibility versus Reduced Inter-stage Saturation Temperature ($T_c = 40°C$)

temperature. With increase in condensation temperature above 30°C, the rate of decrease in θ_{opt} decreases. At 60°C condensation temperature θ_{opt} is about 3.87% lower than the value of θ_{opt} when there was no superheating. The considerable decrease in optimum reduced inter-stage saturation temperature is observed when sub-cooling by 5°C/10°C is carried out (for HCFC22 and R410A)

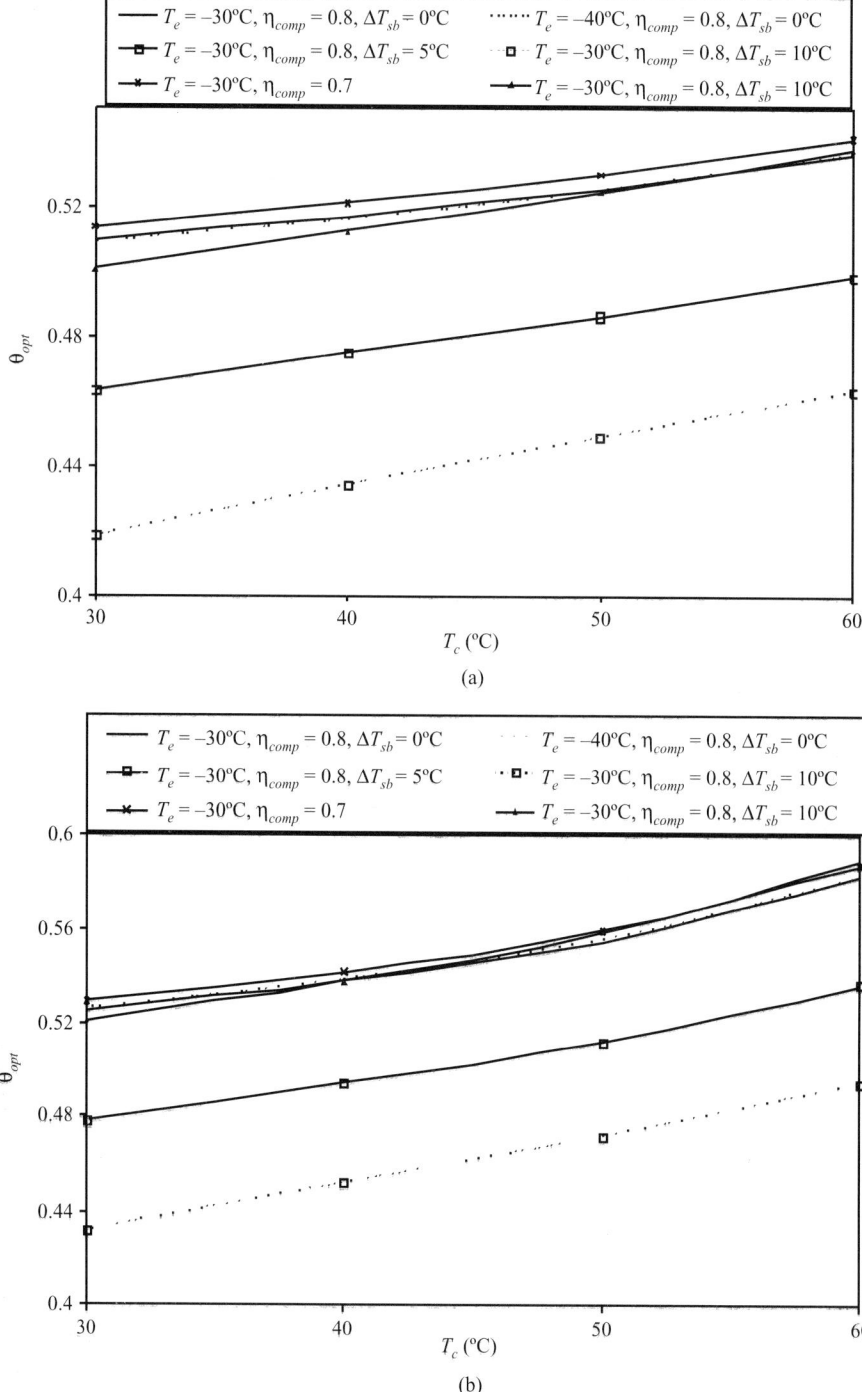

Fig. 2.23 Variation of Reduced Optimum Inter-stage Saturation Temperature versus Condenser Temperature for (a) HCFC22 (b) R410A

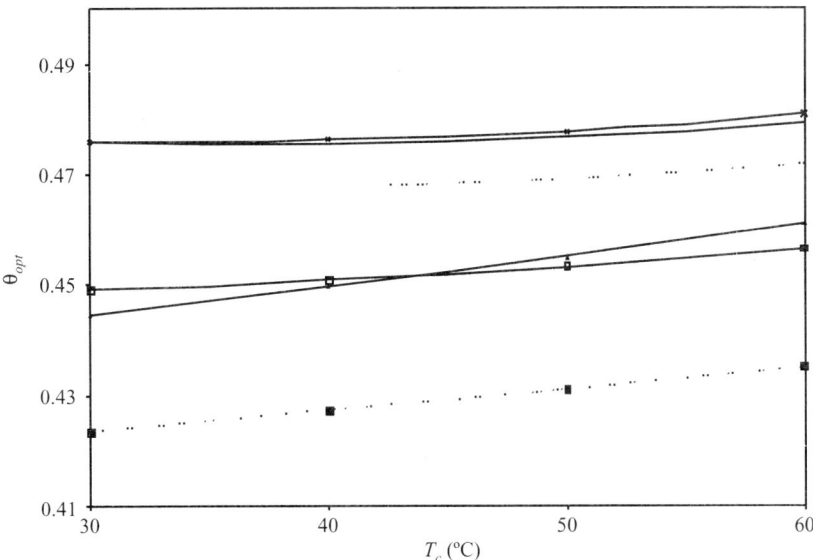

Fig. 2.23(c) Variation of Optimum Reduced Inter-stage Temperature versus Condenser Temperature for R717

which reduces the optimum inter-stage temperature by about 9%/18%. For R717, the value of θ_{opt} reduces by about 4.7%-5.7% for sub-cooling equal to 5°C. The value of θ_{opt} reduces by about 9.2% to 11% with increase in sub-cooling, i.e., sub-cooling by 10°C.

Tables 2.5-2.10 show the comparison of optimum inter-stage saturation temperature with AMT (arithmetic mean of evaporation and condensation temperatures) and GMT (geometric mean of evaporation and condensation temperatures), maximum COP and maximum exergetic efficiency with corresponding values of COP and exergetic efficiency for AMT and GMT respectively corresponding to various operating conditions (fixed operating conditions are: T_e = –30°C, T_c = 30°C, AMT = 0°C, GMT = –1.65°C). Table 2.5 shows the results of an ideal two-stage cycle with isentropic efficiencies of LP and HP compressors as 100%. It is observed that optimum inter-stage saturation temperature for HCFC22 and R410A is higher than AMT (0°C) and GMT (–1.65°C). The optimum inter-stage saturation temperature for R717 is lower than AMT and nearer to GMT. Table 2.6 shows the effect of reducing isentropic efficiencies of compressors from 100% to 80%. It is observed that there is minute increase in optimum inter-stage saturation temperature with decrease in isentropic efficiencies of LP and HP compressors. The corresponding values of maximum COP and maximum exergetic efficiency also reduce. The value of optimum inter-stage saturation temperatures for HCFC22 and R410A is near to AMT whereas for R717, it is near to GMT.

Tables 2.7 and 2.8 illustrate the effect of sub-cooling in condenser on the Ti_{opt}, COP_{max} and maximum exergetic efficiency. The effect of sub-cooling is to reduce Ti_{opt} and increase COP_{max} and maximum exergetic efficiency. The value of Ti_{opt} drops down below GMT due to sub-cooling. With increase in sub-cooling from 5°C to 10°C, the difference between Ti_{opt} and GMT increases. The maximum values of COP and exergetic efficiency also increase by sub-cooling.

Table 2.5 Comparison of Ti_{opt}, AMT and GMT, COP_{max}, COP_{amt} and COP_{gmt} and $\eta_{ex_{max}}$, $\eta_{ex_{amt}}$ and $\eta_{ex_{gmt}}$ for Ideal Two-stage Cycle ($\eta_{comp_{LP}} = \eta_{comp_{HP}} = 1$)

Refrigerant	Ti_{opt} (°C)	COP_{max}	$\eta_{ex_{max}}$ (%)	COP_{amt}	$\eta_{ex_{amt}}$ (%)	COP_{gmt}	$\eta_{ex_{gmt}}$ (%)
HCFC22	0.36	3.52	62.83	3.52	62.82	3.52	62.79
R410A	1.28	3.45	61.49	3.44	61.47	3.44	61.39
R717	−1.53	3.55	63.32	3.54	63.29	3.55	63.31

Table 2.6 Comparison of Ti_{opt}, AMT and GMT, COP_{max}, COP_{amt} and COP_{gmt} and $\eta_{ex_{max}}$, $\eta_{ex_{amt}}$ and $\eta_{ex_{gmt}}$ for Two-stage Cycle ($\eta_{comp_{LP}} = \eta_{comp_{HP}} = 0.8$)

Refrigerant	Ti_{opt} (°C)	COP_{max}	$\eta_{ex_{max}}$ (%)	COP_{amt}	$\eta_{ex_{amt}}$ (%)	COP_{gmt}	$\eta_{ex_{gmt}}$ (%)
HCFC22	0.60	2.77	49.49	2.77	49.48	2.77	49.46
R410A	1.53	2.71	48.42	2.71	48.40	2.71	48.35
R717	−1.52	2.8	49.89	2.79	48.87	2.79	49.88

Table 2.7 Comparison of Ti_{opt}, AMT and GMT, COP_{max}, COP_{amt} and COP_{gmt} and $\eta_{ex_{max}}$, $\eta_{ex_{amt}}$ and $\eta_{ex_{gmt}}$ for Two-stage Cycle ($\eta_{comp_{LP}} = \eta_{comp_{HP}} = 0.8$, $\Delta T_{sb} = 5°C$)

Refrigerant	Ti_{opt} (°C)	COP_{max}	$\eta_{ex_{max}}$ (%)	COP_{amt}	$\eta_{ex_{amt}}$ (%)	COP_{gmt}	$\eta_{ex_{gmt}}$ (%)
HCFC22	−2.21	2.83	50.48	2.83	50.44	2.83	50.47
R410A	−1.3	2.78	49.62	2.77	49.59	2.77	49.61
R717	−3.13	2.83	50.47	2.82	50.41	2.82	50.44

Table 2.8 Comparison of Ti_{opt}, AMT and GMT, COP_{max}, COP_{amt} and COP_{gmt} and $\eta_{ex_{max}}$, $\eta_{ex_{amt}}$ and $\eta_{ex_{gmt}}$ for Two-stage Cycle ($\eta_{comp_{LP}} = \eta_{comp_{HP}} = 0.8$ and $\Delta T_{sb} = 10°C$)

Refrigerant	Ti_{opt} (°C)	COP_{max}	$\eta_{ex_{max}}$ (%)	COP_{amt}	$\eta_{ex_{amt}}$ (%)	COP_{gmt}	$\eta_{ex_{gmt}}$ (%)
HCFC22	−4.91	2.89	51.5	2.88	51.34	2.88	51.43
R410A	−4.1	2.85	50.87	2.84	50.74	2.85	50.82
R717	−4.86	2.87	51.2	2.85	50.93	2.86	51

Table 2.9 shows the effect of superheating of suction vapour on the Ti_{opt}, COP_{max} and maximum exergetic efficiency. The superheating of the suction vapour minutely reduces the Ti_{opt} for HCFC22 and R410A. For R717, Ti_{opt} reduces by 1.85°C. The maximum COP and maximum exergetic

Table 2.9 Comparison of Ti_{opt}, AMT and GMT, COP_{max}, COP_{amt} and COP_{gmt} and $\eta_{ex_{max}}$, $\eta_{ex_{amt}}$ and $\eta_{ex_{gmt}}$ for Two-stage Cycle (($\eta_{comp_{LP}} = \eta_{comp_{HP}} = 0.8$, $\Delta T_{sh} = 10°C$)

Refrigerant	Ti_{opt} (°C)	COP_{max}	$\eta_{ex_{max}}$ (%)	COP_{amt}	$\eta_{ex_{amt}}$ (%)	COP_{gmt}	$\eta_{ex_{gmt}}$ (%)
HCFC22	0.043	2.75	49.09	2.75	49.09	2.75	49.07
R410A	1.25	2.69	48.02	2.69	48.01	2.69	47.96
R717	−3.37	2.76	49.18	2.75	49.12	2.75	49.15

Table 2.10 Comparison of Ti_{opt}, AMT and GMT, COP_{max}, COP_{amt} and COP_{gmt} and $\eta_{ex_{max}}$, $\eta_{ex_{amt}}$ and $\eta_{ex_{gmt}}$ for Two-stage Cycle (($\eta_{comp_{LP}} = \eta_{comp_{HP}} = 0.8$, $\Delta T_{sb} = 10°C$, $\Delta T_{sh} = 20°C$)

Refrigerant	Ti_{opt} (°C)	COP_{max}	$\eta_{ex_{max}}$ (%)	COP_{amt}	$\eta_{ex_{amt}}$ (%)	COP_{gmt}	$\eta_{ex_{gmt}}$ (%)
HCFC22	−6.4	2.85	50.8	2.83	50.56	2.84	50.67
R410A	−5.1	2.81	50.15	2.8	49.97	2.8	50.07
R717	−11.1	2.8	49.91	2.77	49.45	2.78	49.58

efficiency also reduce with superheating of suction vapour. The Ti_{opt} is nearer to AMT for HCFC22 and R410A whereas Ti_{opt} for R717 is nearer to GMT.

The combined effect of reducing isentropic efficiencies of LP and HP compressors, sub-cooling and superheating on Ti_{opt}, coefficients of performances and exergetic efficiencies is presented in Table 2.8. The combined effect of these parameters lowers the value of Ti_{opt} even below than that obtained in previous cases. The maximum values of COP and exergetic efficiency are higher in comparison to the maximum values of COPs and exergetic efficiencies obtained for the cases presented in Tables 2.6 and 2.9. The optimum values of inter-stage saturation temperatures for HCFC22, R410A and R717 are lower than both AMT and GMT.

Table 2.11 shows the difference between the Ti_{opt}, AMT and GMT, percentage change in values of maximum COP and exergetic efficiency with reference to values given in Table 2.6. It is observed that the optimum value of inter-stage temperatures for HCFC22 and R410A are nearer to AMT when isentropic efficiencies of compressors are assumed to be lower than 100% and superheating is considered in evaporator.

The optimum value of inter-stage saturation temperature for HCFC22 and R410A is nearer to GMT when either sub-cooling is considered or in the case of cycle with sub-cooling, superheating and isentropic efficiencies of compressors are less than 100%. The optimum value of inter-stage temperature for R717 is closer to GMT under all conditions considered in this analysis.

Table 2.12 illustrates the comparison of optimum inter-stage temperature, AMT and GMT for the following conditions: $T_e = -30°C$, $\eta_{comp_{LP}} = \eta_{comp_{HP}} = 0.8$, $\Delta T_{sb} = 10°C$, $\Delta T_{sh} = 20°C$ for condensation temperatures 40, 50 and 60°C. It is observed that optimum inter-stage saturation

Table 2.11 Difference Between the Ti_{opt}, AMT and GMT, Percentage Change in Values of Maximum COP and Exergetic Efficiency with Reference to Values Given in Table 2.6.

Table No. 'i'	Refrigerant	Difference between AMT and Ti_{opt}	Difference between GMT and Ti_{opt}	Percentage change in the value of COP_{max} (value corresponding to table no. 'i') with reference to the value given in Table 2.6	Percentage change in the value of $\eta_{ex_{max}}$ (value corresponding to table no. 'i') with reference to the value given in Table 2.6
2.6	HCFC22	*0.6*	2.25	------	------
2.7	HCFC22	−2.21	**−0.56**	1.98	1.98
2.8	HCFC22	−4.91	**−3.26**	4.22	4.06
2.9	HCFC22	*0.043*	1.69	−0.81	−0.81
2.10	HCFC22	−6.4	**−4.74**	2.67	2.65
2.6	R410A	*1.53*	3.18	------	------
2.7	R410A	−1.3	**0.35**	2.45	2.45
2.8	R410A	−4.1	**−2.44**	5.04	5.05
2.9	R410A	*1.25*	2.9	−0.83	−0.83
2.10	R410A	−5.1	**−3.42**	3.57	3.57
2.6	R717	−1.52	**0.13**	------	------
2.7	R717	−3.13	**−1.48**	1.17	1.18
2.8	R717	−4.86	**−3.21**	2.61	2.62
2.9	R717	−3.37	**−1.72**	−1.4	−1.41
2.10	R717	−11.1	**−9.46**	0.064	0.054

The values in **bold** show the proximity of optimum inter-stage temperature to GMT and values in ***italic bold*** show the proximity of optimum inter-stage temperature to AMT.

temperature is nearer to GMT except for the case of R410 at 60°C condensation temperature. In this case the optimum temperature is nearer to AMT.

Table 2.12 Comparison of Optimum Inter-stage Temperature, AMT and GMT at Various Condensation Temperatures ($T_e = -30°C$, $\eta_{comp_{LP}} = \eta_{comp_{HP}} = 0.8$, $\Delta T_{sb} = 10°C$, $\Delta T_{sh} = 20°C$)

Refrigerant	Condensation Temperature (T_c in °C)								
	40			50			60		
	Ti_{opt} (°C)	AMT (°C)	GMT (°C)	Ti_{opt} (°C)	AMT (°C)	GMT (°C)	Ti_{opt} (°C)	AMT (°C)	GMT (°C)
HCFC22	−0.62	5	2.79	5.37	10	7.16	11.6	15	11.46
R410A	1.2	5	2.79	7.86	10	7.16	15.18	15	11.46
R717	−4.25	5	2.79	0.67	10	7.16	12.19	15	11.46

Thus, it can be concluded that optimum inter-stage saturation pressure of an actual two-stage refrigeration system is closer to saturation pressure corresponding to GMT of condensation and evaporation temperatures.

Thus two-stage VCR system is examined with a viewpoint to obtain optimum inter-stage saturation temperature for HCFC22, R410A and R717 and further the effects of various parameters have also been investigated on the optimum inter-stage saturation temperature.

2.4.3 Cascade Refrigeration System

Bansal and Jain (2007) reviewed the literature on cascade refrigeration system. They reported that a cascade refrigeration system is normally required for producing low temperatures ranging from (–)30°C to (–)100°C for various industries such as pharmaceutical, food, chemical, blast freezing and liquefaction of gases. The refrigerants specified for use in high temperature circuit are HCFC22, HFC134a, R507A, ammonia, propane, and propylene whereas carbon dioxide, HFC23 and R508B are suitable for use in low temperature circuit. The refrigerant pairs that have received the most attention in recent years are R717/R744 (ammonia/carbon-dioxide) and R1270/R744 (propylene / carbon dioxide) for applications down to (–) 54°C.

Kanoglu (2002) accomplished the exergy analysis of the multistage cascade refrigeration cycle used for natural gas liquefaction. The relations for the total exergy destruction, exergetic efficiency and minimum work requirement for the liquefaction of natural gas in the cycle were developed. It was shown that the minimum work depends only on the properties of the incoming and outgoing streams of natural gas.

Agnew and Ameli (2004) optimized two-stage cascade refrigeration system using finite time thermodynamics approach for refrigerant R717 and R508b in high temperature and low temperature circuits respectively. This pair was found to exhibit better performance in comparison to R12 and R13 pair.

Nicola et al. (2005) carried out the first law performance of a cascade refrigeration cycle, operating with ammonia in high temperature circuit and blends of CO_2 and HFCs in low temperature circuit, for those applications where temperatures below triple point of CO_2 (216.58 K) are needed. Their results revealed that the R744 blends are an attractive option for the low-temperature circuit of cascade systems operating at temperatures approaching 200 K.

Bhattacharyya et al. (2005) carried out the analysis of a cascade refrigeration system for simultaneous heating and cooling with a CO_2 based high temperature cycle and C_3H_8 (propane) based low temperature cycle. They predicted the optimum performance of the system with variation in the design parameters and operating variables. This cascaded system can operate simultaneously between refrigerating space temperature of –40°C and a heating output temperature of about 120°C. Moreover, propane vindicates itself as a better refrigerant than ammonia due to its non-toxic nature. However, its flammability remains a concern.

Lee et al. (2006) optimised condensing temperature of a two-stage cascade refrigeration system for ammonia and carbon dioxide for maximization of COP and minimization of exergy loss. It was deduced that optimal condensing temperature increased with condensation and evaporation temperatures. The effects of sub-cooling and superheating were not taken into consideration. The computation of exergetic efficiency was also not performed.

Kruse and Rüssmann (2006) investigated the COP of a cascade refrigeration system using N_2O (nitrous oxide) as refrigerant for the low temperature cascade stage and various natural refrigerants like NH_3 (ammonia), C_3H_8, propene, CO_2 and N_2O itself for the high temperature stage. They compared its result with a conventional CFC23/HFC134a cascade refrigeration system for heat rejection temperatures between 25 and 55°C. They concluded that by substituting the lower stage refrigerant CFC23 by N_2O practically achieved the same energetic performance with high stage fluids HFC134a, ammonia and hydrocarbons.

Niu and Zhang (2007) carried out the experimental study of a cascade refrigeration system with R290 in high temperature circuit and a blend of R744/R290 in low temperature circuit. The performance of the blend was compared with R13 in low temperature circuit. The blend showed good cycle performance compared with R13 and is considered a promising alternative refrigerant to R13 when the evaporator temperature is higher than 201 K.

Getu and Bansal (2008) carried out energy analysis of a carbon dioxide-ammonia (R744-R717) cascade refrigeration system. Their study involved the examination of the effects of evaporating, condensing and cascade condenser temperatures, sub-cooling and superheating in both high and low temperature circuits on optimum COP. They employed a multi-linear regression analysis and developed mathematical expressions for maximum COP, the optimum evaporating temperature of R717 and the optimum mass flow ratio of R717 to that of R744 in the cascade system. Their study did not include the exergy analysis approach to achieve maximum exergetic efficiency.

Dopazo et al. (2009) carried out theoretical analysis of a CO_2-NH_3 cascade refrigeration system for cooling applications at low temperatures. The results have been presented for optimization of coefficient of performance in the evaporation temperature range (–)55°C to (–)30°C in low temperature circuit, 25 to 50°C condensation temperature in high temperature circuit and (–)25 to 5°C in cascade condenser. The approach temperature was varied between 3 and 6°C. The effect of compressor isentropic efficiency on system COP is also examined. The results show that, when following both exergy analysis and energy optimization methods, an optimum value of cascade condenser temperature is achieved. However, in this study, effect of sub-cooling and superheating for determining the optimum cascade condenser temperatures is not included.

2.4.3.1 Description of Cascade Vapour Compression Refrigeration System

A cascade refrigeration system is used for applications in temperature range between (–)30°C and (–)100°C for cold storage and liquefaction of gases, etc.

Alternative Refrigerants and Cycles for Compression Refrigeration Systems 69

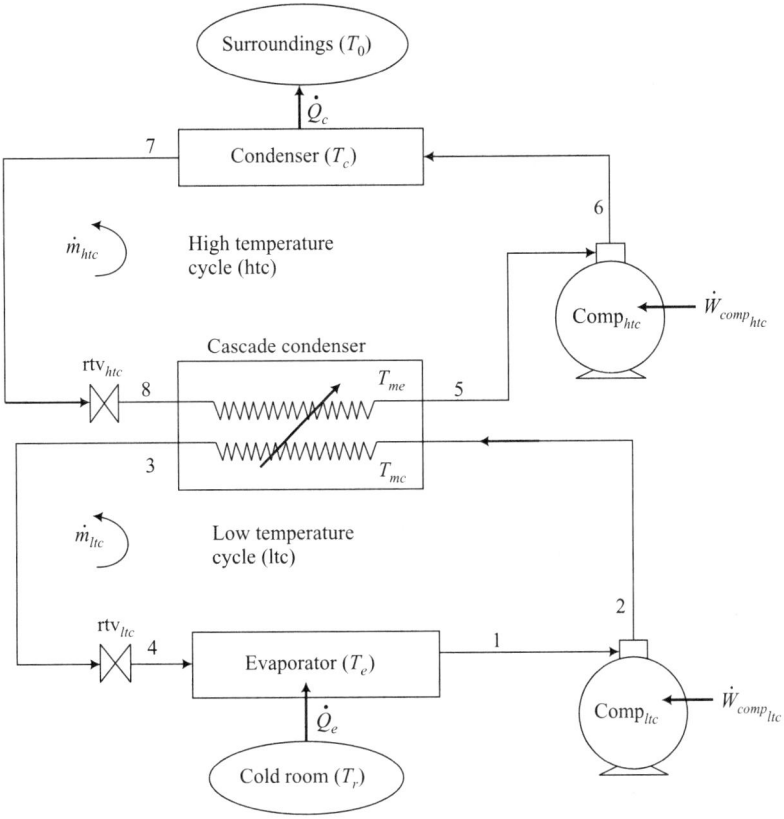

Fig. 2.24(a) Schematic Diagram of a Cascade Refrigeration System

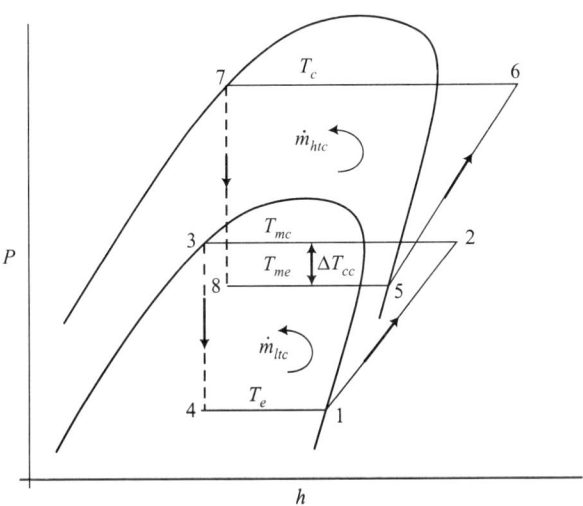

Fig. 2.24(b) P-h Diagram of Cascade Refrigeration System

Figure 2.24(a) schematically represents a two-stage cascade VCR system and Fig. 2.24(b) represents the corresponding pressure enthalpy diagram. This refrigeration system comprises two separate refrigeration cycles—the high-temperature cycle (htc) and the low-temperature cycle (ltc). Each cycle has a different refrigerant suitable for that temperature with lower temperature units progressively using lower boiling point refrigerants.

The lower boiling point refrigerant will have higher saturation pressure at low temperatures that keeps the ingress of air under control and requires a smaller compressor for the same refrigerating effect due to higher density of suction vapours. The cycles are thermally connected to each other through a cascade-condenser, which acts as an evaporator for the 'htc' and a condenser for the 'ltc'. Figure 2.24(a) indicates that the condenser in this cascade refrigeration system rejects heat Q_c from the condenser at condensing temperature T_c to its warm coolant or environment at temperature T_0. The evaporator of the cascade system absorbs refrigeration load $\dot{Q}_e$ from the cold refrigerated space at T_r to the evaporating temperature T_e. The heat absorbed by the evaporator of the ltc plus the work input to the ltc compressor equals the heat absorbed by the evaporator of the htc. T_{mc} and T_{me} represent the condensing and evaporating temperatures of the cascade condenser, respectively. Approach is designated as ΔT_{cc} and it represents the difference between the condensing temperature (T_{mc}) of ltc and the evaporating temperature (T_{me}) of htc. The evaporating temperature (T_e), the condensing temperature (T_c), and the temperature difference in the cascade-condenser (ΔT_{cc}) are three important design parameters of a cascade refrigeration system.

Thermodynamic Analysis of Cascade Refrigeration System

The thermodynamic analysis of the two-stage cascade refrigeration system involves the application of principles of mass conservation, energy conservation and exergy balance as discussed earlier.

The following input data is used (Bansal and Jain (2007)) for computation of results shown in Figs. (2.19) to (2.26).

1. Refrigeration capacity (Q_e) : 1 TR
2. Sub-cooling of refrigerant leaving 'htc' condenser (ΔT_{sb}) : 5°C
3. Superheating of suction vapour in 'ltc' evaporator (ΔT_{sh}) : 10°C
4. Isentropic efficiency of compressors (η_{comp}) : 70%
5. Difference between evaporator and space temperature : $(T_r - T_e) = 10°C$
6. Evaporator temperature (T_e) (in steps of 5°C) : –55°C to –15°C
7. Condenser temperature (T_c) (in steps of 10°C) : 30°C to 60°C
8. Dead state temperature (T_0) and pressure (P_0) : 25°C, 1.01325 bar
9. Reference enthalpy (h_o) and entropy (s_o) of the working fluids have been calculated corresponding to the dead-state temperature (T_0) of 25°C.
10. Heat losses and pressure drops in connecting lines and various components are neglected.

Figure 2.25 (a) and (b) represent the comparison of results obtained using the computer code developed for performance analysis of two-stage cascade VCR system with research of Bansal and Jain (2007). In Fig. 2.25 (a) the variation of COP versus approach in cascade condenser is shown and in Fig. 2.25 (b) variation of COP versus temperature in cascade condenser (T_{cc}) is presented. It

Alternative Refrigerants and Cycles for Compression Refrigeration Systems 71

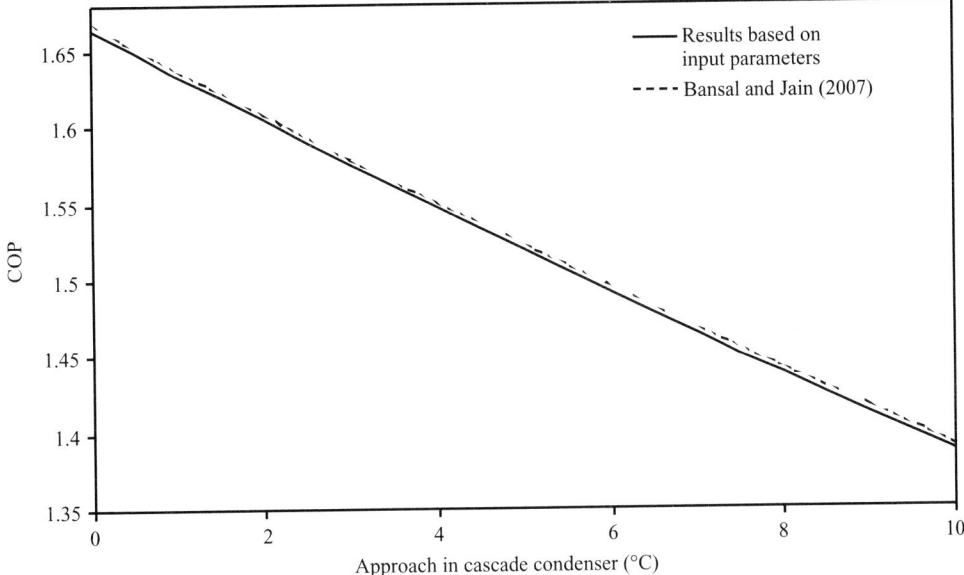

Fig. 2.25(a) Approach versus COP (Comparison of Present Work with Bansal and Jain, 2007) (R-717 in 'htc' and R-744 in 'ltc') ($T_e = -45°C$, $T_c = 30°C$, $T_{cc} = -15°C$)

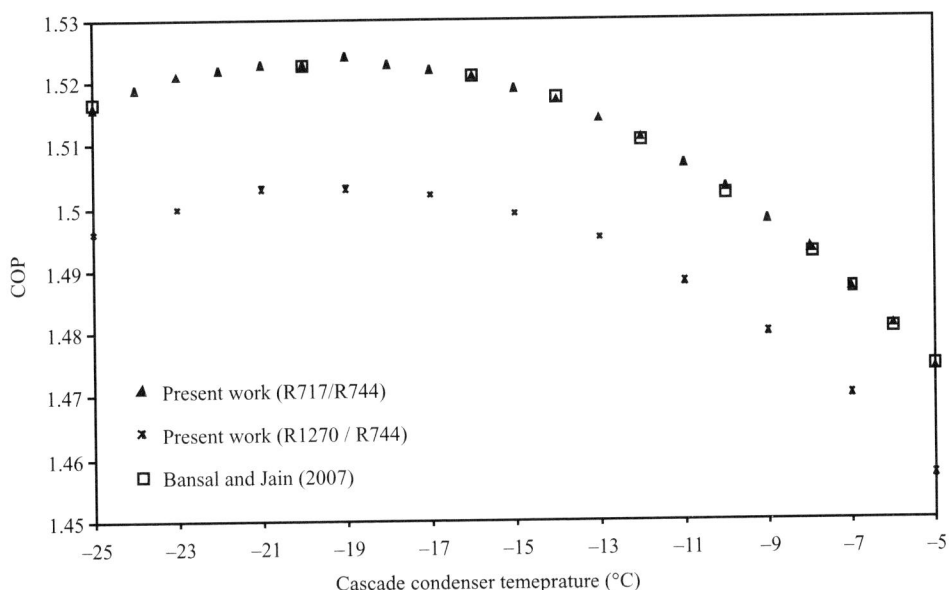

Fig. 2.25(b) Variation of COP versus T_{cc}
($T_e = -45°C$, $T_c = 30°C$, $\Delta T_{sb} = 5°C$, $\Delta T_{sh} = 10°C$, $\eta_{comp} = 0.7$, $\Delta T_{cc} = 5°C$)

is observed that results calculated using the present model are in agreement with the theoretical results reported by Bansal and Jain (2007) for ammonia/carbon dioxide pair and the difference in results is less than 0.5%. The increase in approach causes a drop in COP because of increase in cascade condenser temperature and hence pressure ratio across compressor in 'ltc' increases thereby increasing the compression work in 'ltc'. This enhancement of compression work causes a reduction in COP in 'ltc' and COP of the cascade system also. The variation of COP with temperature in cascade condenser shows that there exists a maximum value of COP corresponding to which cascade condenser temperature is optimum. This happens because the pressure ratio of 'ltc' compressor increases with increase in cascade condenser temperature causing a reduction in COP in 'ltc' because of increase in compressor work in 'ltc' whereas the reverse happens in 'htc' and hence there exists an optimum cascade condenser temperature T_{cc-opt} corresponding to which total compression work is minimum and hence COP is maximum. The results of propylene/carbon-dioxide are also shown in this figure and it can be observed that the COP curve is identical for this pair of refrigerants however the COP offered is lower in comparison to ammonia/carbon-dioxide pair.

Effect of Cascade Condenser Temperature

Figure 2.26 presents the variation of exergetic efficiency and total exergy destruction versus cascade condenser temperature. It is observed that total exergy destruction decreases up to certain cascade condenser temperature and further increases with increase in cascade condenser temperature. The exergetic efficiency shows a reverse trend in comparison to total exergy destruction. The reason for such a behaviour of exergetic efficiency can be explained on the basis of total compressor power required in 'ltc' and 'htc'. The total compressor power required is lowest at a particular

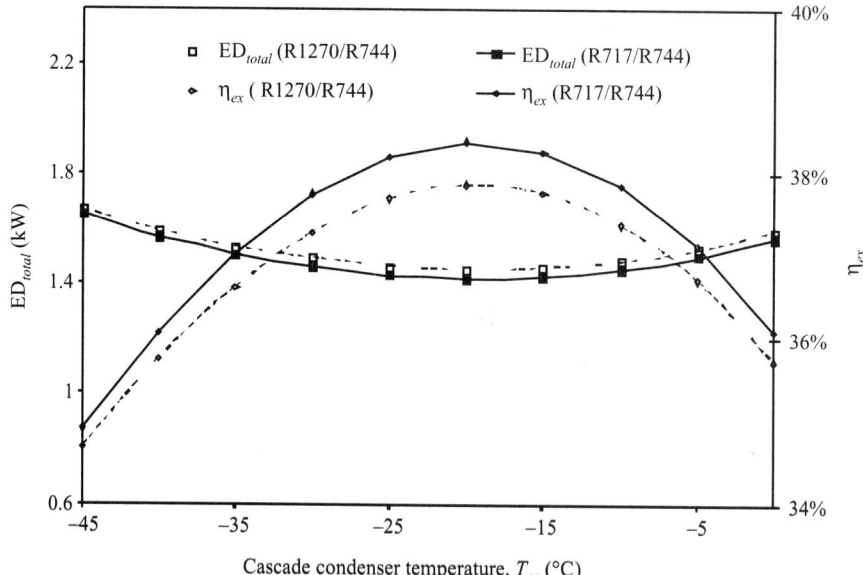

Fig. 2.26 Effect of Cascade Condenser Temperature (T_{cc}) on Total Exergy Destruction and Exergetic Efficiency ($T_e = -45°C$, $T_c = 30°C$, $\Delta T_{sb} = 5°C$, $\Delta T_{sh} = 10°C$, $\eta_{comp} = 0.7$, $\Delta T_{cc} = 5°C$)

cascade condenser temperature. The refrigeration capacity ($\dot{Q}_e$) is constant and the term $\left(1-\dfrac{T_0}{T_r}\right)$ is also constant since both dead state temperature (T_0) and cold room temperature (T_r) are constants. Hence, the value of input exergy expressed by $\dot{Q}_e\left(1-\dfrac{T_0}{T_r}\right)$ is constant. Thus, exergetic efficiency given by $\eta_{e_x}\dfrac{\left|\dot{Q}_e\left(1-\dfrac{T_0}{T_r}\right)\right|}{\dot{W}_{comp_{total}}}$ is highest at a specific cascade condenser temperature corresponding to which total compressor power required is lowest. This specific temperature, corresponding to which the exergetic efficiency is highest, is optimum cascade condenser temperature. It is observed from Figs. 2.25(b) and 2.26 that the optimum cascade condenser temperature corresponding to maximum COP and maximum exergetic efficiency is identical. Fig. 2.26 also depicts that R717/R744 shows better exergetic efficiency as compared to R1270/R744.

The influence of various design parameters which affect the optimum cascade condenser temperature, maximum COP and maximum exergetic efficiency are (i) evaporator temperature; (ii) condenser temperature; (iii) approach in cascade condenser; (iv) isentropic efficiencies of compressors in 'ltc' and 'htc'; (v) sub-cooling of refrigerant leaving condenser in 'htc' and (vi) superheating in evaporator in 'ltc'.

Effect of Evaporator Temperature

Figure 2.27 shows the variation of optimum cascade condenser temperature and maximum COP with evaporator temperature and the influence of isentropic efficiency of 'ltc' and 'htc' compressors on optimum temperature in cascade condenser and maximum COP.

It is evident from Fig. 2.27 that the increase in evaporator temperature increases the optimum temperature in cascade condenser and maximum COP. The increase in optimum cascade condenser temperature is attributed to decrease in overall working temperature range and it also reduces the pressure ratio in 'ltc' and 'htc'. Hence, compressor work decreases and COP of the cascade system increases. It is observed that decrease in isentropic efficiency of the 'ltc' compressor from 1 to 0.8 (keeping the isentropic efficiency of compressor in 'htc' = 1) causes the optimum cascade condenser temperature to decrease whereas the reverse happens in case when isentropic efficiency of the 'htc' compressor decreases to 0.8 from 1(keeping the isentropic efficiency of the 'ltc' compressor =1). This effect is nearly compensated when the efficiencies of both 'ltc' and 'htc' compressors reduce from 1 to 0.8 and the optimum cascade condenser temperature obtained in this particular case is very near (but lower) to the optimum cascade condenser when isentropic efficiencies of both the compressors were assumed to be 1. One more observation that is important to highlight here is that 20% reduction in the efficiency of 'ltc' compressor causes about 10 K drop in optimum cascade condenser temperature as compared to about 8 K rise in cascade condenser temperature for the same decrease in efficiency of 'htc' compressor. On the other hand 20% drop in isentropic efficiency of 'ltc' compressor causes the system COP to drop by 7.3% to 8.9% as compared to a drop of 13-14% in COP for 20% decrease in isentropic efficiency of 'htc' compressor.

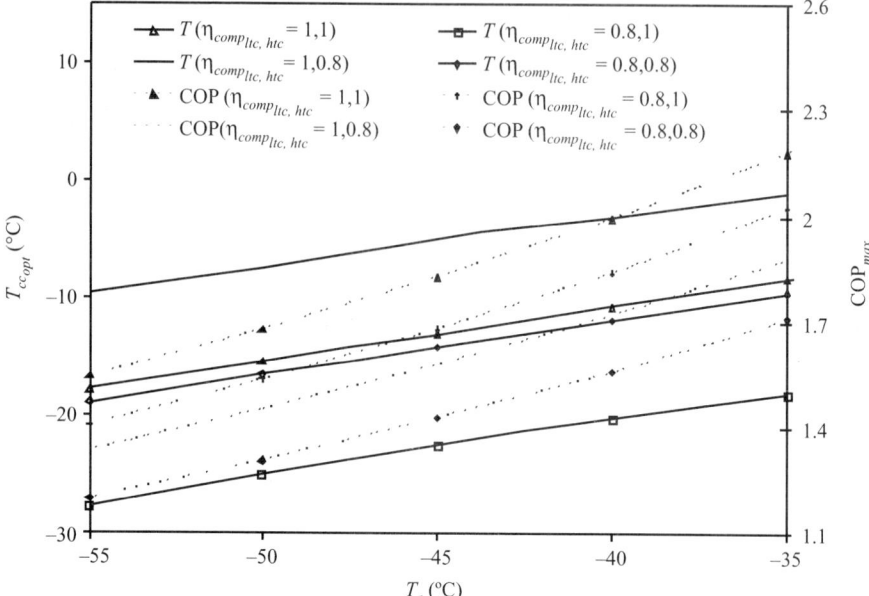

Fig. 2.27 Evaporator Temperature versus Optimum Temperature in Cascade Condenser and COP_{max} for R717/ R744 ($\Delta T_{cc} = 0$, $T_c = 50°C$)

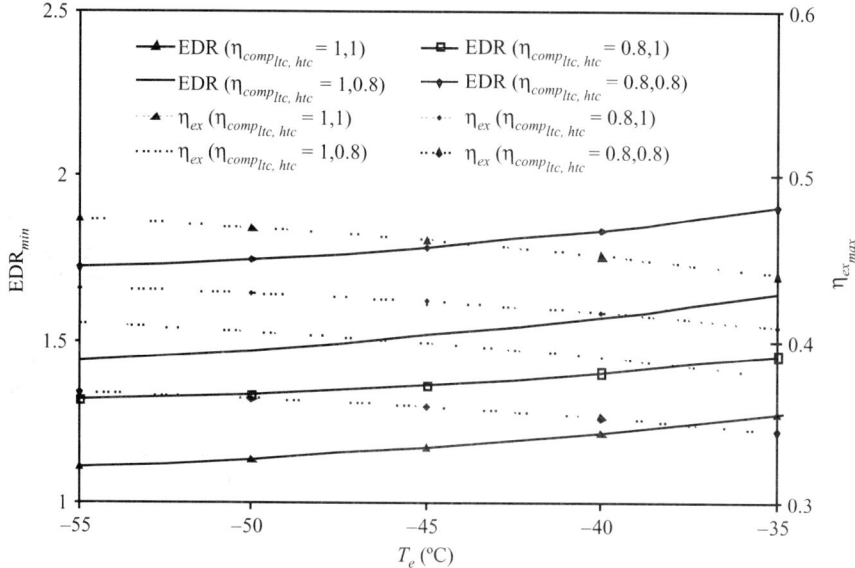

Fig. 2.28 Evaporator Temperature versus Maximum Exergetic Efficiency and Minimum EDR for R717/ R744 ($\Delta T_{cc} = 0$, $T_c = 50°C$)

The variation in EDR_{min} and maximum exergetic efficiency is represented in Fig. 2.28. Two main characteristics of this Figure are decreasing trend of maximum exergetic efficiency with increase in evaporator temperature and decrease in maximum value of exergetic efficiency with decrease in isentropic efficiency of either of the compressors. In this case also, it is crucial to emphasize that the reduction in isentropic efficiency of 'htc' compressor by 20% causes about 13-14% reduction in maximum value of exergetic efficiency as compared to 7.3-9% reduction when the isentropic efficiency of 'ltc' compressor reduces by same amount. Thus once again it is confirmed that lowering of isentropic efficiency of compressor in 'htc' (i.e., compressor for ammonia) has more damaging effect on system performance as compared to carbon dioxide compressor in 'ltc'. The trends of curves of minimum EDR are just opposite to maximum exergetic efficiency curves.

Figure 2.29 illustrates the effect of variation in evaporator temperature on optimum temperature in cascade condenser and maximum COP for R1270/R744 pair.

Figure 2.30 shows the effect of variation in evaporator temperature on minimum EDR and maximum exergetic efficiency. The trends for optimum cascade condenser temperature, maximum COP, maximum exergetic efficiency and minimum EDR are similar to the trends of these parameters presented in Figs. 2.27 and 2.28 for R717/R744. The 20% reduction in isentropic efficiency of 'ltc' compressor (i.e., carbon dioxide compressor) is accountable for lowering both maximum COP and maximum exergetic efficiency by 8.7% and 10.2% corresponding to evaporator temperatures of −35°C and −55°C respectively. Similar to above, the reduction in maximum values of COP and exergetic efficiency is 13.1% and 12.6% for identical temperature conditions when isentropic efficiency of 'htc' compressor reduces by 20%.

Table 2.13 presents the comparison of the two pairs of refrigerants considered for various conditions of isentropic efficiencies of 'ltc' and 'htc' compressors. It is observed that R717/R744 refrigerant pair offers better performance in terms of maximum COP and maximum exergetic efficiency as specified in the table. The values given in braces show the percentage difference by which the values of maximum COP and maximum exergetic efficiency for refrigerant pair R1270/R744 are lower than the corresponding values for R717/R744.

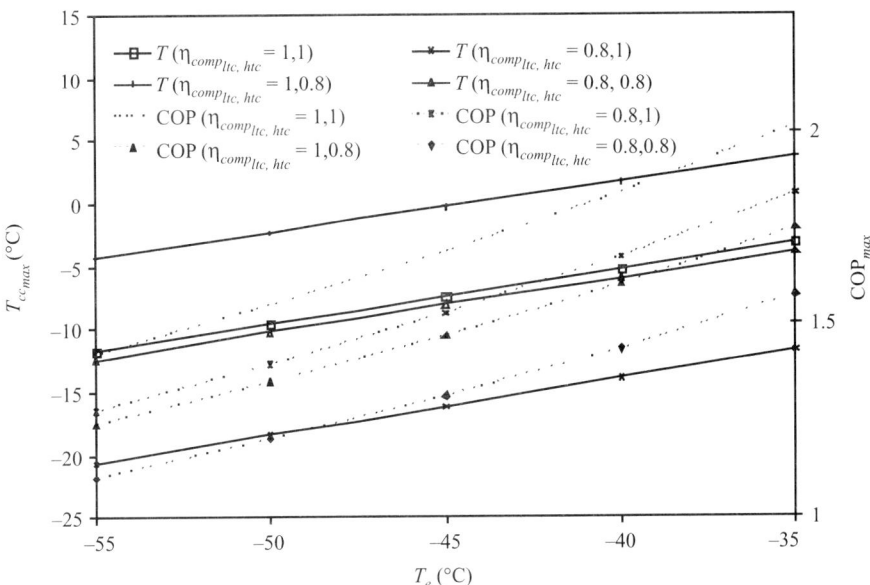

Fig. 2.29 Evaporator Temperature versus Optimum Temperature in Cascade Condenser and Maximum COP for R1270/R744 ($\Delta T_{cc} = 0$, $T_c = 50°C$)

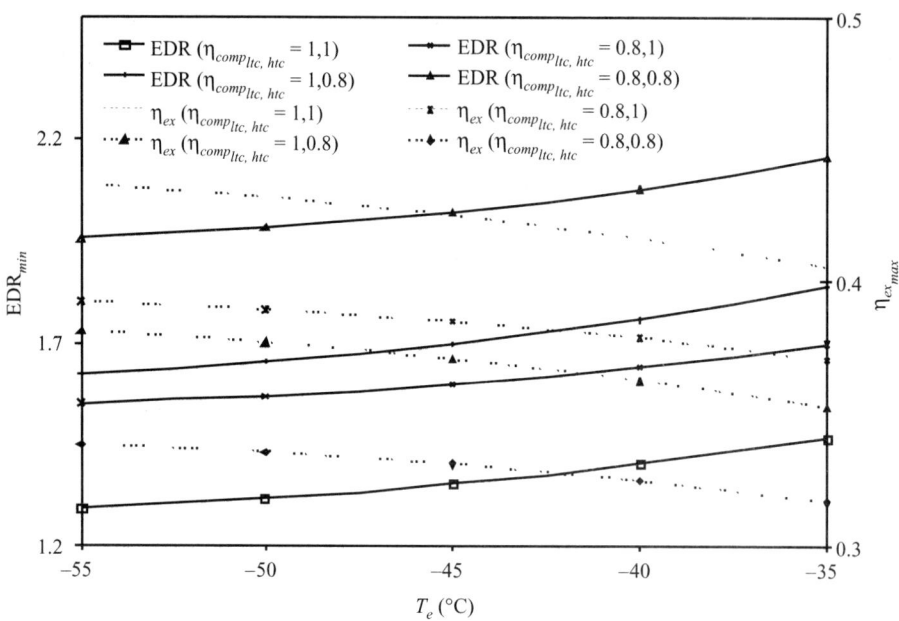

Fig. 2.30 Evaporator Temperature versus Maximum Exergetic Efficiency and Minimum EDR for R1270/R744 ($\Delta T_{cc} = 0$, $T_c = 50°C$)

Table 2.13 Comparison of Optimum Cascade Condenser Temperature, COP_{max} and Maximum Exergetic Efficiency of R717/744 and R1270/R744 Pairs

\multicolumn{8}{c}{$\Delta T_{cc} = 0°C$, $T_c = 50°C$, $T_e = -45°C$}							
η_{comp}		T_{cc} (°C)		COP_{max}		$\eta_{ex_{max}}$	
ltc	htc	R717/R744	R1270/R744	R717/R744	R1270/R744	R717/R744	R1270/R744
1	1	−13.02	−7.43	1.83	1.69 (−7.6%)	0.461	0.425 (−7.8%)
0.8	1	−22.39	−16.18	1.68	1.53 (−8.9%)	0.423	0.385 (−8.98%)
1	0.8	−4.95	−0.32	1.58	1.47 (−6.9%)	0.398	0.371 (−6.8%)
0.8	0.8	−14.16	−8.11	1.43	1.31 (−8.4%)	0.359	0.331 (−7.8%)
0.7	0.7	−15.10	−8.66	1.23	1.13 (−8.13%)	0.31	0.284 (−8.39%)

Effect of Condenser Temperature

Figures 2.31 and 2.32 depict the effect of condenser temperature on optimum cascade condenser temperature, maximum COP, minimum EDR and maximum exergetic efficiency respectively for

R717/R744. Simultaneously, these figures also present the effect of isentropic efficiencies 'ltc' and 'htc' compressors on the above-mentioned parameters. The optimum temperature increases with increase in condenser temperature. This happens because of increase in overall working temperature range. The maximum COP of the system reduces since the increase in condenser temperature

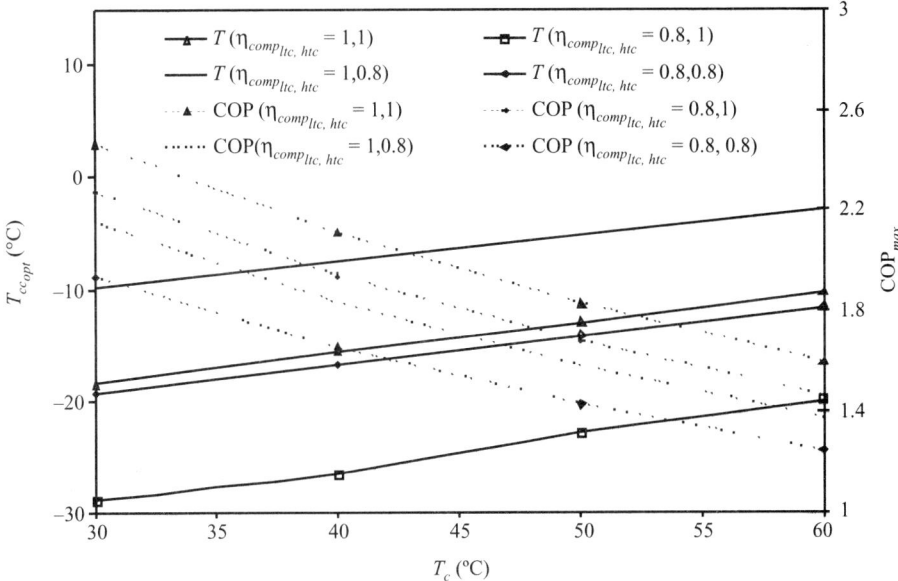

Fig. 2.31 Variation of Optimum Temperature in Cascade Condenser and Maximum COP versus Condenser Temperature for R717/ R744 ($\Delta T_{cc} = 0$, $T_e = -45°C$)

causes the pressure ratios of the 'ltc' and 'htc' compressors to increase and hence the power input increases thereby reducing the maximum COP. The effect of isentropic efficiencies of compressors on optimum temperature in cascade condenser and maximum COP are similar to the trends observed in the case of variation of evaporator temperature depicted in Fig. 2.27 and explained in corresponding para.

Figure 2.32 shows that maximum exergetic efficiency reduces with increase in condenser temperature. This decreasing trend of exergetic efficiency is achieved because of increase in input exergy, i.e., total compressor power required for the same output exergy corresponding to a constant cooling capacity. The effect of isentropic efficiencies of the compressors is similar to the trends that were achieved in case when evaporator temperature was varied (Refer Fig. 2.28 and corresponding para).

The variation of optimum temperature in cascade condenser and maximum COP for R1270/R744 is illustrated in Fig. 2.33. The variation in maximum exergetic efficiency and minimum *EDR* are presented in Fig. 2.34. The explanation of trends of these curves for R1270/R744 is similar to as already has been discussed for R717/R744. The comparison of the optimum temperatures, maximum COP and maximum exergetic efficiencies for a particular set of data is already presented for these two pairs of refrigerants in Table 2.13.

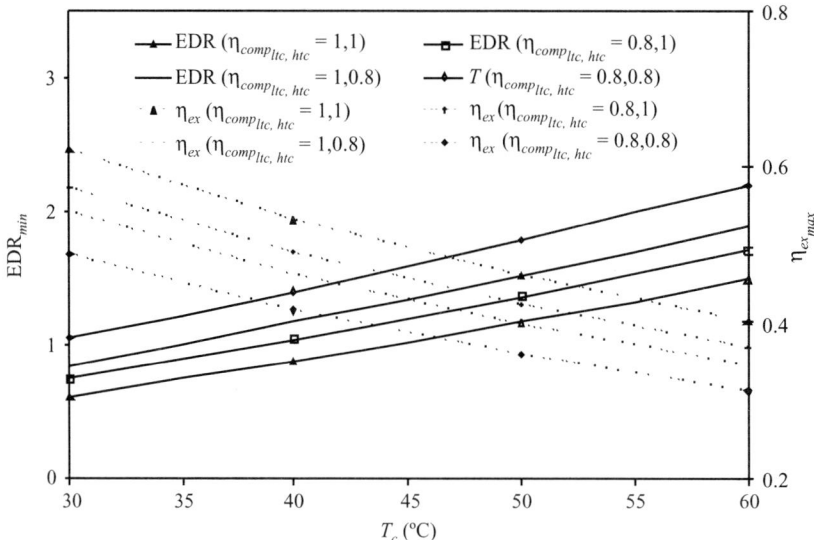

Fig. 2.32 Variation of Maximum Exergetic Efficiency and Minimum EDR versus Condenser Temperature for R717/ R744 ($\Delta T_{cc} = 0$, $T_e = -45°C$, $T_r = T_e + 10\ °C$)

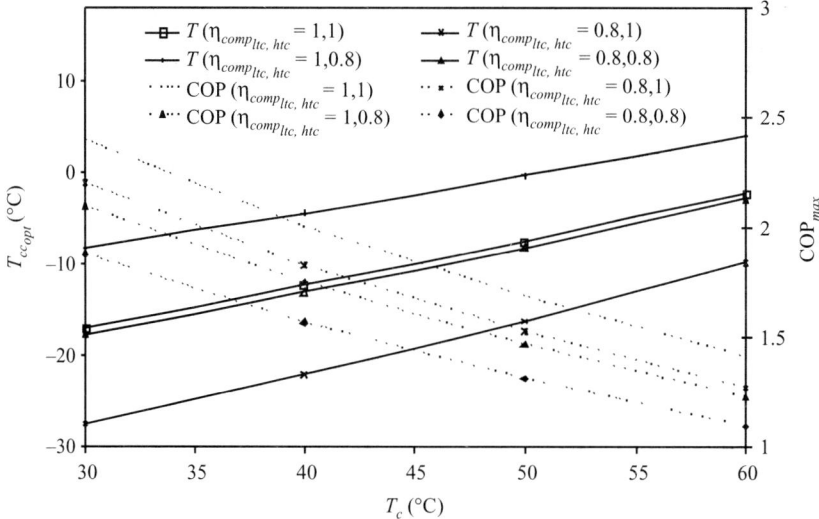

Fig. 2.33 Variation of Optimum Temperature in Cascade Condenser and Maximum COP versus Condenser Temperature for R1270/R744 ($\Delta T_{cc} = 0$, $T_e = -45°C$, $T_r = T_e + 10°C$)

Effect of Approach (ΔT_{cc})

Figures 2.35 and 2.36 represent the effect of approach on optimum temperature in cascade condenser, maximum COP and maximum exergetic efficiency for R717/R744 and R1270/R744. It is observed that the increase in approach causes increase in optimum temperature in cascade condenser. Further

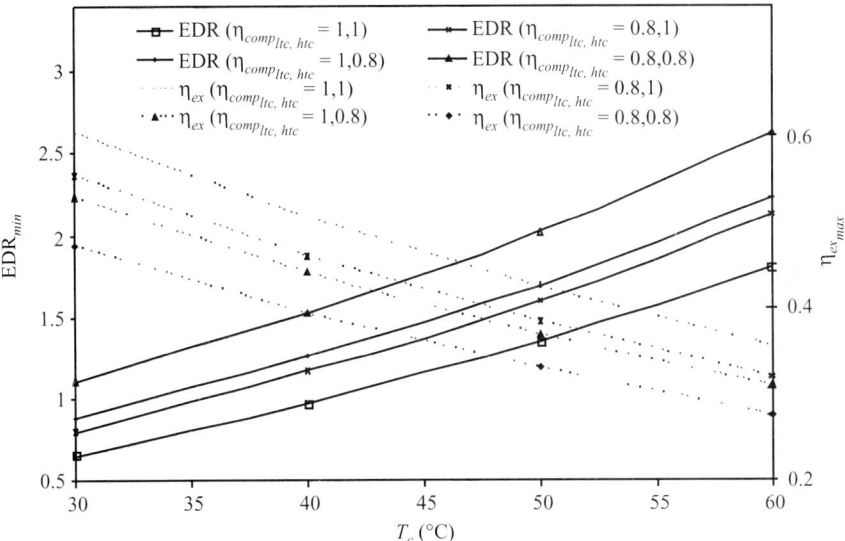

Fig. 2.34 Condenser Temperature versus Maximum Exergetic Efficiency and Minimum EDR for R1270/R744 ($\Delta T_{cc} = 0$, $T_e = -45°C$, $T_r = T_e + 10°C$)

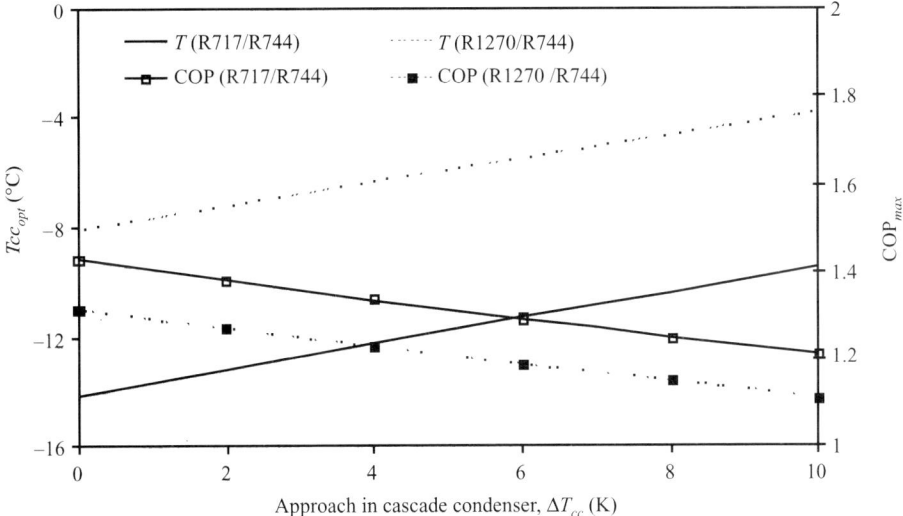

Fig. 2.35 Variation of Optimum Temperature in Cascade Condenser and Maximum COP with Approach (ΔT_{cc}) in Cascade Condenser for R717/R744 and R1270/R744 ($T_e = -45°C$, $T_c = 50°C$, $\eta_{comp_{ltc, htc}} = 0.8$)

approach is responsible for reduction in maximum COP and maximum exergetic efficiencies for both pairs of refrigerants. The increase in approach is responsible for increase in mass flow rates in 'ltc' and 'htc'. This increase in mass flow rates causes the compressor power requirements to increase in 'ltc' and 'htc'. The increase in total power requirement is responsible for reduction of

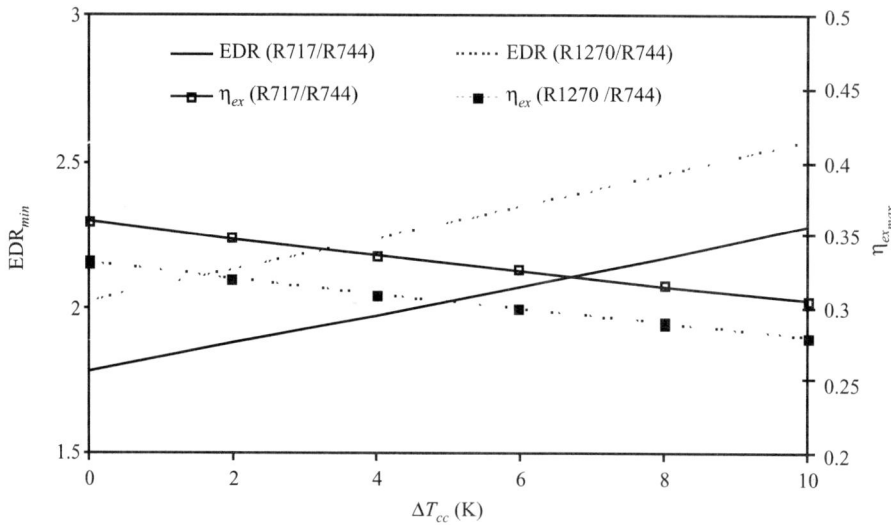

Fig. 2.36 Variation of Minimum EDR and Maximum Exergetic Efficiency with Approach in Cascade Condenser for R717/ R744 and R1270/R744 ($T_e = -45°C$, $T_c = 50°C$, $\eta_{comp_{ltc,\,htc}} = 0.8$)

maximum COP. Simultaneously increase in approach is accountable for increase in exergy destruction due to increase in finite temperature difference in cascade condenser and hence net exergy in product goes down whereas input exergy (total power required for compression) increases. The combined effect of these two factors leads to reduction in exergetic efficiency. The reduction in maximum COP is about 0.021/K (1.5%/K) increase in approach. The decrease in maximum exergetic efficiency is 0.005/K (1.5%/K) rise in approach. The maximum COP and maximum exergetic efficiency for R717/R744 is 7.9 to 8.2% higher (corresponding to 0 K / 10 K approach) in comparison to R1270/R744.

Effect of Sub-cooling of Refrigerant in Condenser in 'htc' (ΔT_{sb}) and Superheating of Refrigerant in Evaporator in 'ltc'(ΔT_{sh})

Figures 2.37, 2.38 and 2.39 represent the effect of sub-cooling of refrigerant in condenser in 'htc' and superheating of refrigerant in evaporator in 'ltc' on optimum temperature in cascade condenser, maximum COP and maximum exergetic efficiency for R717/R744. It is observed that the optimum temperature in cascade condenser is highest for the case when both sub-cooling and superheating are neglected. The sub-cooling of refrigerant by 5 K reduces optimum cascade condenser temperature by approximately 1 K. Consequently maximum COP and maximum exergetic efficiency increase by 1.8% approximately.

The increase in maximum COP is achieved because sub-cooling brings about a reduction in mass flow rate in 'htc' and accordingly the power required to operate 'htc' compressor reduces and hence increase in maximum COP is observed. This reduction in power required in 'htc' compressor translates into rise of exergetic efficiency. The superheating of saturated vapour by 5 K in evaporator reduces the optimum cascade condenser temperature by 0.38 K. This causes the mass flow rate in low temperature cycle and pressure ratio across 'ltc' compressor to reduce. The combined effect

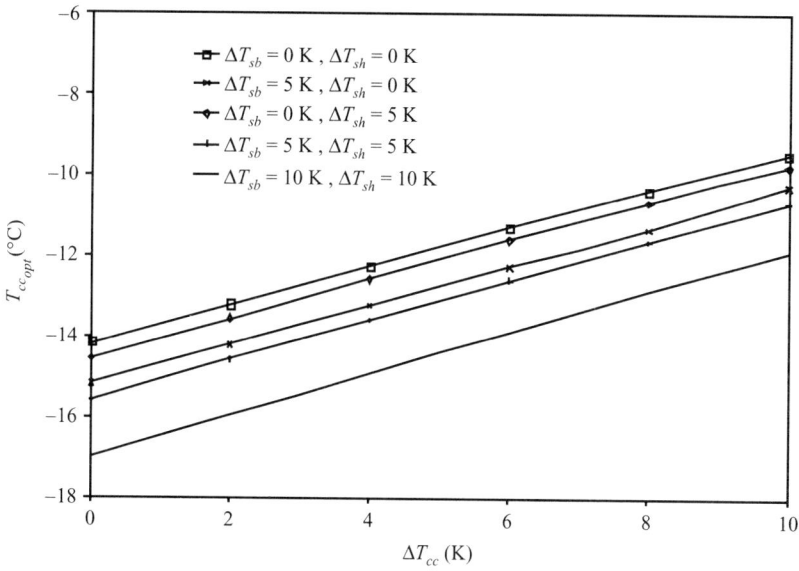

Fig. 2.37 Effect of Approach, Sub-cooling and Superheating on Optimum Temperature in Cascade Condenser for R717/ R744 ($T_c = 50°C$, $T_e = -45°C$, $T_r = T_e + 10°C$, $\eta_{comp_{ltc, htc}} = 0.8$)

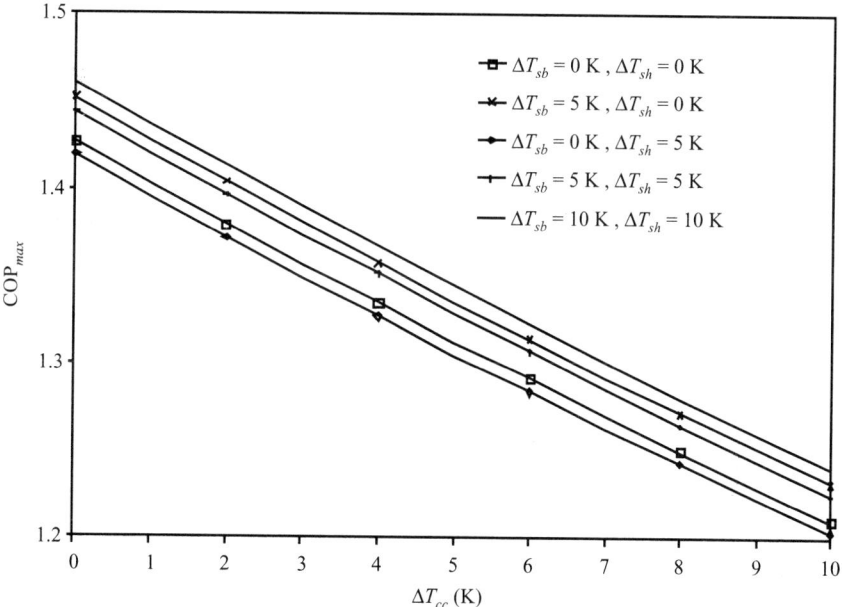

Fig. 2.38 Effect of Approach, Sub-cooling and Superheating on COP_{max} for R717/ R744 ($T_c = 50°C$, $T_e = -45°C$, $T_r = T_e + 10°C$, $\eta_{comp_{ltc, htc}} = 0.8$)

of these two factors is to decrease the power required in 'ltc' compressor. However, lowering of optimum temperature (and consequently the pressure ratio in 'ltc') accounts for an increase in pressure ratio in 'htc' and as a result the power required by the 'htc' compressor increases. The

82 *Alternatives in Refrigeration and Air Conditioning*

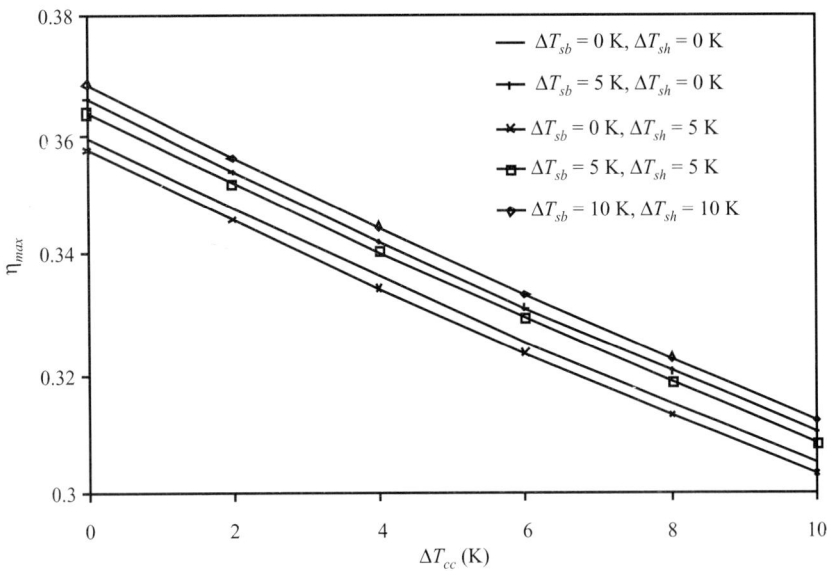

Fig. 2.39 Effect of Approach, Sub-cooling and Superheating on Maximum Exergetic Efficiency for R717/ R744 ($T_c = 50°C$, $T_e = -45°C$, $T_r = T_e + 10°C$, $\eta_{comp_{ltc, htc}} = 0.8$)

overall effect is increase in the total power required and this reduces maximum COP and maximum exergetic efficiency. The drop in maximum COP and maximum exergetic efficiency is about 0.5%.

Further the combined effect of both sub-cooling and superheating (by 5 K and 10 K) is also illustrated in these figures. It is observed that the combined effect of these factors is to reduce the optimum cascade condenser temperature. In this case reduction in optimum cascade condenser temperature is higher in comparison to case when either sub-cooling or superheating is considered to occur in the system. The combined effect of these two factors (for 5 K) is to increase the maximum COP. However the increase in maximum COP is little less in comparison to the case when sub-cooling (of 5 K) alone was considered. However, the combined effect of these factors (for 10 K) is to increase the maximum COP more than each of the previous case assumed. Similar results are obtained for maximum exergetic efficiency and the same is illustrated in Fig. 2.39.

The effect of sub-cooling and superheating on optimum cascade condenser temperature, maximum COP and maximum exergetic efficiency for R1270/R744 pair is illustrated in Figs. 2.40, 2.41 and 2.42 respectively. It is observed that the trends of results obtained for R1270/R744 is similar to results obtained for R717/R744. In this case, the optimum temperature reduces by 2.7 K for a sub-cooling of 5 K whereas a reduction of only 0.23 K is observed for superheating by 5 K. The increase in maximum COP is about 4.2% for 5 K sub-cooling whereas drop in maximum COP is by 0.55% for superheating by 5 K alone. This shows that the effect of sub-cooling on maximum COP is more in comparison to superheating. Similar to maximum COP, maximum exergetic efficiency also follows the analogous trend; it increases by 4.2% for sub-cooling by 5 K and reduces by about 0.6% for superheating in evaporator by 5 K. The explanation of trends of results is specified already in the previous section. Further the combined effect of these two factors shows that maximum COP and maximum exergetic efficiency increase by about 3.6% for 5 K sub-cooling and 5 K

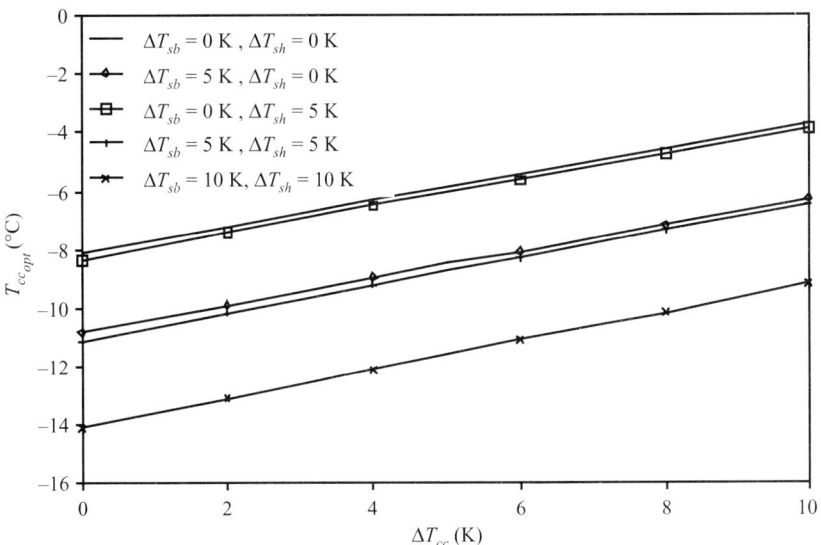

Fig. 2.40 Effect of Variation in Approach, Sub-cooling and Superheating on Optimum Temperature in Cascade Condenser for R1270/R744 ($T_e = -45°C$, $T_c = 50°C$, $T_r = T_e + 10°C$, $\eta_{comp_{ltc,htc}} = 0.8$)

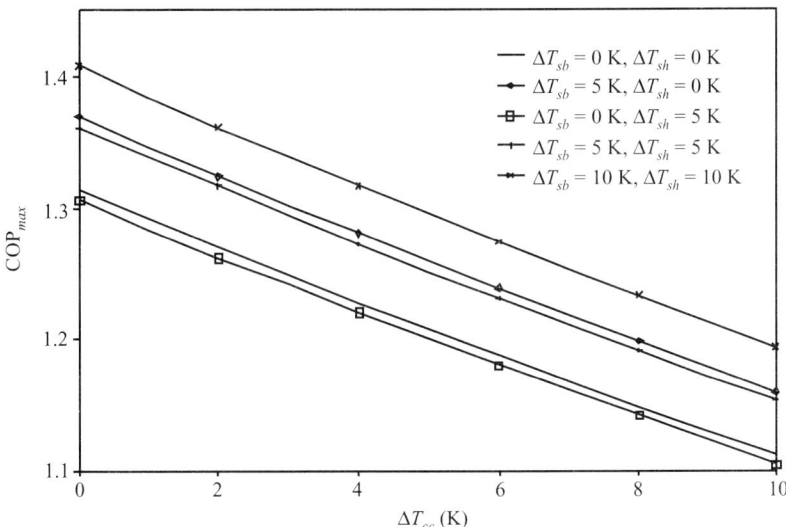

Fig. 2.41 Effect of Variation in Approach, Sub-cooling and Superheating on COP_{max} for R1270/R744 ($T_e = -45°C$, $T_c = 50°C$, $T_r = T_e + 10°C$, $\eta_{comp_{ltc,htc}} = 0.8$)

superheating whereas the corresponding increase in the values of maximum COP and maximum exergetic efficiency is 7.2% for 10 K sub-cooling and superheating.

Table 2.14 shows comparison of effects of sub-cooling and superheating for R717/R744 and R1270/R744. The difference in values of COP_{max} and $\eta_{ex_{max}}$ for the two pairs shows that sub-cooling is more effective for R1270/R744 as compared to R717/R744.

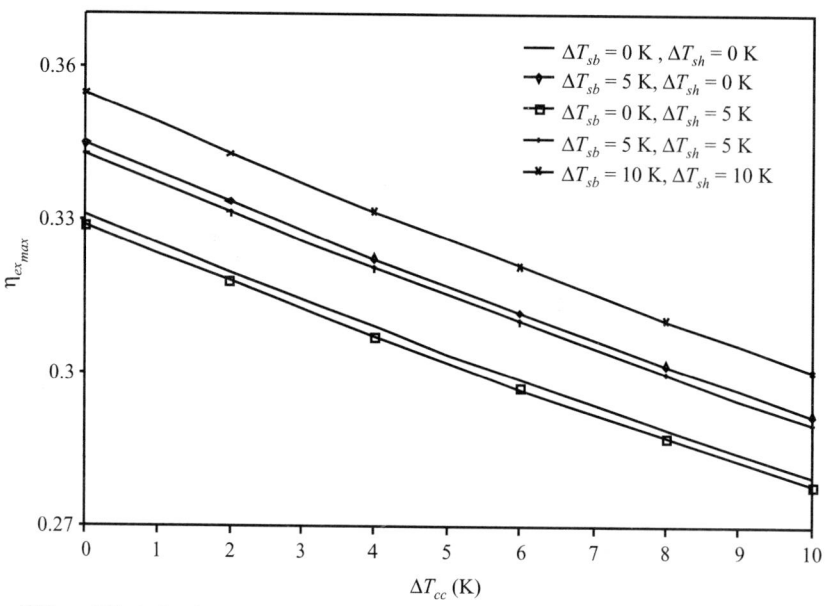

Fig. 2.42 Effect of Variation in Approach, Sub-cooling and Superheating on Maximum Exergetic Efficiency for R1270/R744 ($T_e = -45°C$, $T_c = 50°C$, $T_r = T_e + 10°C$, $\eta_{comp_{ltc,\,htc}} = 0.8$)

Table 2.14 Comparison of Effect of Sub-cooling and Superheating on Maximum COP and Maximum Exergetic Efficiency for R717/R744 and R1270/R744 ($T_e = -45°C$, $T_c = 50°C$, $T_r = T_e + 10°C$, $\eta_{comp_{ltc,\,htc}} = 0.8$, $\Delta T_{cc} = 4°C$)

ΔT_{sb} (K)	ΔT_{sh} (K)	COP_{max}			$\eta_{ex_{max}}$		
		R717/R744 (a)	R1270/R744 (b)	Difference (a-b)/a	R717/R744 (c)	R1270/R744 (d)	Difference (c-d)/c
0	0	1.335	1.227	0.0809	0.3363	0.3092	0.08058
5	0	1.358 (1.74%)*	1.28 (4.3%)*	0.05744	0.3422 (1.75%)¥	0.3224 (4.27%)¥	0.05786
0	5	1.327 (−0.57%)*	1.22 (−0.6%)*	0.08063	0.3344 (−0.57%)¥	0.3074 (−0.58%)¥	0.08074
5	5	1.351 (1.22%)*	1.272 (3.64%)*	0.05848	0.3403 (1.19%)¥	0.3205 (3.65%)¥	0.05818
10	10	1.367 (2.42%)*	1.316 (7.23%)*	0.03731	0.3444 (2.41%)¥	0.3317 (7.28%)¥	0.03688

*The values in percentage in columns (a) and (c) represent the percentage increase or decrease in COP_{max} corresponding to the base value of 0 K sub-cooling and 0 K superheating.
¥Similarly, values in percentage in columns (b) and (d) represent the percentage increase or decrease in $\eta_{ex_{max}}$ corresponding to the base value of 0 K sub-cooling and 0 K superheating.

Figure 2.43 shows the efficiency defects in components of cascade refrigeration system for R717/R744 and R1270/R744. It is evident from figure that for R717/R744 efficiency defect in condenser is highest followed by compressor in 'htc', evaporator, refrigerant throttle valve of 'htc', cascade condenser, 'ltc' compressor and 'ltc' refrigerant throttle valve. For R1270/R744, the

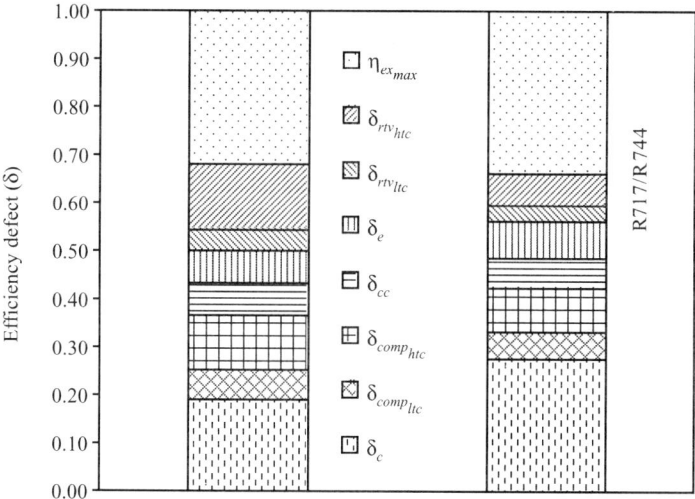

Fig. 2.43 Efficiency Defects in Components of a Cascade Refrigeration System for R717/R744 and R1270/R744 at Optimum Cascade Condenser Temperature
($\Delta T_{cc} = 4$ K, $T_e = -45°C$, $T_c = 50°C$, $T_r = T_e + 10°C$, $\Delta T_{sb} = 5$ K, $\Delta T_{sh} = 5$ K, $\eta_{comp_{ltc,htc}} = 0.8$)

efficiency defect is highest in condenser followed by throttle valve, compressor of 'htc', evaporator, cascade condenser, 'ltc' compressor and throttle valve of 'ltc'.

2.5 CONCLUSION

In this chapter the focus has been given on alternative refrigerants and energy and exergy analysis of single stage, multi-compression and cascade vapour compression refrigeration systems. The importance of factors related to environmental and ecological concerns such as ODP, GWP, TEWI and LCCP has been discussed. Various alternative refrigerants such as azeotropes, zeotropes and pure refrigerants as substitutes to conventional refrigerants for different applications have been specified. The chapter also gives the description of Lorenz cycle when using zeotrope refrigerants. It is a modified form of vapour compression cycle in which the boiling and condensation occurs with temperature glide. Another cycle with alternative refrigerant carbon dioxide is transcritical compression refrigeration cycle (used when the critical temperature of refrigerant is low) is also presented. Subsequently, the focus of the chapter shifts to the description and energy and exergy analysis of vapour compression cycle, multistage vapour compression refrigeration cycles and cascade refrigeration cycle. The parametric study of these systems have been carried out under various operating conditions of the systems. In two-stage system, the optimum reduced inter-stage temperatures and pressures have been computed for HCFC 22, R410A and R717, and the value is found very near to saturation pressure corresponding to Geometric Mean Temperature (GMT) of condensation and evaporation temperatures. In cascade refrigeration system two pairs of refrigerants, viz., R717/R744 and R1270/R744 have been considered and optimum cascade condenser temperature has been computed corresponding to maximum exergetic efficiency. The effects of evaporator and condenser temperatures, approach degree of sub-cooling and super-heating have also been highlighted on optimum cascade condenser temperature, COP and exergetic efficiency of the system.

REFERENCES

Agnew, B. and Ameli, S.M., 2004. A finite time analysis of a cascade refrigeration system using alternative refrigerants. *Applied Thermal Engineering* (24), 2557-2565.

Aprea, C., Mastrullo, R. Renno, C. 2004. An analysis of the performances of a vapour compression plant working both as a water chiller and a heat pump using R22 and R417A. *Applied Thermal Engineering* 24, 487-499.

Aprea, C., Mastrullo R., Rossi F. D., 1996. Behaviour and performances of R502 alternative working fluids in refrigerating plants. *International Journal of Refrigeration* 19(4), 257-63.

Arcaklioglu, E., Cavusoglu, A., Erisen A., 2005. An algorithmic approach towards finding better refrigerant substitutes of CFCs in terms of the second law of thermodynamics. *Energy Conversion and Management* 46, 1595-1611.

Arora, C.P., and Dhar, P.L., 1973. Optimization of Multistage refrigerant compressor, Progress in Refrigeration Science and Technology, *XIII International Congress of Refrigeration (IIR)*. 693-700.

Arora, A., 2009, Energy and Exegy analysis of compression, absorption and combined cycle cooling systems, Ph.D. Thesis, IIT, Delhi.

Arora, A., Arora, B.B., Pathak B.D. and Sachdev H. L., 2007. Exergy analysis of a Vapour Compression Refrigeration System with R-22, R-407C and R-410A, *International Journal of Exergy* 4(4), 441-454.

ASHRAE Handbook (Fundamentals), 1997, Chapter 18.

Bansal, P. K. and Jain, S., 2007. Cascade systems: past, present, and future. *ASHRAE Transactions* 113(1), 245-252.

Baumnann, K., Blass, E., 1961. Beitrag zur Ermittlung des optimalen Mitteldruckes bei zweistuflgen Kaltdampf Verdichter-Kältemaschinen, Kältetechnik 13, 210-216.

Bejan, A., Tsatsaronis, G., Moran, M., 1996. Thermal Design and Optimization. John Wiley and Sons, USA, pp. 143-156.

Bhattacharyya, S., Mukhopadhyay, S., Kumar A., Khurana R.K. and Sarkar J., 2005. Optimization of a CO_2/C_3H_8 cascade system for refrigeration and heating. *International Journal of Refrigeration* 28, 1284-1292.

Bitzer International: Refrigerant report no. 14 (Edition A-501-14), (http://www.bitzer.de/en/etc/download.php?d=doc/a/&f=a-501-14.pdf, 2007).

Bitzer International: Refrigerant report no. 16 (Edition A-501-16) (http://www.bitzer.de/en/etc/download.php?d=doc/a/&f=a-501-16.pdf).

Bivens and Yokozeki, 1996. International Conference on Ozone Protection Technologies, Washington, DC.

Calm, J. M., 2008. The next generation of refrigerants - Historical review, considerations, and outlook. *International Journal of Refrigeration*. 1-11 (In Press).

Calm J.M., Hourahan G.C., 2001. Refrigerant data summary, Engineered Systems 18, 74-78.

Camporese R., Bigolaro G., Bobbo S. and Cortella, G., 1997. Experimental evaluation of refrigerant mixtures as substitutes for CFC12 and R502. *International Journal of Refrigeration*, 20(1), 22-31.

Chen, W., 2008. A comparative study on the performance and environmental characteristics of R410A and R22 residential air conditioners. *Applied Thermal Engineering* 28, 1-7.

Dopazo J.A., Fernández-Seara J., Sieres J., Uhía F. J., 2009. Theoretical analysis of a CO_2-NH_3 cascade refrigeration system for cooling applications at low temperatures. *Applied Thermal Engineering*, 29(8-9), 1577-1583.

Döring R., Buchwald H. and Hellmann, J., 1997. Results of experimental and theoretical studies of the azeotropic refrigerant R507. *International Journal of Refrigeration*, 20(2), 78-84.

Getu, H.M., Bansal P.K., 2008. Thermodynamic analysis of an R744-R717 cascade refrigeration system. *International Journal of Refrigeration* 31(1), 45-54.

Göktun S., 1998. An overview of ozone safe alternatives for R502. Energy 23(5), 379-81.

Goseny, W. B., 1967. Compound compression refrigeration cycles. The Institution of Refrigeration, London, 1-11.

http://www.iiar.org//aaranswers_history.cfm?

Kanoglu, M., 2002. Exergy analysis of multistage cascade refrigeration cycle used for natural gas liquefaction. *International Journal of Energy Research* 26, 763-774.

Katz, S., 1962. Best operating points for staged systems. *Indust. Engg. Chem. Fundamentals* 1(4), 226-240.

Keshwani, H.B. and Rastogi, K.V., 1968. Optimum inter-cooler pressure in Multistage Compression of refrigerants. All India Symposium on Refrigeration, Air-Conditioning and Environmental Control in Cold storage Industry, E6.1-13.

Kilic, M., Kaynakli, O., 2004. Second law-based thermodynamic analysis of water-lithium bromide absorption refrigeration system. *Energy* 32(8), 1505-1512.

Klein, S.A., Alvarado, F., 2005. Engineering Equation Solver, Version 7.441. F-Chart software, Middleton, WI.

Kotas, T.J., 1985. The Exergy Method of Thermal Plant Analysis. Butterworths, London.

Kruse, H. and Rüssmann, H., 2006. The natural fluid nitrous oxide—an option as substitute for low temperature synthetic refrigerants. *International Journal of Refrigeration* 29, 799-806.

Lee T. S., Liu C. H. and Chen T.W., 2006.Thermodynamic analysis of optimal condensing temperature of cascade-condenser in CO_2/NH_3 cascade refrigeration systems. International Journal of Refrigeration, 29,1100-1108.

Lorentzen G., 1995. The use of natural refrigerants: a complete solution to the CFC/HCFC predicament. *International Journal of Refrigeration* 18(3), 190-197.

National Refrigerant Reference Guide, 2011, Fifth Edition. National Refrigerants, Inc.

Nicola G. D., Giuliani, G., Polonara, F. and Stryjek, R., 2005. Blends of carbon dioxide and HFCs as working fluids for the low-temperature circuit in cascade refrigerating systems. *International Journal of Refrigeration* 28, 130-140.

Nikolaidis C. and Probert D., 1998. Exergy-method analysis of a two-stage vapour compression refrigeration-plants performance. *Applied Energy* 60, 241-256.

Niu, B. and Zhang, Y., 2007. Experimental study of the refrigeration cycle performance for the R744/R290 mixtures. *International Journal of Refrigeration*, 30, 37-42.

Ouadha A., En-nacer M., Adjlout L. and Imine O., 2005. Exergy analysis of a two-stage refrigeration cycle using two natural substitutes of HCFC22, *International Journal of Exergy* 2(1), 14-29.

Palm B, 2008. Hydrocarbons as refrigerants in small heat pump and refrigeration systems—A review. *International Journal of Refrigeration* 31, 552-563.

Park, K. J., Jung D., 2007(a). Thermodynamic performance of HCFC22 alternative refrigerants for residential air-conditioning applications. *Energy and Buildings* 39, 675-680.

Park, K. J., Jung D., 2007(b). Thermodynamic performance of R502 alternative refrigerant mixtures for low temperature and transport applications. *Energy Conversion and Management* 48, 3084-3089.

Park, K.J., Shim, Y.B., Jung, D., 2008(a). Experimental performance of R432A to replace R22 in residential air-conditioners and heat pumps. *Applied Energy* xxx, xxx.

Park, K.J., Shim, Y.B., Jung, D., 2008(b). Performance of R433A for replacing HCFC22 used, in residential air-conditioners and heat pumps. *Applied Energy* 85, 896-900.

Pearson, A., 2008. Refrigeration with ammonia. *International Journal of Refrigeration* 31, 545-551.

Prasad, M., 1981. Optimum inter-stage pressure for two-stage refrigeration system. *ASHRAE Journal* 23, 58-60.

Radermacher, R., Hwang Y., 2005. Vapor compression heat pumps with refrigerant mixtures, Taylor and Francis.

Rakhesh B., Venkatarathnam G., Murthy S. S., 2003. Experimental studies on a heat pump operating with R22, R407C and R407A: comparison from an exergy point of view. *Journal of Energy Resources Technology*. 125, 101-112.

Ratts, E.B. and Brown J.S., 2000. A generalized analysis for cascading single fluid vapour compression refrigeration cycles using an entropy generation minimization method. *International Journal of Refrigeration* 23, 353-365.

Riffat, S. B., Afonso C. F., Oliveirat A. C. and Reay D. A., 1997. Natural refrigerants for refrigeration and Air conditioning systems. *Applied Thermal Engineering* 17(I), 33-42.

Said, S.A. and Ismail, B. 1994. Exergetic assessment of the coolants HCFC123, HFC134a, CFC11 and CFC12. *Energy* 19(11), 1181-86.

Sami, S. M. and Desjardins, D. E., 2000a. Performance and comparative study of new alternatives to R502 inside air/refrigerant enhanced surface tubing. *International Journal of Energy Research* 24 (2), 177-186.

Sami, S. M. and Desjardins, D. E., 2000b. Performance enhancement of some alternatives to R502', *International Journal of Energy Research* 24(4), 279-89.

Sencan, A., Yakut, K.A., Kalogirou, S.A., 2005. Exergy analysis of lithium bromide/water absorption systems. *Renewable Energy* 30, 645-657.

Siller, D.A., Anderson K., Shepherd J. J., Strong R., John R., Vallort R. P., Seaton W.W., 2006 Ammonia as a refrigerant. American Society of Heating, Refrigerating and Air-Conditioning Engineers, Inc.www.ashrae.org/content/ASHRAE/ASHRAE/Article AltFormat/200622793710_347.pdf)

Singh, P. 2009. Study of vortex tube integrated alternative refrigeration and air-conditioning options for energy conservation, Ph.D. Thesis, IIT, Delhi.

Stegou-Sagia A., Paignigiannis N., 2005. Evaluation of mixtures efficiency in refrigerating systems. *Energy Conversion and Management* 46, 2787-2802.

Threlkeld, J. L.1966. *Thermal Environmental Engineering*, Prentice-Hall, NJ.

Wang, R.Z.and Li Y., 2007. Perspectives for natural working fluids in China, *International Journal of Refrigeration*. 30, 568-581.

Xuan, Y., Chen, G., 2005. Experimental study on HFC-161 mixture as an alternative refrigerant to R502. *International Journal of Refrigeration* 28 (3), 436-441.

Zubair, S. M., Khan, S. H., 1995. On optimum inter-stage pressure for two-stage and mechanical-subcooling vapour-compression refrigeration cycles. ASME Transactions, *Journal of Solar Energy Engineering* 117 (1), 64-66.

Zubair, S. M., Shaw, D. N., 1986. An Experimental Investigation of Two-Stage Refrigeration Systems, Internal Report, Copeland Corporation, Sidney, OH.

Zubair, S.M., Yaqub, M. and Khan, S.H., 1996. Second-law-based thermodynamic analysis of two-stage and mechanical-sub-cooling refrigeration cycles. *International Journal of Refrigeration* 19(8), 506-516.

CHAPTER 3

Vapour Absorption Cooling and Advanced Absorption Cycles

3.1 INTRODUCTION

In recent years, interest in absorption refrigeration systems is growing because these systems can be operated by solar energy or low temperature waste heat available from various sources such as power plants (Aphornratana and Eames, 1995). Utilization of waste heat helps in reducing global warming. Moreover, refrigerant and absorbent pairs used in absorption refrigeration systems mostly do not deplete ozone layer.

In this chapter, various conventional and advanced vapour absorption systems have been described and their energy and exergy analyses have been presented. A number of researchers (Vliet *et al.*, 1982 ; Kaushik and Chandra, 1985; Gommed and Grossman, 1990; Kumar and Devotta, 1990; Xu *et al.*, 1996; Sun, 1997; Xu and Dai, 1997; Arun *et al.*, 2000, 2001; Kilic and Kaynakli, 2007) investigated single effect and series flow double effect VAR systems using energy analysis approach. The effects of various parameters such as generator temperature, absorber temperature, condenser temperature, solution circulation ratio and solution concentration, etc., had been investigated by these researchers on COP. Their results revealed that COP of double effect VAR system is higher in comparison to single effect system.

Aphornratana and Eames (1995), Talbi and Agnew (2000), and Kilic and Kaynakli (2004) investigated single effect water-lithium bromide system using exergy analysis approach. It was shown that the irreversibility in generator was highest followed by absorber and evaporator. Şencan *et al.* (2005) carried out the exergy analysis of a single effect water-lithium bromide VAR system for the generator temperature equal to 80°C, absorber and condenser temperatures equal to 40°C and evaporator temperature as 7°C and calculated the exergy losses in the system components. The effect of heat source temperature on COP and exergetic efficiency was computed. The effects of variations in absorber and condenser temperatures were not analysed. The results showed that the cooling and heating COP of the system increase slightly when increasing the heat source temperature. However, the exergetic efficiency of the system decreases when increasing the heat source temperature for both cooling and heating applications.

Kaushik *et al.* (1996) and, Kaynakli and Yamankaradeniz (2007) studied the single effect VAR system on the basis of entropy generation method. Kaynakli and Yamankaradeniz (2007) performed calculations for 10 kW cooling load for evaporator temperature equal to 4°C, condenser temperature of 38°C, absorber temperature was taken two degrees higher than condenser temperature, generator temperature equal to 90°C, effectiveness of solution heat exchangers was assumed as 0.5 and efficiency of pump was assumed equal to 0.90. They concluded that entropy generation of the generator is an important fraction of the total entropy generation in the system basically due to the temperature difference between the heat source and the working fluid and in order to decrease the total entropy generation of the system, the generator should be regarded as a system component which should be developed.

Lee and Sherif (2001a) carried out the second law analysis of a single effect water-lithium bromide absorption refrigeration system. The analysis was carried out for chilled water temperature of 7.22°C and cooling water temperatures equal to 29.4°C and 35°C. They evaluated the effect of heat source temperature on COP and exergetic efficiency. However, in this study, the effects of variations in absorber and condenser temperatures were not analyzed. Moreover, the effectiveness of solution heat exchanger was also not specified.

Anand and Kumar (1987) emphasized on calculation of the properties of water-lithium bromide with focus on calculation of entropy of water-lithium bromide. Their other contribution was availability analysis and calculation of irreversibility in system components of single and double effect (series flow) water-lithium bromide absorption systems. The parameters assumed for computation of results were condenser and absorber temperatures equal to 37.8°C, evaporator temperature taken as 7.2°C and generator temperature equal to 87.8°C and 140.6°C for single and double effect systems respectively. They neither carried out the parametric study to evaluate the effects of variations in design parameters nor calculated the second law efficiency of these systems. Lee and Sherif (2001b) presented the second law analysis of single effect and various double effect lithium bromide-water absorption chillers for a chilled water temperature of 7.22°C and cooling water temperatures 29.4°C and 35°C and computed COP and exergetic efficiency. They investigated the effect of heat source temperature on COP and exergetic efficiency. In this study, the effectiveness values of solution heat exchangers considered for analysis was not specified. Moreover, their results are only valid for water cooled systems.

Gomri and Hakimi (2008) presented the exergy analysis of double effect lithium bromide/water absorption refrigeration system. The results show that the performance of the system increases with increasing LP generator temperature, but decreases with increasing HP generator temperature. The highest exergy loss occurs in the absorber and in the HP generator, which therefore makes the absorber and HP generator the most important components of the double effect refrigeration system.

Oh et al. (1994) carried out investigation of a gas-fired, air-cooled $LiBr/H_2O$ double effect (parallel-flow type) absorption heat pump of 2 TR as an air conditioner for domestic use during the summer. The performance of the absorption heat pump in the cooling mode of operation was investigated through cycle simulation to obtain the system characteristics depending on the inlet temperature of air to the absorber, the working solution concentrations, the solution distribution ratio of the mass of solution into the first generator to the total mass of solution from the absorber, and the leaving temperature differences of the heat-exchanging components. Their results established that there exists a critical value of the solution distribution ratio that maximizes the cooling performance of the system.

In the above studies regarding both single and double effect water-lithium bromide systems, the effects of pressure drop between evaporator and absorber, temperature difference between heat source and generator, temperature difference between evaporator and cold room have not been examined. Further, the investigations have been carried out in a narrow range of design parameters.

A series flow triple effect system is one in which the strong solution leaving the absorber is first fed to HP generator then to MP generator followed by LP generator. Since the refrigerant is generated in three generators hence cooling effect produced is nearly three times than that for single effect for the same amount of heat supplied. Thus, the COP obtained is nearly three times as compared to single effect system. Triple effect inherently implies higher temperatures in comparison to double effect systems. The thermodynamic basis of the higher COP values comes from the increased availability of the higher temperature heat input (Herold et al., 1996). Grossman and Wilk (1994) developed a computer code for the simulation of absorption systems at steady state. In their study, the effect of heat source temperature on COP of water-lithium bromide series flow and parallel flow triple effect VAR system has been investigated. Grossman et al. (1994) also carried out simulation and performance analysis of triple effect VAR systems on the basis of energy analysis. They found that COP ranging from 1.272 for the series flow to 1.729 for the parallel flow was achievable at the design point.

Lee and Sherif (1999) examined COP and exergetic efficiency of multi-effect absorption refrigeration systems using water-lithium bromide. It was concluded that triple effect system has a better cooling COP over the single and double effect systems whereas exergetic efficiencies of all the three systems are similar. Their contribution in triple effect cycles is limited to the computation of effect of variation in heat source temperature on COP and exergetic efficiency for generating cooling water at 7.2°C for heat source temperatures between 150°C and 200°C, T_a = 29.4°C and T_c = 29.4 & 35°C. The value of effectiveness of solution heat exchanger(s) chosen by them is not specified in their study. Kaita (2002) carried out the simulation analysis for three kinds of triple-effect absorption cycles of parallel-flow, series-flow and reverse-flow using a newly developed simulation programme. The coefficient of performance, the maximum pressure and the maximum temperature of each cycle were calculated. It was concluded that parallel-flow cycle yields the highest COP among the cycles.

In this chapter energy and exergy analyses of series flow triple effect VAR system has been presented and the effects of HP generator temperature, absorber temperature and evaporator temperature have been discussed.

Other VAR systems like half effect system, resorption, and dual loop system are also described in this chapter. The performance analysis of a half effect system which promises the utilization of LT heat source such as geothermal energy or waste heat available at 60°C to 80°C has been presented.

A half effect system derives its name on the basis of COP achieved which is nearly half that of COP in a single effect VAR system. It comprises two single stage VAR systems connected in series. There are three pressure levels, i.e., LP, intermediate pressure and HP. The LP generator and HP absorber operate at same intermediate pressure. The intermediate pressure has been optimized for maximum exergetic efficiency of the system. The effects of various parameters on energetic and exergetic performance have also been discussed.

92 *Alternatives in Refrigeration and Air Conditioning*

3.2 SINGLE EFFECT AND SERIES FLOW DOUBLE EFFECT GENERATION WATER-LITHIUM BROMIDE VAPOUR ABSORPTION REFRIGERATION (VAR) SYSTEMS

Figure 3.1 (a) shows the schematic diagram of a single effect generation VAR system. The main components of a single effect generation water-lithium bromide VAR system are an evaporator, absorber, generator, condenser, pump, refrigerant and solution throttle valves and a solution heat exchanger. The refrigerant vapours leaving the evaporator (state point 10) are absorbed by the weak (in refrigerant) solution (coming from generator) entering the absorber (state point 6). Thus, the weak solution becomes strong (in refrigerant) solution (at state point 1). It is pumped to the generator via solution heat exchanger where it is raised in temperature by heat transfer from high temperature weak solution returning from generator. An external heat source such as waste heat or solar energy is utilised to drive out refrigerant from the strong solution in generator, leaving behind the weak solution (state point 4). The refrigerant vapours (state point 7) are condensed by means of external cooling using either air or water. The saturated liquid refrigerant (water) so produced is reduced in pressure and temperature by means of a refrigerant throttle valve before its entry to evaporator (state point 9) where it extracts heat from the space to be cooled and thus gets evaporated, thereby producing cooling effect.

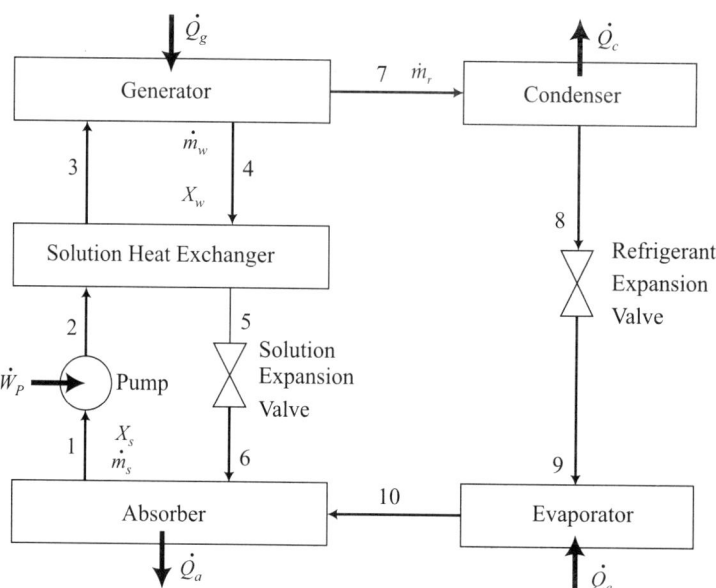

Fig. 3.1(a) Schematic Diagram of Water-Lithium Bromide Single Effect Generation VAR System

Finally, refrigerant vapours leave the evaporator and enter the absorber, where these are absorbed by the weak solution and this completes the cycle. Figure 3.1(b) shows the $\ln P - 1/T$ diagram of the single effect generation VAR system. A single effect generation VAR system is used when heat source temperature lies between 100–110°C (Herold *et al.*, 1996).

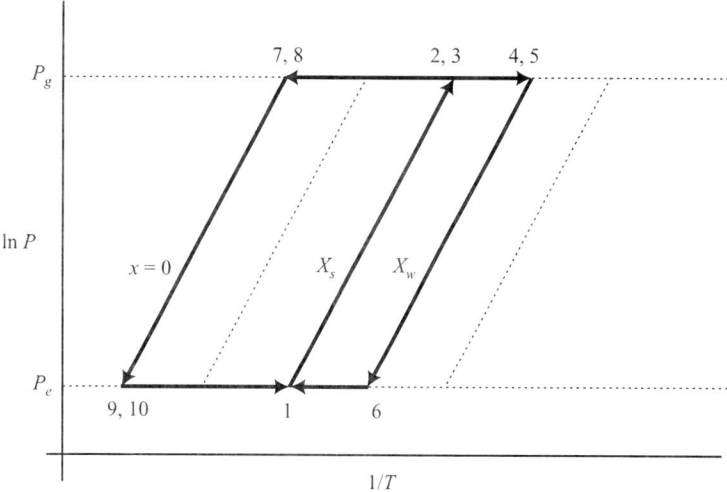

Fig. 3.1(b) ln P-1/T Diagram for Water-lithium Bromide Single Effect Generation VAR System

The term 'double effect' refers to the fact that heat input at high temperature is used twice within the cycle to generate vapours. The 'series flow' or 'parallel flow' in a double effect system refers to the method of delivering the solution (leaving the absorber) to the generators. There are two types of series flow systems. In 'Type –I' series flow system the solution is sent first to high pressure (HP) generator and then to second effect generator (also called low pressure (LP) generator). In 'Type-II' configuration, the solution is sent first to the second effect generator (LP generator) and then to HP generator. In parallel flow, the solution exiting the absorber is distributed in a specific ratio in two streams and delivered separately to HP and LP generators.

Figure 3.2(a) shows the schematic diagram of a series flow double effect generation water-lithium bromide VAR system. It involves three pressure levels, i.e., high, medium and low. The HP generator functions at high pressure and high temperature, the second effect (LP) generator and condenser operate at medium pressure, and the evaporator and absorber work at low pressure level. The strong solution leaving the absorber (state point 1) is pumped to the high temperature generator through solution heat exchangers. A high temperature heat source adds heat in high temperature generator to generate water vapour from the strong solution (state point 15). The weak solution leaving the HP generator (state point 12) enters the second effect generator, via solution throttling valve, where the refrigerant (water) vapours coming from generator (state point 15) are condensed due to low temperature of weak solution and their latent heat is utilised in generating water vapour ($\dot{m}_{r1}$ at state point 7) from the weak solution. The weak solution, therefore, becomes weaker (in refrigerant) (state point 4) and it is delivered to absorber through solution heat exchanger and solution throttle valve. Thus, in a double effect system, the refrigerant is generated in HP generator as well as second effect (LP) generator thereby increasing the quantity of water vapour generated for the same amount of heat input ($\dot{Q}_g$). This results in higher values of COP than those achieved in single effect generation system. Figure 3.2(b) shows the ln P – $1/T$ diagram of the series flow double effect generation VAR system.

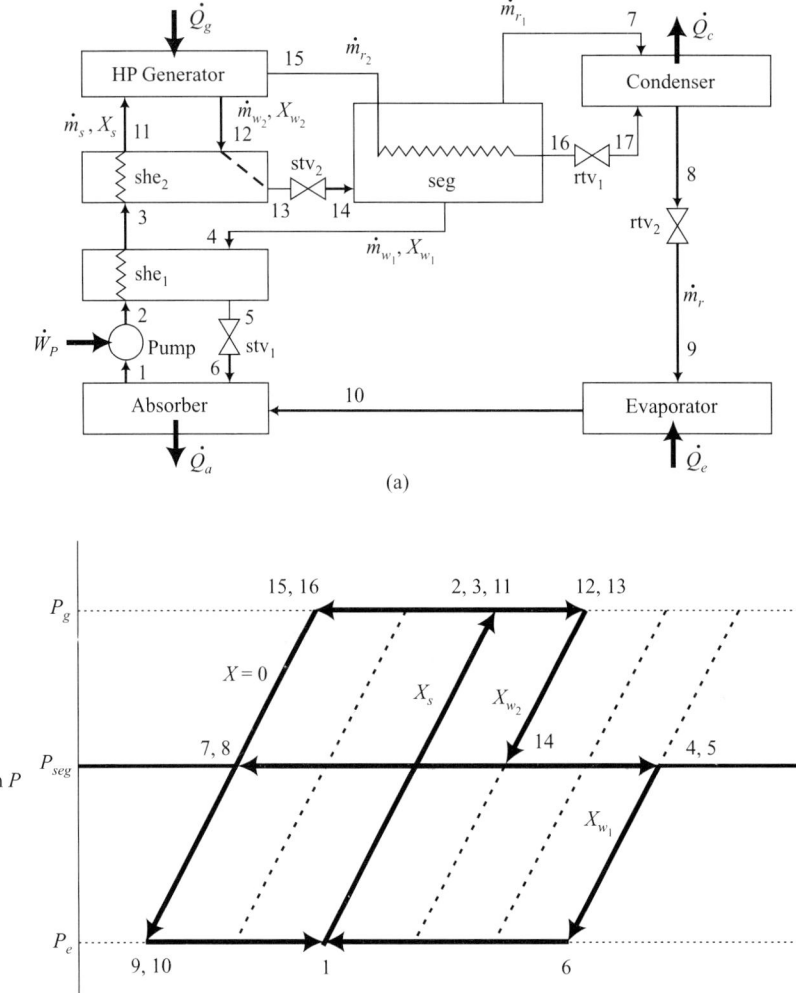

Fig. 3.2 (a) Schematic Diagram of Series Flow Double Effect Generation VAR System
(b) ln P-1/T Diagram for Series Flow Double Effect Generation VAR System

3.2.1 Thermodynamic Analysis

The energy and exergy analyses of absorption systems involve the application of principles of mass conservation, first and second laws of thermodynamics and these have already been presented in Chapter 2. Thus, these are not repeated here again. The principle of species conservation is also applied to absorption systems and is specified below.

Principle of Species Conservation

$$\Sigma \dot{m}_i X_i = \Sigma \dot{m}_e X_e \tag{3.1}$$

Application of these principles to single effect generation absorption refrigeration system is given below:

Mass and Species Conservation

Mass balance at absorber and / or at generator

$$\dot{m}_s = \dot{m}_r + \dot{m}_w \tag{3.2}$$

Mass flow rate through condenser, refrigerant throttle valve and evaporator is m_r.

A balance of the lithium bromide species, assuming that the vapour leaving the generator has zero salt content, yields

$$\dot{m}_s X_s = \dot{m}_w X_w \tag{3.3}$$

Water mass balance can be obtained by subtracting equation (3.3) from equation (3.2) to obtain

$$\dot{m}_s (1 - X_s) = \dot{m}_w (1 - X_w) + \dot{m}_r \tag{3.4}$$

Another mass flow parameter that is very useful is the solution circulation ratio (SCR), it is expressed as:

$$\text{SCR} = \dot{m}_s / \dot{m}_r = (X_w - X_s) \tag{3.5}$$

It actually describes the mass flow rate of solution required for unit refrigerant generation in generator.

3.2.1.1 *Thermodynamic Analysis of Single Effect Generation VAR System*

Energy Balance

The energy balance equations in a single effect generation absorption system are specified below:

Energy balance in absorber

$$\dot{Q}_a = \dot{m}_r h_{10} + \dot{m}_w h_6 - \dot{m}_s h_1 \tag{3.6}$$

Energy balance in generator

$$\dot{Q}_g = \dot{m}_r h_7 + \dot{m}_w h_4 - \dot{m}_s h_3 \tag{3.7}$$

Energy balance in condenser

$$\dot{Q}_c = \dot{m}_r (h_7 - h_8) \tag{3.8}$$

Energy balance in evaporator

$$\dot{Q}_e = \dot{m}_r (h_{10} - h_9) \tag{3.9}$$

Energy balance in solution heat exchanger

$$\dot{Q}_{she} = \dot{m}_s (h_3 - h_2) = \dot{m}_w (h_4 - h_5) \tag{3.10}$$

Pump work is given by equation

$$\dot{W}_p = \dot{m}_s (h_2 - h_1) \tag{3.11}$$

$$\text{Energy input} = \dot{Q}_g + \dot{Q}_e + \dot{W}_p \tag{3.12}$$

$$\text{Energy output} = \dot{Q}_a + \dot{Q}_c \tag{3.13}$$

As per the first law of thermodynamics, energy input into the system must be equal to the energy going out of the system, i.e. $\dot{Q}_g + \dot{Q}_e + \dot{W}_p = \dot{Q}_a + \dot{Q}_c$

Coefficient of performance (COP) of the single effect generation VAR system is defined as the ratio of the cooling capacity of the system to the energy input into the generator and pump. With this definition, COP of single effect generation system is given by:

$$\text{COP} = \frac{\dot{Q}_e}{\dot{Q}_g + \dot{W}_p} \tag{3.14}$$

Exergy Balance

Exergy destruction in each component of a single effect generation VAR system is furnished below:

Exergy destruction in absorber,

$$\dot{ED}_a = \dot{m}_r(h_{10} - T_o s_{10}) + \dot{m}_w(h_6 - T_o s_6) - \dot{m}_s(h_1 - T_o s_1) \tag{3.15}$$

Exergy destruction in generator,

$$\dot{ED}_g = \dot{m}_s(h_3 - T_o s_3) - \dot{m}_w(h_4 - T_o s_4) - \dot{m}_r(h_7 - T_o s_7) + \dot{Q}_g(1 - T_o/T_g) \tag{3.16}$$

Exergy destruction in condenser,

$$\dot{ED}_c = \dot{m}_r((h_7 - h_8) - T_o(s_7 - s_8)) \tag{3.17}$$

Exergy destruction in evaporator,

$$\dot{ED}_e = \dot{m}_r((h_9 - h_{10}) - T_o(s_9 - s_{10})) + Q_e\left(1 - \frac{T_o}{T_r}\right) \tag{3.18}$$

Exergy destruction in solution heat exchanger,

$$\dot{ED}_{she} = \dot{m}_s((h_2 - h_3) - T_o(s_2 - s_3)) + \dot{m}_w((h_4 - h_5) - T_o(s_4 - s_5)) \tag{3.19}$$

Exergy destruction in refrigerant throttle valve,

$$\dot{ED}_{rtv} = \dot{m}_r T_o(s_9 - s_8) \tag{3.20}$$

Exergy destruction in solution throttle valve,

$$\dot{ED}_{stv} = \dot{m}_w T_o(s_6 - s_5) \tag{3.21}$$

Total exergy destruction is the sum of exergy destruction in the system components

$$\dot{ED}_{total} = \dot{ED}_a + \dot{ED}_g + \dot{ED}_c + \dot{ED}_e + \dot{ED}_{she} + \dot{ED}_{rtv} + \dot{ED}_{stv} \tag{3.22}$$

3.2.1.2 Thermodynamic Analysis of Series Flow Double Effect Generation VAR System

Energy Balance

The energy in each component of a series flow double effect VAR system is given by the following equations:

$$\dot{Q}_a = \dot{m}_r h_{10} + \dot{m}_{w_1} h_6 - \dot{m}_s h_1 \tag{3.23}$$

$$\dot{Q}_g = \dot{m}_{r_2} h_{15} + \dot{m}_{w_2} h_{12} - \dot{m}_s h_{11} \tag{3.24}$$

$$\dot{Q}_c = \dot{m}_{r_1} h_7 + \dot{m}_{r_2} h_{17} - \dot{m}_r h_8 \tag{3.25}$$

$$\dot{Q}_e = \dot{m}_r (h_{10} - h_9) \tag{3.26}$$

$$\dot{Q}_{she_1} = \dot{m}_s (h_3 - h_2) = \dot{m}_{w_1} (h_4 - h_5) \tag{3.27}$$

$$\dot{Q}_{she_2} = \dot{m}_s (h_{11} - h_3) = \dot{m}_{w_2} (h_{12} - h_{13}) \tag{3.28}$$

$$\dot{Q}_{seg} = \dot{m}_{r_2} (h_{15} - h_{16}) = \dot{m}_{r_1} h_7 + \dot{m}_{w_1} h_4 - \dot{m}_{w_2} h_{14} \tag{3.29}$$

$$\dot{W}_p = \dot{m}_s (h_2 - h_1) \tag{3.30}$$

$$\text{Energy input} = \dot{Q}_g + \dot{Q}_e + \dot{W}_p \tag{3.31}$$

$$\text{Energy output} = \dot{Q}_a + \dot{Q}_c \tag{3.32}$$

$$\text{COP} = \frac{\dot{Q}_e}{\dot{Q}_g + \dot{W}_p} \tag{3.33}$$

Exergy Analysis

Exergy destruction in each component of a series flow double effect generation VAR system is specified vide equations (3.34) to (3.45).

$$\dot{ED}_a = \dot{m}_r (h_{10} - T_o s_{10}) + \dot{m}_w (h_6 - T_o s_6) - \dot{m}_s (h_1 - T_o s_1) \tag{3.34}$$

$$\dot{ED}_g = \dot{m}_s (h_{11} - T_o s_{11}) - \dot{m}_{w_2} (h_{12} - T_o s_{12}) - \dot{m}_{r_2} (h_{15} - T_o s_{15}) + \dot{Q}_g (1 - T_0/T_g) \tag{3.35}$$

$$\dot{ED}_c = \dot{m}_{r_1} (h_7 - T_o s_7) + \dot{m}_{r_2} (h_{17} - T_o s_{17}) - \dot{m}_r (h_8 - T_o s_8) \tag{3.36}$$

$$\dot{ED}_e = \dot{m}_r ((h_9 - h_{10}) - T_o (s_9 - s_{10})) + \dot{Q}_e \left(1 - \frac{T_o}{T_r}\right) \tag{3.37}$$

$$\dot{ED}_{she_1} = \dot{m}_s((h_2 - h_3) - T_o(s_2 - s_3)) + \dot{m}_{w_1}((h_4 - h_5) - T_o(s_4 - s_5)) \quad (3.38)$$

$$\dot{ED}_{she_2} = \dot{m}_s((h_3 - h_{11}) - T_o(s_3 - s_{11})) + \dot{m}_{w_2}((h_{12} - h_{13}) - T_o(s_{12} - s_{13})) \quad (3.39)$$

$$\dot{ED}_{rtv_1} = \dot{m}_{r_2} T_o(s_{17} - s_{16}) \quad (3.40)$$

$$\dot{ED}_{rtv_2} = \dot{m}_r T_o(s_9 - s_8) \quad (3.41)$$

$$\dot{ED}_{stv_1} = \dot{m}_{w_1} T_o(s_6 - s_5) \quad (3.42)$$

$$\dot{ED}_{stv_2} = \dot{m}_{w_2} T_o(s_{14} - s_{13}) \quad (3.43)$$

$$\dot{ED}_{seg} = \dot{m}_{w_2}(h_{14} - T_o s_{14}) + \dot{m}_{r_2}((h_{15} - h_{16}) - T_o(s_{15} - s_{16})) \quad (3.44)$$
$$- \dot{m}_{w_1}(h_4 - T_o s_4) - \dot{m}_{r_1}(h_7 - T_o s_7)$$

$$\dot{ED}_{total} = \dot{ED}_a + \dot{ED}_g + \dot{ED}_c + \dot{ED}_e + \dot{ED}_{she_1} + \dot{ED}_{she_2} + \dot{ED}_{rtv_1} + \dot{ED}_{rtv_2} \quad (3.45)$$
$$+ \dot{ED}_{stv_1} + \dot{ED}_{stv_2} + ED_{seg}$$

Exergetic Efficiency

Exergetic efficiency is given by the equation (3.46).

$$\eta_{ex} = \frac{\left| \dot{Q}_e \left(1 - \frac{T_0}{T_r}\right) \right|}{\dot{Q}_g \left(1 - \frac{T_0}{T_g}\right) + \dot{W}_p} \quad (3.46)$$

where T_r is the temperature of the space to be cooled. The exergetic efficiency can also be expressed in terms of total exergy destruction and exergy supplied to the system, i.e.:

$$\eta_{ex} = 1 - \frac{\dot{ED}_{total}}{\dot{Q}_g \left(1 - \frac{T_0}{T_g}\right) + \dot{W}_p} \quad (3.47)$$

The exergetic efficiency shows the percentage of the fuel exergy provided to a system that is found in the product exergy.

Efficiency Defect

Efficiency defect is given by:

$$\delta_i = \frac{\dot{ED}_i}{E\dot{Q}_g + \dot{W}_p} = \frac{\dot{ED}_i}{\dot{Q}_g \left(1 - T_0/T_g\right) + \dot{W}_p} \quad (3.48)$$

The total of efficiency defects is linked to the exergetic efficiency of the system by means of the relation (3.49):

$$\eta_{ex} = \left(1 - \Sigma \delta_i\right) \qquad (3.49)$$

Assumptions

The exergy analysis presented here is based on the following assumptions:
1. Heat losses through the system components are negligible.
2. Solution leaving the absorber and the generator are assumed to be saturated in equilibrium conditions at their respective temperatures and concentrations.
3. The refrigerant states leaving the condenser and evaporator are assumed to be saturated.
4. Refrigerant vapour leaving the generator is considered to be superheated. Non-equilibrium states at the inlet to generator, and absorber and states at outlet to the solution pump and solution heat exchanger are taken to be at their actual conditions.
5. The temperature in high temperature heat source, medium temperature heat sink and low temperature heat source are assumed to be constant while the fluid temperature varies in non-isothermal components due to different inlet/outlet solution concentrations.
6. The reference enthalpy (h_0) and entropy (s_0) used for calculating exergy of the working fluid are the values for water at an environment temperature and pressure of 25°C and 1.01325 bar respectively.

3.2.2 Results and Discussion

The properties of water-lithium bromide solution are taken from Pátek and Klomfar (2006).

Table 3.1 Maximum Differences in Values of Enthalpy and Entropy Calculated Using the Formulations Developed by Pátek and Klomfar (2006) Respectively, from Values Calculated According to Formulations of Other Authors

Author	$(\Delta h)_{max}$, kJ kg^{-1}	$(\Delta s)_{max}$, kJ kg^{-1}K^{-1}
Mcneely (1979)	14	–
Herold and Moran (1997)	16	–
Feurecker et al. (1993)	4	0.016
Chua et al. (2000)	10	0.030
Kaita (2001)	8	0.012

Table 3.1 illustrates the maximum difference between the enthalpy values and entropy values calculated using formulations developed by Pátek and Klomfar (2006) and others researchers respectively. The base parameters assumed for computation of results are specified in Tables 3.2 and 3.3.

3.2.2.1 *Validation of the Simulation*

The simulation results of energy analysis of the single effect and series flow double effect generation VAR systems (present work) have been compared with the numerical data reported by Anand and

Kumar (1987) (Refer Tables 3.2 and 3.3). It is observed that difference in heat transfer rates in all components are within ±1% of the reported data of Anand and Kumar (1987).

The COP values for single effect and series flow double effect systems are 0.7609 and 1.26 respectively and these are also matching with the corresponding values of Anand and Kumar (1987). Figure 3.3 presents the validation of simulation results (of energy analysis) with the reported experimental results of Aphornratana and Sriveerakul (2007) for single effect generation water-lithium bromide absorption refrigeration system. It is clear that the simulation results (for COP and SCR) are in good agreement with theoretical and experimental results of Aphornratana and Sriveerakul (2007).

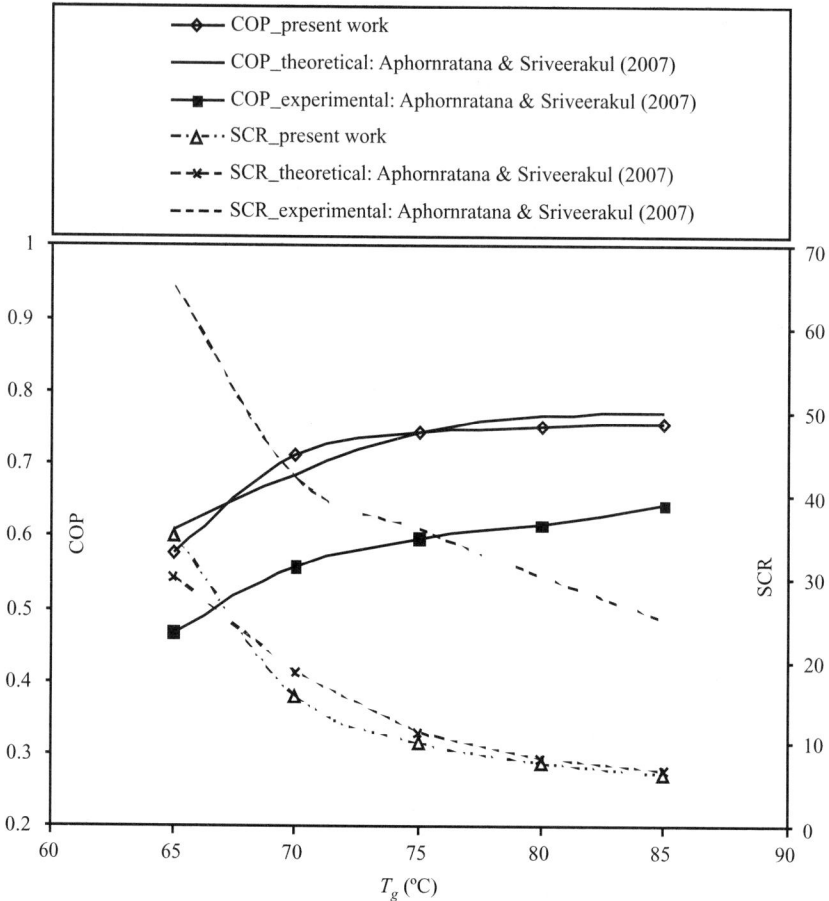

Fig. 3.3 Validation of the Single Effect System

Tables 3.2 and 3.3 also exhibit the results of exergy analysis of single effect system and series flow double effect system respectively. The exergy destruction in case of series flow double effect system is 451.17 kW as compared to 538.65 kW in case of single effect system. The reason for reduction of exergy destruction in double effect system is provision of one more solution heat exchanger which improves the performance of HP generator by regenerative heating of strong

Table 3.2 Comparison of Results of Energy Analysis of Present Work with Numerical Data Given in Anand and Kumar (1987) and Exergy Analysis Results of the Present Work (Single Effect System)

Single Effect Generation 'VAR' System			
Parameters: T_g = 87.8 °C, T_e = 7.2 °C, $T_c = T_a$ = 37.8 °C, effectiveness of solution heat exchanger = 0.7, mass flow rate of refrigerant (water) = 1 kg/s			
Energy Analysis Results			
Component	**Anand and Kumar (1987)** $\dot{Q}$, kcal kg^{-1} s (kW)	**Present work** $\dot{Q}$, (kW)	**Difference (%)**
Generator	734 (3073.11)	3095.70	0.735
Absorber	698 (2922.39)	2945.27	0.783
Condenser	599 (2507.89)	2505.91	−0.079
Evaporator	563 (2357.17)	2355.45	−0.0729
Solution heat exchanger	125 (523.25)	518.72	−0.86
Solution throttle valve	0	0	–
Refrigerant throttle valve	0	0	–
Pump	0	0.0314	–
Energy input	1297 (5430.51)	5451.184	0.385
COP (dimensionless)	0.76703	0.7609	
Exergy Analysis Results			
$\dot{ED}_a$ = 191.72 kW	$\dot{ED}_{rtv}$ = 6.94 kW	Exergy input = $\dot{Q}_g(1-T_0/T_g)+\dot{W}_p$	
$\dot{ED}_c$ = 109.76 kW	$\dot{ED}_{stv}$ = 0.024 kW	= 538.654 kW	
$\dot{ED}_e$ = 86.28 kW	$\dot{ED}_{she}$ = 25.08 kW	Exergy output = $\dot{Q}_e\left|(1-T_0/T_r)\right|$	
		= 63.28 kW	
$\dot{ED}_g$ = 55.57 kW	$\dot{ED}_{total}$ = 475.374 kW	**Exergetic efficiency (η_{ex}) = 11.75 %**	

solution by the weak solution returning from HP generator. The exergy destruction is highest in absorber followed by condenser, evaporator and solution heat exchangers. Hence, more attention should be paid to improve the design of the absorber.

3.2.2.2 Effect of Generator Temperature

Figure 3.4 (a) and Fig.3.4 (b) show the effect of generator temperature on the coefficient of performance (COP) and exergetic efficiency in single and double effect systems.

For $T_a = T_c$ = 37.8°C: The COP of single effect system varies between 0.6 and 0.75 and the maximum value is achieved at about 91°C and further increasing the temperature does not increase the COP. The value of the COP nearly remains constant between 91–110°C. The COP for double effect system varies between 1–1.28 and it is about 60–70% higher than the COP for single effect system. The maximum value occurs at about 150°C, subsequently the curve flattens. The reason for the better COP in double effect system is additional amount of refrigerant produced in the second effect generator because of the heat available from condensation of vapour refrigerant

Table 3.3 Comparison of Energy Analysis of Series Flow Double Effect System of Present Work with Numerical Data Given in Anand and Kumar (1987) and Results of Exergy Analysis of Series Flow Double Effect Generation VAR System for the Present Work

Double Effect 'VAR' System			
Parameters : T_g= 140.6°C , T_e = 7.2 °C, T_c = T_a = 37.8 °C, effectiveness of solution heat exchangers 1 and 2 = 0.7, mass flow rate of refrigerant = 1 kg/s			
Energy Analysis Results			
Component	Anand and Kumar (1987) Q, kcal kg^{-1} s (kW)	Present work Q, (kW)	Difference (%)
Generator	444 (1858.94)	1868.711	0.525
Second effect generator	303 (1268.6)	1272.479	0.3057
Absorber	698 (2922.39)	2942.175	0.677
Condenser	308 (1289.53)	1282.052	−0.58
Evaporator	563 (2357.17)	2355.45	−0.073
Solution heat exchanger 1	125 (523.35)	518.596	−0.908
Solution heat exchanger 2	207 (866.67)	861.206	−0.63
Solution throttle valves 1,2	0	0	–
Refrigerant throttle valves 1,2	0	0	–
Pump	0	0.3598	–
Energy input	1007(4216.65)	4224.227	0.193
Coefficient of performance (COP)	1.268	1.26	
Exergy Analysis Results			
$\dot{ED}_a$ = 188.66 kW	$\dot{ED}_{she2}$ = 43.1 kW	Exergy input = $\dot{Q}_g(1 - T_0/T_g) + \dot{W}_p$	
$\dot{ED}_c$ = 55.87 kW	$\dot{ED}_{rtv1}$ = 7.81 kW	= 514.45 kW	
$\dot{ED}_e$ = 86.28 kW	$\dot{ED}_{rtv2}$ = 6.94 kW	Exergy output = $\dot{Q}_e |(1 - T_0/T_r)|$	
$\dot{ED}_g$ = 21.72 kW	$\dot{ED}_{stv1}$ = 0.024 kW	= 63.28 kW	
$\dot{ED}_{seg}$ = 12.48 kW	$\dot{ED}_{stv2}$ = 0.23 kW	**Exergetic efficiency = η_{ex}= 12.3 %**	
$\dot{ED}_{she1}$ = 28.06 kW	$\dot{ED}_{total}$ = 451.17 kW		

coming from the high temperature generator (i.e., external heat is not supplied in the second effect generator).

It is observed that the COP increases initially with increase in generator temperature, tends to level off rather than continue to increase, and with a further increase in generator temperature, even drops somewhat. The increase in the generator temperature results in an increase in temperatures of refrigerant and the solution exiting the generator which are higher than before. Thus, increase in the average temperatures of both condenser and the absorber is observed and it accounts for

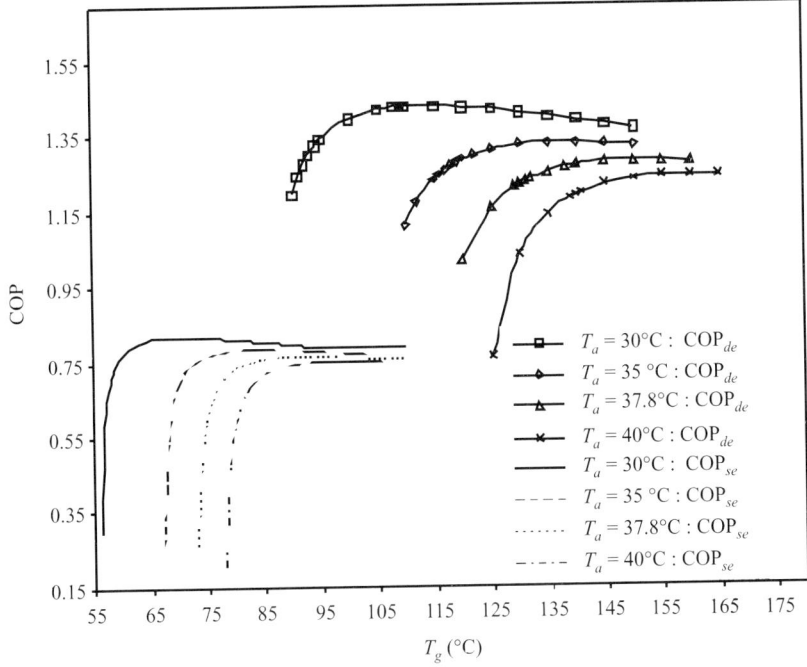

Fig. 3.4(a) Variation of COP with Generator Temperature in Single and Double Effect Generation VAR Systems ($T_e = 7.2\,°C$, $T_c = T_a$, $\varepsilon_{she(s)} = 0.7$, $T_r = 17.2°C$)

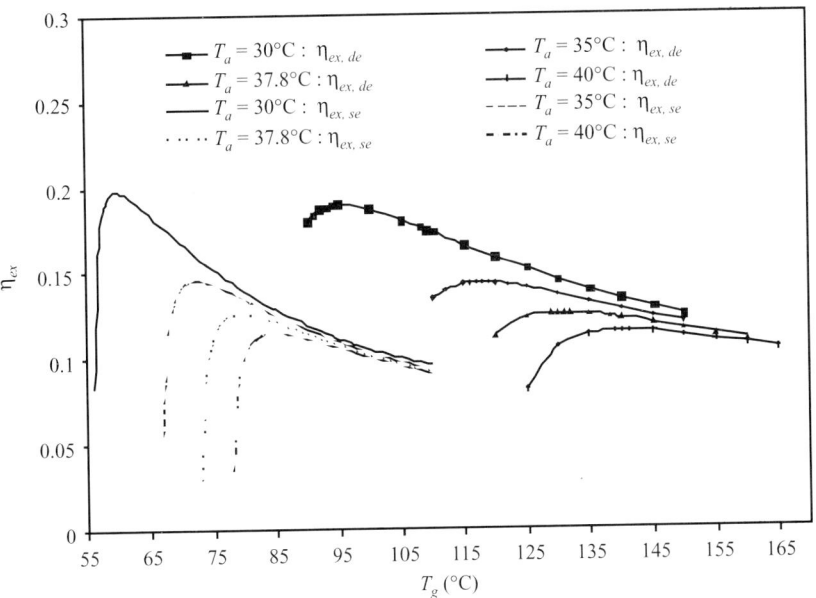

Fig. 3.4(b) Variation of Exergetic Efficiency with Generator Temperature in Single and Double Effect VAR Systems ($T_e = 7.2\,°C$, $T_c = T_a$, $\varepsilon_{she(s)} = 0.7$, $T_r = 17.2°C$)

increased irreversibility in these components. Thus, the positive effect of increase in the COP by virtue of increase in the generator temperature is offset by the degradation of the COP due to increase in the condenser and absorber temperatures. The solution circulation ratio (i.e., ratio of mass flow rate of strong solution required per unit mass flow rate of refrigerant) is another important factor for this behaviour of COP. The increase in generator temperature causes the solution circulation ratio to decrease and consequently the heat duties in generator, condenser and absorber decrease. Thus, for constant cooling load, the reduction in generator heat duty causes initial increase in the COP. At higher generator temperatures the rate of reduction in solution circulation ratio drops but the temperature difference between the generator and sub-cooled solution entering the generator increases causing an increase in the irreversibility occurring in generator. Thus, the positive effect of decrease in the solution circulation ratio is balanced by negative effect of increase in the temperature difference between generator and sub-cooled solution entering the generator. Thus, the generator heat duty levels off and the COP curve turns nearly flat showing marginal drop in COP. The exergetic efficiency of each system increases considerably initially and then declines continuously as the generator temperature increases. The reason already discussed above for the COP curve variations can be attributed for such behaviour of the variation in exergetic efficiency. In single effect system, ascend and descend of the exergetic efficiency with increase in the generator temperature is higher than that in double effect system. The thermodynamically optimum value of the generator temperature, corresponding to given temperature conditions of other components, is 130°C for double effect system and 80°C for single effect system. The maximum value is about 0.3% higher in case of single effect system. Lee and Sherif (1999) also concluded that single effect system has relatively higher exergetic efficiency within its low heat source temperature range, when compared to double effect system.

It is observed that decreasing the absorber and condenser temperatures below 37.8°C cause COP, exergetic efficiency and their maximum values to increase and this happens because of reduction in heat duty of generator, irreversibilities associated with absorber, generator and condenser due to reduction in their average temperatures. However, increasing the absorber and condenser temperatures above 37.8°C reduces COP, exergetic efficiency and their maximum values also. The detailed reasoning is specified in next paragraph.

3.2.2.3 Effect of Absorber Temperature

The effect of the variation in the absorber temperature on the COP and exergetic efficiency is illustrated in Fig. 3.5. It is important to discuss the effect of absorber temperature on solution circulation ratio to understand its significance. The solution circulation ratio is defined by the expression, $SCR = X_w/(X_w - X_s)$. With increase in the absorber temperature (at constant evaporator temperature and constant absorber pressure) the concentration X_s increases whereas concentration X_w remains constant since generator temperature and generator pressure remain constant. Thus, the numerator remains constant whereas the denominator reduces and hence solution circulation ratio increases. Thus, increase in the absorber temperature increases the solution circulation ratio which increases the generator heat duty and pump work. The evaporator load remains constant because mass flow rate of refrigerant is assumed constant. Thus, the COP reduces with increase in the absorber temperature. The increased heat duty in the generator causes the increase in the input exergy while the exergy of the cooling effect remains constant and hence the exergetic efficiency decreases. It is appropriate to mention here that increase in the solution circulation ratio is the foremost reason for degradation of system performance in both the systems. The rate of decrease in the COP and the exergetic efficiency is higher in double effect system. The COP for double effect system is approximately 50% to 70% higher than that for single effect system. However, the exergetic efficiency for double effect system is higher than that in single effect system at absorber temperatures between 25–40°C.

At absorber temperatures higher than 40°C, the solution circulation ratio for series flow double effect system is more than that for single effect system and hence the circulation losses are higher in double effect system. Hence, exergetic efficiency is higher for single effect system than that for the series flow double effect system.

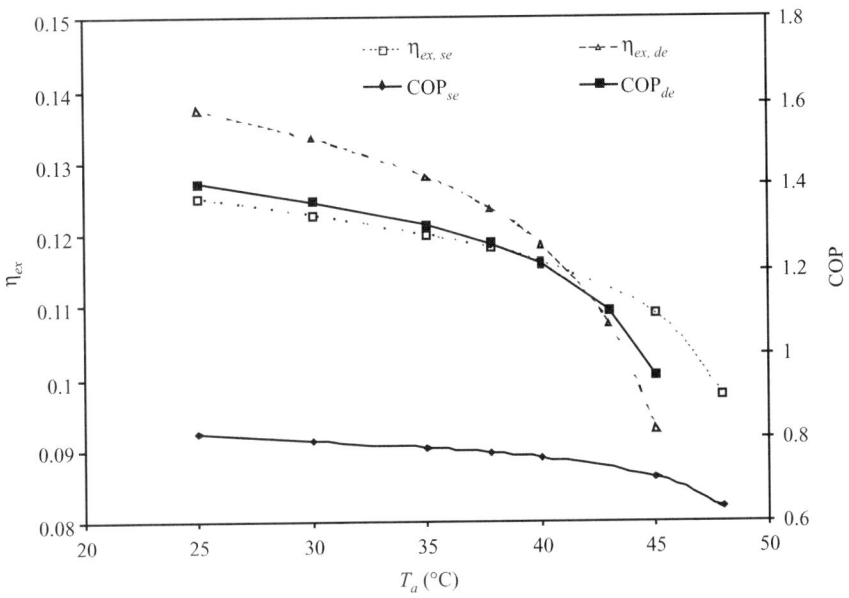

Fig. 3.5 Effect of Absorber Temperature on COP and η_{ex} (T_g = 87.8 °C for Single Effect, T_g = 140.6 °C for Double Effect, T_c = 37.8 °C, T_e = 7.2 °C, T_r = 17.2°C, $\varepsilon_{she(s)}$ = 0.7)

3.2.2.4 *Effect of Difference in Absorber and Condenser Temperatures*

Figure 3.6 shows the effect of $T_a \geq T_c$ and $T_c \geq T_a$ on COP and exergetic efficiency in a single effect system. With increase in either T_a or T_c, the decrease in COP and exergetic efficiency are observed.

Case 1: $T_c \geq T_a$: In the previous para it is specified that with increase in absorber temperature solution circulation ratio increases and the exergetic efficiency depends on solution circulation ratio. Higher the solution circulation ratio, higher are circulation losses and lesser is exergetic efficiency. This is the reason for decrease in exergetic efficiency with increase in absorber temperature.

Case 2: $T_c \geq T_a$: (for $T_a = 30°C$), since absorber temperature remains constant X_s remains constant whereas X_w reduces since generator temperature is constant but generator pressure increases because of increase in condenser temperature. This also causes an increase in solution circulation ratio and reduction in exergetic efficiency. In the first case (i.e., $T_a \geq T_c$) the increase in solution circulation ratio is larger in comparison to the later case (i.e., $T_c \geq T_a$), thus reduction in exergetic efficiency is higher for earlier case. In both circumstances, the decrease in COP and exergetic efficiency is in the range of 3-4%.

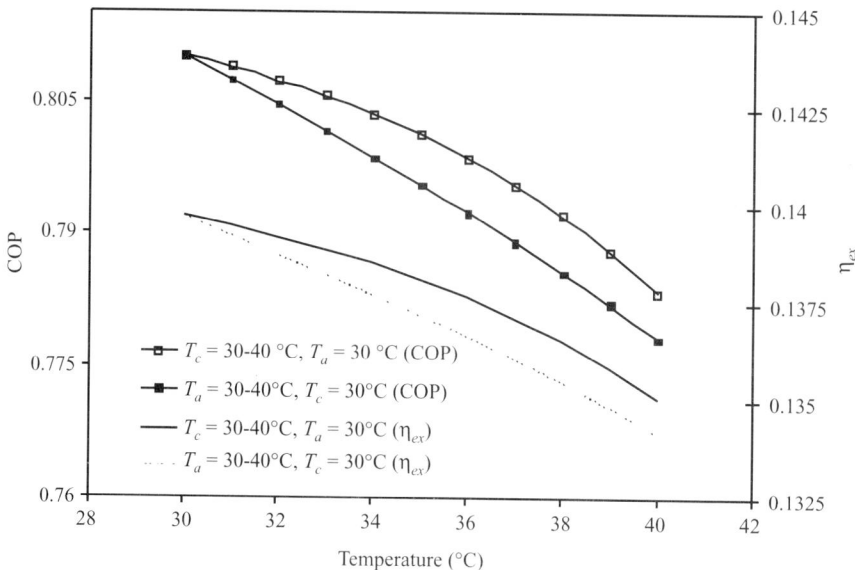

Fig. 3.6 Effect of Difference in Absorber and Condenser Temperatures on COP and η_{ex} in a Single Effect System ($T_g = 80°C$, $\varepsilon_{she} = 0.7$, $T_e = 7.2°C$ and $T_r = 17.2°C$)

The effect of $T_a \geq T_c$ and $T_c \geq T_a$ on COP and exergetic efficiency in a series flow double effect system is presented in Fig. 3.7. For series flow double effect system, solution circulation ratio is defined as $Xw_1/(Xw_1 - Xs)$ where Xw_1 is the concentration of solution leaving the second effect generator. In the first case, i.e., $T_a \geq T_c$, the increase in absorber temperature increases X_s whereas Xw_1 reduces since temperature of the second effect generator reduces whereas the pressure in it remains constant. The overall effect is to increase the solution circulation ratio from 5.24 to 13 and hence exergetic efficiency reduces. In the later case, i.e., $T_c \geq T_a$, X_s remains constant since absorber temperature and pressure remain constant however, 'Xw_1' reduces and the solution circulation ratio increases marginally from 5.24 to 7.15, therefore the reduction in exergetic efficiency is lower. The reduction in COP and exergetic efficiency is about 11.3% when absorber temperature is 10°C higher than condenser temperature. For the later case: $T_c \geq T_a$, for condenser temperature 10°C higher than absorber temperature, the reduction in COP and exergetic efficiency is about 4.3%.

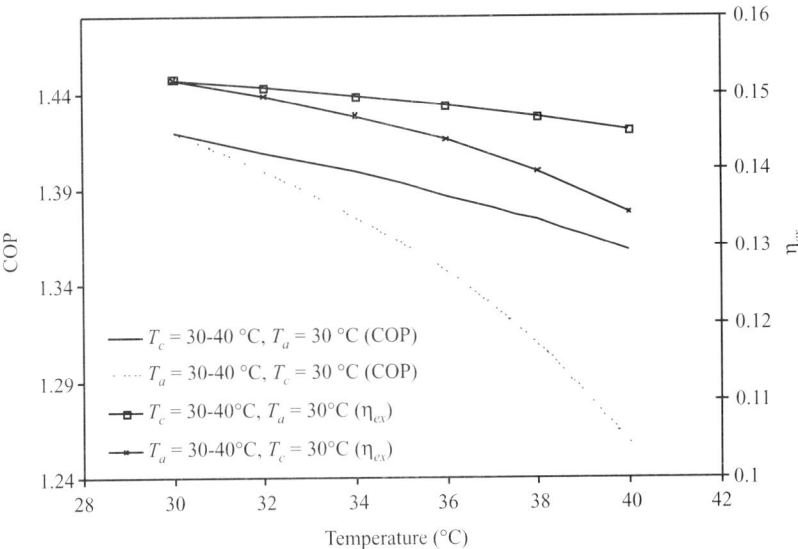

Fig. 3.7 Effect of Difference in Absorber and Condenser Temperatures on COP and Exergetic Efficiency in a Series Flow Double Effect System ($T_g = 125°C$, $\varepsilon_{she(s)} = 0.7$, $T_e = 7.2\ °C$, $T_r = 17.2°C$)

3.2.2.5 Effect of Variation in Evaporator Temperature

Figure 3.8 shows the effect of evaporator temperature on COP and exergetic efficiency in the two systems. The COP of both the systems increases with increase in evaporator temperature. It is observed that the rate of increase in COP of series flow double effect system varies between 3.4 and 13.3% whereas for single effect it lies between 1.9 and 3.8%. Moreover, COP values for series flow double effect system are 44 to 70% higher in comparison to single effect system. The exergetic efficiency for single effect system is about 0-9% higher for evaporator temperatures between 2 and 4°C in comparison to double effect system whereas above 4°C, the exergetic efficiency of series flow double effect system is 3.5 – 7.7% higher than the corresponding values in a single effect system.

3.2.2.6 Effect of Effectiveness of Solution Heat Exchangers

Figure 3.9 represents the effect of solution heat exchanger effectiveness on the COP and exergetic efficiency of a series flow double effect system. It is observed that the effectiveness of solution heat exchanger '2' (ε_{she2}) has greater effect on both COP and exergetic efficiency of the series flow double effect system than effectiveness of solution heat exchanger '1' (ε_{she1}). The values of COP and exergetic efficiency of the system increase by about 15.7 % when effectiveness of she_2 is

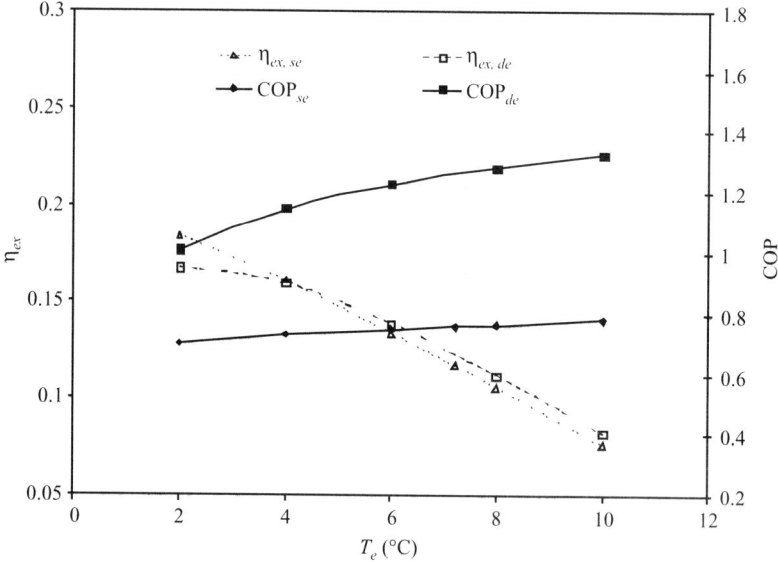

Fig. 3.8 Effect of Evaporator Temperature on COP and η_{ex} ($T_a = T_c = 37.8$ °C, $T_g = 87.8$ °C for Single Effect and 140.6 °C for Double Effect System, $\varepsilon_{she} = 0.7$)

increased from 0.1 to 1 whereas these values increase by 13% for identical increase in effectiveness of she_1. Moreover, it is also observed that increase in ε_{she1} is responsible for reduction in temperature of the weak solution leaving the she_1 to such a level that chances of crystallization are enhanced. Thus, in a double effect system ε_{she1} should be chosen prudently.

3.2.2.7 *Effect of Pressure Drop between Evaporator and Absorber*

The pressure drop between evaporator and absorber varies between 0 and 0.5 kPa (Herold *et al.*, 1996). It can be ascertained from Fig. 3.10 that pressure drop between evaporator and absorber severely affects the COP and exergetic efficiency in a series flow double effect system as compared to a single effect system. In a single effect system, the COP and exergetic efficiency drops by about 31% for pressure drop equal to 0.5 kPa whereas in a double effect system COP and exergetic efficiency drops by 57.8% for pressure drop of 0.45 kPa. The rate at which COP and exergetic efficiency decreases in a single effect system varies between 0.02 and 0.08 per kPa for pressure drop between 0 and 0.39 kPa whereas this value varies between 0.08–0.26 per kPa for pressure drop between 0.39 and 0.5 kPa. In a series flow double effect system, the rate of variation of COP and exergetic efficiency lies between 0.04 and 0.1 per kPa for pressure drop between 0 and 0.3 kPa whereas further increase in pressure drop from 0.3 to 0.45 kPa causes the corresponding value to vary between 0.1 and 0.51 per kPa. Thus, it is evident that pressure drop between evaporator and absorber is a very important parameter in both the systems and it should be reduced.

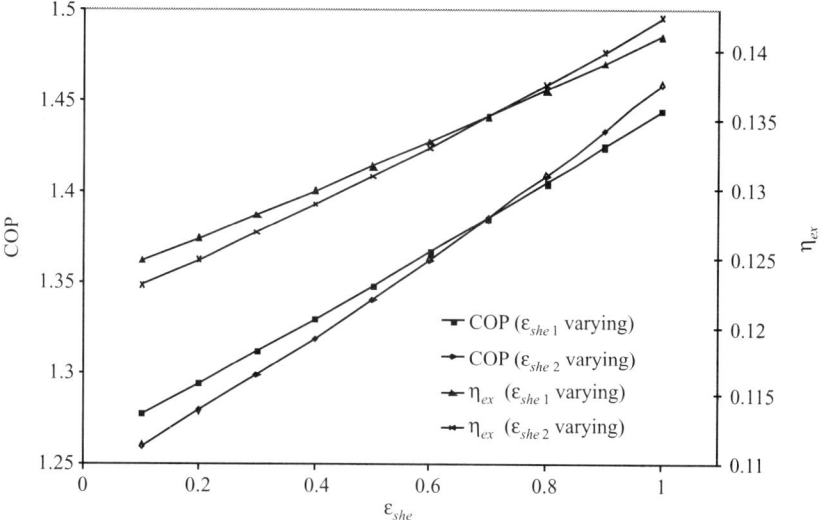

Fig. 3.9 Effect of Effectiveness of Solution Heat Exchangers on COP and η_{ex} in Series Flow Double Effect System ($T_a = T_c = 30\ °C$, $T_g = 140.6\ °C$, $T_e = 7.2\ °C$, $T_r = 17.2\ °C$)

3.2.2.8 Effect of Temperature Difference in Heat Source and Generator on Exergetic Efficiency

The difference in heat source temperature and generator temperature is an important parameter and its effect on exergetic efficiency of absorption systems is shown in Fig. 3.11. The generator temperatures assumed for this particular case are 130°C for series flow double effect system and 80°C for single effect system. These temperatures have been assumed because corresponding to these values of generator temperatures the exergetic efficiency was found to be maximum corresponding to given conditions. It can be observed that with increase in difference between generator temperature and heat source temperature the exergetic efficiency falls. This occurs due to increase in exergy destruction because of increase in finite temperature difference. The rate of decline of exergetic efficiency in single effect system is higher in comparison to series flow double effect system. For a particular amount of heat to be transferred through the same temperature difference, the entropy generation will be larger when heat source temperature is lower. That is why in case of single effect system the drop in exergetic efficiency is higher than in double effect for the same temperature difference. The drop in exergetic efficiency for single effect system is 0.00052°C temperature difference between the source and generator. The corresponding value for double effect system is 0.000367/°C temperature difference between source and generator.

If both single and double effect systems are operated by same heat source, say engine exhaust at a temperature of about 300°C, then the exergetic efficiency of single effect is 4.1% and that for series flow double effect is 6.84% (corresponding to the base data assumed in the present analysis). This means that a double effect system is better than a single effect system if both systems are operated by same heat source.

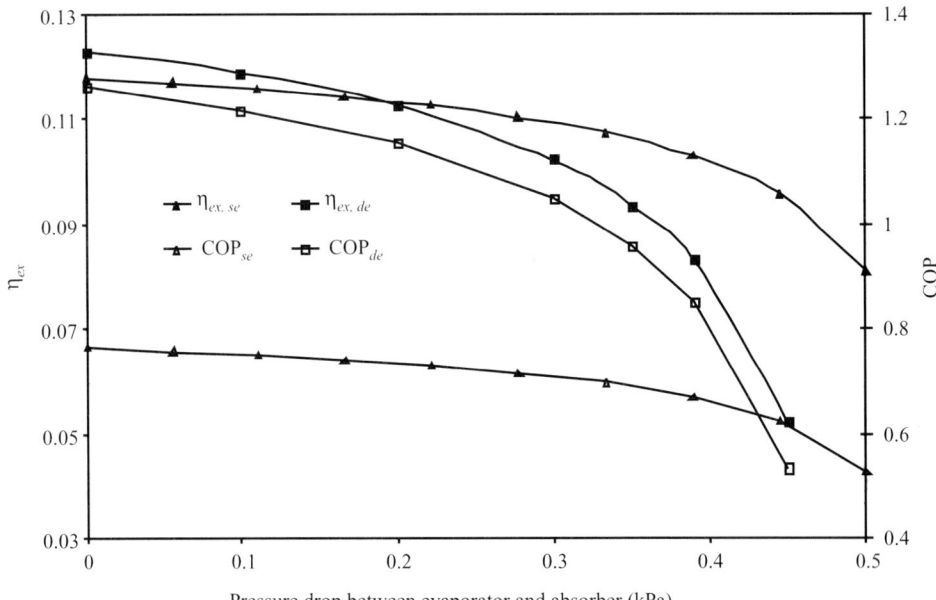

Fig. 3.10 Effect of Pressure Drop Between Evaporator and Absorber on COP and η_{ex} ($T_a = T_c = 37.8°C$, $T_g = 87.8°C$ for Single Effect and $140.6°C$ for Double Effect System, $T_e = 7.2°C$, $\varepsilon_{she} = 0.7$)

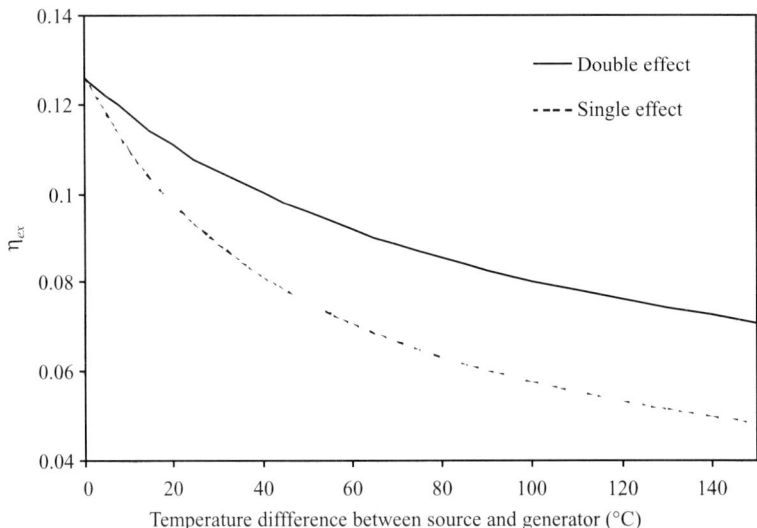

Fig. 3.11 Effect of Temperature Difference Between Source and Generator on Exergetic Efficiency ($T_g = 130°C$ (Double Effect)/$80°C$ (Single Effect), $T_e = 7.2°C$, $T_c = T_a = 37.8°C$, $\varepsilon_{she(s)} = 0.7$)

3.2.2.9 Effect of Temperature Difference in Cold Room and Evaporator on Exergetic Efficiency

In Fig. 3.12, the effect of temperature difference between cold room and evaporator is shown. The slopes of curves of decline in exergetic efficiency for both the systems are same for the assumed temperature difference between cold room and evaporator. This is because the evaporator temperature and temperature difference between the cold room and evaporator are same for both systems. However, one important point to note here is that the rate of fall of exergetic efficiency is about 0.017/°C for the difference between cold room and evaporator temperatures.

Thus, from above it can be concluded that the temperature difference in cold room and evaporator is of greater importance in comparison to the temperature difference between heat source and generator and hence the earlier should be minimized.

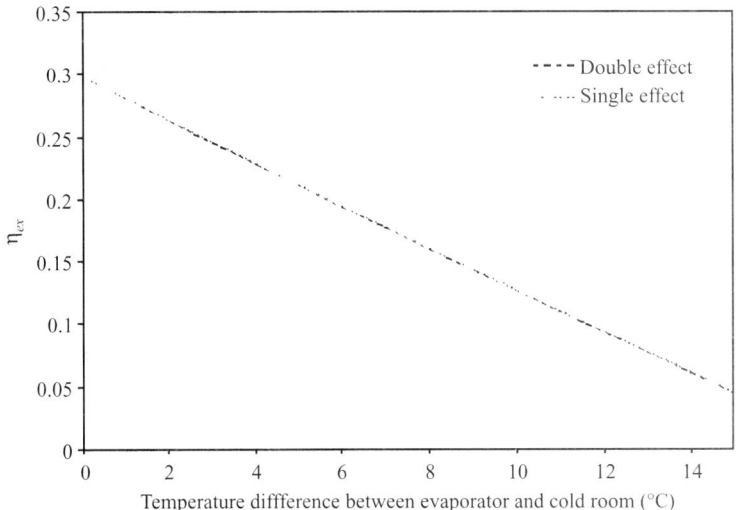

Fig. 3.12 Effect of Temperature Difference Between Evaporator and cold Space on Exergetic Efficiency (T_g = 130°C (Double Effect)/80 °C (Single Effect), T_e = 7.2°C, $T_c = T_a$ = 37.8°C, $\varepsilon_{she(s)}$ = 0.7)

3.2.2.10 Effect of Generator Temperature on Efficiency Defects

Figures 3.13 and 3.14 represent the variation in efficiency defects with generator temperature in the components of the single effect and series flow double effect systems. The absorber is the key contributor towards drop in the exergetic efficiency since efficiency defect in it is largest in comparison to other components. The main cause for the high efficiency defect in the absorber is entropy generation due to mixing of streams of weak solution and refrigerant at different temperatures. With increase in generator temperature, the temperature of the solution stream leaving the solution heat exchanger at (low temperature) increases and hence more entropy generation takes place due to mixing of streams with larger temperature differences. Hence, it is obvious that improvement in design of the absorber is a must for enhancing the exergetic efficiency of both the systems.

At the optimum generator temperature equal to 80 °C for single effect system, the components arranged in the descending order of efficiency defects are absorber, condenser, evaporator, solution heat exchanger and generator ($\delta_a > \delta_c > \delta_e > \delta_{she} > \delta_g$). The efficiency defects in expansion valves are least. The efficiency defect in generator is on the lower side since while computing efficiency defects in system components it is assumed that heat source and generator are at same temperature.

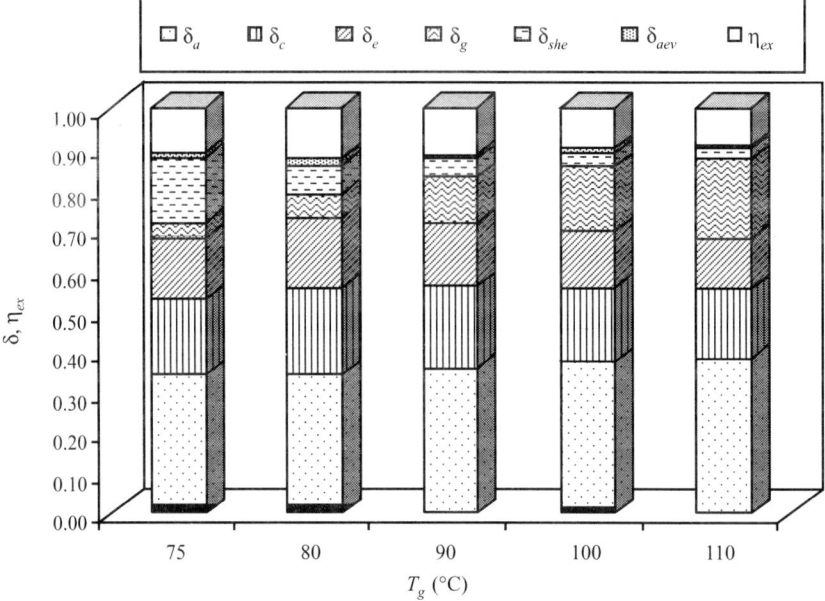

Fig. 3.13 Variation of Efficiency Defects in Components of a Single Effect System with Generator Temperature ($T_a = T_c = 37.8°C$, $T_e = 7.2°C$, $\varepsilon_{she} = 0.7$)

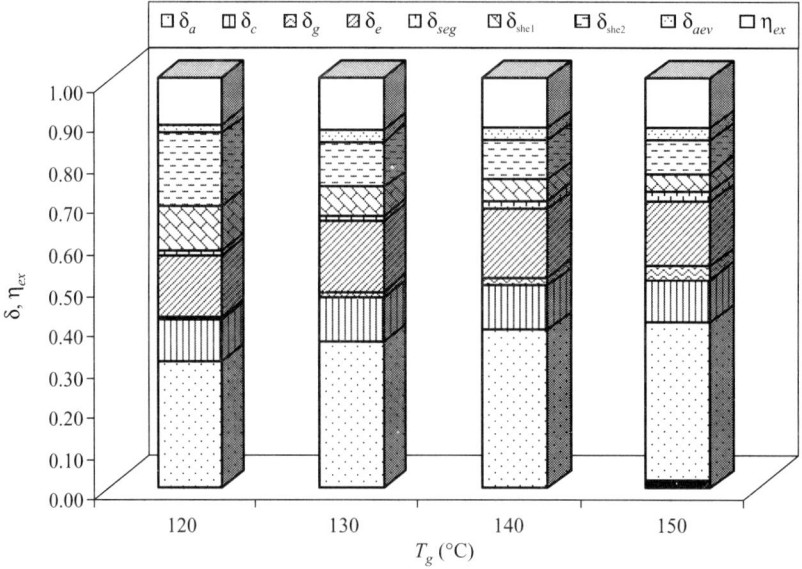

Fig. 3.14 Variation of Efficiency Defects in Components of a Series Flow Double Effect System with Generator Temperature ($T_a = T_c = 37.8°C$, $T_e = 7.2°C$, $\varepsilon_{she(s)} = 0.7$)

Vapour Absorption Cooling and Advanced Absorption Cycles 113

Similar to above, the descending order of components based on efficiency defects in a series flow double effect system is absorber, evaporator, she$_2$, condenser, she$_1$, all expansion valves, second effect generator and HP generator.

3.3 PARALLEL FLOW DOUBLE EFFECT GENERATION 'VAR' SYSTEM

Figure 3.15 (a) shows scheme of a LiBr/H$_2$O parallel-flow, double effect VAR system and Fig. 3.15 (b) represents its *ln P- 1/T* diagram. It includes the most important components of the double effect system: absorber, evaporator, HP generator, second effect generator (LP generator), condenser, HT heat exchanger (she$_2$), LT heat exchanger (she$_1$) and a solution pump. The strong (in absorbent) solution produced in the absorber is separated at the outlet of the LT heat exchanger and distributed separately to the HP generator and the LP generator. This is the main feature of the parallel-flow double effect generation absorption system. The ratio of solution entering the HP generator to the

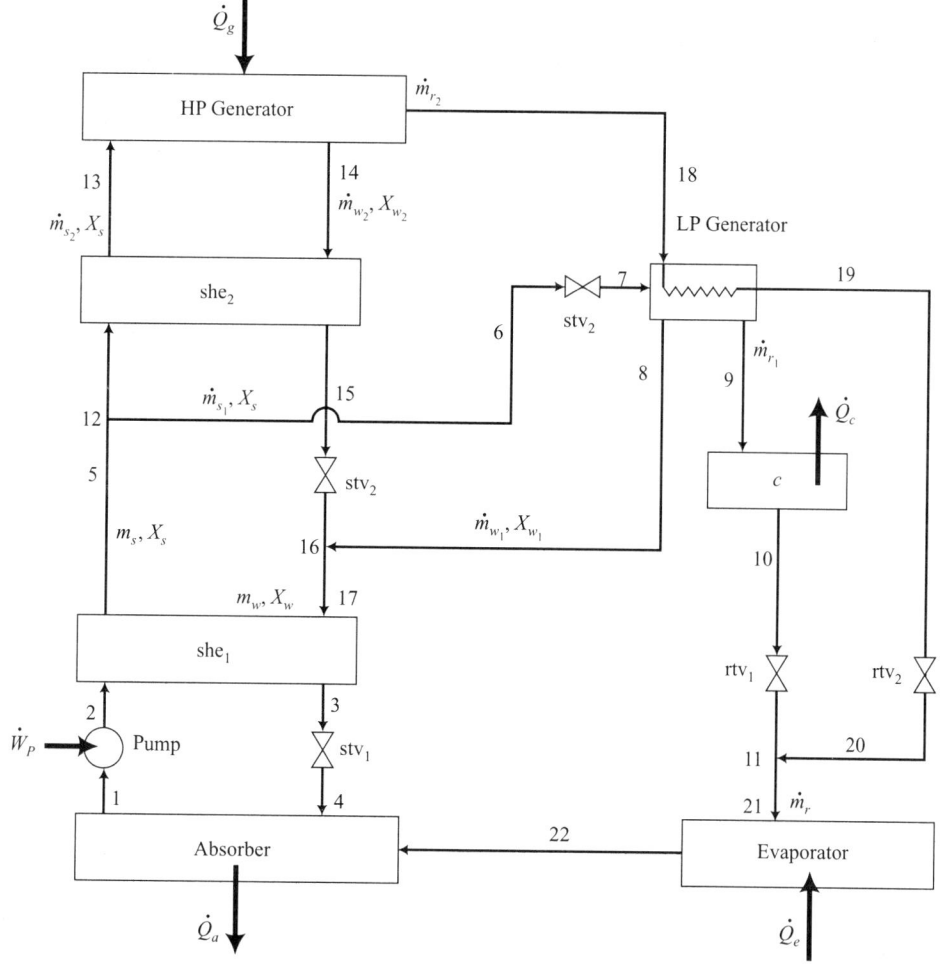

Fig. 3.15(a) Schematic Diagram of Parallel Flow Double Effect Generation VAR System

114 *Alternatives in Refrigeration and Air Conditioning*

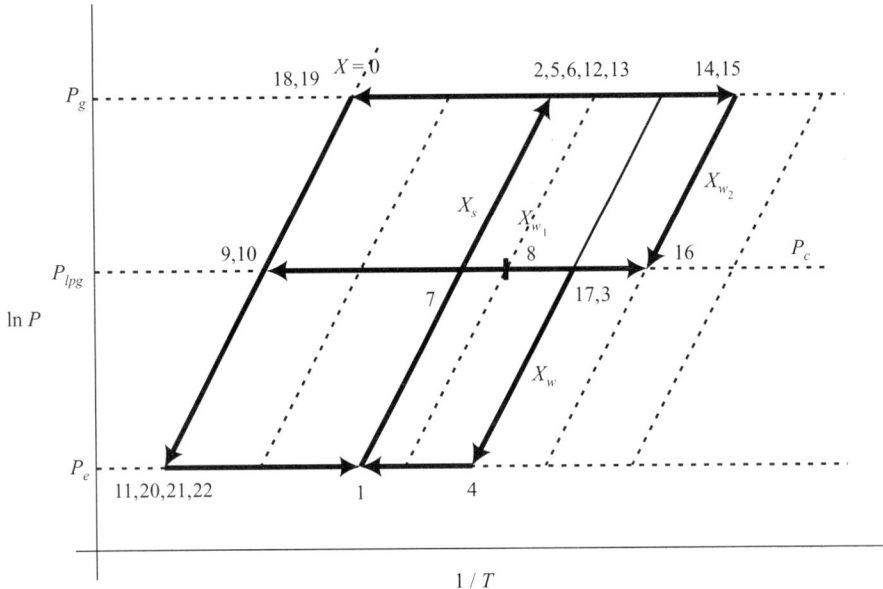

Fig 3.15(b) ln P – $1/T$ Diagram for Parallel Flow Double Effect Generation VAR System

solution leaving the absorber is known as solution distribution ratio and it is represented by R. The solution in the HP generator is heated externally from solar energy or waste heat. The refrigerant vapour (steam) produced in the HP generator is used as a heat source for the LP generator. The weak (in refrigerant) solution from the HP generator passes through the high-temperature heat exchanger and mixes with the weak solution from the LP generator at the inlet of the low-temperature heat exchanger. The solution then enters the absorber and is diluted by absorbing the refrigerant vapour coming from evaporator. The refrigerant vapour produced by the LP generator is condensed in the condenser, and flows to the evaporator. The refrigerant generated in LP and HP generators mix before entering the evaporator. In evaporator, liquid refrigerant extracts heat from the space to be cooled and evaporates, thereby producing cooling effect.

3.3.1 Results and Discussion of Thermodynamic Analysis

The mass, energy and exergy balances for a parallel flow double effect system can be applied in similar manner as has been applied for single effect and series flow double effect generation absorption refrigeration systems. The only difference for this particular case is solution distribution ratio as has been explained in system description and it is expressed as $R = \dot{m}_{s2}/\dot{m}_s$.

The energy analysis results computed for the parallel flow double effect generation absorption refrigeration system are compared with the results reported by Riffat and Shankland (1993) (Refer Table 3.4). The difference in the values of heat transfer rates in various components is less than 4%. The difference in the value of COP is less than 0.5 %.

The following parameters are assumed for analysis:-
1. High pressure generator temperature (T_{hpg}) = 120°C – 170°C
2. Evaporator temperature (T_e) = 7.2°C
3. Condenser and absorber temperatures are equal $(T_c = T_a)$ = 29.4°C/37.8°C
4. Effectiveness of solution heat exchanger(s), $(\varepsilon) = 0.7$
5. Temperature of the space to be cooled $(T_r) = 17.2°C$
6. Cooling capacity of the system $(\dot{Q}_e) = 100$ kW

Figure 3.16 illustrates the effect of variation in solution distribution ratio and HP generator temperature on the COP (for $T_c = T_a = 29.4°C$). The COP increases with reduction in HP generator

Table 3.4 Comparison of Present Results with the Results of Riffat and Shankland (1993)

Results of Energy Analysis of Parallel Flow Double Effect Generation VAR System			
Parameters for computation of results: $T_e = 5°C, T_a = T_c = 30°C$, HP generator temperature $(T_{hpg}) = 115$ °C, LP generator temperature $(T_{hpg}) = 70°C$ (Riffat and Shankland, 1993), Solution distribution ratio(R) = 0.5			
Parameter	Results reported by Riffat and Shankland (1993)	Computed results	Difference in reported and computed values
Heat supplied in HP generator, $(\dot{Q}_{hpg})$, kW	71.2	70.81	– 0.55 %
Cooling effect $(\dot{Q}_e)$, kW	100	100	—
Heat rejected by the absorber $(\dot{Q}_a)$, kW	119.99	121.5	1.26%
Heat rejected by the condenser $(\dot{Q}_c)$, kW	51.20	49.32	-3.67%
Pump work $(\dot{W}_p)$, kW	0.0	0.0083	—
COP	1.404	1.41	0.43%

temperature. The reason is that (at same value of solution distribution ratio) the increase in HP generator temperature causes an increase in exergy destruction in HP generator, absorber, condenser, LP generator and the total exergy destruction. The maximum value of COP varies between 1.39 and 1.42. The maximum value of maximum of COP is observed to occur at 135°C HP generator temperature. Further, with increase in solution distribution ratio, it is observed that total exergy destruction first decreases up to the optimum value of solution distribution ratio and then again starts increasing. The optimum value of solution distribution ratio for maximum COP varies between 0.19 and 0.28 depending upon the HP generator temperature, indicating best results are achieved when most of the solution is distributed to the LP generator. It is also observed that minimum total

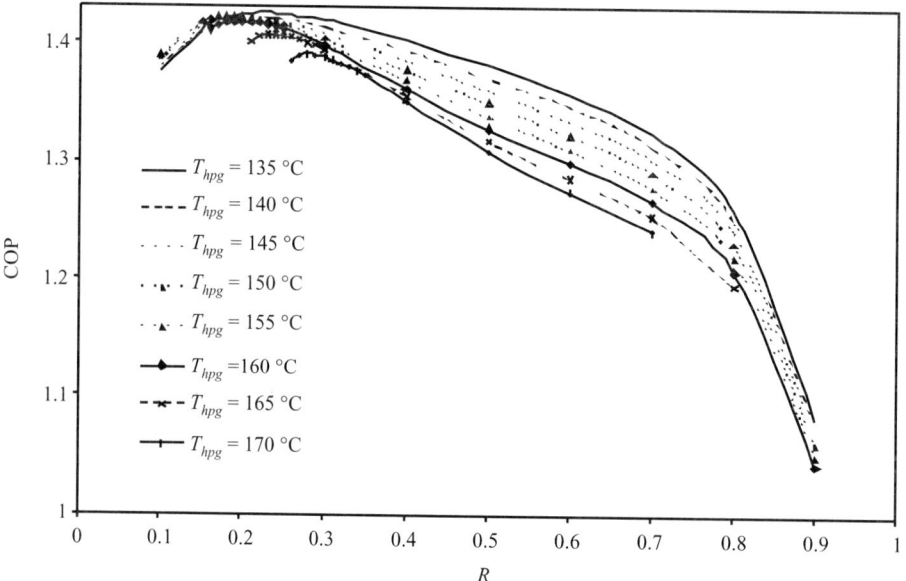

Fig. 3.16 Effect of Solution Distribution Ratio (R) and High Pressure Generator Temperature on COP ($T_e = 7.2$ °C, $T_a = T_c = 29.4$°C, $\varepsilon_{she} = 0.7$, $T_r = 17.2$°C)

exergy destruction is observed corresponding to optimum solution distribution ratio. Secondly, it is observed that with increase in HP generator temperatures (i.e., between 135°C and 160°C) the optimum solution distribution ratio decreases from 0.23 to 0.19 and for temperatures between 165°C and 170°C, it lies between 0.24 and 0.28. Moreover, lower values of R_{opt} show that circulation losses are also less when lesser amount of solution is entering the HP generator.

Figure 3.17 presents the effect of increasing the absorber and condenser temperatures on the COP and optimum solution distribution ratio. It is observed that with increase in the absorber and condenser temperatures (from 29.4 to 37.8°C), the COP and its maximum value drops for the range of HP generator temperatures considered. It happens because of increase in solution circulation ratio, with increase in absorber and condenser temperatures, and it causes increase in circulation losses and hence reduction in COP. Further, the optimum solution distribution ratio is achieved at higher values of solution distribution ratio than before. It indicates that more solution is required to be pumped to HP generator at higher absorber and condenser temperatures to achieve maximum value of the COP. However, increasing the HP generator temperature causes the optimum distribution ratio to shift toward lower values of solution distribution ratio.

Figure 3.18 depicts the effect of solution distribution ratio and HP generator temperature on exergetic efficiency. It is observed that increase in the HP generator temperature brings down the exergetic efficiency. The same reasons can be attributed for such a behaviour as already explained for the trend of COP curve. Secondly, with increase in solution distribution ratio, there is increase in exergetic efficiency up to optimum value of solution distribution ratio and beyond which exergetic efficiency decreases. The optimum value of solution distribution ratio lies between 0.23 and 0.19 for the HP generator temperatures between 135°C and 155°C, and for temperatures above 155°C and up to 170°C, it increases to 0.28.

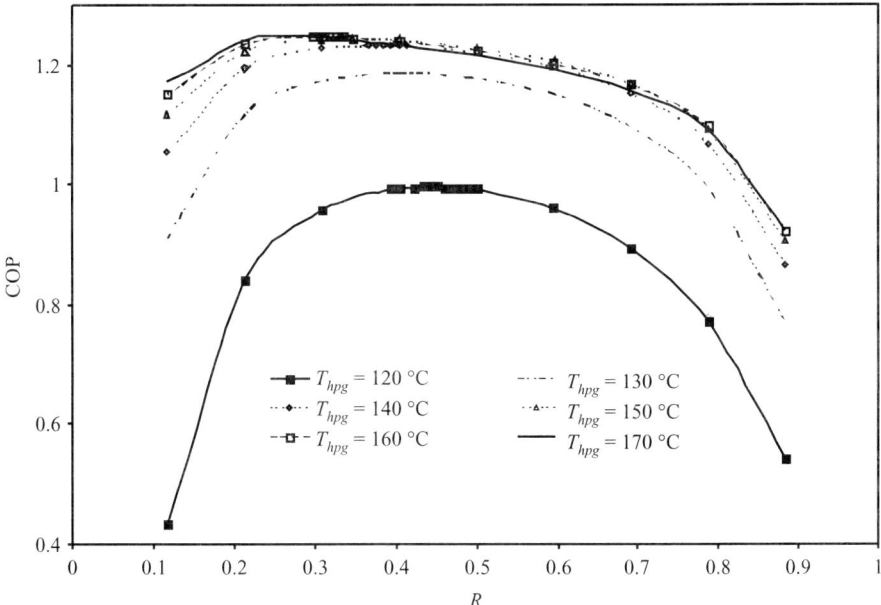

Fig. 3.17 Effect of Solution Distribution Ratio (R) and High Pressure Generator Temperature on COP (T_e = 7.2°C, $T_a = T_c$ = 37.8°C, ε_{she} = 0.7, T_r = 17.2°C)

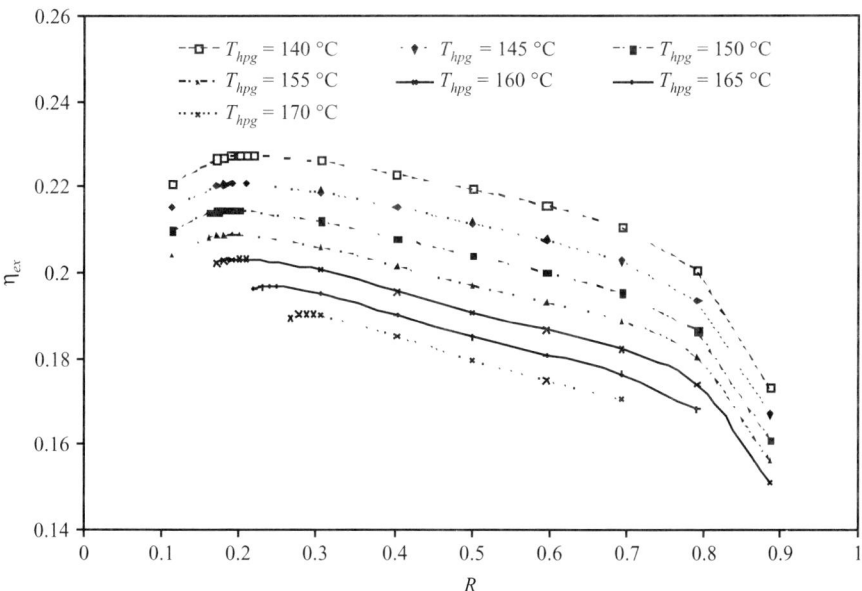

Fig. 3.18 Effect of Solution Distribution Ratio (R) and High Pressure Generator Temperature on Exergetic Efficiency (T_e = 7.2°C, $T_a = T_c$ = 29.4°C, ε_{she} = 0.7, T_r = 17.2 °C)

118 *Alternatives in Refrigeration and Air Conditioning*

The effects of increase in absorber and condenser temperatures on exergetic efficiency and optimum solution distribution ratio are represented in Fig. 3.19. The increase in the absorber and condenser temperatures is responsible for reduction of exergetic efficiency. It happens because of overall increase in solution circulation ratio (with increase in absorber and condenser temperatures) causes an increase in circulation losses and irreversibility in various components of the system increase, thereby reducing exergetic efficiency. The higher values of the exegetic efficiency are achieved at lower values of HP generator temperature for solution distribution ratio varying between 0.2 and 0.77. The optimum value of solution distribution ratio shifts towards higher side as compared to the case when absorber and condenser temperatures are lower.

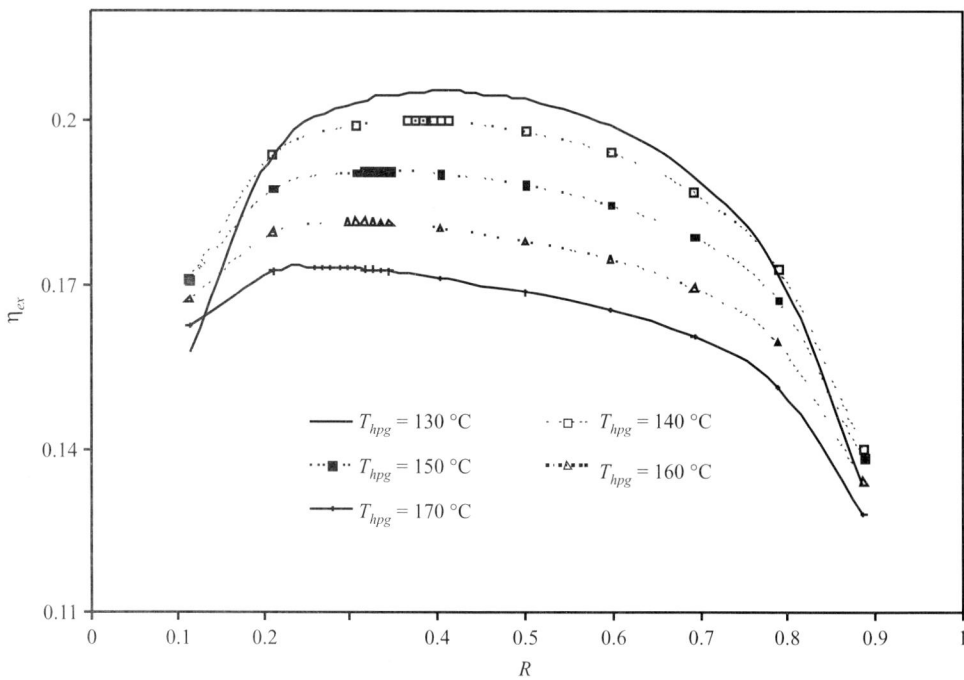

Fig. 3.19 Effect of Solution Distribution Ratio (R) and High Pressure Generator Temperature on Exergetic Efficiency ($T_e = 7.2°C$, $T_a = T_c = 37.8°C$, $\varepsilon_{she} = 0.7$, $T_r = 17.2$)

Figure 3.20 depicts the variation of the COP and exergetic efficiency with solution distribution ratio. It is observed that the maximum value of the COP and exergetic efficiency occur for same solution distribution ratio. This happens because minimum irreversibility occurs at optimum solution distribution ratio. This fact is also proved by considering the different values of absorber and condenser temperatures as shown in Fig 3.21.

Figure 3.22 shows the effect of variation in the evaporator temperature on the optimum solution distribution ratio, maximum COP and maximum exergetic efficiency. The overall system irreversibility increases with reduction in evaporator temperature and hence maximum value of the COP and the exergetic efficiency reduce.

Vapour Absorption Cooling and Advanced Absorption Cycles 119

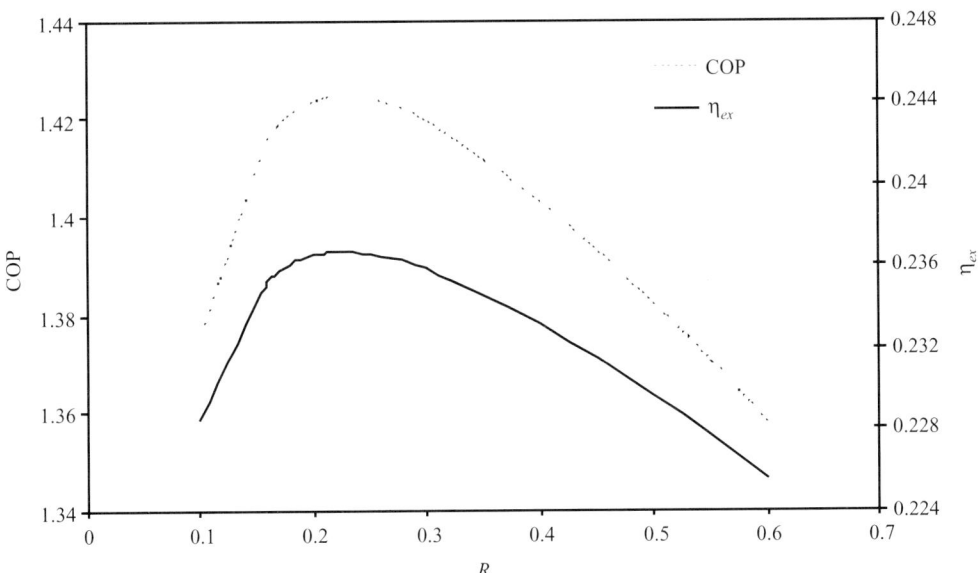

Fig. 3.20 Variation of COP and Exergetic Efficiency versus Solution Distribution Ratio (R)
(T_{hpg} = 135°C, T_e = 7.2°C, T_r = 17.2°C, T_a = T_c = 29.4°C, T_0 = 298.15 K, ε_{she} = 0.7)

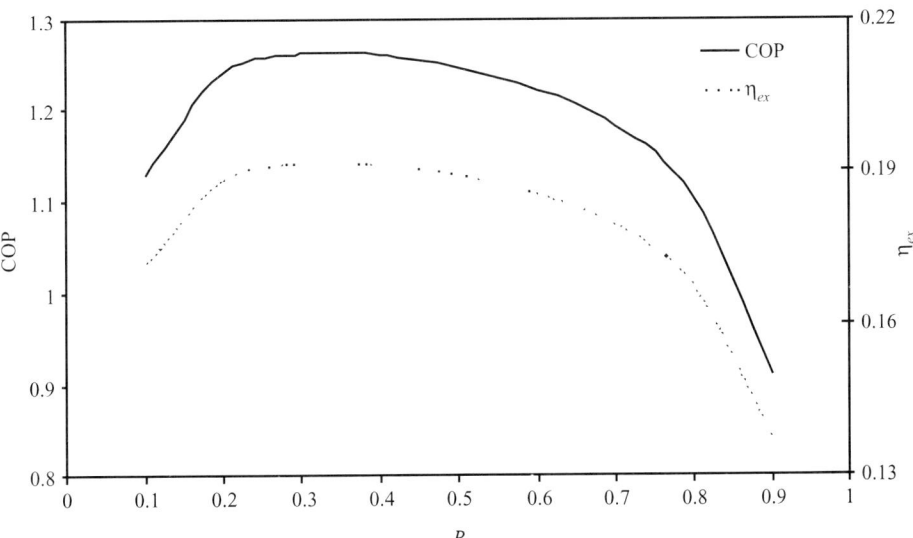

Fig. 3.21 Variation of COP and Exergetic Efficiency versus Solution Distribution Ratio (R)
(T_{hpg} = 135°C, T_e = 7.2 °C, T_r = 17.2°C, T_a = T_c = 37.8°C, T_0 = 298.15 K, ε_{she} = 0.7)

120 *Alternatives in Refrigeration and Air Conditioning*

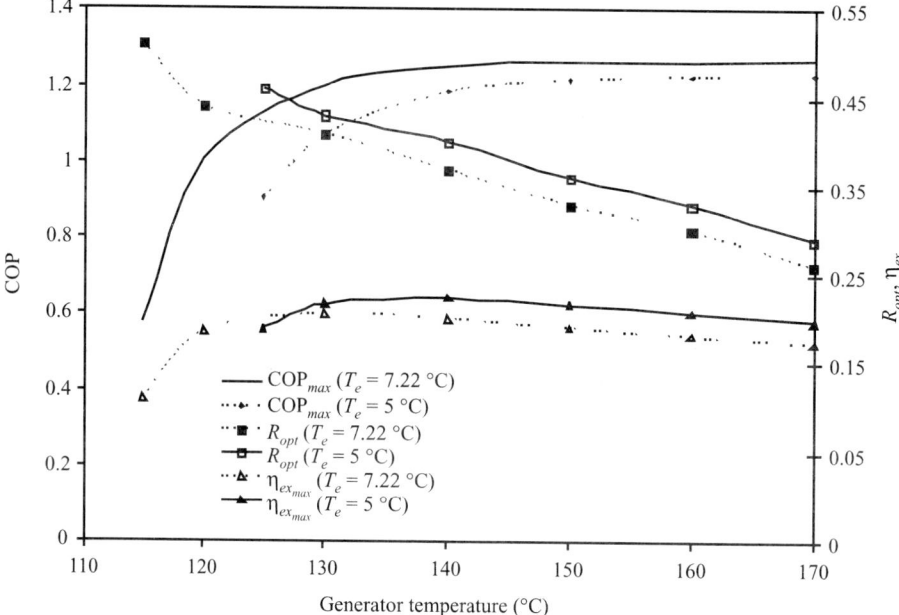

Fig. 3.22 Variation of Maximum COP, R_{opt} and $\eta_{ex_{max}}$ versus T_{hpg} ($T_a = T_c = 37.8°C$, $\varepsilon_{she} = 0.7$) for Varying Evaporator Temperatures

Figure 3.23 shows the efficiency defects in various components of a parallel flow double effect system.

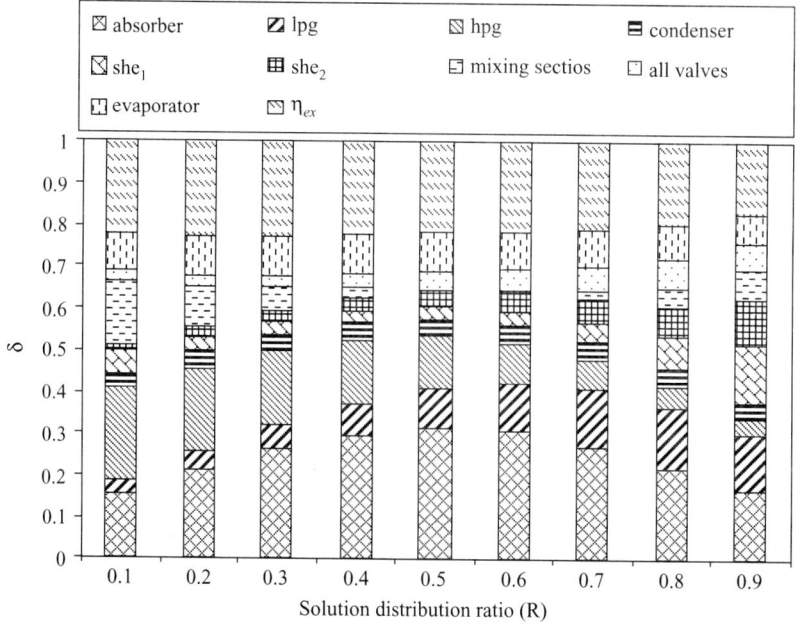

Fig. 3.23 Variation of Efficiency Defects versus Solution Distribution Ratio ($T_{hpg} = 140°C$, $T_a = T_c = 29.4°C$, $T_e = 7.2°C$ Effectiveness of Solution Heat Exchanger (s) = 0.7)

3.4 TRIPLE EFFECT GENERATION WATER-LITHIUM BROMIDE VAPOUR ABSORPTION COOLING SYSTEMS

Figure 3.24 (a) shows the schematic diagram of a triple effect series flow VAR system. It is an extension of the double effect series flow VAR system. It is a four-pressure system against three pressure double effect system. The additional components required are generator at medium pressure, solution heat exchanger, solution and refrigerant throttle valves. This cycle therefore includes two internal heat exchange processes between condensing refrigerant and weak solution. Thus, each unit of heat is used in three different generators to generate refrigerant vapour and hence the name triple effect. The solution leaving the absorber flows in series from HP generator to MP generator and finally to LP generator before entering the absorber via solution throttle valve. This also enhances the COP since heat input is reduced for the same cooling capacity. Figure 3.24(b) shows the ln $P - 1/T$ diagram of the triple effect series flow VAR system.

3.4.1 Thermodynamic Analysis

The thermodynamic analysis involves the application of principles of mass, energy and exergy balances. The application of these concepts is illustrated below.

Mass Balance

The equations from (3.49) to (3.53) denote mass balance.

HP generator
$$\dot{m}_s = \dot{m}_{w3} + \dot{m}_{r3} \tag{3.50}$$

MP generator
$$\dot{m}_{w3} = \dot{m}_{w2} + \dot{m}_{r2} \tag{3.51}$$

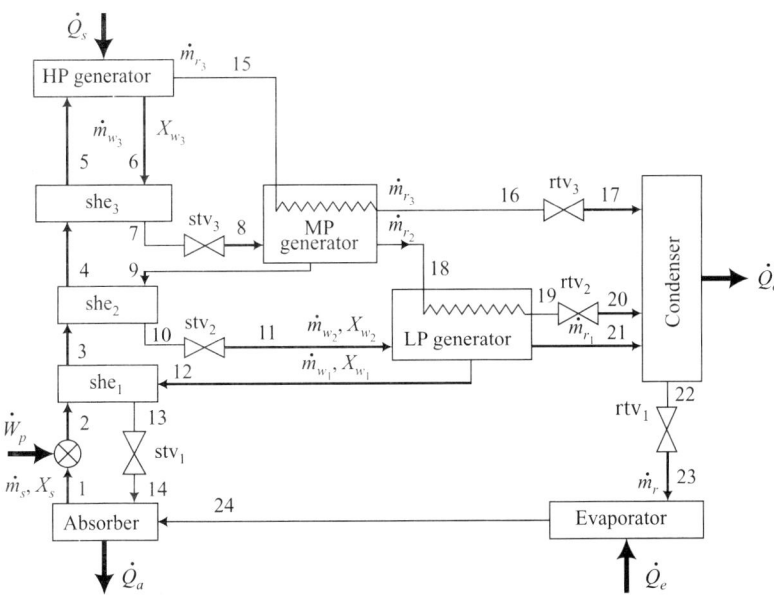

Fig. 3.24(a) Schematic Diagram of Series Flow Triple Effect Generation H_2O/LiBr VAR System

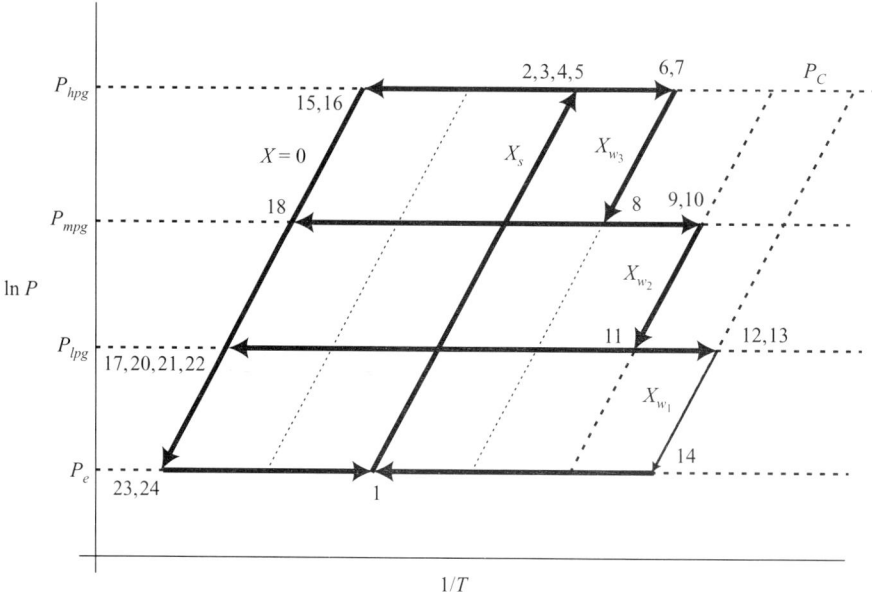

Fig. 3.24(b) ln $P - 1/T$ Diagram of a Series Flow Triple Effect Generation H_2O/LiBr VAR System

LP generator
$$\dot{m}_{w2} = \dot{m}_{w1} + \dot{m}_{r1} \tag{3.52}$$
Condenser
$$\dot{m}_r = \dot{m}_{r3} + \dot{m}_{r2} + \dot{m}_{r1} \tag{3.53}$$

Energy Balance

The energy balance in each component of a series flow triple effect generation VAR system cycle is given by the following equations:

$$\dot{Q}_a = \dot{m}_r h_{24} + \dot{m}_{w1} h_{14} - \dot{m}_s h_1 \tag{3.54}$$

$$\dot{Q}_{hpg} = \dot{m}_{r3} h_{15} + \dot{m}_{w3} h_6 - \dot{m}_s h_5 \tag{3.55}$$

$$\dot{Q}_c = \dot{m}_{r1} h_{21} + \dot{m}_{r2} h_{20} + \dot{m}_{r3} h_{17} - \dot{m}_r h_{22} \tag{3.56}$$

$$\dot{Q}_e = \dot{m}_r (h_{24} - h_{23}) \tag{3.57}$$

$$\dot{Q}_{she_1} = \dot{m}_s (h_3 - h_2) = \dot{m}_{w1} (h_{12} - h_{13}) \tag{3.58}$$

$$\dot{Q}_{she_2} = \dot{m}_s (h_4 - h_3) = \dot{m}_{w3} (h_9 - h_{10}) \tag{3.59}$$

$$\dot{Q}_{she_3} = \dot{m}_s (h_5 - h_4) = \dot{m}_{w3} (h_6 - h_7) \tag{3.60}$$

$$\dot{Q}_{mpg} = \dot{m}_{r3} (h_{15} - h_{16}) = \dot{m}_{r2} h_{18} + \dot{m}_{w2} h_9 - \dot{m}_{w3} h_8 \tag{3.61}$$

$$\dot{Q}_{lpg} = \dot{m}_{r2} (h_{18} - h_{19}) = \dot{m}_{r1} h_{21} + \dot{m}_{w1} h_{12} - \dot{m}_{w2} h_{11} \tag{3.62}$$

$$\dot{W}_p = \dot{m}_s (h_2 - h_1) \tag{3.63}$$

$$\text{Energy in} = \dot{Q}_{hpg} + \dot{Q}_e + \dot{W}_p \tag{3.64}$$

Vapour Absorption Cooling and Advanced Absorption Cycles 123

$$\text{Energy out} = \dot{Q}_a + \dot{Q}_c \tag{3.65}$$

$$\text{COP} = \frac{\dot{Q}_e}{\dot{Q}_{hpg} + \dot{W}_p} \tag{3.66}$$

Exergy Balance

Exergy destruction in each component of a series flow triple effect generation VAR system is furnished below:

$$\dot{ED}_a = \dot{m}_r(h_{24} - T_o s_{24}) + \dot{m}_{w1}(h_{14} - \dot{m}_{14} T_o s_{14}) - \dot{m}_s(h_1 - T_o s_1) \tag{3.67}$$

$$\dot{ED}_{hpg} = \dot{m}_s(h_5 - T_o s_5) - \dot{m}_{w3}(h_6 - T_o s_6) - \dot{m}_{r3}(h_{15} - T_o s_{15}) + \dot{Q}_{hpg}\left(1 - \frac{T_0}{T_{hpg}}\right) \tag{3.68}$$

$$\dot{ED}_c = \dot{m}_{r1}((h_{21} - T_o s_{21}) + \dot{m}_{r2}(h_{20} - T_o s_{20}) + \dot{m}_{r3}(h_{17} - T_o s_{17}) - \dot{m}_r(h_{22} - T_o s_{22}) \tag{3.69}$$

$$\dot{ED}_e = \dot{m}_r((h_{23} - h_{24}) - T_o(s_{23} - s_{24})) + \dot{Q}_e\left(1 - \frac{T_0}{T_r}\right) \tag{3.70}$$

$$\dot{ED}_{she1} = \dot{m}_s((h_2 - h_3) - T_o(s_2 - s_3)) + \dot{m}_{w1}((h_{12} - h_{13}) - T_o(S_{12} - S_{13})) \tag{3.71}$$

$$\dot{ED}_{she2} = \dot{m}_s((h_3 - h_4) - T_o(s_3 - s_4)) + \dot{m}_{w2}((h_9 - h_{10}) - T_o(s_9 - s_{10})) \tag{3.72}$$

$$\dot{ED}_{she3} = \dot{m}_s((h_4 - h_5) - T_o(s_4 - s_5)) + \dot{m}_{w3}((h_6 - h_7) - T_o(s_6 - s_7)) \tag{3.73}$$

$$\dot{ED}_{rtv1} = \dot{m}_r T_o(s_{23} - s_{22}) \tag{3.74}$$

$$\dot{ED}_{rtv2} = \dot{m}_r T_o(s_{20} - s_{19}) \tag{3.75}$$

$$\dot{ED}_{rtv3} = \dot{m}_r T_o(s_{17} - s_{16}) \tag{3.76}$$

$$\dot{ED}_{stv1} = \dot{m}_{w1} T_o(s_{14} - s_{13}) \tag{3.77}$$

$$\dot{ED}_{stv2} = \dot{m}_{w2} T_o(s_{11} - s_{10}) \tag{3.78}$$

$$\dot{ED}_{stv3} = \dot{m}_{w3} T_o(s_8 - s_7) \tag{3.79}$$

$$\dot{ED}_{mpg} = \dot{m}_{w3}(h_8 - T_o s_8) + \dot{m}_{r3}((h_{15} - h_{16}) - T_o(s_{15} - s_{16})) \\ - \dot{m}_{w2}(h_9 - T_o s_9) - \dot{m}_{r2}(h_{18} - T_o s_{18}) \tag{3.80}$$

$$\dot{ED}_{lpg} = \dot{m}_{w2}(h_{11} - T_o s_{11}) + \dot{m}_{r2}((h_{18} - h_{19}) - T_o(s_{18} - s_{19})) - \dot{m}_{w1}(h_{12} - T_o s_{12}) \\ - \dot{m}_{r1}(h_{21} - T_o s_{21}) \tag{3.81}$$

$$\dot{ED}_{total} = \dot{ED}_a + \dot{ED}_{hpg} + \dot{ED}_{mpg} + \dot{ED}_{lpg} + \dot{ED}_c + \dot{ED}_e + \dot{ED}_{she1} + \dot{ED}_{she2} \\ + \dot{ED}_{she3} + \dot{ED}_{rtv1} + \dot{ED}_{rtv2} + \dot{ED}_{rtv3} + \dot{ED}_{stv1} + \dot{ED}_{stv2} + \dot{ED}_{stv3} \tag{3.82}$$

Exergetic Efficiency

$$\eta_{ex} = \frac{\dot{Q}_e |(1 - T_0/T_r)|}{\dot{Q}_{hpg}(1 - T_0/T_{hpg}) + \dot{W}_p} \tag{3.83}$$

3.4.2 Results and Discussion

The triple effect system operates at heat supply temperatures above 175°C (Grossman and Zaltash, 2001). Accordingly, HP generator temperature is varied between 175°C and 200°C.

Table 3.5 State Points for the Series Flow Triple Effect Generation Water-Lithium Bromide VAR System (Parameters: HP Generator Temperature, T_{hpg} = 185°C, T_e = 7.2°C, $T_c = T_a$ = 37.8°C, $\varepsilon_{she(s)}$ = 0.7, Mass Flow Rate of Refrigerant (Water) = 1 kg/s)

State	T_i (°C)	P_i (kPa)	X_i (kg of absorbent/kg sol)	m_i (kg/s)	h_i (kJ/kg)	s_i (kJ/kg/K)
1	37.8	1.016	0.5542	12.567	91.88	0.2288
2	37.8	287.5	0.5542	12.567	92.06	0.2288
3	65.38	287.5	0.5542	12.567	147.7	0.4018
4	107.7	287.5	0.5542	12.567	236.7	0.6481
5	159.3	287.5	0.5542	12.567	347.6	0.9203
6	185	287.5	0.5713	12.19	403.6	1.018
7	131.2	287.5	0.5713	12.19	289.1	0.7535
8	131.2	53.67	0.5713	12.19	289.1	0.7539
9	132.11	53.67	0.5865	11.88	294.1	0.7399
10	85.74	53.67	0.5865	11.88	200	0.4941
11	85.74	6.558	0.5865	11.88	200	0.49419
12	83.11	6.558	0.6021	11.567	201.6	0.4679
13	51.56	6.558	0.6021	11.567	141.1	0.2901
14	51.56	1.016	0.6021	11.567	141.1	0.2901
15	185	287.5	0	0.3096	2835	7.265
16	132.11	287.5	0	0.3096	555.4	1.657
17	37.8	6.558	0	0.3096	555.4	1.82
18	132.11	53.67	0	0.3149	2745	7.822
19	83.11	53.67	0	0.3149	348	1.112
20	37.8	6.558	0	0.3149	348	1.153
21	83.11	6.558	0	0.3755	2655	8.555
22	37.8	6.558	0	1	158.3	0.5428
23	7.2	1.016	0	1	158.3	0.5661
24	7.2	1.016	0	1	2514	8.968

The other parameters assumed are taken from Anand and Kumar (1987). These are evaporator temperature (T_e) = 7.2°C, condenser and absorber temperatures ($T_c = T_a$) = 37.8°C, effectiveness of solution heat exchangers = 0.7, mass flow rate of refrigerant (water) = 1 kg/s. Table 3.5 presents the state point property values and Table 3.6 represents the details of heat transfer and exergy destruction in various components of series flow triple effect system.

Table 3.6 Results of Energy and Exergy Analyses of Series Flow Triple Effect Water-Lithium Bromide VAR System (Parameters: HP Generator Temperature, T_g = 185°C, T_e = 7.2°C, $T_c = T_a$ = 37.8°C, $\varepsilon_{she(s)}$ = 0.7, Mass Flow Rate of Refrigerant (Water) = 1 kg/s)

Results of energy analysis		
$\dot{Q}_a$ = −2991 kW $\dot{Q}_c$ = −981.9 kW $\dot{Q}_e$ = 2355.45 kW $\dot{Q}_{hpg}$ = 1615 kW $\dot{W}_p$ = 2.26 kW	$\dot{Q}_{she1}$ = 699.7 kW $\dot{Q}_{she2}$ = 1117 kW $\dot{Q}_{she3}$ = 1394 kW $\dot{Q}_{mpg}$ = 855.8 kW $\dot{Q}_{lpg}$ = 754.8 kW	Energy input = $\dot{Q}_e + \dot{Q}_g + \dot{W}_p$ = 3972.71 kW Energy output = $\dot{Q}_a + \dot{Q}_c$ = 3972.9 kW **COP = 1.456**
Results of exergy destruction in system components		
$\dot{ED}_{she2}$ = 52.46 kW $\dot{ED}_{she3}$ = 56.85 kW $\dot{ED}_{rtv1}$ = 6.936 kW	$\dot{ED}_c$ = 42.12 kW $\dot{ED}_e$ = 86.28 kW $\dot{ED}_{hpg}$ = 15.99 kW	$\dot{ED}_{stv3}$ = 1.333 kW $\dot{ED}_{total}$ = 502.98 kW Exergy input = $\dot{Q}_{hpg}(1 - T_0/T_{hpg}) + \dot{W}_p$ = 566.27 kW
$\dot{ED}_{rtv2}$ = 3.823 kW	$\dot{ED}_{mpg}$ = 9.476 kW	Exergy output = $\dot{Q}_e \|1 - T_0/T_r\|$ = 63.28 kW
$\dot{ED}_{rtv3}$ = 18.25 kW $\dot{ED}_{stv1}$ = 0.03511 kW $\dot{ED}_a$ = 173.7 kW	$\dot{ED}_{lpg}$ = 0.0402 kW $\dot{ED}_{she1}$ = 35.4 kW $\dot{ED}_{stv2}$ = 0.2855 kW	Exergetic efficiency, η_{ex} = **11.17 %**

It is observed that the heat supplied in HP generator ($\dot{Q}_{hpg}$ = 1615 kW) in triple effect VAR system is lower in comparison to heat supplied in HP generator (1868.71 kW) in series flow double effect system for the same amount of refrigerant generated (1 kg/s) and same amount of cooling effect ($\dot{Q}_e$ = 2355.45 kW) produced in both the systems (Refer Table 3.3) corresponding to identical state conditions. Therefore, the COP of the triple effect system (COP = 1.456) is higher than the COP for the series flow double effect system (COP = 1.26). The exergetic efficiency of the series flow triple effect system is lower in comparison to exergetic efficiency of the series flow double effect system. This happens because of increase in input exergy in series flow triple effect system for constant exergy output given by $\dot{Q}_e \left|\left(1 - \frac{T_0}{T_r}\right)\right|$. The input exergy = $\left[\dot{Q}_{hpg}\left(1 - \frac{T_0}{T_{hpg}}\right) + \dot{W}_p\right]$ is dependent on two factors, viz., HP generator temperature and heat supplied in generator. Though the heat supplied is reduced yet the increase in temperature of HP generator causes an increase in the input exergy in comparison to series flow double effect system. The contribution of pump work has negligible effect on increase of the input exergy.

Figure 3.25 illustrates the effect of HP generator temperature on COP of series flow triple effect system. It can be observed that, for an absorber temperature of 35°C, increase in HP generator temperature causes increase in COP initially and further increase in generator temperature brings slight decrease in COP.

126   *Alternatives in Refrigeration and Air Conditioning*

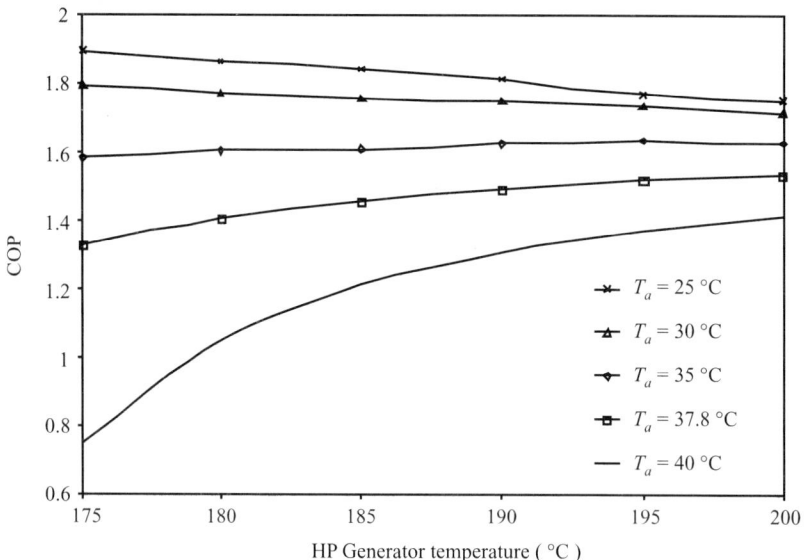

Fig. 3.25 Effect of HP Generator Temperature on COP at Various Absorber Temperatures
($T_e = 7.2°C$, $T_a = T_c$, $\varepsilon_{she\,(s)} = 0.7$)

3.4.2.1 *Effect of HP Generator Temperature*

The increase in the HP generator temperature results in an increase in temperatures of refrigerant and the leaving generator. Thus, increase in the average temperature of both condenser and the absorber is observed and it accounts for increased irreversibility in these components. Initially, the positive effect of increase in the COP by virtue of increase in the HP generator temperature overcomes the degradation of the COP due to increase in irreversibility in the condenser and absorber temperatures, hence COP increases. However, at higher HP generator temperatures, the reduction in COP, due to irreversibility in condenser and absorber, exceeds the increase in COP due to increase of generator temperature and hence a marginal drop in COP is observed. The solution circulation ratio (i.e., ratio of mass flow rate of strong solution required per unit mass flow rate of refrigerant) is another important factor for this behaviour of COP. The increase in HP generator temperature causes the solution circulation ratio to decrease and consequently the heat duties in HP generator, condenser and absorber decrease. Thus, for constant cooling load, the reduction in generator heat duty causes initial increase in the COP. At higher HP generator temperatures, the rate of reduction in solution circulation ratio drops but the temperature difference between the generator and sub-cooled solution entering the HP generator increases causing an increase in the irreversibility occurring in generator. Thus, the positive effect of decrease in the solution circulation ratio is balanced by negative effect of increase in the temperature difference between HP generator and sub-cooled solution entering the HP generator. Thus, the HP generator heat duty levels off and the COP curve turns nearly flat showing marginal drop in COP. For the absorber temperatures below 35°C, the trend of the curves is that of reducing COP achieved after maximum value of COP. However, above 35°C absorber temperatures, the trend of the curves is that of increasing COP achieved before maximum value of COP. Moreover, with increase in absorber temperature, corresponding to a particular HP generator temperature, the COP reduces. This happens because of increase in overall solution circulation ratio (defined by m_s/m_r). The increase of solution circulation ratio causes the heat supply in HP generator to increase for the constant cooling load, hence COP reduces. Figure 3.26 shows the effect of generator temperature on exergetic efficiency. The trends of exergetic efficiency curves are similar to trends of COP curves discussed above. The reason already explained above for the behaviour of COP curves can be attributed for the variation of exergetic efficiency curves also. Neglecting pump work, it is observed that exergetic efficiency is a function of COP and the expression $(1 - T_0/T_{hpg})$. The expression $(1 - T_0/T_r)$ is a constant term. With increase in T_{hpg}, the value of expression $(1 - T_0/T_{hpg})$ increases. However, the variation in its value is small. The exergetic efficiency is thus directly proportional to COP. Hence, the increase in COP causes exergetic efficiency to increase and vice versa.

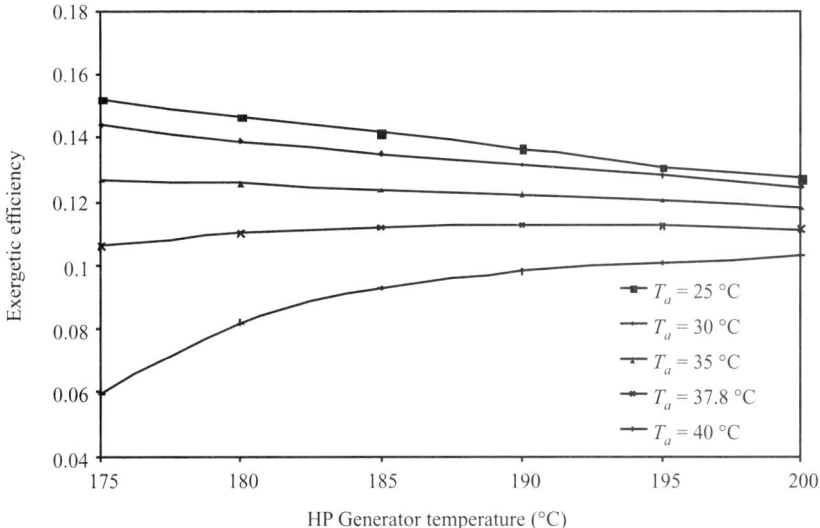

Fig. 3.26 Effect of HP Generator Temperature on Exergetic Efficiency for Varying Absorber Temperature ($T_e = 7.2°C$, $T_r = T_e + 10°C$, $T_a = T_c$, $\varepsilon_{she\,(s)} = 0.7$)

3.4.2.2 Effect of Absorber Temperature

Figures 3.27(a) and 3.27(b) show the effect of absorber temperature on solution circulation ratio and COP respectively for different HP generator temperatures. The foremost reason for decrease in COP with increase in absorber temperature is increase in heat duty of generator and pump work with increase in solution circulation ratio. It is observed that solution circulation ratio increases with increase in absorber temperature and for a particular value of absorber temperature the solution circulation ratio increases with decrease in high pressure generator temperature.

The rate of rise in solution circulation ratio is highest at HP generator temperature of 175°C in comparison to higher generator temperatures. Further, at absorber temperatures above 37°C, the rate of rise of solution circulation ratio increases sharply. The increase in solution circulation ratio is accountable for increase in HP generator heat duty and consequently COP decreases. The sharp rise in solution circulation ratio at absorber temperatures above 37°C causes the COP to decrease sharply

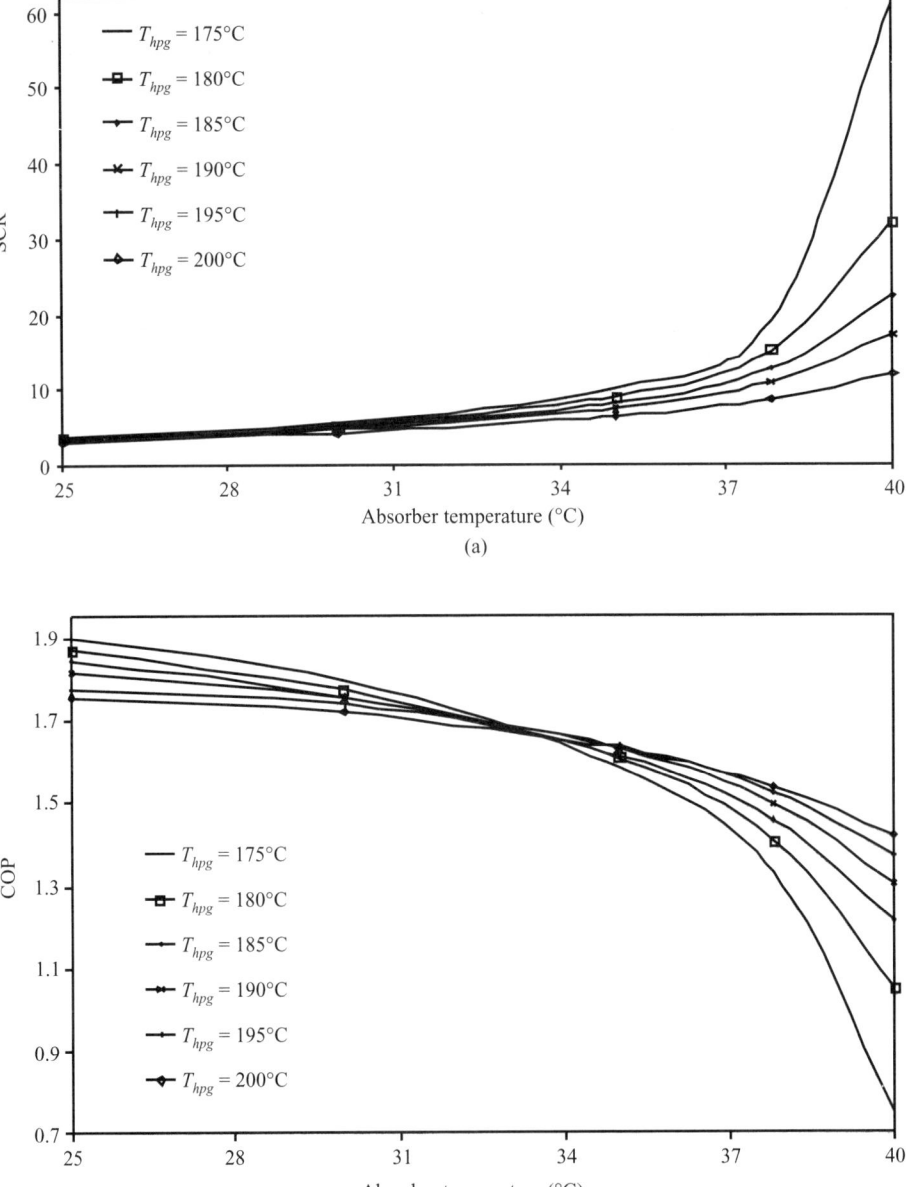

Fig. 3.27 Effect of Absorber Temperature on (a) SCR and (b) COP at Different High Pressure Generator Temperatures ($T_e = 7.2°C$, $\varepsilon_{she\,(s)} = 0.7$)

at higher values of absorber temperature. The similar trend is observed at higher generator temperatures; however the rate of decrease of COP is low because of lower rate of rise in solution circulation ratio.

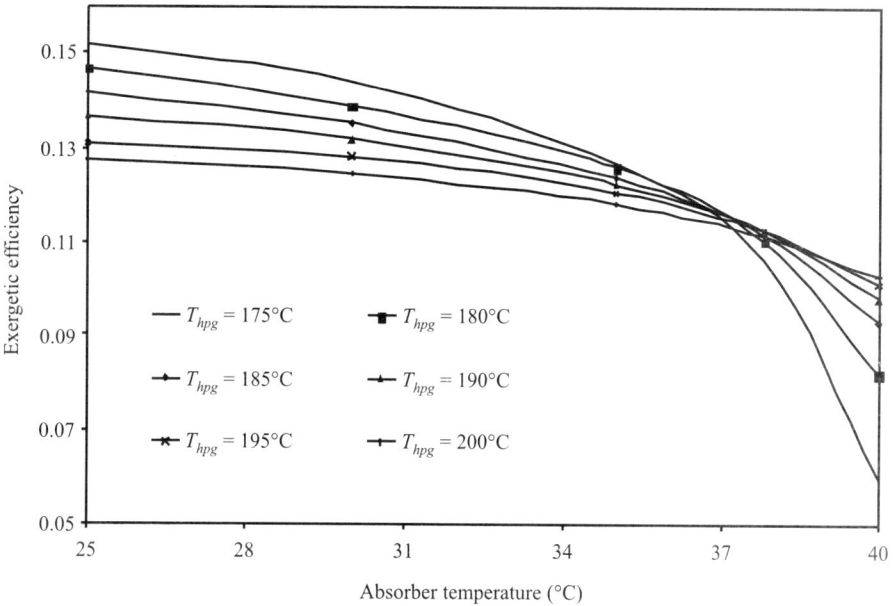

Fig. 3.28 Effect of Absorber Temperature on Exergetic Efficiency
($T_a = T_c$, $T_e = 7.2°C$, $T_r = T_e + 10°C$, $\varepsilon_{she(s)} = 0.7$)

Figure 3.28 shows the results of variation in exergetic efficiency with HP generator temperature. It is observed that exergetic efficiency decreases with increase in absorber temperature. It has already been explained in the previous paragraph that solution circulation ratio increases with increase in absorber temperature. The increase in solution circulation ratio is accountable for an increased input exergy required for the operation of the system because of increase in generator heat duty. The output exergy (i.e., exergy of the cooling effect) remains constant since the refrigerant mass flow rate and evaporator temperature are assumed constant. Thus, exergetic efficiency, which is the ratio of output exergy to the input exergy, decreases. One important feature of this figure that needs clarification is why the exergetic efficiency curve corresponding to HP generator temperature of 175°C shows largest drop in exergetic efficiency at absorber temperatures above 37°C. The reason for decrease in exergetic efficiency with increase in HP generator temperatures is the increase in finite temperature difference between the sub-cooled solution leaving the solution heat exchanger 3(she_3) and the HP generator temperature. The increase in finite temperature difference increases the irreversibility in HP generator. However, the irreversibility in HP generator is least for 175°C but the rate of rise in solution circulation ratio is highest (refer Fig. 3.27(a)). The latter is accountable for larger drop in exergetic efficiency.

3.4.2.3 Effect of Evaporator Temperature

The variation in COP with evaporator temperature is shown in Fig. 3.29. It is observed that with increase in evaporator temperature COP increases. The increase in evaporator temperature brings about an increase in evaporator pressure. Therefore, absorber pressure also increases since it is assumed equal to evaporator pressure. The increase in absorber pressure causes the strong solution concentration Xs to decrease whereas the decrease in weak solution concentration Xw_1 is less. The overall effect is to decrease the solution circulation ratio ($SCR_{hpg} = Xw_1/(Xw_1 - Xs)$) across HP generator. Hence, the HP generator heat duty decreases. The cooling effect produced in evaporator also increases due to increase in evaporator temperature. Thus, COP improves. The increase in absorber temperature for a constant evaporator temperature is observed to decrease the COP. This happens because with increase in absorber temperature the solution circulation ratio across the HP generator increases, consequently HP generator heat duty increases. The condenser temperature also increases since it is assumed equal to absorber temperature.

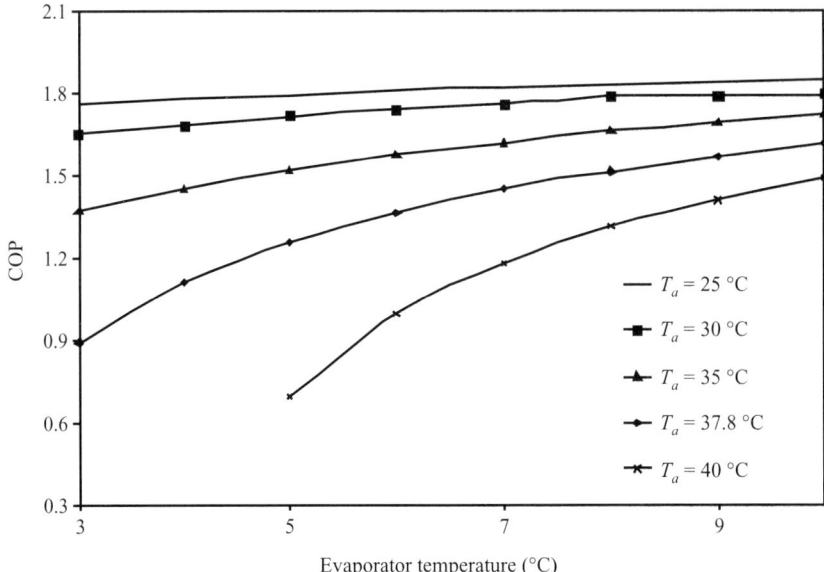

Fig. 3.29 Effect of Evaporator Temperature on COP for Varying Absorber Temperatures
($T_g = 185\ °C$, $T_a = T_c$, $\varepsilon_{she(s)} = 0.7$)

This increase in condenser temperature is accountable for increase in flashing of refrigerant vapour at the exit of refrigerant throttle valve and hence cooling effect decreases. The combined effect of increase in HP generator heat duty and decrease of cooling effect in evaporator accounts for decrease in the COP.

Figure 3.30(a) illustrates the effect of evaporator temperature on exergetic efficiency. It is observed that exergetic efficiency decreases with increase in evaporator temperature. It has been explained in the previous paragraph that HP generator heat duty decreases with increase in evaporator temperature, thus input exergy given by the expression $\left[\dot{Q}_{hpg}\left(1-T_0/T_{hpg}\right)+\dot{W}_p\right]$ also reduces. The output exergy given by the expression $\dot{Q}_e\left(1-T_0/T_r\right)$ also reduces.

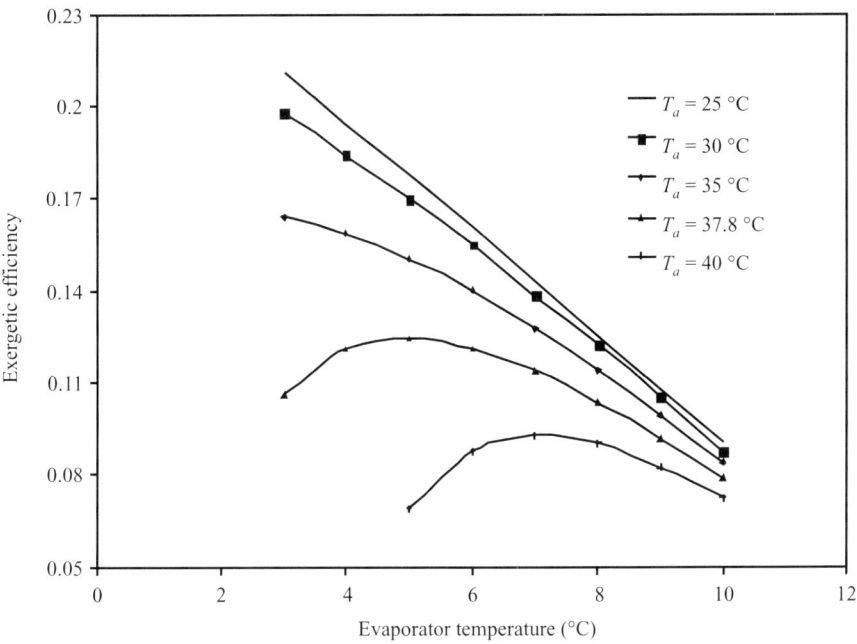

Fig. 3.30(a) Effect of Evaporator Temperature on Exergetic Efficiency for Varying Absorber Temperatures (T_{hpg} = 185°C, $T_a = T_c$, $T_r = T_e$ + 10°C $\varepsilon_{she(s)}$ = 0.7)

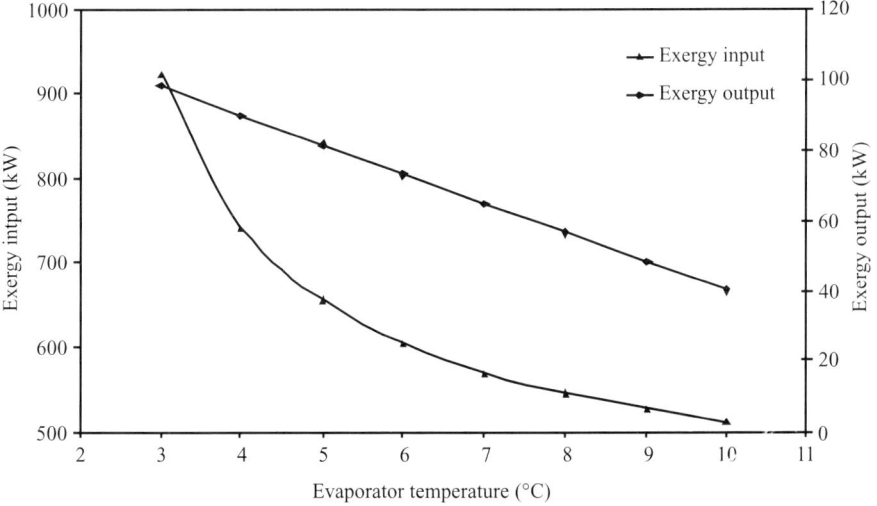

Fig. 3.30(b) Variation of Input and Output Exergies with Evaporator Temperature (T_{hpg} = 185°C, $T_a = T_c$ =37.8°C, $T_r = T_e$ + 10°C, $\varepsilon_{she(s)}$= 0.7)

It is observed that at absorber temperatures between 25°C and 35°C, the ratio of output to input exergy causes the exergetic efficiency to reduce since the rate at which output exergy reduces is higher in comparison to reduction in input exergy. However, at absorber temperatures higher than 35°C, the exergetic efficiency increases initially achieves a maximum value and then decreases. The reason for this behaviour can be explained with the help of Fig. 3.30(b) which depicts the variation of input and output exergies with increase in evaporator temperature. It is evident that the rate at which input exergy reduces is higher in comparison to rate of drop in output exergy (which is nearly constant) at evaporator temperatures between 3°C and 5°C, hence exergetic efficiency increases. However, for evaporator temperatures above 5°C, the rate of fall in input exergy drops considerably and hence exergetic efficiency reduces.

Figure 3.31 shows the effect of absorber temperature on COP at varying evaporator temperatures. The results show that with increase in absorber temperature the COP drops for all values of evaporator temperatures considered in this analysis. The increase in absorber temperature results in an increase in solution circulation ratio which causes the increase in heat duty in HP generator. The cooling effect reduces since condenser temperature is assumed equal to absorber temperature which results in increase in flashing of refrigerant vapours at exit from refrigerant throttle valve. The combined effect of both these factors is to reduce the COP.

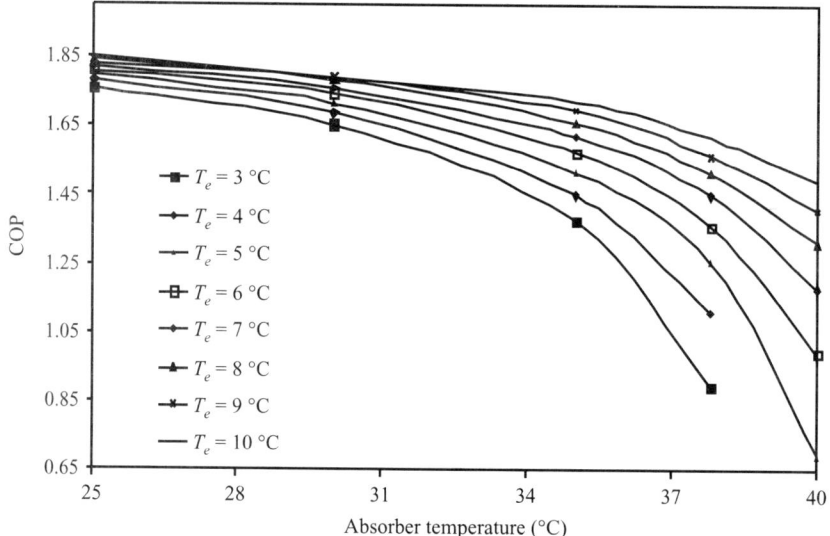

Fig. 3.31 Effect of Absorber Temperature on COP for Varying Evaporator Temperature
$(T_{hpg} = 185\ °C,\ T_a = T_c,\ \varepsilon_{she\,(s)} = 0.7)$

Figure 3.32 represents the effect of absorber temperature on exergetic efficiency for varying evaporator temperature. It is clear that exergetic efficiency decreases with increase in absorber temperature at any particular evaporator temperature. As specified in the previous paragraph that generator heat duty increases with increase in absorber temperature which further results in increase in input exergy whereas decrease in cooling effect is accountable for reduction in output exergy. Thus, exergetic efficiency reduces with increase in absorber temperature. The reason for higher drop in exergetic efficiency at lower evaporator temperatures (and at higher absorber temperatures) is increase in rate of rise of solution circulation ratio.

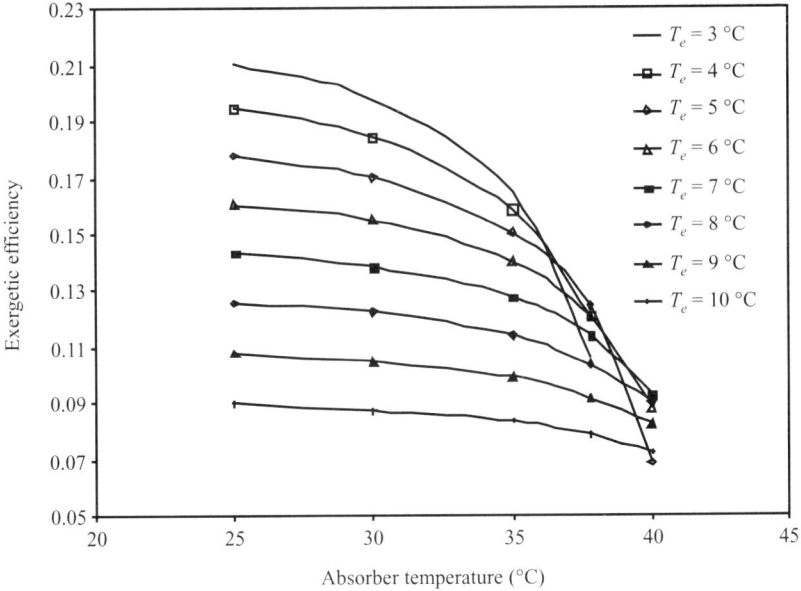

Fig. 3.32 Effect of Absorber Temperature on Exergetic Efficiency for Varying Evaporator Temperatures ($T_{hpg} = 185°C$, $T_a = T_c$, $T_r = T_e + 10°C$, $\varepsilon_{she(s)} = 0.7$)

Figure 3.33 represents the effect of HP generator temperature on exergy destruction in the components of a series flow triple effect absorption system. The percentage exergy destruction is highest in absorber followed by evaporator, she_3, she_2, she_1 and condenser. All generators taken together, all refrigerant throttle valves taken together and all solution throttle valves taken together follow the condenser.

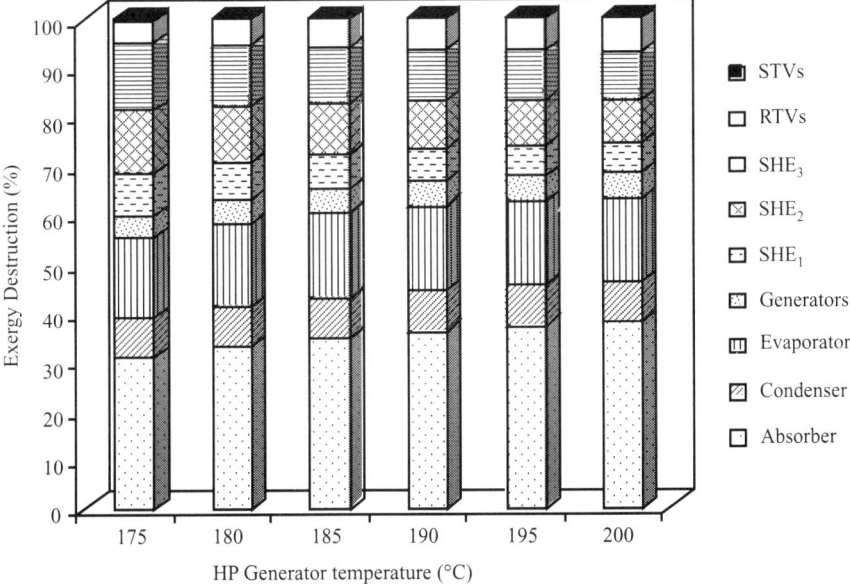

Fig. 3.33 Variation of Percentage Exergy Destruction with HP Generator Temperature ($T_e = 7.2°C$, $T_a = T_c = 37.8°C$, $T_r = T_e + 10°C$, $\varepsilon_{she(s)} = 0.7$)

3.5 HALF EFFECT GENERATION WATER-LITHIUM BROMIDE VAPOUR ABSORPTION COOLING SYSTEM

The half effect water-lithium bromide absorption refrigeration system shown schematically in Fig. 3.34 comprises an evaporator, LP & HP absorbers, LP & HP generators, LP and HP solution heat exchangers, condenser, solution pumps and solution and refrigerant throttle valves. The condenser and HP generator operate at same pressure which is the highest system pressure. The HP absorber and LP generator operate at same intermediate pressure whereas the LP absorber and evaporator operate at same lowest system pressure. Following is the relationship among the pressures in these components:

$$(P_{g_h} = P_c) > (P_{g_l} = P_{a_h}) > (P_{a_l} = P_e)$$

The refrigerant water is circulated through the evaporator, LP absorber, LP generator, HP absorber, HP generator and condenser. After water vapour has condensed in the condenser, it returns to the evaporator through an expansion valve. However, the absorbent lithium bromide aqueous solution is circulated within two separate stages, i.e., an LP stage between the LP absorber and the LP generator, and a HP stage between the HP absorber and the HP generator.

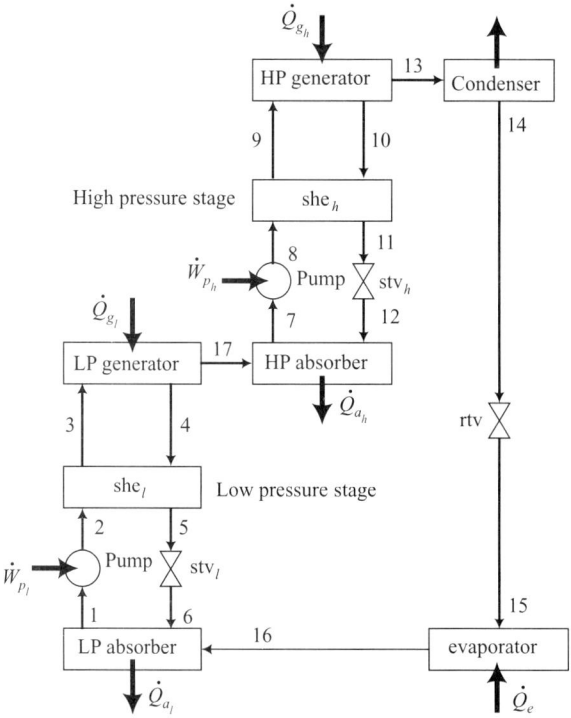

Fig. 3.34 Schematic Diagram of a Half Effect Generation Water-Lithium Bromide VAR System

Figure 3.35 shows the half effect generation water-lithium bromide VAR cycle on $\ln P\text{-}1/T$ diagram. As compared with a single stage absorption refrigeration system, there are two additional

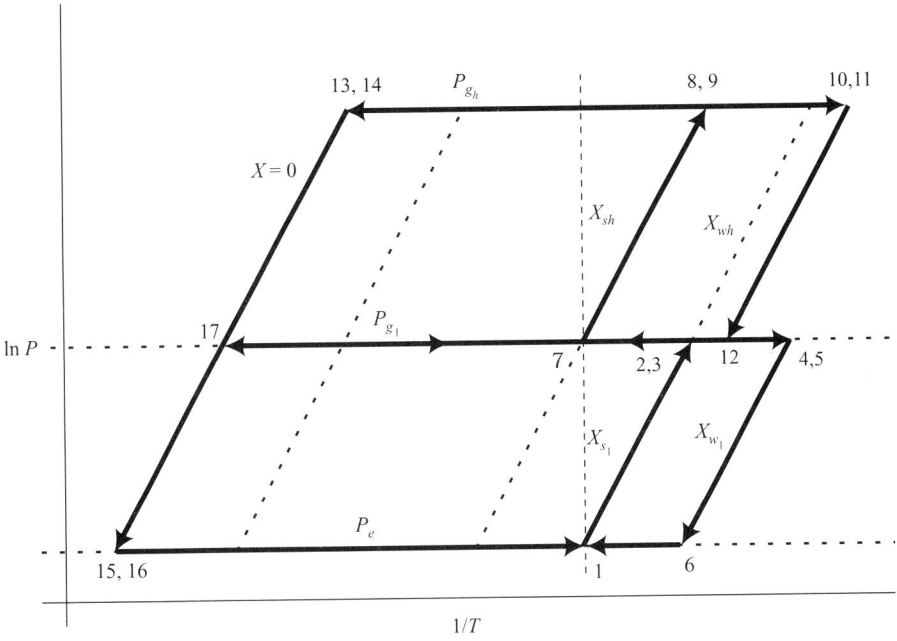

Fig. 3.35 ln P-1/T Diagram of a Half Effect Generation H_2O/LiBr Refrigeration System

components, i.e., the HP absorber and LP generator, in a half effect system. These are used to concentrate the lithium bromide aqueous solution in the LP stage cycle.

3.5.1 Thermodynamic Analysis

The thermodynamic analysis of the half effect generation VAR system is given below.

Mass Balance

The equations of mass balance in a half effect system are specified below.

LP generator and LP absorber

$$\dot{m}_{s_l} = \dot{m}_{w_l} + \dot{m}_r \tag{3.84}$$

HP generator and HP absorber

$$\dot{m}_{s_h} = \dot{m}_{w_h} + \dot{m}_r \tag{3.85}$$

The mass flow rate of refrigerant through evaporator is $\dot{m}_r$.

Energy Balance

The energy in each component of a half effect system is given by the following equations:

$$\dot{Q}_{a_l} = \dot{m}_r h_{16} + \dot{m}_{w_l} h_6 - \dot{m}_{s_l} h_1 \tag{3.86}$$

$$\dot{Q}_{g_l} = \dot{m}_r h_{17} + \dot{m}_{w_l} h_4 - \dot{m}_{s_l} h_3 \tag{3.87}$$

$$\dot{Q}_{a_h} = \dot{m}_r h_{17} + \dot{m}_{w_h} h_{12} - \dot{m}_{s_h} h_7 \tag{3.88}$$

$$\dot{Q}_{g_h} = \dot{m}_r h_{13} + \dot{m}_{w_h} h_{10} - \dot{m}_{s_h} h_9 \tag{3.89}$$

$$\dot{Q}_c = \dot{m}_r (h_{13} + h_{14}) \tag{3.90}$$

$$\dot{Q}_e = \dot{m}_r (h_{16} + h_{15}) \tag{3.91}$$

$$\dot{Q}_{she_l} = \dot{m}_{s_l}(h_3 + h_2) = \dot{m}_{w_l}(h_4 - h_5) \tag{3.92}$$

$$\dot{Q}_{she_h} = \dot{m}_{s_l}(h_9 + h_8) = \dot{m}_{w_h}(h_{10} - h_{11}) \tag{3.93}$$

$$\dot{W}_{p_l} = \dot{m}_{s_l}(h_2 + h_1) \tag{3.94}$$

$$\dot{W}_{p_h} = \dot{m}_{s_h}(h_8 + h_7) \tag{3.95}$$

$$\text{Energy in} = \dot{Q}_{g_l} + \dot{Q}_{g_h} + \dot{Q}_e + \dot{W}_{p_l} + \dot{W}_{p_h} \tag{3.96}$$

$$\text{Energy in} = \dot{Q}_{a_l} + \dot{Q}_{a_h} + \dot{Q}_c \tag{3.97}$$

The coefficient of performance (COP) of the half effect VAR system is defined as the ratio of the cooling capacity obtained at the evaporator divided by energy input into the high and low pressure generators and pumps. Accordingly, COP is given by equation (3.97).

$$\text{COP} = \frac{\dot{Q}_e}{\dot{Q}_{g_l} + \dot{Q}_{g_h} + \dot{W}_{p_l} + \dot{W}_{p_h}} \tag{3.98}$$

Exergy Balance

Exergy destruction in each component of a half effect generation VAR system is furnished below:

$$\dot{ED}_{a_l} = \dot{m}_r(h_{16} - T_0 s_{16}) + \dot{m}_{w_l}(h_6 - T_0 s_6) - \dot{m}_{s_l}(h_1 - T_0 s_1) \tag{3.99}$$

$$\dot{ED}_{g_l} = \dot{m}_{s_l}(h_3 - T_0 s_3) - \dot{m}_{w_l}(h_4 - T_0 s_4) - \dot{m}_r(h_{17} - T_0 s_{17}) + \dot{Q}_{g_l}(1 - T_0/T_{g_l}) \tag{3.100}$$

$$\dot{ED}_{a_h} = \dot{m}_r(h_{17} - T_0 s_{17}) + \dot{m}_{w_h}(h_{12} - T_0 s_{12}) - \dot{m}_{s_h}(h_7 - T_0 s_7) \tag{3.101}$$

$$\dot{ED}_{g_h} = \dot{m}_{s_h}(h_9 - T_0 s_9) + \dot{m}_{w_h}(h_{10} - T_0 s_{10}) - \dot{m}_r(h_{13} - T_0 s_{13}) + \dot{Q}_{g_h}(1 - T_0/T_{g_h}) \tag{3.102}$$

$$\dot{ED}_c = \dot{m}_r((h_7 - T_8) - T_0(s_7 - s_8)) \tag{3.103}$$

$$\dot{ED}_e = \dot{m}_r((h_9 - T_{10}) - T_0(s_9 - s_{10})) + \dot{Q}_e\left(1 - \frac{T_0}{T_r}\right) \tag{3.104}$$

$$\dot{ED}_{she_l} = \dot{m}_{s_l}((h_2 - T_3) - T_0(s_2 - s_3)) + \dot{m}_{w_l}((h_4 - h_5) - (s_4 - s_5)) \tag{3.105}$$

$$\dot{ED}_{she_l} = \dot{m}_{s_h}((h_8 - T_9) - T_0(s_8 - s_9)) + \dot{m}_{w_l}((h_{10} - h_{11}) - T_0(s_{10} - s_{11})) \tag{3.106}$$

$$\dot{ED}_{rtv} = \dot{m}_r T_o (s_{15} - s_{16}) \tag{3.107}$$

$$\dot{ED}_{stv_l} = \dot{m}_{w_l} T_o (s_6 - s_5) \tag{3.108}$$

$$\dot{ED}_{stv_h} = \dot{m}_{w_h} T_o (s_{12} - s_{11}) \tag{3.109}$$

$$\dot{ED}_t = \dot{ED}_{a_l} + \dot{ED}_{a_h} + \dot{ED}_{g_l} + \dot{ED}_{g_h} + \dot{ED}_c + \dot{ED}_e + \dot{ED}_{she_l} + \dot{ED}_{she_h} + \dot{ED}_{rtv} + \dot{ED}_{stv_l} + \dot{ED}_{stv_h} \tag{3.110}$$

Exergetic Efficiency

$$\eta_{ex} = \frac{\dot{Q}_e |(1 - T_0 / T_r)|}{\dot{Q}_{g_l}(1 - T_0 / T_{g_l}) + \dot{W}_{P_l} + \dot{Q}_{g_h}(1 - T_0 / T_{g_h}) + \dot{W}_{P_h}} \tag{3.111}$$

3.5.2 Results and Discussion

The unique feature of the half effect machine is that the required heat input temperature is lower than that for a single effect machine for the same evaporator and heat rejection temperatures. Accordingly, minimum LP and HP generator temperatures are assumed as 54°C (Herold et al., 1996). The other base parameters assumed are evaporator temperature (T_e) = 7°C, condenser temperature (T_c) = 37.8°C, absorber temperature in low and high pressure stages ($T_{a_l} = T_{a_h}$) = 37.8°C, effectiveness of solution heat exchangers = 0.7, mass flow rate of refrigerant (water) = 1 kg/s.

Figure 3.36 presents the effect of variation in LP and HP generator temperatures on COP and exergetic efficiency of the half effect generation VAR system. The HP and LP generator temperatures are assumed equal. The analysis is performed for 2 kPa pressure in the HP absorber. The initial value of the HP and LP generator temperatures assumed is 54°C. However, the first result is obtained for HP and LP generator temperatures of 60.5°C.

A small increase in the generator temperatures above 60.5°C, causes the COP and exergetic efficiency values to increase abruptly. With further increase in HP and LP generator temperatures, the COP becomes constant whereas exergetic efficiency shows a decreasing trend. The initial values of COP and exergetic efficiency are nearly zero since the solution circulation ratio in HP stage is very high, the same can be observed in Fig. 3.37, and consequently heat duty rate in HP generator is high. The maximum values of COP and exergetic efficiency obtained are about 0.41 and 9.5% respectively.

Figure 3.38 shows the effect of HP absorber pressure on COP and exergetic efficiency. It is observed that there exists an optimum intermediate pressure corresponding to which the COP and exergetic efficiency are maximum. Accordingly, rest of the analysis is carried out for the optimum intermediate pressure conditions in HP absorber or LP generator.

Figure 3.39 illustrates the effect of generator temperature on optimum intermediate pressure. It is observed that with increase in generator temperature the optimum intermediate pressure increases. This can be explained on the basis of change in concentration in HP stage.

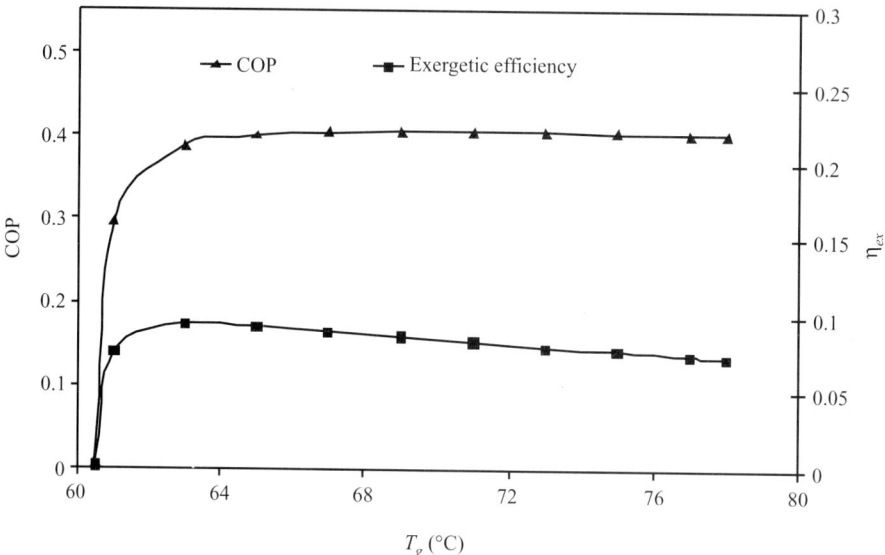

Fig. 3.36 Variation of COP and Exergetic Efficiency with Generator Temperature ($T_e = 7°C$, $T_{a_l} = T_{a_h} = T_c = 37.8°C$, $T_{g_l} = T_{g_h}$, $P_{g_l} = P_{a_h} = 2$ kPa, Effectiveness of Solution Heat Exchangers = 0.7)

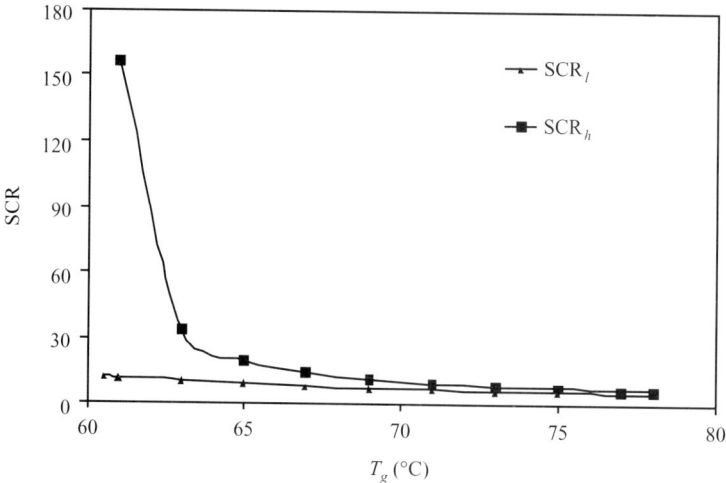

Fig. 3.37 Variation of Solution Circulation Ratio in HP and LP Stages ($T_e = 7°C$, $T_{a_l} = T_{a_h} = T_c = 37.8°C$, $T_{g_l} = T_{g_h}$, $P_{g_l} = P_{a_h} = 2$ Bar, Effectiveness of Solution Heat Exchangers = 0.7)

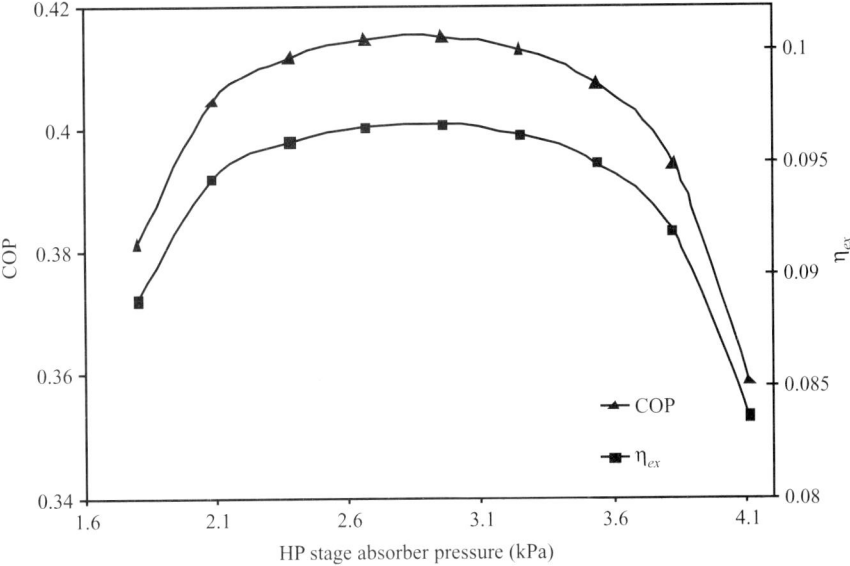

Fig. 3.38 Variation of COP and Exergetic Efficiency with HP Stage Absorber Pressure ($T_e = 7°C$, $T_{a_l} = T_{a_h} = T_c = 37.8°C$, $T_{g_l} = T_{g_h} = 65°C$, $P_{g_l} = P_{a_h}$, Effectiveness of Solution Heat Exchangers = 0.7)

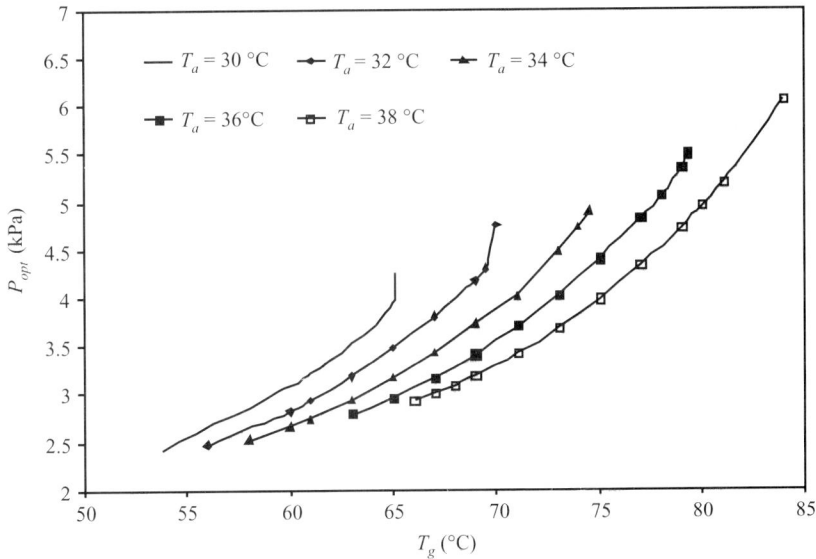

Fig. 3.39 Variation of Optimum Intermediate Pressure
($T_e = 7°C$, $T_{a_l} = T_{a_h} = T_c$, $T_{g_l} = T_{g_h}$, $\varepsilon_{she_l} = \varepsilon_{she_h} = 0.7$)

In HP stage the concentration of the solution exiting the HP generator is governed by HP generator pressure and temperature. The HP generator pressure is constant since the condenser pressure is also constant being a function of the condenser temperature which is assumed constant.

When the HP generator temperature is increased, then the concentration of solution leaving the HP generator X_{w_h} increases. This will increase the solution circulation ratio in HP stage and consequently the heat supply required in HP generator will increase. The heat supply in HP generator also increases because of increase in difference between HP generator temperature and temperature of the solution entering the HP generator. Thus, the combined effect of these two factors is to reduce the COP and exergetic efficiency. Similar increase in heat supply in LP generator and solution circulation ratio in LP stage will also be observed in the lower stage because of increase of LP generator temperature. Thus, overall COP will reduce. The increase in LP and HP generator temperatures, cause the solution temperature and refrigerant temperature exiting the generators to increase. The increase in average temperature of the condenser is also observed. Hence, exergy destruction in absorbers, generators, condenser and solution heat exchangers will increase. To overcome these ill effects, the intermediate pressure (i.e., $P_{g_l} = P_{a_h}$) should be increased. By increasing intermediate pressure, the X_{s_h} reduces and consequently the SCR_h goes down thereby nullifying the effect of increase in heat supply in HP generator due to increase in temperature in HP generator. This increase in intermediate pressure in LP stage will cause X_{w_l} to reduce and hence SCR_l will go down and reduce the negative effect of increase in heat supply required in LP stage. Thus, optimum conditions will be restored. The above paragraph explains in detail why optimum LP generator or HP absorber pressure is necessary to be increased to have maximum values of COP and exergetic efficiencies. For a constant value of generator temperature, the increase in absorber temperature is responsible for increase in solution circulation ratios and this causes the heat supply to increase in HP and LP generators. In order to maximize the COP and exergetic efficiency, the optimum pressure should be reduced. In this case, the optimum pressure is achieved at higher values of solution circulation ratios in both HP and LP stages in comparison to solution circulation ratios at lower absorber temperatures. Thus, maximum COP and maximum exergetic efficiency reduce.

Figure 3.40 represents the effect of generator temperature on maximum value of COP. It is observed that the maximum COP nearly remains constant with increase in generator temperature.

This happens because the optimum pressure is adjusted in such a way that small reduction in heat supply in HP stage generator is observed whereas there is small increase in heat supply in LP generator. These two effects negate each other and hence maximum COP remains nearly constant. As specified in the previous paragraph that with increase in absorber temperature the optimum intermediate pressure is achieved for higher values of solution circulation ratios, hence more heat is to be supplied for the same generator temperature. Further, it is also observed that increase in condenser temperature causes more flashing during throttling in refrigerant throttle valve, hence cooling effect reduces marginally. Thus, increase in heat duty in generators and decrease in cooling effect in evaporator result in a decrease of COP when absorber and condenser temperatures increase. The maximum value of maximum COP achieved is about 0.438 (corresponding to $T_{a_h} = T_{a_l} = T_c = 30°C$) whereas minimum value of maximum COP achieved is about 0.415 (corresponding to $T_{a_h} = T_{a_l} = T_c = 38°C$).

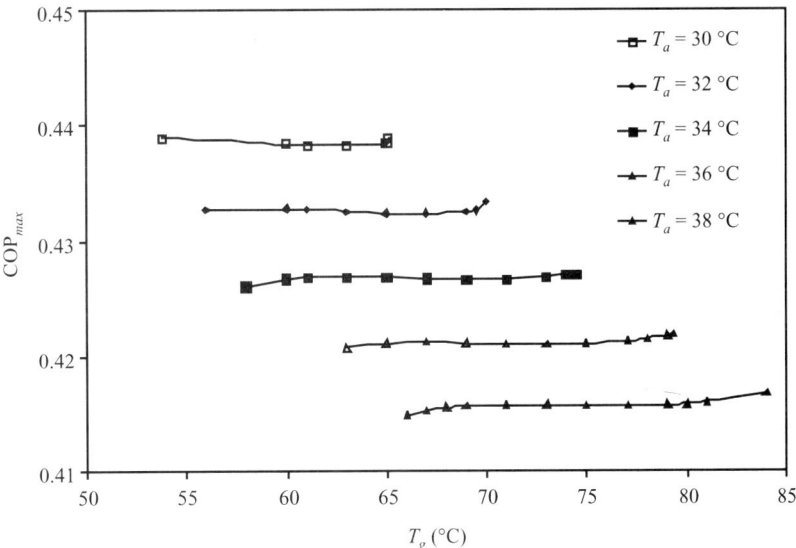

Fig. 3.40 Variation of Maximum COP with Generator Temperature
($T_e = 7°C$, $T_{a_l} = T_{a_h} = T_c$, $T_{g_l} = T_{g_h}$, $\varepsilon_{she_l} = \varepsilon_{she_h} = 0.7$, $P_{g_l} = P_{a_h} = P_{opt}$)

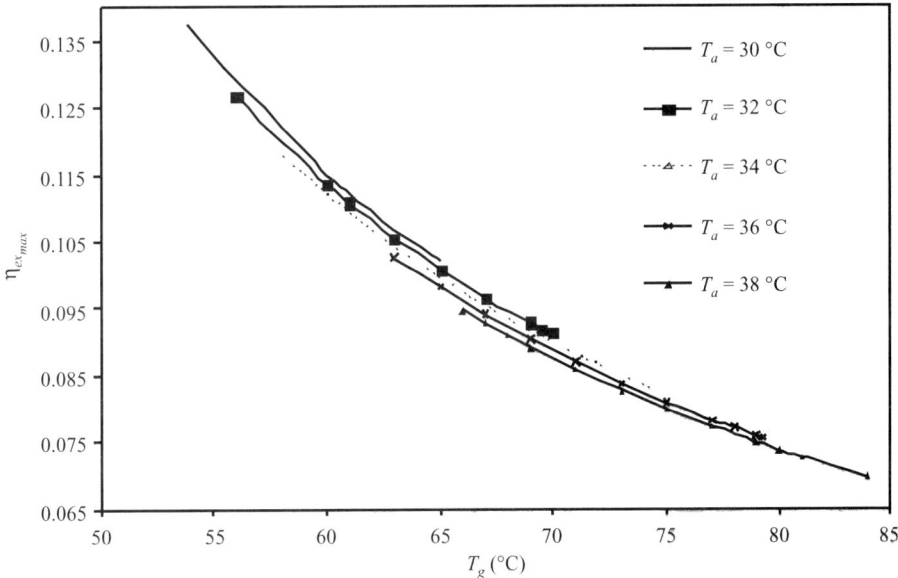

Fig. 3.41 Variation of Maximum Exergetic Efficiency with Generator Temperature for Optimum Intermediate Pressure ($T_e = 7°C$, $T_{a_l} = T_{a_h} = T_c$, $T_{g_l} = T_{g_h}$, $\varepsilon_{she_l} = \varepsilon_{she_h} = 0.7$)

The variation of maximum exergetic efficiency is shown in Fig. 3.41. It is observed that with increase in HP and LP generator temperatures, maximum exergetic efficiency reduces. This happens because of increase in total irreversibility. The irreversibility in generators increase, since the temperature difference between the solution leaving the solution heat exchanger and generator increases. The irreversibility in absorbers also increases because of the weak solution which leaves the solution heat exchangers is at a temperature higher than before and hence the temperature difference between refrigerant and weak solution increases causing an increase in entropy generation and consequently higher exergy destruction is observed in absorbers. The irreversibility in condenser also increases because of increase in average temperature of condenser. The change in exergy destruction in valves is negligible. Thus, increase in total exergy destruction is observed for the same output exergy and hence maximum exergetic efficiency decreases. The increase in absorber temperature causes increase in solution circulation ratio in LP and HP stages and consequently exergy destruction increases causing the maximum exergetic efficiency to reduce. The maximum and minimum values of maximum exergetic efficiency obtained are 13.74% and 6.96% under identical set of conditions as specified in previous paragraph.

Figures 3.42 and 3.43 illustrate the variation in optimum solution circulation ratio in LP and HP stages with generator temperature. It is observed that the optimum solution circulation ratio reduces with increase in generator temperature in HP stage up to a specific value of generator temperature and further increase in generator temperature causes the solution circulation ratio to drop suddenly.

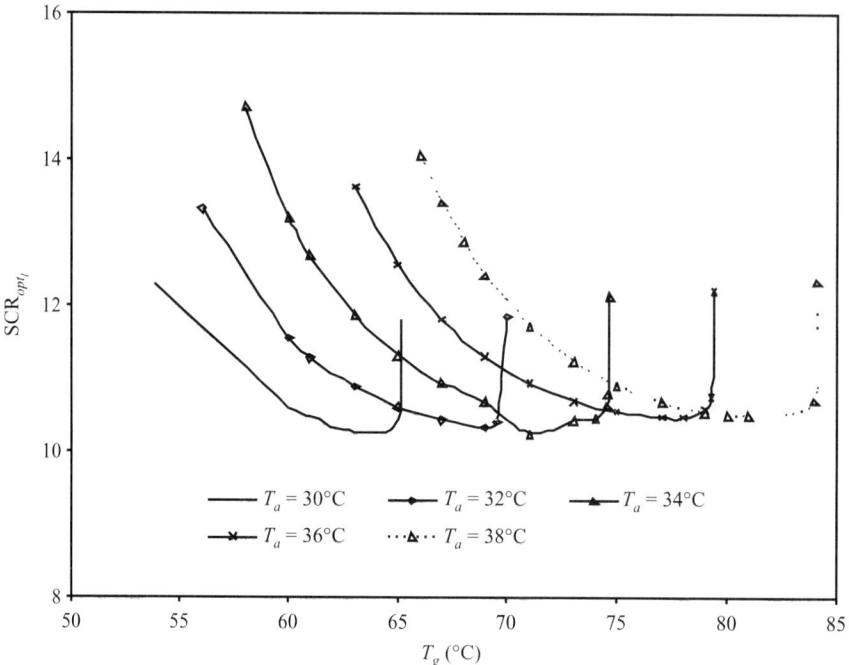

Fig. 3.42 Variation of Optimum Solution Circulation Ratio in LP Stage with Generator Temperature
($T_e = 7\ °C, T_{a_l} = T_{a_h} = T_c, T_{g_l} = T_{g_h}, \varepsilon_{she_l} = \varepsilon_{she_h} = 0.7$)

This can be explained on the basis that with increase in HP generator temperature the X_{w_h} keeps on increasing whereas X_{w_l} keeps on decreasing and correspondingly mass flow rates of weak as well as strong solution keeps on reducing in HP stage. Finally, a limit reaches where the mass flow rate of weak solution is negligible and the mass of strong solution leaving the absorber is unity. Thus, the solution circulation ratio is nearly unity. Hence, system cannot be operated above this

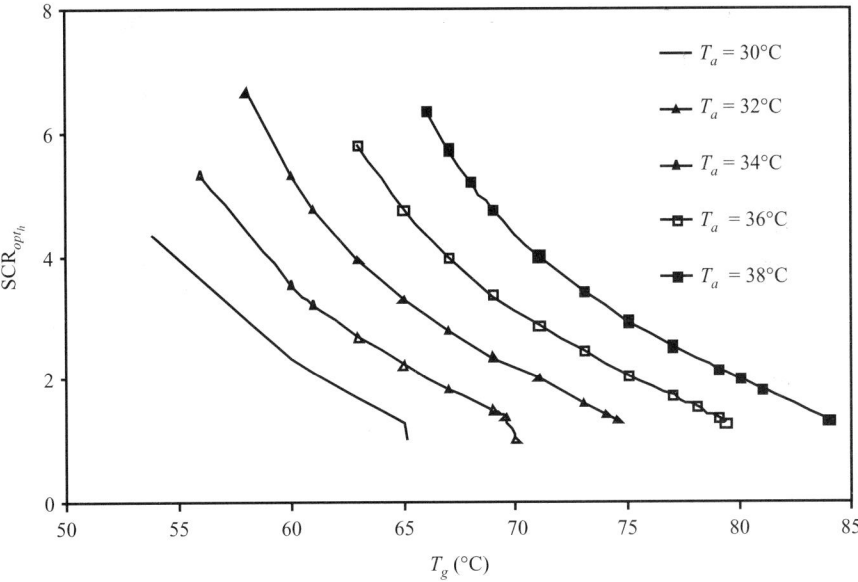

Fig. 3.43 Variation of Optimum Solution Circulation Ratio in HP Stage with Generator Temperature ($T_e = 7\ °C$, $T_{a_l} = T_{a_h} = T_c$, $T_{g_l} = T_{g_h}$, $\varepsilon_{she_l} = \varepsilon_{she_h} = 0.7$)

particular value of the solution circulation ratio since this is the minimum value which can be attained. Further, it is also observed that at this specific generator temperature, the optimum intermediate pressure (i.e., optimum pressure in HP absorber or LP generator) approaches condenser pressure. This is the reason for abrupt increase in the value of solution circulation ratio in LP stage.

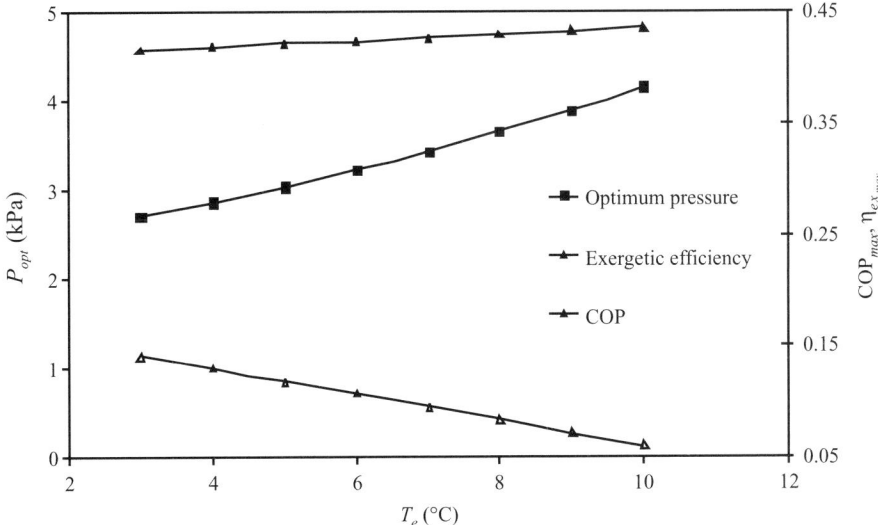

Fig. 3.44 Variation of Optimum Pressure, Maximum COP and Maximum Exergetic Efficiency with Evaporator Temperature ($T_{a_h} = T_{a_l} = T_c = 34°C$, $T_{g_h} = T_{g_l} = 67°C$, $\varepsilon_{she_l} = \varepsilon_{she_h} = 0.7$)

Figure 3.44 illustrates the effect of evaporator temperature on optimum intermediate pressure, COP_{max} and maximum exergetic efficiency. It is observed that with increase in evaporator temperature, evaporator pressure increases and consequently optimum intermediate pressure also increases. The COP max increases with increase in evaporator temperature because of reduction in solution circulation ratio in both LP and HP stages which causes heat supply rates to decrease in generators. The maximum exergetic efficiency reduces with increase in evaporator temperature. The results show a drop in both input and output exergies. However, the rate of drop in output exergy per unit increase in evaporator temperature is found to be higher in comparison to the rate of drop in input exergy per unit increase in evaporator temperature, hence exergetic efficiency, being a ratio of output exergy to input exergy, reduces.

Figure 3.45 shows the variation of optimum intermediate pressure in LP generator or HP absorber, COP_{max} and maximum exergetic efficiency versus effectiveness of solution heat exchangers. It is seen that the optimum intermediate pressure, COP_{max} and maximum exergetic efficiency increase with increase in effectiveness of solution heat exchangers.

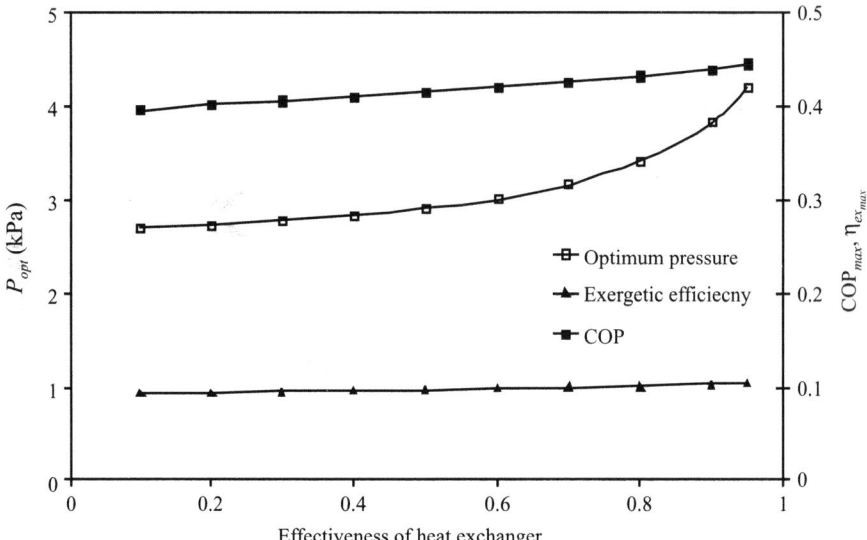

Fig. 3.45 Variation of Optimum Pressure, Maximum COP and Maximum Exergetic Efficiency with Effectiveness of Heat Exchangers ($T_{a_h} = T_{a_l} = T_c = 34°C$, $T_{g_h} = T_{g_l} = 65°C$, $T_e = 7°C$)

The COP_{max} and maximum exergetic efficiency increase because with increase in effectiveness of solution heat exchangers the heat supply at generators is reduced. Secondly, the irreversibility reduces as the temperature difference between the solution leaving the solution heat exchangers and generators reduce and it contributes in increasing the exergetic efficiency.

The variation of optimum intermediate pressure with variation in effectiveness of one of the solution heat exchangers (when the effectiveness of other heat exchanger is kept constant) is indicated in Fig. 3.46. The results indicate that when effectiveness of LP heat exchanger is assumed constant while effectiveness of HP heat exchanger is varied, the optimum intermediate pressure

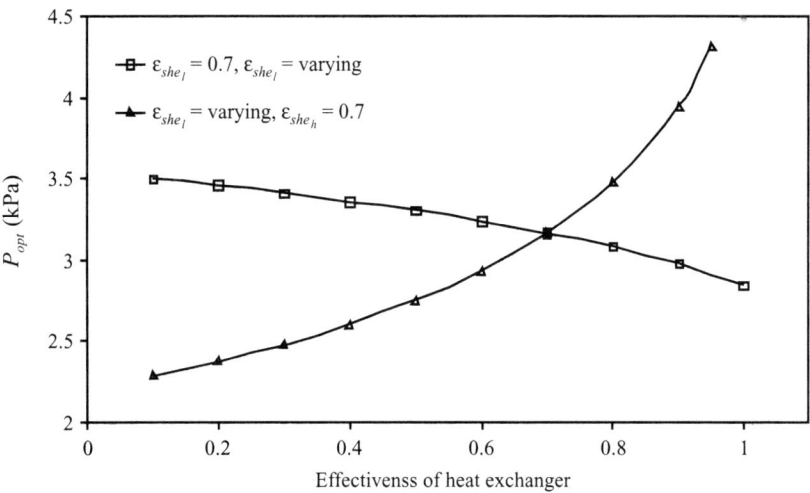

Fig. 3.46 Variation of Optimum Pressure with Effectiveness of Heat Exchangers Keeping the Effectiveness of One Heat Exchanger Constant ($T_{a_h} = T_{a_l} = T_c = 34°C$, $T_{g_h} = T_{g_l} = 65°C$, $T_e = 7°C$)

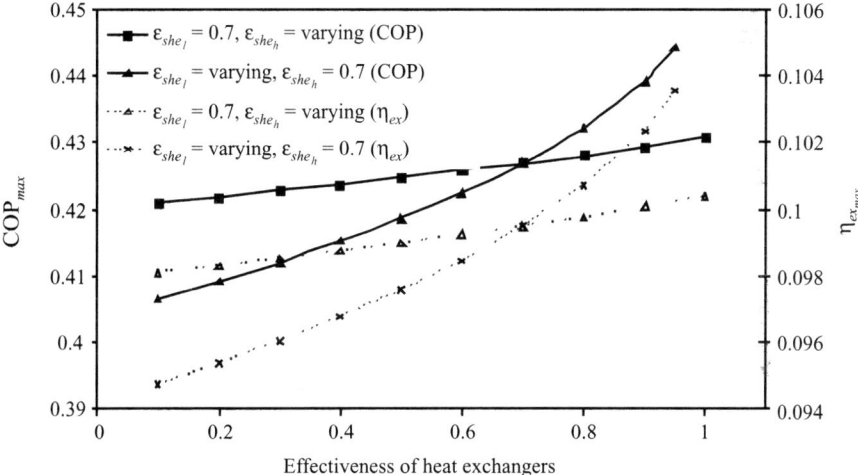

Fig. 3.47 Variation of Maximum COP with Effectiveness of Heat Exchangers ($T_{a_h} = T_{a_l} = T_c = 34°C$, $T_{g_h} = T_{g_l} = 65°C$, $T_e = 7°C$)

reduces and trends reverse when effectiveness of HP heat exchanger is assumed constant while effectiveness of LP heat exchanger is varied. Moreover, the effect of varying the effectiveness of LP heat exchanger is higher on both COP_{max} and maximum exergetic efficiency since the rate of rise of COP_{max} and maximum exergetic efficiency with increase in effectiveness of LP heat exchanger is higher in comparison to corresponding effect of increase in effectiveness of HP heat exchanger as shown in Fig. 3.47.

Figure 3.48 shows the effect of LP generator temperature on maximum COP and maximum exergetic efficiency. The results indicated in this figure show that it is better to have HP generator temperature lower than the LP generator temperature since maximum COP and maximum exergetic efficiency are higher when LP generator temperature is higher than HP generator temperature.

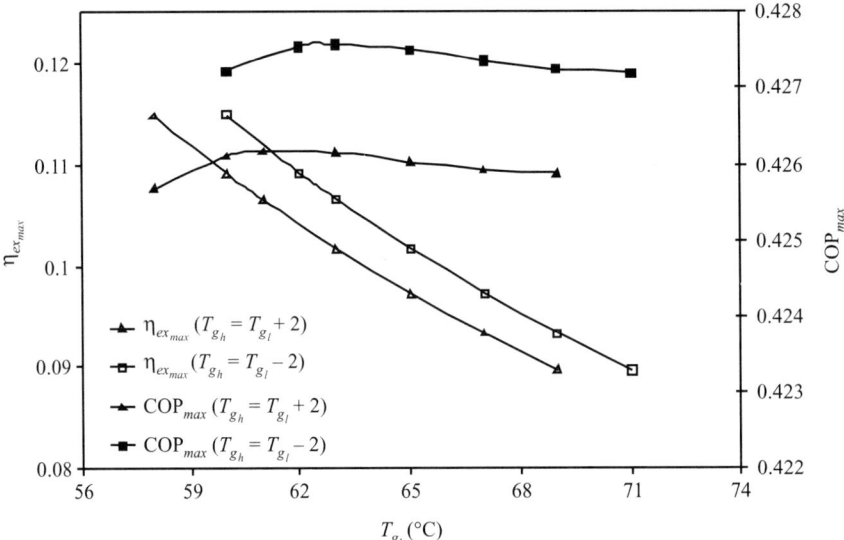

Fig. 3.48 Variation of Maximum COP and Maximum Exergetic Efficiency with LP Generator Temperature ($T_{a_l} = T_{a_h} = T_c = 34°C$, $T_e = 7°C$, $\varepsilon_{she_l} = \varepsilon_{she_h} = 0.7$, $P_{g_l} = P_{g_h} = P_{opt}$)

3.6 RESORPTION SYSTEM

A simple resorption system is shown in Fig. 3.49. The cycle employs two solution circuits instead of one. The condenser/expansion valve/evaporator section of a conventional single effect cycle are replaced by a solution circuit comprising an absorber, solution heat exchanger, pump and desorber. The absorber takes the role of condenser and rejects heat whereas the desorber performs the role of evaporator. One aspect of the resorption cycle that complicates the modeling slightly is that the mass fraction of the aqueous lithium bromide in the new solution circuit tends to be much lower than that encountered in a conventional cycle. The cooling COP of resorption cycle is of the order of 0.55-0.6. This value is lower than that found for the conventional single effect cycle due to primarily increased irreversibilites in the new solution circuit. The ln P-1/T diagram of the resorption cycle is shown in Fig. 3.50.

3.7 DUAL LOOP FLOW TYPE DOUBLE EFFECT VAPOUR ABSORPTION REFRIGERATION SYSTEM

In a dual loop flow type double effect vapour absorption system (Fig. 3.51), two individual absorption systems are cascaded and operate together as a dual loop, double effect system as shown in Duhring diagram (ln P – 1/T diagram). One of the two cycles is operated at relatively low pressure

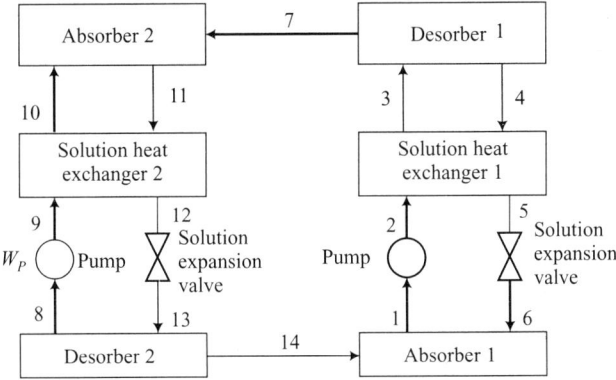

Fig. 3.49 Schematic Diagram of a Resorption System

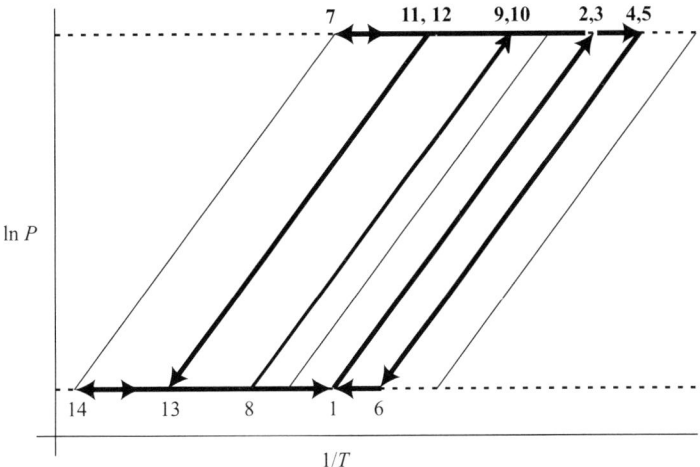

Fig. 3.50 Resorption Cycle

and temperature levels and is labelled as low stage cycle. The other is operated at high pressure and temperature levels and is designated as a high stage cycle or loop. Each of the two cycles operates independently like the single effect system (Lee, 1999) except that there are heat transfer interactions between these two cycles.

In dual loop, double effect absorption system, a high temperature heat source is employed in the generator of the high stage loop (or cycle) to generate water vapour from the LiBr/H_2O solution from the absorber. The heat rejected from the condenser of the high stage cycle is recovered and introduced into generator of the low stage cycle to generate water vapour from the solution. Cooling water is distributed to the condenser and absorber of the low stage cycle as well as absorber of the high stage cycle. If the operating conditions for the high stage cycle are carefully selected, the evaporators of both the high stage and low stage cycles in this operating type can create cooling capacity simultaneously.

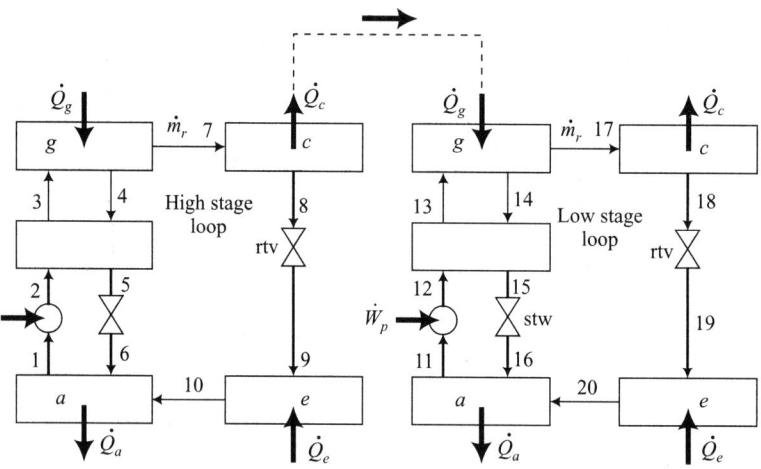

Fig. 3.51 Schematic Diagram of Dual Loop Flow Type Double Effect Absorption System

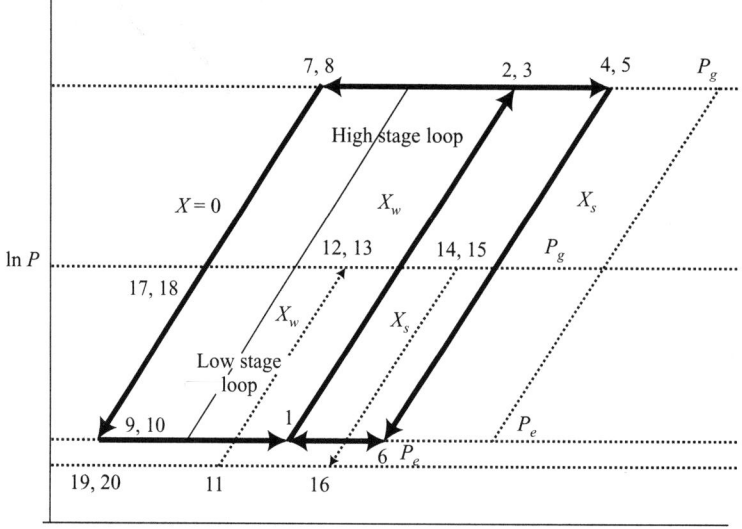

Fig. 3.52 ln $P - 1/T$ Diagram of Dual Loop Flow Type Double Effect Absorption System

The advantage of the dual loop, double effect absorption system is that the system can employ different working fluid pairs for each loop or cycle since they are operated independently. System designers may select suitable working fluid pairs for each loop of the dual loop system.

3.8 CONCLUSION

In this chapter, the various basic and advanced vapour absorption cooling cycles have been described. The basic cycles described are single effect and double effect (series and parallel flow and dual loop) generation water-lithium bromide vapour absorption refrigeration systems. The advanced cycles discussed are triple effect half effect and resorption systems. The detailed energy and exergy analysis of single effect, double effect series flow, double effect parallel flow, triple effect and half effect have been carried out. The source temperature required for the operation of each of these cycles is different, i.e., in the range of 65-85°C for half effect system, 90-120°C for single effect system, 130-170°C for double effect systems and 180-220°C for triple effect systems. The COP of these systems lies in the range of 0.3 to 0.4 for half effect system, 0.6 to 0.8 for single effect system, 1.2 to 1.4 for the double effect systems and 1.4 to 1.8 for the triple effect systems. The exergetic efficiency of vapour absorption systems is highly dependent upon the operating parameters.

Nomenclature

COP	Coefficient of performance (Non-dimensional)
$\dot{E}$	Exergy rate (kW)
ED	Rate of exergy destruction (kW)
h	Specific enthalpy (kJ kg^{-1})
HP	High pressure
LiBr	Lithium bromide
ln	Natural logarithmic
LP	Low pressure
LT	Low temperature
$\dot{m}$	Mass flow rate (kg s^{-1})
MP	Medium pressure
P	Pressure (kPa)
Q	Heat transfer rate (kW)
R	Refrigerant-solution distribution ratio
rtv	Refrigerant throttle (expansion) valve
s	Specific entropy (kJ kg^{-1}K^{-1})
SCR	Solution circulation ratio (non-dimensional)
Seg	Second effect generator
stv	Solution throttle (expansion) valve
T	Temperature (K)
TR	Tons of refrigeration
VAR	Vapour absorption refrigeration
W	Work transfer rate (kW)
X	Mass fraction of Lithium bromide in solution

Greek

η	Efficiency
Δ	Difference

δ	Efficiency defect
ε	Effectiveness
Σ	Summation

Subscripts

0	Represents dead state
a	Absorber
c	Condenser
comp	Compressor
e	Evaporator; exit
ex	Exergetic
g	High pressure generator; generator
hpg	High pressure generator
i	Inlet; represents any component of the system under consideration
lpg	Low pressure generator
max	Maximum
opt	Optimum
p	Pump
r	Refrigerant, space to be cooled
rtv	Refrigerant throttle (expansion) valve
s	Strong
seg	Second effect generator
she	Solution heat exchanger
stv	Solution throttle (expansion) valve
total	Addition of all
w	Weak
1,2,3,......	State points

REFERENCES

Agnew, B. and Ameli, S.M., 2004. A finite time analysis of a cascade refrigeration system using alternative refrigerants. *Applied Thermal Engineering*, (24), 2557–2565.

Anand, D.K., Kumar, B., 1987. Absorption machine irreversibility using new entropy calculations. Solar Energy 39 (3), 243–256.

Aphornratana, S., Eames, I.W., 1995. Thermodynamic analysis of absorption refrigeration cycles using the second law of thermodynamics method. *International Journal of Refrigeration* 18 (4), 244–252.

Aphornratana, S., Sriveerakul, T., 2007. Experimental studies of a single-effect absorption refrigerator using aqueous lithium bromide: Effect of operating condition to system performance. Experimental Thermal and Fluid Science 32 (2), 658–669.

Arivazhagan S., Murugesan S.N., Saravanan R., Renganarayanan S., 2005. Simulation studies on R134a-DMAC based half effect absorption cold storage systems, *Energy Conversion and Management* 46, 1703–1713.

Arivazhagan S., Saravanan R., Renganarayanan S., 2006. Experimental studies on HFC based two-stage half effect vapour absorption cooling system. *Applied Thermal Engineering*, 26, 1455–1462.

Arun, M.B., Maiya, M.P., Murthy, S.S., 2000. Equilibrium low pressure generator temperatures for double effect series flow absorption refrigeration systems. Applied Thermal Engineering 20, 227–242.

Arun, M.B., Maiya, M.P., Murthy, S.S., 2001. Performance comparison of double effect parallel flow and series flow water-lithium bromide absorption systems. *Applied Thermal Engineering*, 21, 1273–1279.

Arzu Şencan, Kemal A. Yakut, Soteris A. Kalogirou, 2005. Exergy analysis of lithium bromide/water absorption systems, *Renewable Energy* 30, 645–657.

Chua, H.T., Toh, H.T., Malek, A., Ng, K.C., Srinivasan, K., 2000. Improved thermodynamic property fields of LiBr-H_2O solution. *International Journal of Refrigeration* 23 (6), 412–429.

Feurecker, G., Scharfe, J., Greiter, I., Frank, C., Alefeld, G., 1993. Measurement of thermophysical properties of aqueous LiBr solutions at high temperatures and concentrations. Proceedings International Absorption Heat Pump Conference, ASME 31, 493–499.

Gommed, K., Grossman, G., 1990. Performance analysis of staged absorption heat pumps: water–lithium bromide systems. ASHRAE Transactions 30 (6), 1590–1598.

Gomri, R., Hakimi, R., (2008). Second law analysis of double effect vapour absorption cooler system. *Energy Conversion and Management*, 49(11), 3343–3348.

Grossman G., Wilk M., 1994. Advanced modular simulation of absorption systems. International *Journal of Refrigeration*, 17(4), 231–244.

Grossman G., Wilk M., DeVault, R.C., 1994. Simulation and performance analysis of triple-effect absorption cycles. *ASHRAE Transactions* 100(2):452–462.

Grossman, G., Zaltash, A., 2001. ABSIM-modular simulation of advanced absorption systems, *International Journal of Refrigeration*, 24, 531–543.

Herold, K.E., Moran, M.J., 1997. Thermodynamic properties of lithium bromide/water solutions. *ASHRAE Transactions* 93, 35–48.

Herold, K.E., Radermacher, R., Klein, S.A., 1996. Absorption chillers and heat pumps. CRC Press, USA.

Kaita, Y., 2001. Thermodynamic properties of lithium bromide water solutions at high temperatures. *International Journal of Refrigeration*, 24, 374–390.

Kaita, Y., 2002. Simulation results of triple-effect absorption cycles. *International Journal of Refrigeration*, 25, 999–1007.

Kaushik, S.C., Chandra, S., 1985. Computer modeling and parametric study of a double effect generation absorption refrigeration cycle. *Energy Conversion and Management* 25 (1), 9–14.

Kaushik S. C., Neveu P., Meunier F., 1996. An entropic model for liquid vapour absorption cooling system: Second law analysis. International Journal of Ambient Energy, 17, 101–109.

Kaynakli, O., Yamankaradeniz, R., 2007. Thermodynamic analysis of absorption refrigeration system based on entropy generation. *Current Science*, 92 (4), 472–479.

Kilic, M., Kaynakli, O., 2004. Second law-based thermodynamic analysis of water–lithium bromide absorption refrigeration system. *Energy* 32(8), 1505-1512.

Kilic, M., Kaynakli, O., 2007. Theoretical study on the effect of operating conditions on performance of absorption refrigeration system. Energy Conversion and Management 48, 599–607.

Klein, S.A., Alvarado, F., 2005. Engineering Equation Solver, Version 7.441. F-Chart software, Middleton, WI.

Kumar, P., Devotta, S., 1990. Study of an absorption refrigeration system for simultaneous cooling and heating. *ASHRAE Transactions*, 92 (2), 291–298.

Lee, S.F., Sherif, S.A., 2001a. Thermodynamic analysis of a lithium bromide/water absorption system for cooling and heating applications. *International Journal of Energy Research* 25, 1019–1031.

Lee S.F., Sherif S.A., 1999. Second law analysis of multi-effect lithium bromide/water absorption chillers. *ASHRAE Transactions*, Ch-99-23-3, 1256–1266.

Lee, S.F., Sherif, S.A., 2001b. Second law analysis of various double effect Lithium Bromide / Water absorption chillers. *ASHRAE Transactions* 9 (5), 664–673.

Ma W. B., Deng S.M., (1996). Theoretical analysis of low-temperature hot source driven two-stage LiBr/H_2O absorption refrigeration system. *International Journal of Refrigeration*. 19(2), 141–146.

Mc-Neely, L.A., 1979. Thermodynamic properties of aqueous solutions of lithium bromide. *ASHRAE Transactions*, 85, 413–434.

Oh M. D., Kim S. C., Kim Y. L. and Kim Y. I., 1994. Cycle analysis of an air-cooled LiBr/H_2O absorption heat pump of parallel-flow type. *International Journal of Refrigeration*, 17, 555–565.

Pa´tek, J., Klomfar, J., 2006. A computationally effective formulation of the thermodynamic properties of water–lithium bromide solutions from 273 to 500 K over full composition range. *International Journal of Refrigeration* 29, 566–578.

Riffat, S. B. and Shankland, N., 1993. Integration of absorption and vapour-compression systems. *Applied Energy* 46(4), 303–316.

Şencan, A., Yakut, K.A., Kalogirou, S.A., 2005. Exergy analysis of lithium bromide/water absorption systems. *Renewable Energy* 30, 645–657.

Sözen A., 2001. Effect of heat exchangers on performance of absorption refrigeration systems. *Energy Conversion and Management*, 42, 1699–1716.

Sumathy K., Huang Z. C. and Li Z. F., 2002. Solar absorption cooling with low grade heat source - a strategy of development in south China. *Solar Energy*, 72(2), 155–165.

Sun, D.W., 1997. Thermodynamic design data and optimum design maps for absorption refrigeration systems. *Applied Thermal Engineering*, 17 (3), 211–221.

Talbi, M.M., Agnew, B., 2000. Exergy analysis: an absorption refrigerator using lithium bromide and water as the working fluids. Applied Thermal Engineering 20, 619–630.

Vliet, G.C., Lawson, M.B., Lithgow, R.A., 1982. Water–lithium bromide double effect absorption cooling system analysis. *ASHRAE Transactions*, 5 (2), 811–823.

Xu, G.P., Dai, Y.Q., 1997. Theoretical analysis and optimization of a double-effect parallel flow type absorption chiller. *Applied Thermal Engineering* 17 (2), 157–170.

Xu, G.P., Dai, Y.Q., Tou, K.W., Tso, C.P., 1996. Theoretical analysis and optimization of a double-effect series-flow-type absorption chiller. *Applied Thermal Engineering*, 16 (12), 975–987.

CHAPTER 4

Compression-Absorption Combined Cooling Systems

4.1 INTRODUCTION

This chapter deals with the analysis of compression-absorption combined cooling systems for LiBr–H$_2$O absorbent and refrigerant combination. The first configuration discussed in this chapter is Absorption Recompression Refrigeration (ARR) system. In ARR system, a compressor is used after the generator and the superheated refrigerant leaving the compressor is utilized as a source of heat in the generator to generate the refrigerant vapours. The condensed refrigerant leaves the generator and it is throttled further. The remaining system is similar to single effect VAR system (Riffat and Shankland, 1993). Herold *et al.* (1991) analysed hybrid refrigeration cycles which combined a mechanical compressor and an absorption cycle in such a way that they shared a single evaporator. The work and the heat output of an engine were used an input for the compressor and generator respectively.

Kim *et al.* (2002) carried out the simulation study of the compressor-assisted triple-effect H$_2$O/LiBr absorption cooling cycles. The triple-effect absorption cooling machines using the lithium bromide-based working fluid is strongly limited by the corrosion problem caused by the high generator temperature. In this study four compressor-assisted H$_2$O/LiBr cooling cycles were suggested to solve the problem by lowering the generator temperature of the basic theoretical triple-effect cycle.

The second configuration discussed here is Compression Absorption Refrigeration (CAR) system which is also similar to single effect VAR system. In this system, the desorber (generator) operates at low pressure and also performs the function of an evaporator and the absorber operates at high pressure and performs the function of condenser. The refrigerant leaving the desorber is sent to the compressor where it is compressed and sent to the absorber at high pressure. The thermodynamic feasibility analysis of this system was carried out by Hulten and Berntsson (1999) and Pratihar *et al.* (2001) for NH$_3$-H$_2$O (refrigerant absorbent pair). Zhou and Radermacher (1997) investigated a vapour compression cycle with a solution circuit and desorber/absorber heat exchange (DAHX) experimentally using the ammonia/water mixture. COPs and cooling capacities in the range of 1.2 and 1.8 & 7 and 12 kW respectively were obtained experimentally for a temperature lift between 60 and 80°C. The experimental results are compared to that of the single-stage and two-stage cycles.

The two-stage system had the highest temperature lift (110-120°C) and the lowest COP (0.69-1.04). The single-stage system has the highest COP (2.2-3.5) but the lowest temperature lift (40°C). Also, a solution bypass between the Absorber I outlet and Desorber II inlet was proposed to improve the cycle performance. The experimental results showed that the COP varied in the range of 1-2%, while the temperature lift increased by the range between 0 and 6°C. Tarique and Siddiqui (1999) carried out the performance study and economic analysis of the combined absorption/compression cycle using NH_3-NaSCN solution and compared it with pure ammonia in the compression cycle under various operating conditions. The capital and running costs of the compressors in the NH_3-NaSCN are highly reduced as compared to the cycle using only pure ammonia.

Brunin et al. (1997) defined, plotted and discussed the working domains of a model of a compression heat pump using different fluids and a model of a compression-absorption heat pump using water-ammonia mixtures. They reported that the disappearance from use of CFC and HCFC fluids leaves only one alternative for the implementation of high temperature electric heat pumps: hydrocarbons in compression devices or water-ammonia mixtures in compression absorption devices.

Bourouis et al. (2000) studied the thermodynamic performance of a single-stage absorption/compression heat pump using the ternary working fluid Trifluoroethanol + Water + Tetraethylenglycol dimethylether (TFE-H_2O-TEGDME) for upgrading waste heat. The results obtained show that the operation of the cycle with this ternary system is still more advantageous than the TFE + TEGDME binary working pair.

Pratihar et al. (2010) carried out a detailed simulation of a 400 kW ammonia water compression absorption refrigeration system for water chilling application for summer air conditioning for three different configurations of system having 17%, 23%, and 30% relative solution heat exchanger areas. The effect of relative solution heat exchanger area and mass flow rate of weak solution on the COP, cooling capacity and absorber heat load was reported. The results show that the COP of the system can be increased by maintaining low mass flow rate of weak solution and large relative solution heat exchanger area. Increase in the relative solution heat exchanger area from 10 to 30% resulted in an increase in COP by about 16%. Further increase in relative SHX area revealed that there exists an optimum area at which COP becomes maximum.

The third configuration explained in this chapter is a cascade refrigeration system. The cascade system is suitable for low temperature applications in the temperature range from −30°C to −100°C in various industries such as pharmaceutical, food, chemical, blast freezing and liquefaction of gases. The disadvantage of the vapour compression refrigeration systems in this type of applications is their high electricity consumption. This disadvantage can be overcome by a novel cascade refrigeration system comprising VCR system in low temperature stage and a VAR system in high temperature stage. The VAR system can be operated by waste heat and therefore electricity consumption can be reduced. Fernandez-Seara et al. (2006) carried out the energy analysis of a compression absorption cascade refrigeration system with CO_2 and NH_3 as refrigerants in the compression stage and the pair of NH_3–H_2O in the absorption stage. Kairouani and Nehdi (2006) studied the possibility of using geothermal energy (temperature source in the range 343–349 K to supply vapour absorption system cascaded with conventional compression system. They selected three working fluids (R717, R22, and R134a) for the conventional compression system and the ammonia-water pair for the absorption system. Their results showed that the COP of a combined system can be improved by 37–54%, compared with the conventional cycle, under the same operating conditions, that is an evaporation temperature at 263 K and a condensation temperature of 308 K.

Garimella et al. (2011) conceptualized and analysed a novel cascaded absorption/vapour-compression cycle with a high temperature lift for a naval ship application. A single effect LiBr-H_2O absorption cycle and a subcritical CO_2 vapour-compression cycle were coupled together to provide low-temperature refrigerant (– 40°C) for high heat flux electronics applications, medium-temperature refrigerant (5°C) for space conditioning and other low heat flux applications, and as an auxiliary benefit, provide medium-temperature heat rejection (48°C) for water heating applications.

Cimsit and Ozturk (2012) studied compression-absorption cascade refrigeration cycles. While LiBr-H_2O and NH_3-H_2O are used as fluid pair in cascade absorption section, R134a, R-410A and NH_3 fluids were used in the vapour compression section of cascade cycle. It was reported that electrical energy consumption in the cascade refrigeration cycle is 48-51% lower than classical vapour compression refrigeration cycles that use R134a, R-410A and NH_3 as working fluids under the same operating conditions, that are an evaporator temperature of 263 K and a condenser temperature of 313 K. Separately, the results show that by using LiBr-H_2O pair for absorption section the thermal energy consumption of cascade refrigeration cycle could be reduced by 35% and also general coefficient of performance could be improved by 33% compared to the NH_3-H_2O pair.

4.2 ABSORPTION RECOMPRESSION REFRIGERATION (ARR) SYSTEM

An absorption recompression system is shown in Fig. 4.1. It comprises an evaporator, absorber, generator, compressor, solution heat exchanger, pump, refrigerant and solution throttle valves. The description of system is similar to the description of a single effect cycle as explained in Chapter 3. The only difference is the use of the compressor in place of condenser. The refrigerant leaving the generator is compressed to high pressure and the superheated refrigerant vapour is condensed in the generator and the heat of condensation is utilized to generate refrigerant vapour in the generator.

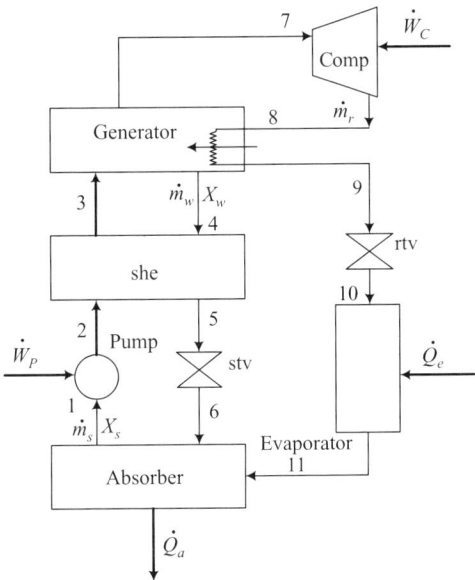

Fig. 4.1 Schematic Diagram of an Absorption Recompression Refrigeration System

4.2.1 Thermodynamic Analysis of 'ARR' System

The thermodynamic analysis of the absorption recompression refrigeration system involves the application of mass, energy and exergy balances.

Mass Balance

Mass balance at absorber or generator

$$\dot{m}_s = \dot{m}_r + \dot{m}_w \tag{4.1}$$

Mass flow rate through compressor, refrigerant throttle valve and evaporator is m_r.

Energy Balance

The energy balance in each component of a single effect absorption recompression system is given by the following equations:

$$\dot{Q}_a = \dot{m}_r h_{11} + \dot{m}_w h_6 - \dot{m}_s h_1 \tag{4.2}$$

$$\dot{Q}_g = \dot{m}_r h_7 + \dot{m}_w h_4 - \dot{m}_s h_3 \tag{4.3}$$

$$\dot{W}_{comp} = \dot{m}_r (h_7 - h_8) \tag{4.4}$$

$$\dot{Q}_e = \dot{m}_r (h_{11} - h_{10}) \tag{4.5}$$

$$\dot{Q}_{she} = \dot{m}_s (h_3 - h_2) = \dot{m}_w (h_4 - h_5) \tag{4.6}$$

$$\dot{W}_p = \dot{m}_s (h_2 - h_1) \tag{4.7}$$

$$\text{Energy in} = \dot{W}_{comp} + \dot{Q}_e + \dot{W}_p \tag{4.8}$$

$$\text{Energy output} = \dot{Q}_a \tag{4.9}$$

Coefficient of performance (COP) of the single effect ARR system is defined as the ratio of the cooling capacity obtained at the evaporator to the energy input into the compressor and pump. Accordingly, COP is given by equation (4.10).

$$\text{COP} = \frac{\dot{Q}_e}{\dot{W}_{comp} + \dot{W}_p} \tag{4.10}$$

Exergy Balance

Exergy destruction in each component of a single effect absorption recompression refrigeration cycle is given below:

$$\dot{ED}_a = \dot{m}_r (h_{11} - T_o s_{11}) + \dot{m}_w (h_6 - T_o s_6) - \dot{m}_s (h_1 - T_o s_1) \tag{4.11}$$

$$\dot{ED}_g = \dot{m}_s (h_3 - T_o s_3) - \dot{m}_w (h_4 - T_o s_4) - \dot{m}_r (h_7 - T_o s_7) + \dot{m}_r (h_8 - T_o s_8)$$
$$- \dot{m}_r (h_9 - T_o s_9) \tag{4.12}$$

$$\dot{ED}_{comp} = \dot{m}_r (h_7 - T_o s_7) - \dot{m}_r (h_8 - T_o s_8) + \dot{W}_{comp} \tag{4.13}$$

$$\dot{ED}_e = \dot{m}_r((h_{10} - h_{11}) - T_o(s_{10} - s_{11})) + \dot{Q}_e\left(1 - \frac{T_o}{T_r}\right) \quad (4.14)$$

$$\dot{ED}_{she} = \dot{m}_s((h_2 - h_3) - T_o(s_2 - s_3)) + \dot{m}_w((h_4 - h_5) - T_o(s_4 - s_5)) \quad (4.15)$$

$$\dot{ED}_{rtv} = \dot{m}_r T_o(s_{10} - s_9) \quad (4.16)$$

$$\dot{ED}_{stv} = \dot{m}_w T_o(s_6 - s_7) \quad (4.17)$$

$$\dot{ED}_{total} = \dot{ED}_a + \dot{ED}_g + \dot{ED}_{comp} + \dot{ED}_e + \dot{ED}_{she} + \dot{ED}_{rtv} + \dot{ED}_{stv} \quad (4.18)$$

Exergetic efficiency is given by

$$\eta_{ex} = \frac{\dot{Q}_e\left|\left(1 - \frac{T_0}{T_r}\right)\right|}{\dot{W}_{comp} + \dot{W}_p} = COP\left|\left(1 - \frac{T_0}{T_r}\right)\right| \quad (4.19)$$

Efficiency defect (δ_i) in a particular component is given by:

$$\delta_i = \frac{\dot{ED}_i}{\dot{W}_{comp} + \dot{W}_p} \quad (4.20)$$

where $\dot{ED}_i$ represents the rate of exergy destruction occurring in a process occurring in the component 'i' under consideration.

4.2.2 Results and Discussion

The parameters assumed for computation of results presented in Table 4.1 are obtained from Riffat and Shankland (1993). The parameters used in Figs. 4.2–4.5 are mentioned below:

1. Cooling capacity ($\dot{Q}_e$) : 100 kW
2. Isentropic efficiency of compressor, (η_{comp}) : 80 %
3. Evaporator temperature, (T_e) : 3°C to 10°C
4. Generator temperature (T_g) : 55-90°C
5. Absorber temperature (T_a) : 25-40°C
6. Effectiveness of solution heat exchanger, (ε_{she}) : 0.7
7. Condenser temperature (T_c) assumed only for single effect absorption refrigeration system : 25-40°C
8. Temperature difference between condensed refrigerant and generator temperature (i.e., $T_9 - T_g$) : 10°C

158 *Alternatives in Refrigeration and Air Conditioning*

Table 4.1 represents the comparison of results of energy analysis of present work with Riffat and Shankland (1993). It is observed that the difference between the present results and theoretical results of Riffat and Shankland (1993) is less than 3.08%.

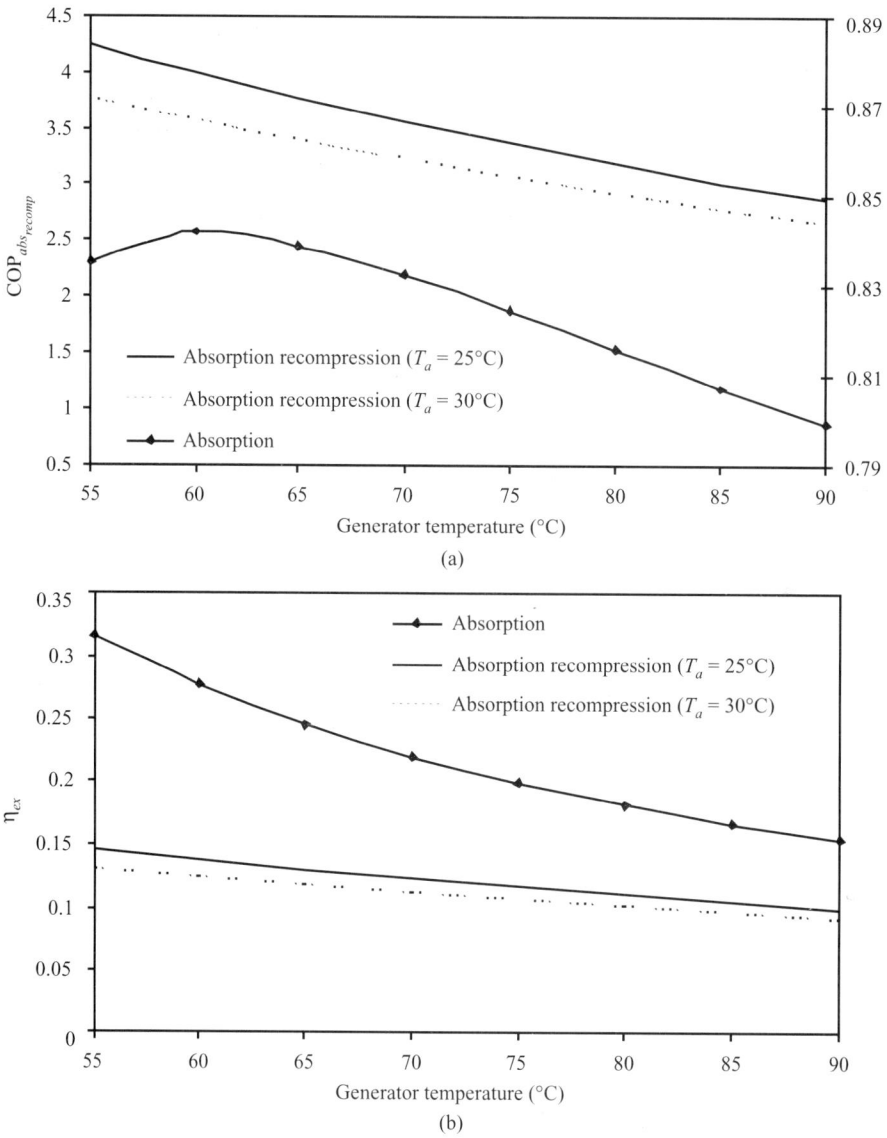

Fig. 4.2 Comparison of (a) COP and (b) Exergetic Efficiency of Single Effect Absorption and Absorption Recompression Systems ($T_a = T_c = 25°C$ (for Single Effect), $T_e = 5°C$, Effectiveness of Solution Heat Exchanger = 0.7, Efficiency of Compressor = 0.8)

Figures 4.2(a) and (b) illustrate variation of COP and exergetic efficiency with variation in generator temperature for single effect absorption and absorption recompression systems. It is observed that the COP of single effect system first increases and then decreases with increase in generator temperature. The reason for this behaviour of rise and fall of COP has already been explained in chapter 3. The COP of absorption recompression system decreases with increase in generator temperature. The generator temperature is dependent on compressor discharge temperature T_8 and temperature T_9. To increase the generator temperature it is necessary to increase the compressor discharge temperature which increases on increasing the pressure ratio across compressor. Thus, increase in pressure ratio leads to increase in power requirement and hence COP and exergetic efficiency decrease. The COP and exergetic efficiency decrease with increase in absorber temperature. The increase in absorber temperature raises the concentration of solution leaving the absorber (X_s).

Table 4.1 Comparison of Results of Energy Analysis with Results of Riffat and Shankland (1993)

Single Effect Absorption Recompression System Results			
Parameters: T_g = 65°C, T_e = 5°C, T_a = 25°C, effectiveness of solution heat exchanger = 0.7, efficiency of compressor = 0.8, Cooling capacity = 100 kW			
Component	Riffat and Shankland (1993) $\dot{Q}$(kW)	Present work $\dot{Q}$(kW)	Difference (%)
Generator	127.76	131.7	3.08
Absorber	126.48	126.47	0
Compressor	26.47	26.57	0.38
Evaporator	100	100	0
Solution Heat Exchanger	---	18.36	---
Solution throttle valve	0	0	0
Refrigerant throttle valve	0	0	0
Pump	0	0.000737	----
Energy input	126.47	126.5707	0.08
COP	**3.78**	**3.764**	**–0.42**

Thus, the solution circulation ratio increases. The increase in solution circulation ratio causes the generator heat duty to increase. This increase in generator heat duty is achieved from the heat of condensation of superheated refrigerant vapours leaving the compressor. Since the generator temperature is kept constant therefore temperature T_9 is also constant. Thus the only option to increase the heat of condensation is to increase the compressor discharge temperature T_8. The compressor discharge temperature is increased by increasing the pressure ratio across the compressor and it accounts for increase in heat of condensation and the compressor work. The overall effect is to reduce the COP. The increase in pressure ratio enhances exergy destruction in compressor.

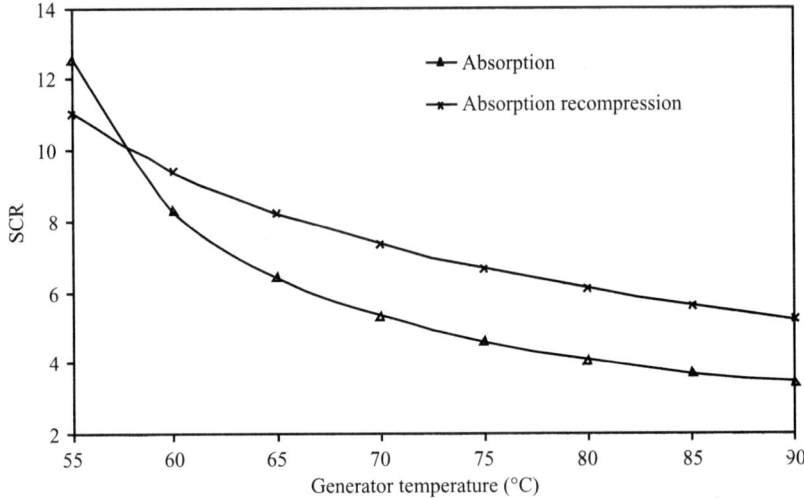

Fig. 4.3 Comparison of Solution Circulation Ratio in Single Effect VAR and Vapour Absorption Recompression Refrigeration Systems ($T_a = T_c = 25°C$, $T_e = 5°C$, Effectiveness of Solution Heat Exchanger = 0.7, Efficiency of Compressor = 0.8)

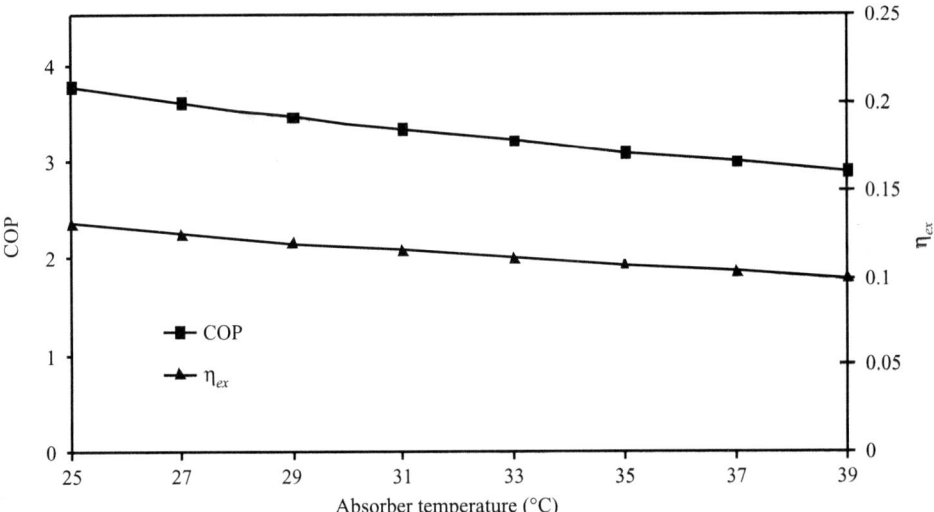

Fig. 4.4 Absorber Temperature Versus COP and Exergetic Efficiency in Absorption Recompression Systems ($T_g = 65°C$, $T_c = 25°C$, $T_e = 5°C$, Effectiveness of Solution Heat Exchanger = 0.7, Efficiency of Compressor = 0.8)

Moreover, the increase in discharge temperature of compressor is liable for increase in temperature difference between generator temperature and the refrigerant being condensed, thus exergy destruction in generator increases. Thus, there is increase in total exergy destruction and hence exergetic efficiency decreases. The solution circulation ratio decreases with increase in generator temperature, as shown in Fig. 4.3. The solution circulation ratio is lower in absorption system in comparison to absorption recompression system. Figure 4.4 shows the variation of COP and exergetic efficiency with increase in absorber temperature. The increase in absorber temperature lowers the COP and exergetic efficiency. The reason for decrease in COP and exergetic efficiency with increase in absorber temperature has already been explained in the previous section.

The increase in evaporator temperature increases the COP and exergetic efficiency as shown in Fig. 4.5. This happens because with increase in evaporator temperature the absorber pressure increases and hence the concentration of strong solution X_s leaving the absorber reduces. The fall in value of concentration of strong solution X_s is accountable for decrease of solution circulation ratio (solution circulation ratio = $X_w / (X_w - X_s)$). Thus, generator heat duty decreases and this drop in generator heat brings down the pressure ratio across the compressor (at constant generator temperature). Hence, power required to compress the refrigerant vapour drops thereby increasing the COP and exergetic efficiency.

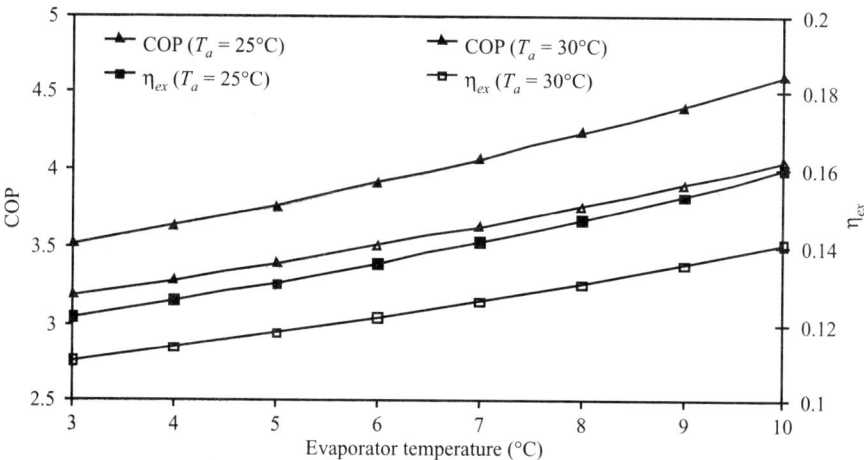

Fig. 4.5 Variation of COP and Exergetic Efficiency with Evaporator Temperature ($T_g = 65°C$, $T_a = T_c = 25°C$, Effectiveness of Solution Heat Exchanger = 0.7, Efficiency of Compressor = 0.8)

4.3 COMPRESSION ABSORPTION REFRIGERATION (CAR) SYSTEM

The schematic diagram of a compression absorption system is shown in Fig. 4.6. In this system, the evaporator is replaced by a generator which functions at low pressure whereas the absorber operates at higher pressure. The superheated refrigerant (water) generated in desorber, is compressed from the generator pressure to the absorber pressure in the compressor. The remaining refrigerant-absorbent solution (weak in water), leaving the desorber at state point 1, is pumped through the solution heat exchanger to the absorber pressure P_a, where mixing of this solution with the superheated water vapour discharged from the compressor, takes place. The refrigerant-absorbent mixture, after absorbing the water vapour (becomes strong in refrigerant), passes through the solution heat exchanger where the cold solution pumped by the pump and hot solution coming from the absorber exchange heat. The strong solution is then throttled to the generator pressure at state point 6. This completes the cycle.

4.3.1 Thermodynamic Analysis of Compression Absorption Refrigeration System

Mass balance

Mass balance at the absorber or desorber is given by equation (4.21).

$$\dot{m}_s = \dot{m}_r + \dot{m}_w \tag{4.21}$$

Mass flow rate through the compressor is $\dot{m}_r$.

Energy Balance

The energy balance in each component of a single effect compression absorption refrigeration system is given by the following equations:

$$\dot{Q}_a = \dot{m}_r h_8 + \dot{m}_w h_3 - \dot{m}_s h_4 \tag{4.22}$$

$$\dot{Q}_g = \dot{m}_r h_7 + \dot{m}_w h_1 - \dot{m}_s h_6 \tag{4.23}$$

$$\dot{W}_{comp} = \dot{m}_r (h_8 - h_7) \tag{4.24}$$

$$\dot{Q}_{she} = \dot{m}_w (h_3 - h_2) = \dot{m}_s (h_4 - h_5) \tag{4.25}$$

$$\dot{W}_p = \dot{m}_w (h_2 - h_1) \tag{4.26}$$

$$\text{Energy in} = \dot{W}_{comp} + \dot{W}_p \tag{4.27}$$

$$\text{Energy out} = \dot{Q}_a \tag{4.28}$$

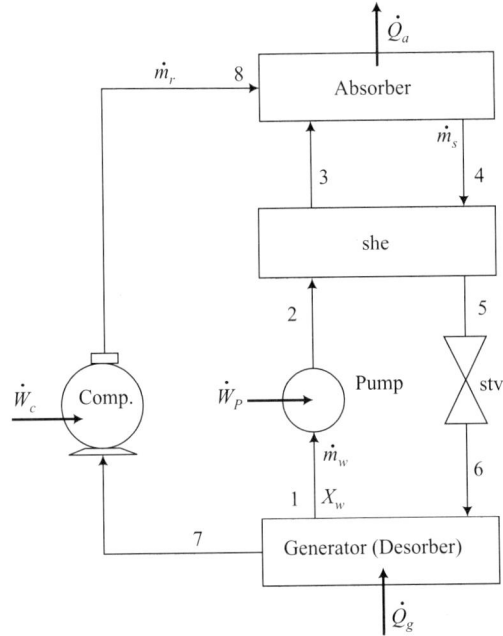

Fig. 4.6 Schematic Diagram of a Compression Absorption Refrigeration System

The coefficient of performance (COP) of the single effect compression absorption refrigeration system is defined as the ratio of the cooling capacity obtained at the evaporator to the energy input into the compressor and pump. With this definition of the cycle COP is given as:

$$\text{COP} = \frac{\dot{Q}_g}{\dot{W}_{comp} + \dot{W}_p} \quad (4.29)$$

Exergy Balance

Exergy destruction in each component of a single effect compression absorption refrigeration cycle is given below:

$$\dot{ED}_a = \dot{m}_r(h_8 - T_o s_8) + \dot{m}_w(h_3 - T_o s_3) - \dot{m}_s(h_4 - T_o s_4) \quad (4.30)$$

$$\dot{ED}_g = \dot{m}_s(h_6 - T_o s_6) - \dot{m}_w(h_1 - T_o s_1) - \dot{m}_r(h_7 - T_o s_7) + \dot{Q}_g\left(1 - \frac{T_o}{T_r}\right) \quad (4.31)$$

$$\dot{ED}_{comp} = \dot{m}_r(h_7 - T_o s_7) - \dot{m}_r(h_8 - T_o s_8) + \dot{W}_{comp} \quad (4.32)$$

$$\dot{ED}_{she} = \dot{m}_w((h_2 - h_3) - T_o(s_2 - s_3)) + \dot{m}_s((h_4 - h_5) - T_o(s_4 - s_5)) \quad (4.33)$$

$$\dot{ED}_{stv} = \dot{m}_s T_o(s_6 - s_5) \quad (4.34)$$

$$\dot{ED}_{total} = \dot{ED}_a + \dot{ED}_g + \dot{ED}_{comp} + \dot{ED}_{she} + \dot{ED}_{stv} \quad (4.35)$$

Exergetic efficiency is given by

$$\eta_{ex} = \frac{\dot{Q}_g\left(1 - \frac{T_0}{T_r}\right)}{\dot{W}_{comp} + \dot{W}_p} = \text{COP}\left(1 - \frac{T_0}{T_r}\right) \quad (4.36)$$

4.3.2 Results and Discussion

The input parameters used for obtaining results as shown in Figs. 4.7–4.13 are mentioned below:
1. Pressure ratio across compressor and pump : 5-10
2. Isentropic efficiency of compressor, (η_{comp}) : 80%
3. Generator temperature (T_g) : 2-11°C
4. Absorber temperature (T_a) : 25-30°C
5. Effectiveness of solution heat exchanger, (ε_{she}) : 0.7
6. Temperature difference between generator and cold room (i.e., $T_r - T_g$) : 10°C
7. Mass flow rate of the refrigerant through the evaporator : 1 kg/s

4.3.2.1 Effect of Variation in Generator Temperature

The variation of cooling effect (Q_g), compressor power and solution circulation ratio with generator temperature are shown in Fig. 4.7(a). Figure 4.7(b) and (c) show the effect of generator temperature

164 *Alternatives in Refrigeration and Air Conditioning*

on COP and exergetic efficiency. For a particular value of weak solution concentration, it is observed that as the generator temperature increases from 5°C to 6°C, the rise in COP curve is steep whereas with further increase in generator temperature the increase in COP is meager and the curve becomes nearly horizontal. The variation in solution circulation ratio is accountable for large change in COP (refer Fig. 4.7(a)). It is observed that at lower values of generator temperature, the compressor discharge pressure (for a fixed pressure ratio) is low and hence the strong solution concentration X_s leaving the absorber is high. This accounts for higher values of solution circulation ratio (SCR = $X_w/(X_w - X_s)$). At higher solution circulation ratios, the cooling effect produced at the generator is low and it increases when the solution circulation reduces. However, the compressor power input is dependent on pressure ratio and mass flow rate of refrigerant (which is constant) and is not affected with variations in solution circulation ratio. Since the pressure ratio is assumed constant, hence the no variation in compressor power input is observed. Thus, at lower generator temperatures, the COP is low. As the generator temperature increases from 5°C to 6°C, a large drop in solution circulation ratio is observed which results in sharp rise in cooling effect and COP. With further increase in generator temperature, the rate of drop in solution circulation ratio is small and corresponding rise in cooling effect is also small. This accounts for nearly horizontal COP curve.

It is also observed that as the concentration of the weak solution X_w increases (refer Fig. 4.7(a) and (b)), the rate of rise in COP increases. This happens because of rate of drop in solution circulation ratio increases with increase in generator temperature at higher values of weak solution concentration. Lower the value of weak solution concentration higher is the COP. The COP ranges between 5.69 (at $T_g = 5°C$) to 7.5 (at $T_g = 11°C$) corresponding to $X_w = 34\%$ and the corresponding values of COP at $X_w = 40\%$ varies between 3.7 and 7.46.

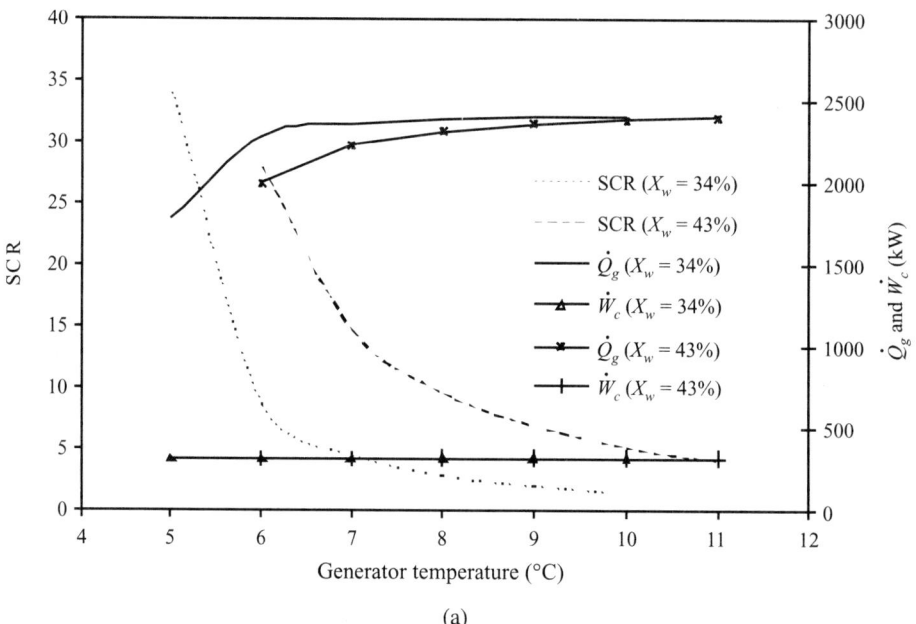

Fig. 4.7 (a) Effect of Solution Concentration Leaving the Generator on SCR, Cooling Effect and Compressor Power ($T_a = 30°C$, Pressure Ratio Across Pump and Compressor = 5)

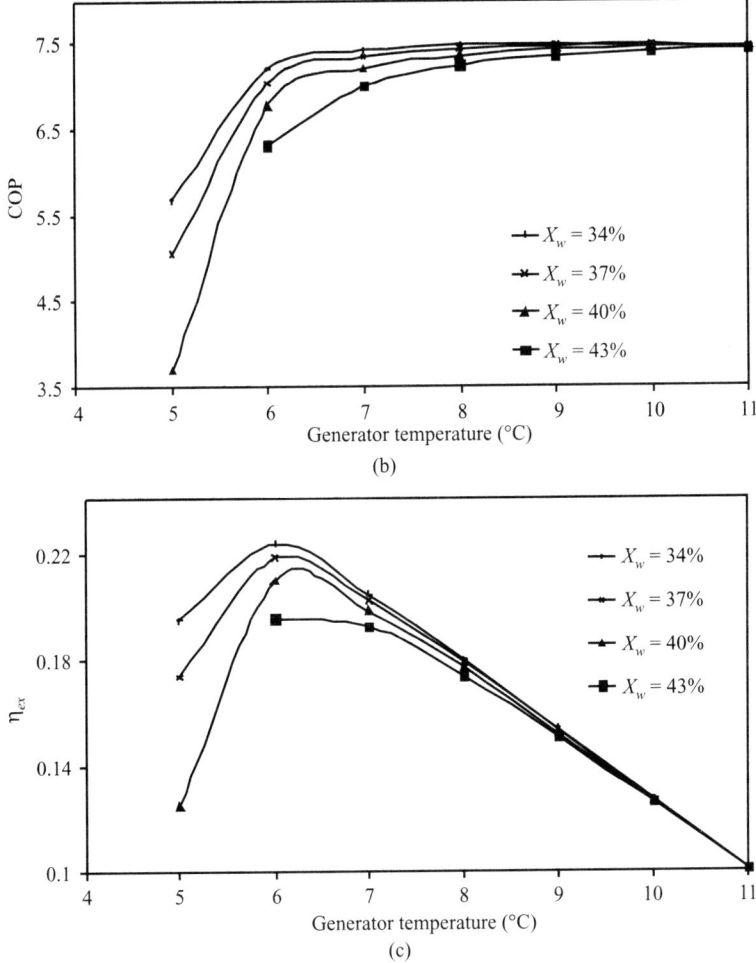

Fig. 4.7 Effect of Solution Concentration Leaving the Generator on (b) COP and (c) η_{ex} ($T_a = 30°C$, Pressure Ratio Across Pump and Compressor = 5)

Figure 4.7 (c) represents the effect of generator temperature on exergetic efficiency. It is observed that exergetic efficiency increases with increase in generator temperature from 5°C to 6°C and decreases with further increase in generator temperature. The exergetic efficiency is the product of two terms, i.e., COP and $\left|\left(1-\dfrac{T_0}{T_r}\right)\right|$. The COP increases sharply between 5°C and 6°C whereas second term reduces. The combined effect of these two factors is to increase the exergetic efficiency. As stated earlier, with increase in generator temperature, the COP curve nearly becomes horizontal whereas the second term has a decreasing trend. The combined effect of these factors accounts for decrease of the exergetic efficiency. The exergetic efficiency also reduces with increase in weak solution concentration. The maximum value of exergetic efficiency varies between 22.36% (corresponding to $X_w = 34\%$) and 19.55% (corresponding to $X_w = 43\%$).

166 *Alternatives in Refrigeration and Air Conditioning*

Figures 4.8 (a), (b) and (c) illustrate the effect of generator temperature on COP for varying absorber temperatures. The effect of generator temperature on COP has already been explained. It has already been shown in Fig. 4.7 (a) that with increase in solution circulation ratio, cooling effect at the generator reduces. As shown in Fig. 4.8(c) that with increase in absorber temperature, solution circulation ratio increases. Thus, COP reduces with increase in absorber temperature. With increase in generator temperature, the COP curves at different absorber temperatures converge to same value of COP ~ 7.5 (at $T_g = 11°C$). The trend of exergetic efficiency is dependent on COP and

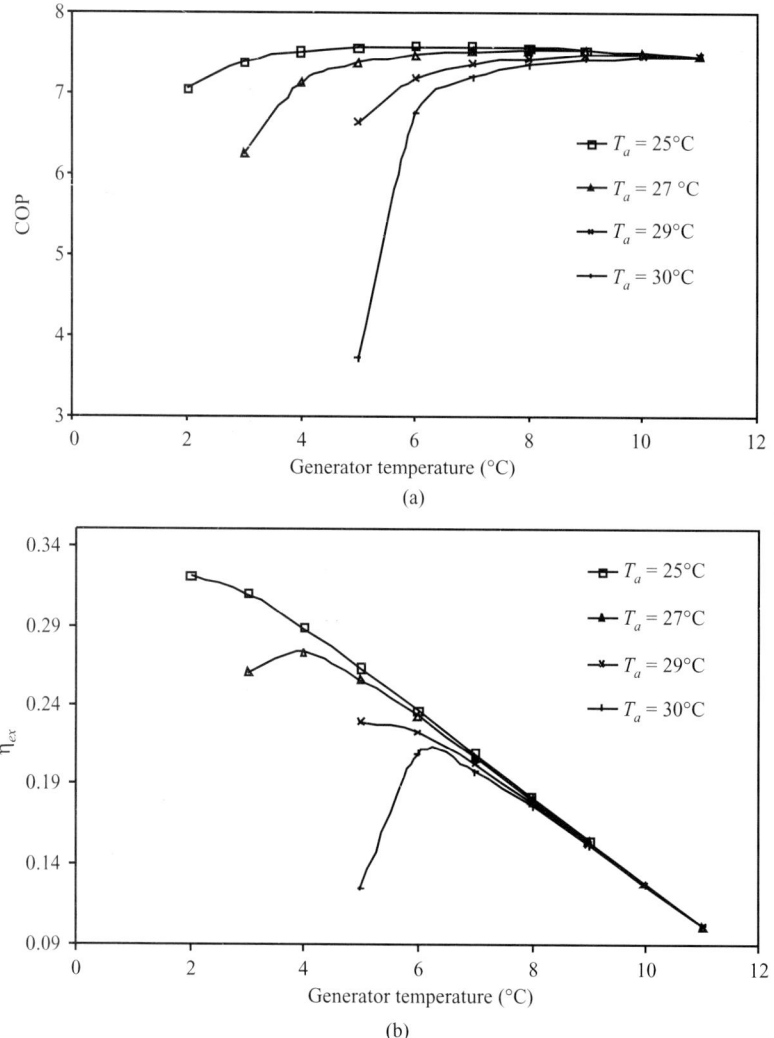

Fig. 4.8 Variation of (a) COP and (b) η_{ex} versus Generator Temperature for Varying Absorber Temperature (Effectiveness of Solution Heat Exchanger = 0.7, Pressure Ratio Across Pump and Compressor = 5, $X_w = 40\%$, Compressor Efficiency = 0.8)

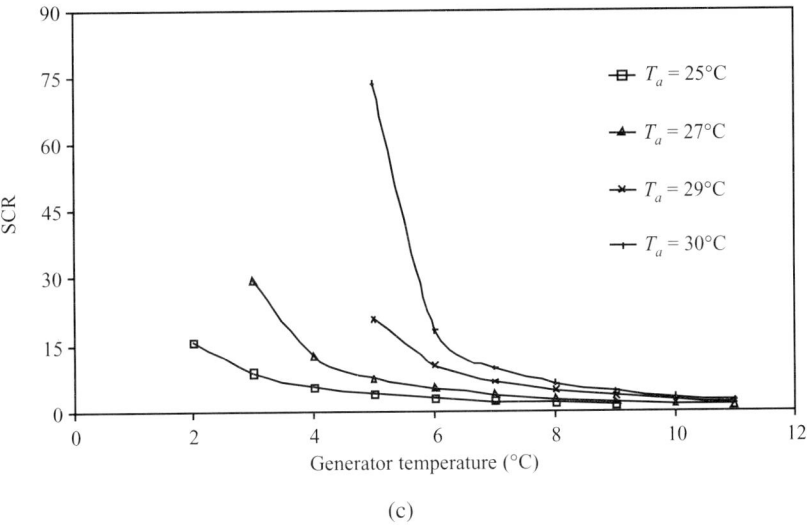

Fig. 4.8 (c) Variation of Solution Circulation Ratio versus Generator Temperature for Varying Absorber Temperature (Effectiveness of she = 0.7, Pressure Ratio Across Pump and Compressor = 5, X_w = 40%, Compressor Efficiency = 0.8)

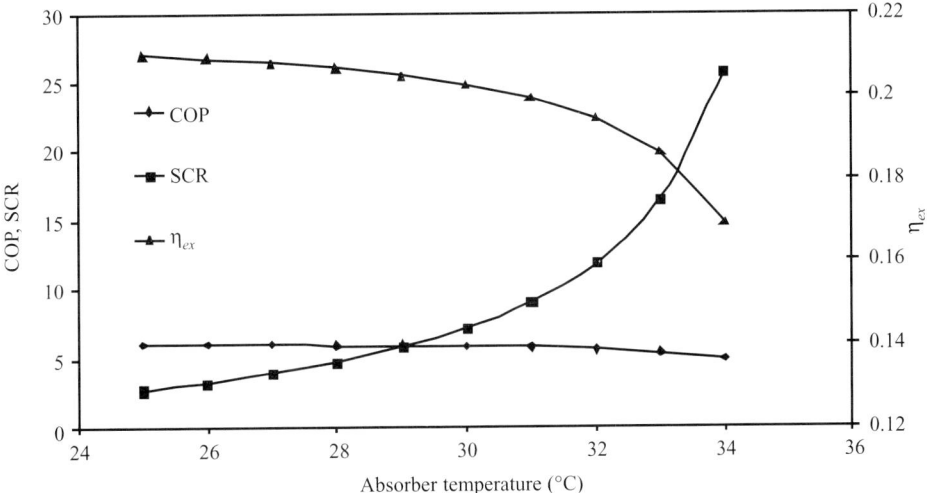

Fig. 4.9 COP, SCR and Exergetic Efficiency versus T_a (T_g = 5°C, X_w = 45%, Pressure Ratio = 7, Effectiveness of Solution Heat Exchanger = 0.7 and Efficiency of Compressor = 0.8)

$\left|\left(1-\dfrac{T_0}{T_r}\right)\right|$. The effect of these two parameters on exergetic efficiency has been explained. The exergetic efficiency values also increase with reduction in absorber temperature. The maximum value of exergetic efficiency obtained is about 32% (at T_g = 2°C) for T_a = 25°C whereas the minimum value observed is 12.5% (at T_g = 5°C) for T_a = 30°C.

4.3.2.2 Effect of Variation in Absorber Temperature

Figure 4.9 shows the variation of COP, exergetic efficiency and solution circulation ratio with absorber temperature. It has already been explained that with increase in absorber temperature, the solution circulation ratio increases and it accounts for decrease in cooling effect at the generator. The compressor power remains constant and hence COP reduces. The exergetic efficiency also reduces since COP and the term $|(1 - T_0/T_r)|$ reduce.

The effect of generator temperature on COP and exergetic efficiency at varying pressure ratios is shown in Figs. 4.10 and 4.11. The increase in pressure ratio from 5 to 10 causes the COP to drop

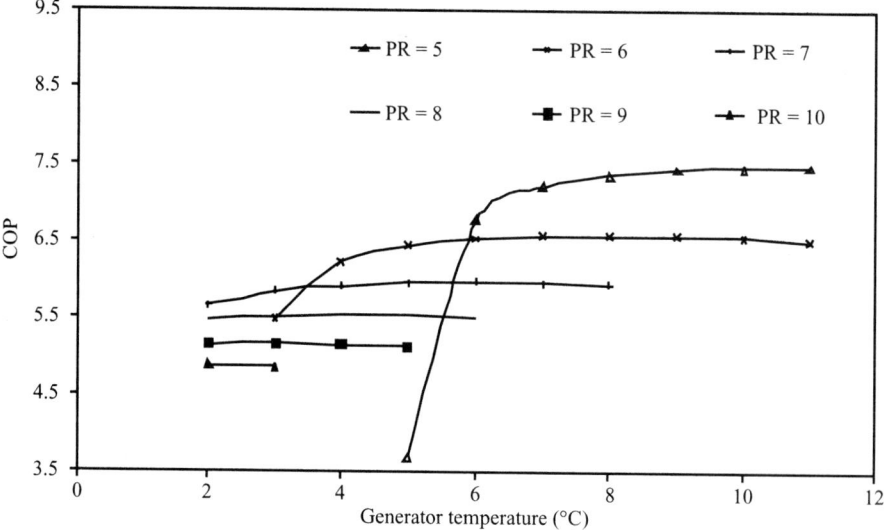

Fig. 4.10 COP versus Generator Temperature for Varying Pressure Ratio ($T_a = 30°C$, $X_w = 40\%$)

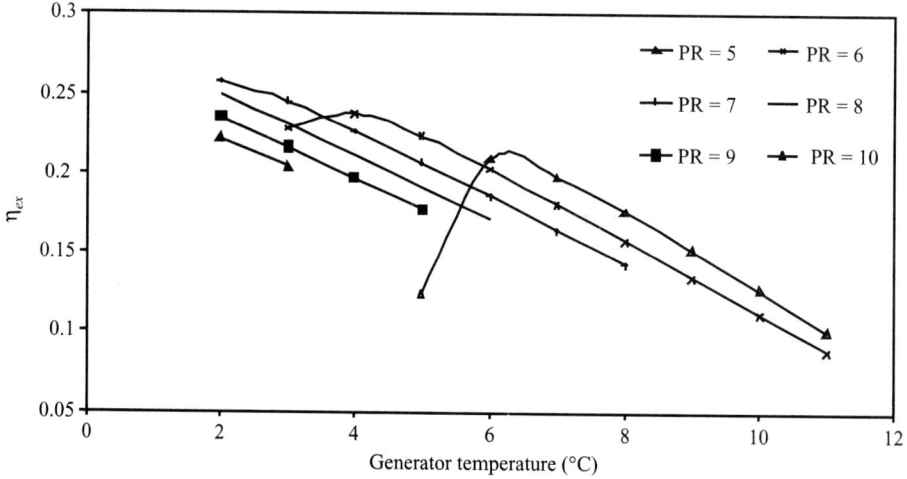

Fig. 4.11 η_{ex} versus Generator Temperature for Varying Pressure Ratio ($T_a = 30°C$, $X_w = 40\%$)

from 7.46 to 4.86. Further, it is observed that as the pressure ratio increases, the system seizes to operate at higher values of generator temperature. The exergetic efficiency also reduces with increase in pressure ratio at any specific generator temperature. However, the maximum value of exergetic efficiency achieved increases with increase in pressure ratio and also shifts towards lower values of generator temperature.

In Fig. 4.12, it is shown that the increase in pressure ratio causes solution circulation ratio to decrease. The effect of decrease in solution circulation ratio is to increase the cooling effect at

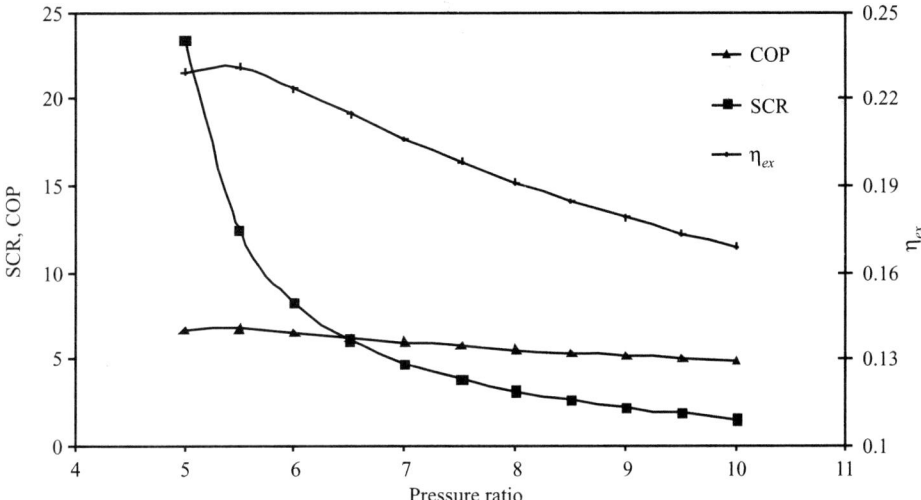

Fig. 4.12 Variation of COP, Exergetic Efficiency and Solution Circulation Ratio versus Pressure Ratio ($X_w = 45\%$, $T_g = 5°C$, $T_a = 28°C$)

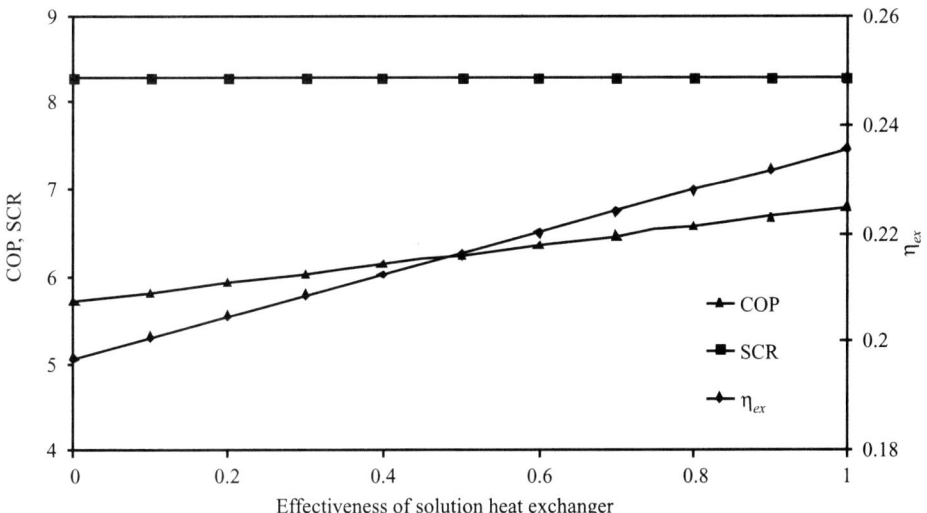

Fig. 4.13 COP, Solution Circulation Ratio and Exergetic Efficiency versus Effectiveness of Solution Heat Exchanger ($T_g = 5°C$, $T_a = 28 °C$, $X_w = 45\%$, Pressure Ratio = 6)

170 Alternatives in Refrigeration and Air Conditioning

evaporator. However simultaneously, the compressor work and pump work increase because of increase in pressure ratio. The rate of increase in compressor work is found to be higher than the rate of increase in cooling effect. Therefore, COP decreases with increase in pressure ratio. At a particular generator temperature, the effect of increase in pressure ratio is to reduce COP, hence exergetic efficiency also reduces.

4.3.2.3 Effect of Variation in Effectiveness of Solution Heat Exchanger

Figure 4.13 represents the effect of solution circulation ratio, COP and effectiveness of solution heat exchanger on exergetic efficiency. It is observed that the effectiveness of solution heat exchanger does not have any effect on solution circulation ratio. As the effectiveness increases, the temperature of the strong solution leaving the solution heat exchanger reduces and enthalpy h_6 also reduces. The cooling effect at generator is given by equation $\dot{Q}_g = \dot{m}_r h_7 + \dot{m}_w h_1 - \dot{m}_s h_6$. The right-hand side shows that only variable is h_6 which decreases with increase in effectiveness of solution heat exchanger and hence cooling effect increases whereas power input to the compressor remains constant. Thus, COP increases. Since COP increases and the expression $\left| \left(1 - \dfrac{T_0}{T_r} \right) \right|$ is constant, therefore exergetic efficiency also increases.

4.4 COMPRESSION-ABSORPTION CASCADE REFRIGERATION (CACR) SYSTEM

A compression-absorption cascade refrigeration system, comprising a VCR system in LT stage and a VAR system in the HT stage, is analysed. CO_2 and NH_3 have been considered as refrigerants in the compression stage and the H_2O-LiBr refrigerant absorbent pair in the absorption stage. The analysis has been realized by means of a mathematical model of the refrigeration system implemented using an EES program. The study presents the results obtained regarding the performance of the refrigeration system based on energy and exergy analysis.

4.4.1 Description of the Compression Absorption Cascade Refrigeration System

The compression-absorption cascade system being considered is shown schematically in Fig. 4.14. It consists of a single-stage compression system in the low temperature stage and a single effect absorption system in the high temperature stage. The compression system comprises the evaporator, compressor, condenser and an expansion device. The absorption system uses the pair water-lithium bromide and its major components are the absorber, generator, condenser, evaporator, solution heat exchanger, pump, solution throttle valve and a refrigerant throttle valve. Both systems share a common heat exchanger (cascade condenser), which operates simultaneously as the condenser of the compression system and as the evaporator of the absorption system. The compression system is used for the generation of the cooling at low temperature and the absorption system rejects heat to the surroundings, as shown in Fig. 4.14. This refrigeration system would decrease the electricity consumption compared to the two stages of compression system, since it is only required to operate the compression system at the low stage; whereas the absorption system is driven by heat.

4.4.2 Thermodynamic Analysis of 'CACR' System

The thermodynamic analysis of compression absorption system involves the principles of mass conservation, energy conservation and exergy balance.

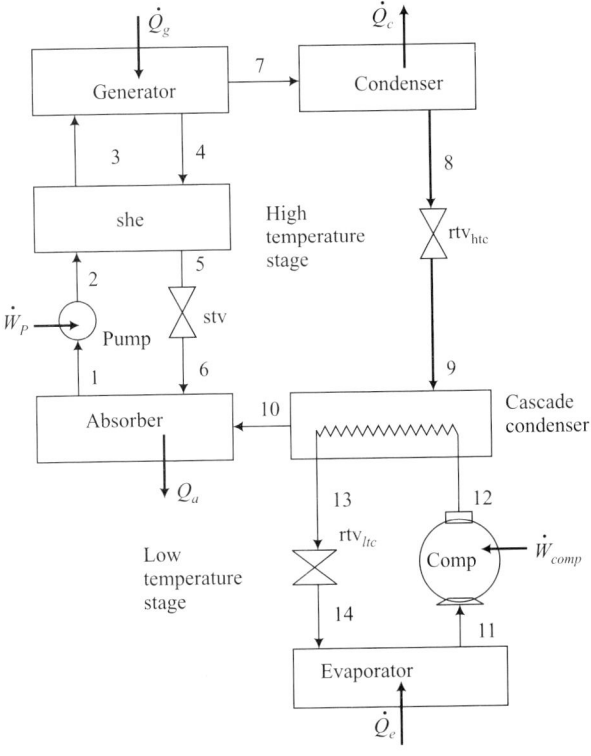

Fig. 4.14 Schematic Diagram of Compression-Absorption Cascade Refrigeration System

Mass Balance

The mass flow rate through each component of low temperature stage is $\dot{m}_{r_{vcr}}$. It is calculated using equation (4.37).

$$\dot{Q}_e = \dot{m}_{r_{vcr}}(h_{11} - h_{14}) \tag{4.37}$$

Mass balance equations in high temperature stage are specified below.

Mass balance at absorber or generator

$$\dot{m}_s = \dot{m}_r + \dot{m}_w \tag{4.38}$$

Mass flow rate through condenser and evaporator is $\dot{m}_r$.

Energy Balance

The energy balance equations for the system shown in Fig. 4.14 are given below:

Energy balance in high temperature stage

$$\dot{Q}_a = \dot{m}_r h_{10} + \dot{m}_w h_6 - \dot{m}_s h_1 \tag{4.39}$$

$$\dot{Q}_g = \dot{m}_r h_7 + \dot{m}_w h_4 - \dot{m}_s h_3 \tag{4.40}$$

172 *Alternatives in Refrigeration and Air Conditioning*

$$\dot{Q}_c = \dot{m}_r (h_7 - h_8) \tag{4.41}$$

$$\dot{Q}_{she} = \dot{m}_s (h_3 - h_2) = \dot{m}_w (h_4 - h_5) \tag{4.42}$$

$$\dot{W}_p = \dot{m}_s (h_2 - h_1) \tag{4.43}$$

Energy balance in low temperature stage

$$\dot{W}_{comp} = \dot{m}_{r_{vcr}} (h_{12} - h_{11}) \tag{4.44}$$

$$\dot{Q}_e = \dot{m}_{r_{vcr}} (h_{11} - h_{14}) \tag{4.45}$$

$$\text{Power input} = \dot{W}_{comp} + \dot{W}_p \tag{4.46}$$

Energy input

$$\dot{E}_{in} = \dot{Q}_g + \dot{Q}_e + \dot{W}_p + \dot{W}_{comp} \tag{4.47}$$

Energy output

$$\dot{E}_{out} = \dot{Q}_a + \dot{Q}_c \tag{4.48}$$

Coefficient of performance

$$\text{COP} = \frac{\dot{Q}_e}{\dot{Q}_g + \dot{W}_p + \dot{W}_{comp}} \tag{4.49}$$

Exergy Balance

Exergy destruction in components of the cascade system is given below:

Exergy analysis in high temperature stage

$$\dot{ED}_a = \dot{m}_r (h_{10} - T_o s_{10}) + \dot{m}_w (h_6 - T_o s_6) - \dot{m}_s (h_1 - T_o s_1) \tag{4.50}$$

$$\dot{ED}_g = \dot{m}_s (h_3 - T_o s_3) - \dot{m}_w (h_4 - T_o s_4) - \dot{m}_r (h_7 - T_o s_7) + \dot{Q}_g \left(1 - \frac{T_o}{T_g}\right) \tag{4.51}$$

$$\dot{ED}_c = \dot{m}_r ((h_7 - h_8) - T_o (s_7 - s_8)) \tag{4.52}$$

$$\dot{ED}_{cc} = \dot{m}_r ((h_9 - h_{10}) - T_o (s_9 - s_{10})) + \dot{m}_{r_{vcr}} ((h_{12} - h_{13}) - T_o (s_{12} - s_{13})) \tag{4.53}$$

$$\dot{ED}_{she} = \dot{m}_s ((h_2 - h_3) - T_o (s_2 - s_3)) + \dot{m}_w ((h_4 - h_5) - T_o (s_4 - s_5)) \tag{4.54}$$

$$\dot{ED}_{rtv_{htc}} = \dot{m}_r T_o (s_9 - s_8) \tag{4.55}$$

$$\dot{ED}_{stv} = \dot{m}_w T_o (s_6 - s_5) \tag{4.56}$$

Exergy analysis in low temperature stage

$$\dot{ED}_{comp} = \dot{m}_{r_{vcr}} T_0 (s_{12} - s_{11}) \tag{4.57}$$

$$\dot{ED}_e = \dot{m}_{r_{vcr}} (h_{14} - T_0 s_{14}) + \dot{Q}_e \left(1 - \frac{T_0}{T_r}\right) - \dot{m}_{r_{vcr}} (h_{11} - T_0 s_{11}) \tag{4.58}$$

$$\dot{ED}_{rtv_{ltc}} = \dot{m}_{r_{vcr}}(h_{13} - T_0 s_{13}) - \dot{m}_{r_{vcr}}(h_{14} - T_0 s_{14}) = \dot{m}_{r_{vcr}} T_0 (s_{14} - s_{13}) \qquad (4.59)$$

$$\dot{ED}_{total} = \dot{ED}_a + \dot{ED}_g + \dot{ED}_c + \dot{ED}_{cc} + \dot{ED}_{she} + \dot{ED}_{rtv_{htc}} + \dot{ED}_{stv} + \dot{ED}_{rtv_{ltc}}$$
$$+ \dot{ED}_{comp} + \dot{ED}_e \qquad (4.60)$$

Exergetic Efficiency

$$\eta_{ex} = \frac{\text{Exergy in product}}{\text{Exergy of fuel}} = \frac{\dot{EP}}{\dot{EF}} = \frac{\dot{EF} - \dot{ED}_{total}}{\dot{EF}} = 1 - \frac{\dot{ED}_{total}}{\dot{EF}} \qquad (4.61)$$

$$\dot{EF} = \dot{W}_{comp} + \dot{W}_p + \dot{Q}_g \left(1 - \frac{T_0}{T_g}\right) \qquad (4.62)$$

$$\dot{EP} = \dot{Q}_e \left|\left(1 - \frac{T_0}{T_r}\right)\right| \qquad (4.63)$$

$$\eta_{ex} = \frac{\dot{Q}_e \left|\left(1 - \frac{T_0}{T_r}\right)\right|}{\dot{W}_{comp} + \dot{W}_p + \dot{Q}_g \left(1 - \frac{T_0}{T_g}\right)} \qquad (4.64)$$

4.4.3 Results and Discussion

The analysis has been carried out considering two different natural refrigerants, carbon dioxide and ammonia in the compression system. The absorption system uses the water-lithium bromide pair. The parameters assumed for computation of results are mentioned below:

1. Cooling capacity (Q_e) : 100 kW
2. Isentropic efficiency of compressor, (η_{comp}) : 60–80 %
3. Evaporator temperature, ($T_{e_{vcr}}$) : –45°C to –35°C
4. Cascade condenser temperature, (T_{cc}) : 2–11°C
5. Generator temperature (T_g) : 50–115°C
6. Absorber temperature (T_a) : 25–40°C
7. Effectiveness of solution heat exchanger, (ε_{she}) : 0.6–0.8
8. Condenser temperature (T_c) : 25–40°C
9. Approach in cascade condenser (A) : 0–10°C

The cascade condenser temperature is varied between 2 and 11°C because the lowest temperature in water-lithium bromide system depends upon the freezing point of water which can't be below 0°C.

4.4.3.1 Effect of Cascade Condenser Temperature on COP and Exergetic Efficiency

Figure 4.15 shows the effect of cascade condenser temperature on COP and exergetic efficiency for ammonia and carbon-dioxide in low temperature stage. It is observed that both COP and exergetic efficiency reduce with increase in cascade condenser temperature for carbon dioxide cycle. In ammonia cascade system, the COP first increases, achieves a maximum value and then reduces. However, the difference between maximum and minimum values is very small. The carbon-dioxide cycle cascade system has the lower values of COP and exergetic efficiency in comparison to ammonia cycle cascade system. The increase in cascade condenser temperature results in increase in pressure ratio across compressor, increase in mass flow rate in compression stage and increase in absorber pressure. The first two factors account for increase in compressor power and the last factor, i.e., increase in absorber pressure accounts for decrease in strong solution concentration X_s. The weak solution concentration X_w remains constant. Thus, the solution circulation ratio (given by $X_w/(X_w - X_s)$) decreases. The reduction in solution circulation ratio reduces the generator heat duty. Thus, COP of the system may increase or decrease depending upon the increase in compressor power requirement and reduction in generator heat duty.

Since the selection of cascade condenser temperature depends upon the freezing point of refrigerant (water), hence for a given set of parameters, viz., evaporator temperature (depends upon application), condenser and absorber temperatures (depends upon ambient conditions) there must exist a unique generator temperature corresponding to which COP is maximum. Similar to this, the exergetic efficiency will be maximum corresponding to some other value of generator temperature. This point is highlighted in Fig. 4.16(a) for ammonia cascade system. It can be observed that maximum COP and maximum exergetic efficiency occur corresponding to different generator

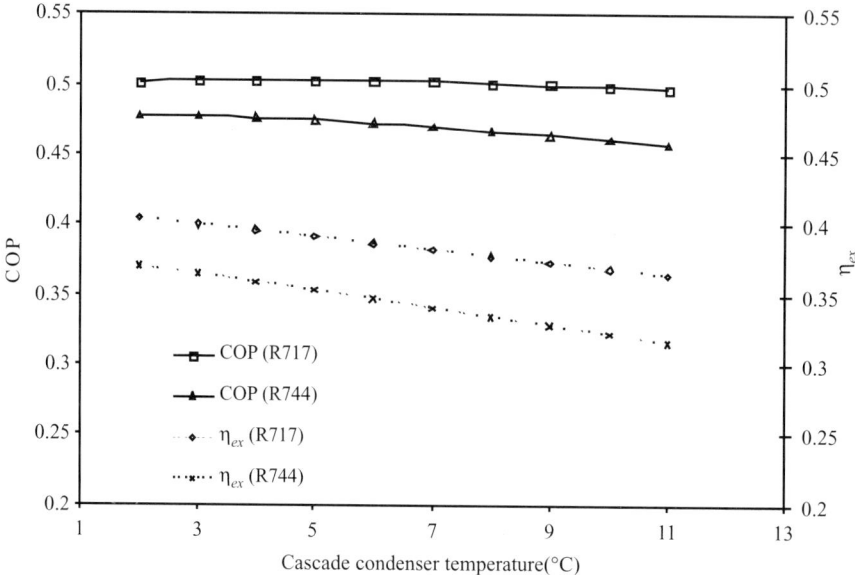

Fig. 4.15 Effect of Cascade Condenser Temperature on COP and Exergetic Efficiency ($T_a = T_c = 35°C$, $T_{e_{vcr}} = -40°C$, $T_g = 85°C$, $\varepsilon_{she} = 0.7$, $\eta_{she} = 0.7$, $\eta_{comp} = 0.8$, $\dot{Q}_{e_{vcr}} = 100$ kW, A = 0°C, $T_o = 298.15$ K)

temperatures. The maximum COP occurs corresponding to generator temperature of 83.5°C and maximum exergetic efficiency occurs corresponding to 72.25°C.

It is observed that the optimum generator temperature corresponding to maximum COP is higher in comparison to optimum generator temperature corresponding to maximum exergetic

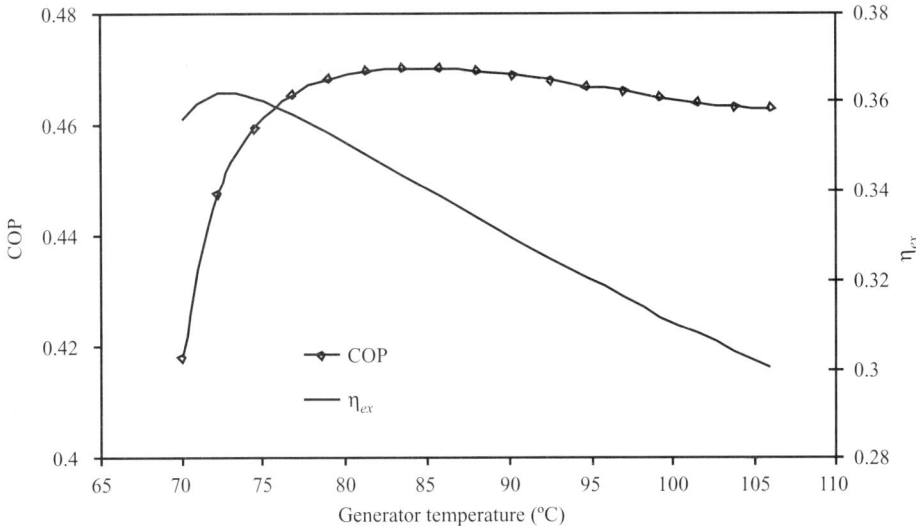

Fig. 4.16(a) Variation of COP and η_{ex} versus Generator Temperature for Ammonia ($T_a = T_c = 35$ °C, $T_{e_{vcr}} = -40$ °C, $\varepsilon_{she} = 0.7$, $\varepsilon_{comp} = 0.8$, $Q_{e_{vcr}} = 100$ kW, $A = 0$ °C, $T_o = 298.15$ K)

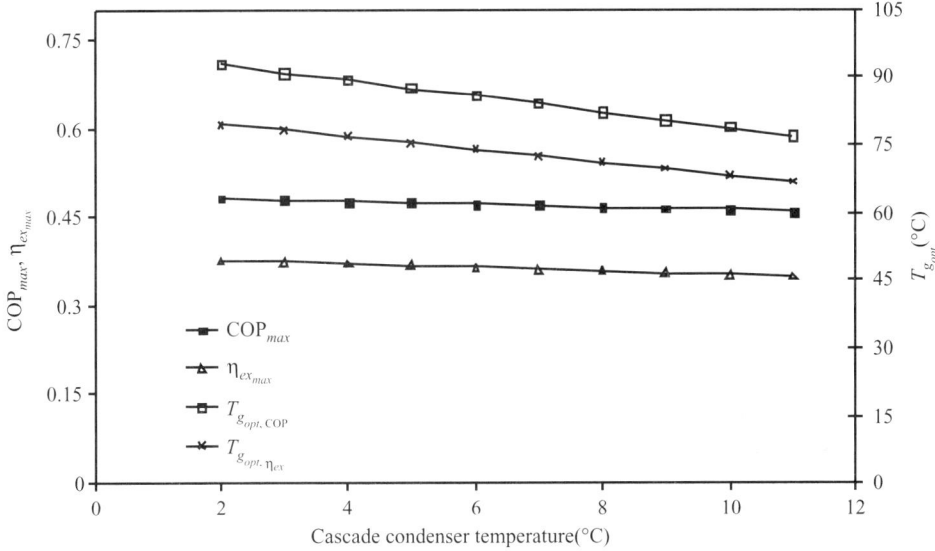

Fig. 4.16(b) COP_{max} and $\eta_{ex_{max}}$ versus Cascade Condenser Temperature and Corresponding $T_{g_{opt}}$ for Ammonia ($T_a = T_c = 35$°C, $T_{e_{vcr}} = -40$°C, $\varepsilon_{she} = 0.7$, $\eta_{comp} = 0.8$, $Q_{e_{vcr}} = 100$ kW, $A = 0$°C, $T_o = 298.15$ K)

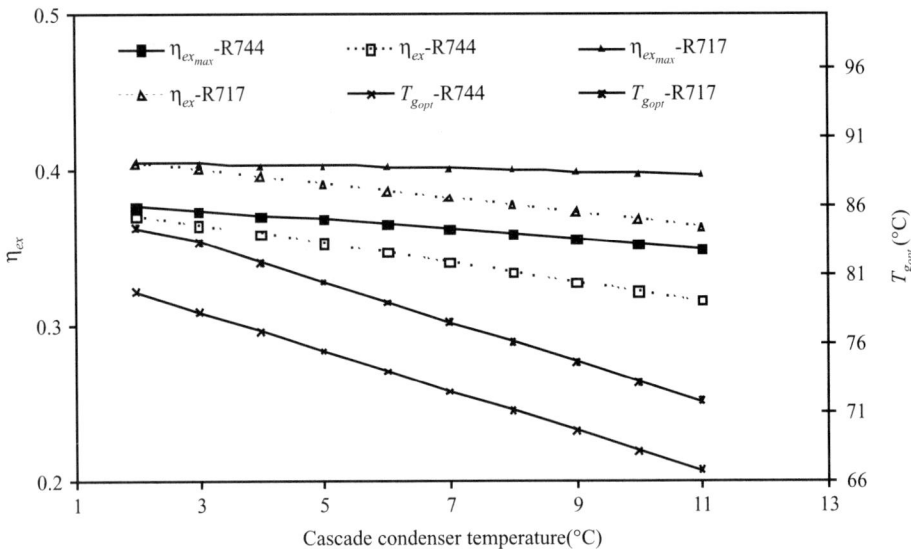

Fig. 4.16(c) Exergetic Efficiency versus Cascade Condenser Temperature and $T_{g_{opt}}$ Corresponding Maximum Exergetic Efficiency ($T_a = T_c = 35°C$, $T_{e_{vcr}} = -40°C$, $T_g = 85°C$, $\varepsilon_{she} = 0.7$, $\eta_{comp} = 0.8$, $Q_{e_{vcr}} = 100$ kW, $A = 0$ °C, $T_o = 298.15$ K)

efficiency. Figure 4.16(b) illustrates the comparison of maximum exergetic efficiencies of ammonia and carbon dioxide and corresponding optimum generator temperatures. The values of exergetic efficiency are also calculated for generator temperature equal to 85°C. It is observed that if generator temperature is not optimum then exergetic efficiency continues to decrease with increase in cascade condenser temperature. The maximum exergetic efficiency and corresponding generator temperature also decrease with increase in cascade condenser temperature. The maximum exergetic efficiency and optimum generator temperature both are lower for carbon dioxide cycle cascade system in comparison to ammonia cycle cascade system. The cascade condenser temperature corresponding to the optimum generator temperature is certainly optimum cascade condenser temperature. However, the maximum exergetic efficiency is not the global maximum exergetic efficiency rather it is localized maximum value of exergetic efficiency corresponding to given conditions.

4.4.3.2 Effect of Generator Temperature

Figure 4.17(a) and (b) illustrate the effect of generator temperature on COP and exergetic efficiency at different absorber temperatures and different evaporator temperatures. It is observed that with increase in generator temperature (for a specific absorber and specific evaporator temperature), COP increases initially, attains a maximum value and with further increase in generator temperature COP decreases marginally. The trend followed by exergetic efficiency curves is similar to trend followed by COP curves with the exception that the fall of exergetic efficiency is significant after it attains maximum value. At particular evaporator and cascade condenser temperatures, the compressor power required for a constant cooling load remains constant. This implies that COP of the low temperature stage is constant. Thus, the variation in COP of the cascade system is dependent on

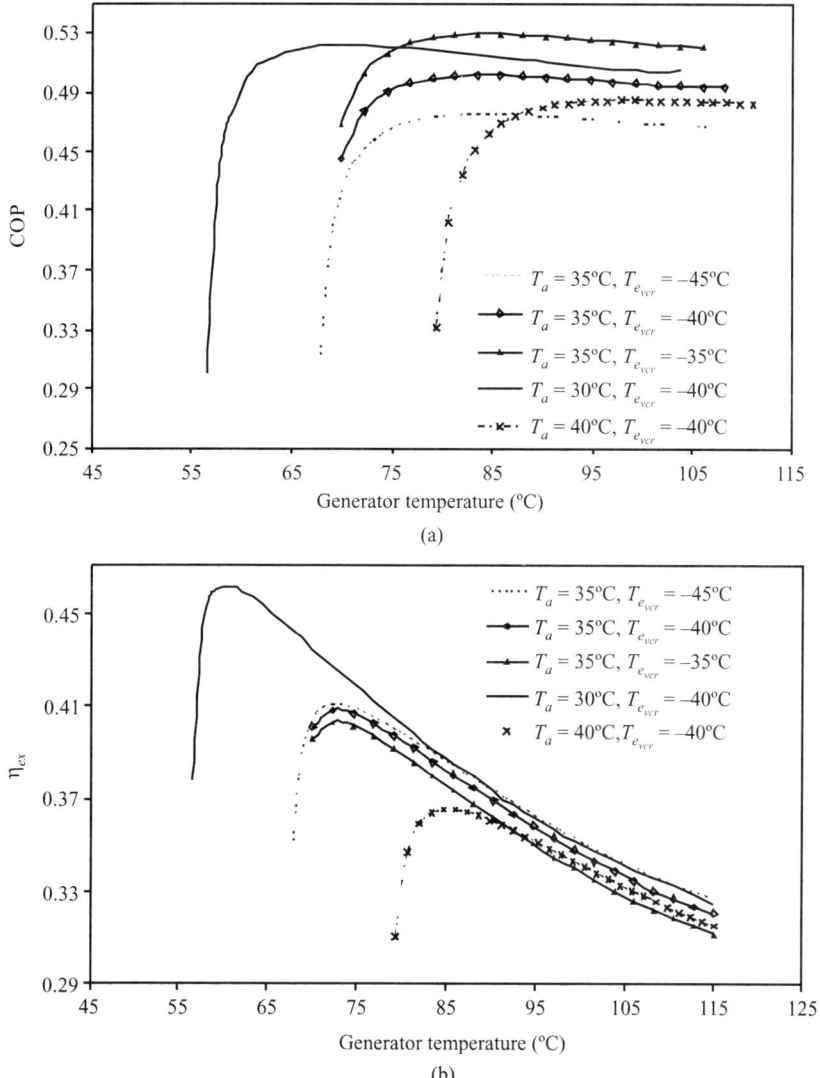

Fig. 4.17 Effect of Generator Temperature on (a) COP and (b) Exergetic Efficiency for Range of Evaporator and Absorber Temperatures (Refrigerant Ammonia, $T_a = T_c$, $T_{cc} = 7°C$, Effectiveness = 0.7, $\dot{Q}_{e_{vcr}} = 100$ kW, Compressor Efficiency = 0.8, $A = 0$ °C)

variations in COP of the high temperature stage, i.e., COP of the single effect absorption system. The increase in generator temperature causes the solution circulation ratio to reduce and the heat duty of the generator reduces. The reduction in generator heat duty increases the COP of the high temperature stage and consequently the COP of the cascade system increases initially. However, with further increase in generator temperature, the temperature difference between generator and sub-cooled solution entering the generator increases. Thus, the irreversibility in generator increases. The increase in generator temperature also accounts for increase in temperatures of weak solution and the refrigerant leaving the generator.

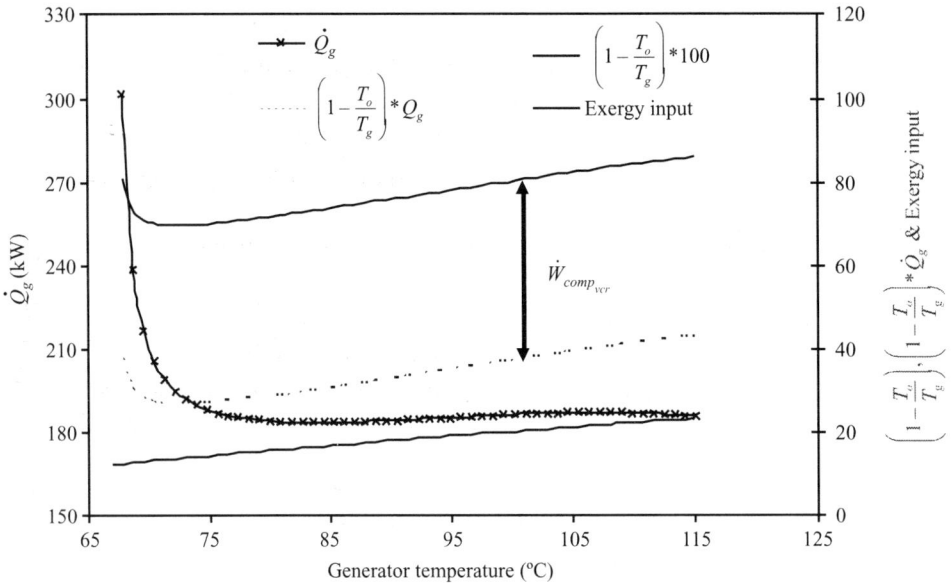

Fig. 4.17 (c) Variation of Parameters Factors which Constitute Input Exergy (Refrigerant Ammonia, $T_a = T_c = 35°C$, $T_{cc} = 7°C$, Effectiveness = 0.7, $\dot{Q}_{e_{vcr}} = 100$ kW, Compressor Efficiency = 0.8, $A = 0$ °C)

Thus, irreversibility in absorber and condenser increases, hence the positive effect of increase of generator temperature on COP of high temperature stage is counterbalanced by negative effects of increase in irreversibility in generator, absorber and condenser. Hence, COP of the cascade system becomes constant and even reduces marginally with increase in generator temperature. The reduction in generator heat duty brings down the exergy input initially due to fall in solution circulation ratio. The exergetic efficiency, therefore, increases and achieves a maximum value. The rate of decrease of solution circulation ratio reduces with further increase in generator temperature and therefore exergy input (given by $\dot{Q}_g \left(1 - \frac{T_0}{T_g}\right) + \dot{W}_p + \dot{W}_{c_{vcr}}$) increases whereas output exergy (given by $\dot{Q}_e \left| \left(1 - \frac{T_0}{T_r}\right) \right|$) remains constant and hence the exergetic efficiency drops sharply. The trend of various parameters which constitute input exergy is shown in Fig. 4.17 (c). Figs. 4.17 (a) and (b) also show that with increase in evaporator temperature the COP increases whereas exergetic efficiency decreases. The COP increases because increase in evaporator temperature reduces the pressure ratio across the compressor and hence the compressor power required reduces which, in turn, reduces the total input energy required. The exergetic efficiency reduces because of reduction in output exergy. The input exergy required also reduces however the rate of decrease in input exergy is lower in comparison to rate of decrease of output exergy. Figure 4.17(a) and (b) also show the effect of absorber temperature on COP and exergetic efficiency. The reduction in absorber temperature accounts for reduction in solution circulation ratio which reduces generator heat duty and increases COP. The reduction in generator heat duty also accounts for reduction in input exergy, it results in increase in exergetic efficiency. Simultaneously, the optimum generator temperature also reduces with reduction in absorber temperature. This indicates that an absorption compression cascade refrigeration system requires lower temperature heat source when absorber temperature is reduced.

Figure 4.18 (a) and (b) illustrates the variation of COP and exergetic efficiency for carbon-dioxide cascade system. The comparison of Figs. (4.17) and (4.18) illustrates that the COP and exergetic efficiency values for carbon-dioxide cascade cycle are lower in comparison to ammonia cascade cycle.

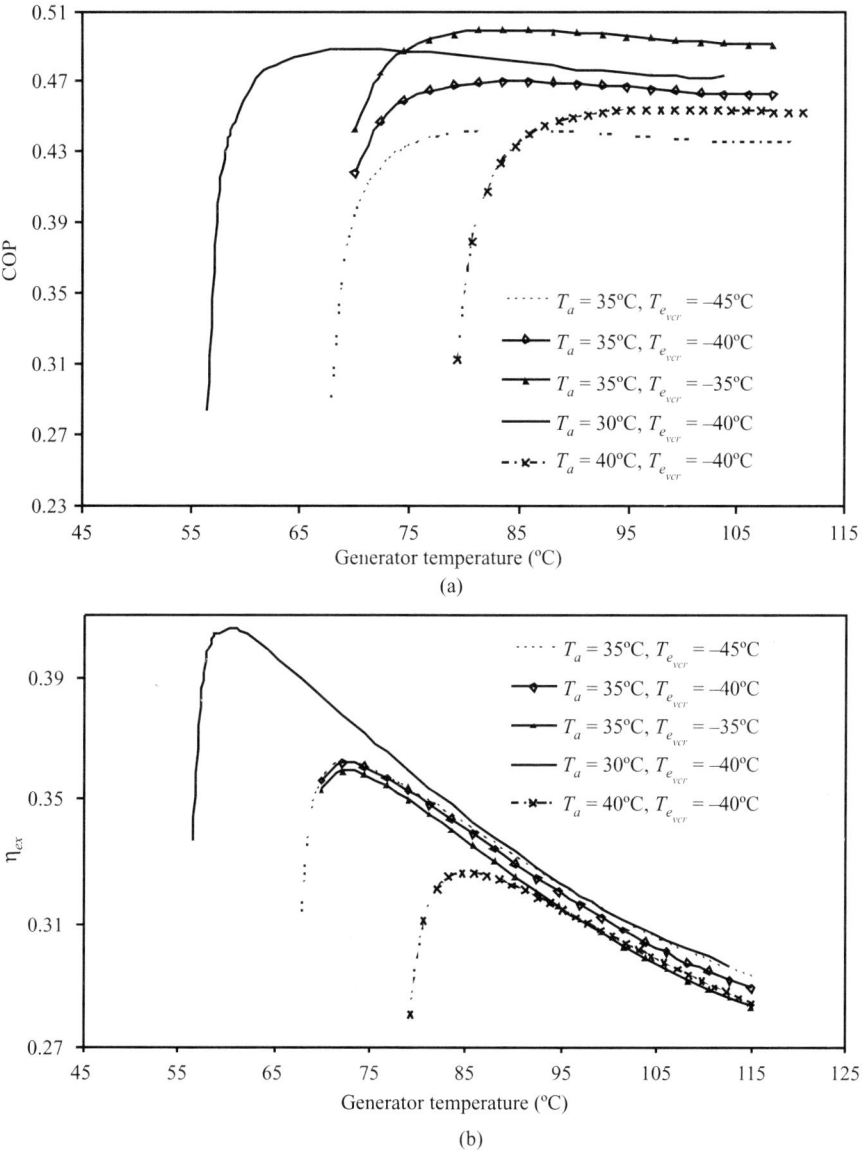

Fig. 4.18 Effect of Generator Temperature on (a) COP and (b) Exergetic Efficiency for Range of Evaporator and Absorber Temperatures (Refrigerant Carbon Dioxide, $T_a = T_c$, $T_{cc} = 7°C$, Effectiveness = 0.7, $\dot{Q}_{e_{vcr}}$ = 100 kW, Compressor Efficiency = 0.8, A = 0 °C)

4.4.3.3 *Effect of Effectiveness of Solution Heat Exchanger, Approach in Cascade Condenser and Efficiency of Compressor in Low Temperature Stage*

Figures 4.19 and 4.20 show the effect of effectiveness of solution heat exchanger, approach in cascade condenser and efficiency of compressor in low temperature stage on the COP and exergetic efficiency. The reduction in effectiveness of solution heat exchanger is responsible for decrease in COP and exergetic efficiency. The reduction in effectiveness of solution heat exchangers causes

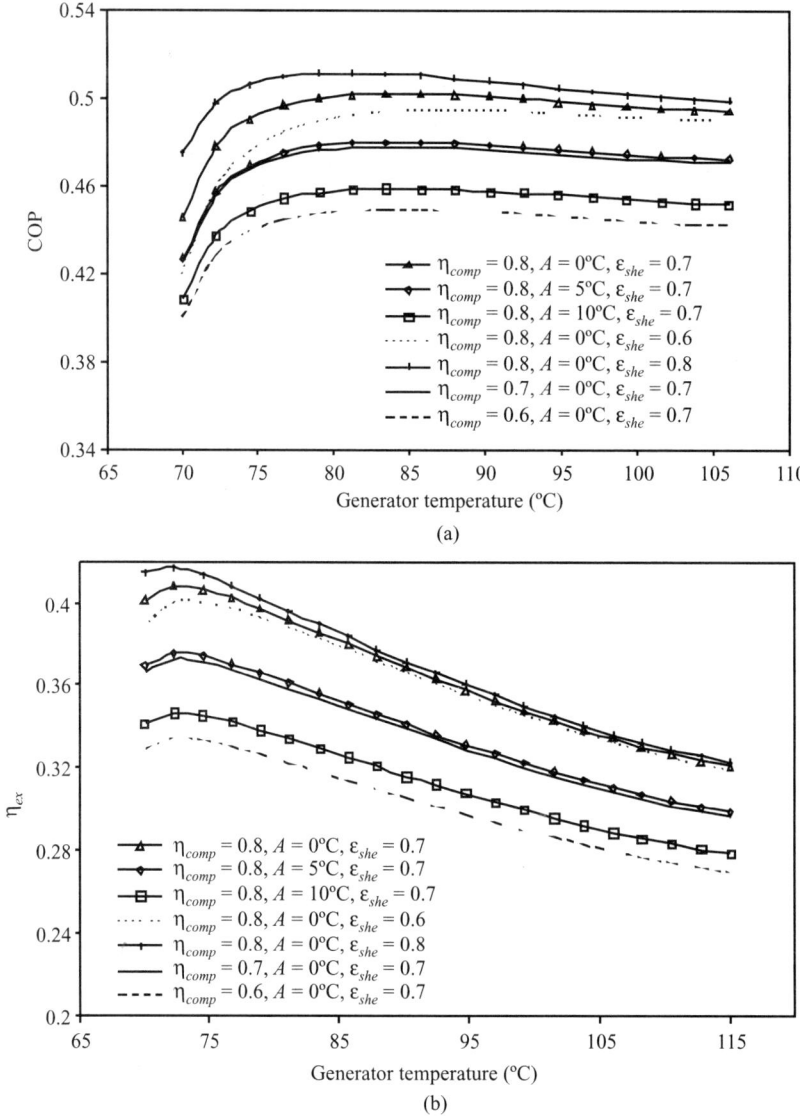

Fig. 4.19 Effect of Generator Temperature on (a) COP and (b) Exergetic Efficiency for Varying Approach, Effectiveness of 'she' and Compressor Efficiency (Refrigerant Ammonia, $T_a = T_c = 35°C, T_{cc} = 7°C, T_{e_{vcr}} = -40°C, \dot{Q}_{e_{vcr}} = 100 \text{ kW}$)

the temperature difference to increase between the strong solution leaving the solution heat exchanger and the generator temperature. Thus, the heat duty and irreversibility in generator increase. Moreover at lower values of effectiveness of solution heat exchanger, the temperature difference between

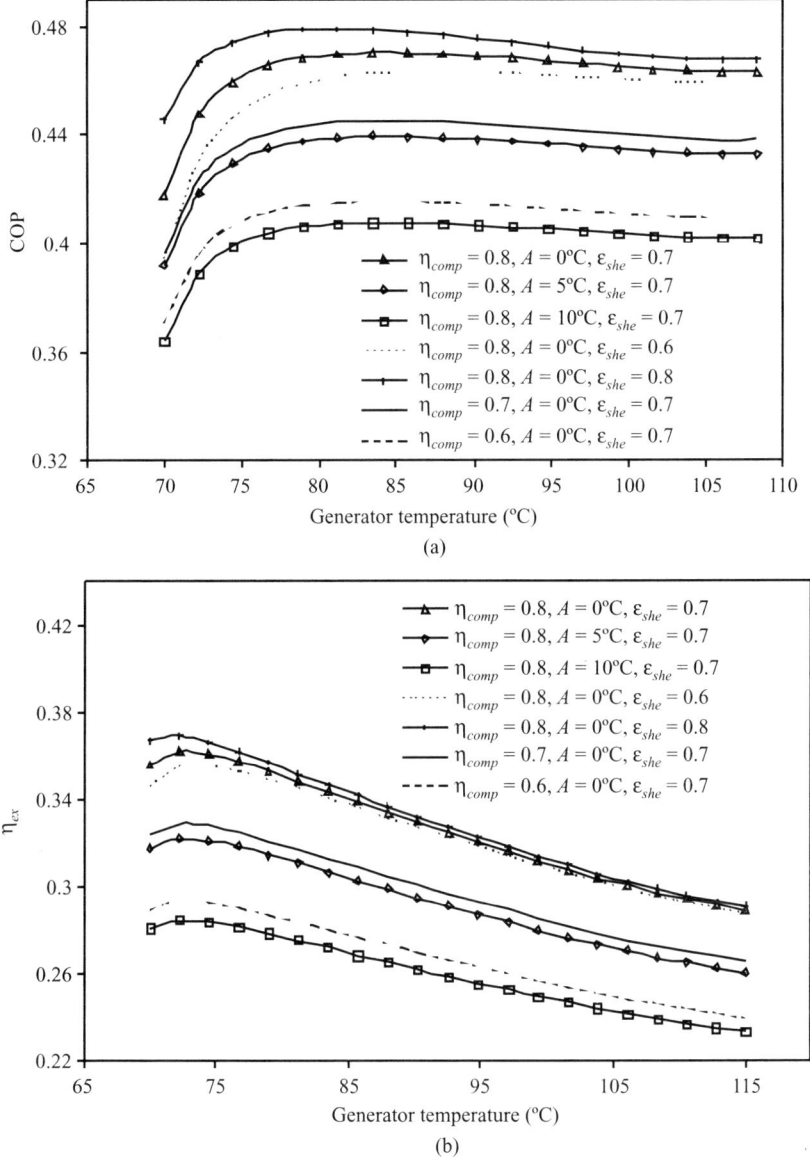

Fig. 4.20 Effect of Generator Temperature on (a) COP and (b) Exergetic Efficiency for Varying Approach, Effectiveness of 'she' and Compressor Efficiency (Refrigerant Carbon Dioxide,

$T_a = T_c = 35°C$, $T_{cc} = 7°C$, $T_{e_{vcr}} = -40°C$, $\dot{Q}_{e_{vcr}} = 100$ kW)

entering and leaving streams of strong and weak solutions is larger in comparison to when the effectiveness is higher. Thus, at lower values of effectiveness, the irreversibility in solution heat exchanger is more. The above factors lead to increase the total exergy destruction in the cycle and the COP and exergetic efficiency reduce.

The increase in approach from 0°C to 10°C causes the absorber pressure to increase which increases the solution circulation ratio. The increase in solution circulation ratio increases the generator heat duty which brings down the COP. It also increases the irreversibility in generator and cascade condenser. Thus, total exergy destruction increases and hence exergetic efficiency reduces.

The increase in isentropic efficiencies of compressor reduces the compressor power required and hence input energy and input exergy reduce thereby increasing the COP and the exergetic efficiency.

4.5 CONSCLUSION

This chapter highlights the concepts and analysis of three combined compression absorption systems, i.e., absorption recompression refrigeration system, compression absorption refrigeration system and compression absorption cascade refrigeration system. The combined systems seem to be very promising on the basis of their high COP values lying in the range of 3 to 4 for absorption recompression refrigeration system and 3.5 to 7.5 for compression absorption refrigeration system. However, the exergetic efficiency values offered by the above-mentioned systems are lower than the conventional single effect vapour absorption refrigeration system. The compression absorption compression cascade refrigeration system is a novel concept in refrigeration systems and its analysis shows that though the exergetic efficiency of compression absorption cascade system is almost same as that of compression cascade system yet its primary energy (electrical energy) consumption is much lower than that of compression cascade system.

Nomenclature

A	Approach (K, °C)
ARR	Absorption Recompression Refrigeration
CAR	Compression Absorption Refrigeration
comp	Compressor
CO_2	Carbon dioxide
COP	Coefficient of performance (Non dimensional)
ED	Rate of exergy destruction (kW)
EF	Exergy rate of fuel (kW)
EL	Exergy loss rate (kW)
EP	Exergy rate of product (kW)
h	Specific enthalpy (kJ kg^{-1})
LiBr	Lithium bromide
m	Mass flow rate (kg s^{-1})
P	Pressure (kPa)

Q	Heat transfer rate (kW)
s	Specific entropy (kJ kg^{-1}K^{-1})
she	Solution heat exchanger
SCR	Solution circulation ratio (non dimensional)
T	Temperature (K)
VAR	Vapour absorption refrigeration
VCR	Vapour compression refrigeration
W	Work transfer rate (kW)
X	Mass fraction of lithium bromide in solution

Greek

η	Efficiency
δ	Efficiency defect
ε	Effectiveness

Subscripts

0	Represents dead state
a	Absorber
c	Condenser
cc	Cascade condenser
$comp$	Compressor
e	Evaporator, exit
e_{vcr}	Evaporator of VCR system
ex	Exergetic
g	Generator
htc	High temperature cycle
i	Inlet, represents any component of the system under consideration
in	Input
ltc	Low temperature cycle
out	Output
p	Pump
r	Refrigerant; space to be cooled
rtv	Refrigerant throttle (expansion) valve
s	Strong
she	Solution heat exchanger
stv	Solution throttle (expansion) valve
$total$	Addition of all

r_{vcr}	Refrigerant in vapour compression refrigeration system
rtv_{htc}	Refrigerant throttle valve of high temperature circuit
rtv_{ltc}	Refrigerant throttle valve of low temperature circuit
w	Weak
1,2,3,……	State points

REFERENCES

Fernandez-Seara J., Sieres, J., Vazquez M, 2006. Compression-absorption cascade refrigeration system. *Applied Thermal Engineering* 26, 502-512.

Hulte´n M. and Berntsson T., 1999. The compression/absorption cycle-influence of some major parameters on COP and a comparison with the compression cycle. *International Journal of Refrigeration* 22, 91-106.

Klein, S.A., Alvarado, F., 2005. *Engineering Equation Solver*, Version 7.441. F-Chart software, Middleton, WI.

Pa´tek, J., Klomfar, J., 2006. A computationally effective formulation of the thermodynamic properties of water-lithium bromide solutions from 273 to 500 K over full composition range. *International Journal of Refrigeration* 29, 566-578.

Pratihar, A.K., Kaushik S.C. and Agarwal R.S., 2001. Thermodynamic modeling and feasibility analysis of compression-absorption refrigeration system. Proceedings of the International Conference on Emerging Technologies in Air-conditioning and Refrigeration, Delhi, India, 207-215.

Riffat, S. B. and Shankland, N., 1993. Integration of absorption and vapour-compression systems. *Applied Energy* 46(4), 303-316.

Herold K E., Howe L. A. and Radermacher R., 1991, Analysis of a hybrid compression-absorption cycle using lithium bromide and water as the working fluid. *International Journal of Refrigeration* Vol. 14, 264-272.

Kairouani L., Nehdi E. 2006. Cooling performance and energy saving of a compression-absorption refrigeration system assisted by geothermal energy. *Applied Thermal Engineering*. 26, 288-294.

Zhou Q. and Radermacher R. 1997. Development of a vapor compression cycle with a solution circuit and desorber/absorber heat exchange. *International Journal of Refrigeration*, 20(2), 85-95.

Tarique S. M, Siddiqui M. A. (1999). Performance and economic study of the combined absorption/compression heat pump. *Energy Conversion & Management* 40, 575-591.

Kim J.S., Ziegler F., Lee H. (2002). Simulation of the compressor-assisted triple-effect H_2O/LiBr absorption cooling cycles. *Applied Thermal Engineering*. 22, 295-308.

Garimella S., Brown A. M., Nagavarapu A.K.(2011). Waste heat driven absorption/vapor-compression cascade refrigeration system for megawatt scale, high-flux, low-temperature cooling, *International Journal of Refrigeration*, 34,1776-1785.

Brunin O., Feidt M. and Hivet B. 1997. Comparison of the working domains of some compression heat pumps and a compression-absorption heat pump. *International Journal of Refrigeration*. 20(5), 308-318, 1997.

Bourouis M., Nogue M., Boer D., Coronas A. (2000). Industrial heat recovery by absorption/compression heat pump using TFE-H2O-TEGDME working mixture. *Applied Thermal Engineering* 20, 355-369.

Pratihar A.K., Kaushik S.C., Agarwal R.S.. (2010). Simulation of an ammonia water compression absorption refrigeration system for water chilling application. International Journal of Refrigeration, 33, 1386-1394.

Cimsit C., Ozturk I.T. (2012). Analysis of compression absorption cascade refrigeration cycles *Applied Thermal Engineering* 40, 311-317.

CHAPTER 5

Use of Ejector in Refrigeration and Air Conditioning

5.1 INTRODUCTION

There are many sources from which low grade thermal energies are available directly or in the form of waste heat such as automobile engines, air conditioners, refrigerators, fuel cell stacks, geothermal, industrial processes and solar radiations, etc. These low grade energies can be effectively recovered through heat operated refrigeration systems using ejectors. It will not only reduce problem related to ozone depletion but also help in reducing the problem of global warming because these systems do not involve the use of electrical energy. In the past, ejector based refrigeration systems were not paid much attention by the scientific community. But the Montreal Protocol, Kyoto Protocol, Earth Summit and similar other efforts and agreements at world level, once again attracted lot of researchers towards research on ejector refrigeration systems due to its simple design and being heat driven systems. The ejector refrigeration system comprises (1) evaporator, (2) condenser, (3) ejector, (4) generator, (5) pump and (6) expansion valve. The ejector refrigeration system is different from vapour compression refrigeration system as the mechanical compressor is replaced by three components, viz., ejector, generator and pump which act as a thermo-compressor. Since the mechanical compressor is replaced by thermo-compressor therefore this system can be easily operated using low grade thermal energy, viz., solar energy, waste heat, geothermal energy, etc. When water is used as a refrigerant then ejector refrigeration system is called steam jet refrigeration system. Steam jet refrigeration system is mostly used for air conditioning because it is not possible to produce temperatures below 0°C (freezing point of water). To obtain temperatures below 0°C refrigerants like R-123, R-142b, R-134a and natural refrigerants like ammonia, carbon dioxide, etc., can be used. The ejector refrigeration technology is one of the reliable refrigeration technologies because of the low installation and operating costs and low noise and vibrations. However, their growth was hampered in past because of very low COP values in comparison to compression and absorption refrigeration technologies. The low performance of these systems is due to low entrainment performance of ejector which depends on both operating parameters and geometry of the ejector. Further, if the ejector refrigeration system without any use of mechanical pump could be developed then the problems associated with pump and additional devices coming with pump could be avoided.

In recent past many researchers have focused on improving the energy efficiency of these systems by (i) improving the design of ejector, (ii) choosing a suitable refrigerant, (iii) the optimizing operating parameters and (iv) adding different components like pre-cooler and regenerator to the basic system. The use of ejector integration in other types of refrigeration and air conditioning systems to enhance their performance and to make them eco-friendly and affordable for vast applications, is another area where researchers have achieved lot of success and still it is the promising area which needs extensive research. The present chapter deals with all these issues.

5.2 HISTORICAL BACKGROUND OF EJECTOR AND ITS APPLICATIONS

The history of ejector goes back to 1838 when a patent was granted to the Frenchman Pelletan for compressing the steam using jet of motive steam. Later, in 1858, Henry Giffard developed injector using the principle of condensation and used it for replenishing the steam engine boiler reservoir with water. In 1869, an engineer named Schau introduced the converging-diverging motive nozzle. The first supersonic steam nozzle experiments was carried out by de Laval in 1890. Since twentieth century a number of studies have been carried out on use of ejectors in various applications. In 1901, an Englishman Charles Parsons utilized the ejector's vacuum creating capability for removing non- condensable gases present in steam condensers. In 1910, Maurice Leblanc, a French engineer, developed steam jet ejector refrigeration system (SJERS) for producing steam jet having velocities of the order of 1200 m/s. Later in 1910 Westinghouse designed the first commercial ejector system in Paris. From 1910 onwards, the SJERS became famous in air conditioning of large buildings, breweries, chemical factories, railroad cars, and warships, etc. Presently, SJERS are being utilized to used to harness solar energy or other low-grade energy sources. In 1926, the French engineer Follain improved the steam jet refrigeration machine by introducing multiple stages of vaporization and condensation of the suction steam. Gay (1931) patented a two-phase ejector and used it for minimizing throttling losses of the expansion valve in refrigeration systems. With the use of two phase ejector, the specific refrigerating effect increases, compression work reduces because of reduced compression ratio thereby increasing COP of the system. Around 1955, the first closed vapour jet refrigeration system was developed by I.S. Badylkes. Addy et al. (1981) conducted research on ejectors used in high energy chemical lasers. Lorentzen (1983) presented a transcritical CO_2 refrigeration cycle using an ejector. Huang et al. (1985) conducted experimental studies on vapour jet ejector system using R113. The cooling capacity and COP varied between 0.4-2.2 kW and 0.02-0.26 respectively. Kornhauser (1990) presented a 1-D iterative model with two-phase ejector using R12 as working fluid and showed that theoretical COP improved upto 21% in comparison to the conventional cycle with expansion valve. Sokolov and Hershgal (1990a) discussed conventional and improved ejector refrigeration cycles. Special emphasis is given to the adaptation of these cycles for the utilization of low grade or waste heat. A compression enhanced ejector system was suggested as a mechanically efficient way to improve the ejector cycle. It was demonstrated that a combination of mechanical and thermal energies may provide a wide range of design alternatives which should yield a competitive refrigeration system. This research paper provides an over-all view of the systems by discussing their principle of operation, expected performance and design considerations. Sokolov and Hershgal (1990b) described the development of procedures which will enable system design, optimization and control of operation. Special attention was given to the ejector design and the recommended modification of conventional system components such as the evaporator and generator. A multi-ejector system was introduced in order to expand the range in which the system may operate efficiently. Sokolov and Hershgal (1991) designed and constructed an experimental, double ejector, compression enhanced refrigeration cycle. Their results indicated that the system is capable of smooth adaptive operation when the demands and environment vary within the design limitation. Many of the two-phase ejector models available in the open literature are based on this numerical approach. Kornhauser's study triggered intensive ejector research effort in his workgroup. Several of his students worked on improving the ejector after initial experimental results obtained by Menegay (1991) showed COP improvements of only a few percent. Eames et al. (1995) described an experimental refrigerator designed on the jet-pump cycle and tested it for boiler and evaporator temperatures varying between 120-140°C and 5-10°C respectively and obtained COP values greater than 0.5. Tomasek and Radermacher (1995) numerically investigated how an ejector can be utilized to improve the performance of a domestic household refrigerator-freezer having two evaporators. Their results showed COP enhancement up to 12% in comparison to conventional refrigerator-freezer systems. Domanski (1995) concluded that the ejector efficiency was very sensitive to the theoretical COP of the ejector expansion refrigeration cycle. Harrell et al. (1995) used a R-134a two-phase ejector and the test rig to estimate the COP of the refrigeration cycle. It was found that the COP improvement ranged from 3.9% to 7.6%. Menegay et al. (1996) developed a bubbly flow tube to reduce the thermodynamic non-equilibrium in the motive nozzle with R-12 as the refrigerant. This device was installed upstream of the motive nozzle. The COP of the system using the bubbly flow tube can be improved up to 3.8% over the conventional cycle under standard conditions. However, they reported that the result was not as good as was expected and study of the ejector expansion refrigeration cycle should be extended. Sun (1998) described a novel refrigeration cycle based on the integration of ejector and vapor compression cycles. The results showed that the new cycle has a significant increase in system performance over the conventional systems, and has similar COP values as that of absorption systems. The combined cycle can operate with a single refrigerant or dual refrigerants. Nakagawa et al. (1998) showed that the longer divergent part provided a longer period of time for the two-phase flow to achieve equilibrium. He concluded that the longer the length of divergent part of the motive nozzle, the higher motive nozzle efficiency could be obtained. Riffat and Holt (1998) presented a numerical study to simulate the performance of a heat pipe with an integrated ejector. Their cycle was actually a modification of the vapour jet ejector cycle, but instead of using a pump, capillary action and a wick were used to transport the liquid to the vapour generator. A basic one-dimension modeling approach yielded COPs of up to 0.7 with environmentally friendly working fluids such as methanol. Huang et al. (1985) and Garris et al. (1998) cited work carried out on automotive air conditioning systems using the hot exhaust gases from the combustion engine as an energy source. In a subsequent study, Huang et al. (1998) were able to show experimentally that the COP of a vapour jet ejector cycle is comparable to that of absorption systems. They also showed that the ejector performance decreased for unchoked flow conditions. According to Chen et al. (1998), working fluids for an ejector refrigeration cycle can be categorized as wet vapour (R11, R12, R134a or Steam) and dry vapour (R113, R123 or R141b). For the dry vapour fluid, there is no phase change during the expansion process through the converging–diverging nozzle of the ejector. On the other hand, the wet vapour fluid will partly condense in the nozzle and small droplets may be formed. Sherif et al. (1998) presented thermodynamic and economic analyses of a steam-jet refrigeration system. The COP was found to increase with an increase in the evaporator temperature and decrease in condenser temperature, whereas the total irreversibility increased with an increase in the condenser temperature and decrease in evaporator temperature. The COP of a steam-jet system ($\approx 0.1 - 0.3$) was found to fall a little short when compared to vapour compression ($\approx 2 - 3$) and vapour absorption ($\approx 0.5 - 1$) systems. Sun (1999) compared the theoretical performance of eleven refrigerants used in an ejector refrigeration system. These refrigerants include water (R718), halocarbon compounds, i.e., CFCs (R11, R12, R113), HCFCs (R21, R123, R142b) and HFCs (R134a, R152a), a cyclic organic compound (RC318), and an azeotrope (R500). The results showed that a steam jet refrigeration cycle has the lowest COP value. For CFCs, R12 gives better performance; for HCFCs, R142b gives high COP value; the HFC refrigerants tested have comparative performance, with R152a giving the best performance among all the other refrigerants. Huang et al. (1999) carried out a 1-D analysis for the prediction of ejector performance at critical-mode operation by assuming constant-pressure mixing and entrained flow at choking condition. They conducted experiments to determine the coefficients, η_p, η_s, ϕ_p and ϕ_m defined in the 1-D model by matching the experimental data with the analytical results. They had shown that the 1-D analysis using the empirical coefficients could accurately predict the performance of the 11 ejectors using R141b as the working fluid. Huang and Chang (1999) further pointed out that the vapour jet cycle does not require any lubrication, thereby reducing the negative impact on performance caused by lubrication oil in conventional vapour compression systems. Eames et al. (1999) examined the effects of ejector geometry on the performance of steam jet-pump refrigerators, using two primary nozzles and three diffusers with mixing chamber. It is seen from their experimental results that the entrainment ratio increases almost linearly with the ejector ratio area, if the primary pressure ratio (p_g/p_c) and the ratio of the primary nozzle exit area to throat area (A_{ne}/A_{nt}) are held constant. Water or steam driven ejectors are being used to provide emergency cooling water to nuclear reactors, as reported by Beithou and Aybar (2000). Chang and Chen (2000) used a petal nozzle to enhance the performance of a steam-jet refrigeration system. According to their experimental results, when the system is operated at larger area ratios, the performance of system with a petal nozzle is better than that with a conical nozzle. Eames (2002) introduced a new method for designing ejectors to be used in ejector refrigeration systems. It is assumed in the method that the momentum of flow changes at a constant rate within the diffuser passage of a supersonic ejector. The theoretical method produces a diffuser geometry that removes the thermodynamic shock process within the diffuser at the design-point operating conditions. Elgozali et al. (2002) investigated a gas-liquid reactor with an ejector-type gas distributor. In this application, the ejector was basically used to enhance the desired mixing process of two different fluid streams. Butrymowicz (2003) chose a different approach in his efforts to model two-phase ejector systems. He argued that the iterative one-dimensional modeling routine suggested by Kornhauser (1990) did not explicitly take into account mixing shock waves. Therefore, he constructed an ejector performance curve by relating the suction pressure ratio to the mass entrainment ratio. More research on two-phase ejector systems was reported by Takeuchi et al. (2004). For the transportation refrigeration system investigated, it was claimed that the use of a two-phase ejector simultaneously improved the cooling capacity and COP by 25% to 45% and 45% to 65%, respectively. Ozaki et al. (2004) presented more details related to the same two-phase ejector research efforts. Their study included the presentation of limited experimental results showing COP improvements of 20% over conventional transcritical R744 automotive systems with expansion valve at an outdoor temperature of 35°C. Chunnanond and Aphornratana (2004a) contributed a comprehensive review paper in which they gave a detailed overview of vapour jet ejector refrigeration systems. They summarized basic theories regarding fundamental ejector flow features and discussed important design issues. Chunnanond and Aphornratana (2004b) examined the effects of the nozzle geometry and position on the performance of a steam ejector refrigerator with a conical mixing chamber. Based on their tests, they expressed that decreasing the generator pressure, using a nozzle with smaller throat area (hence higher area ratio) and retracing the nozzle out of the mixing chamber can increase the COP and cooling capacity of the refrigerator, provided that the critical condenser pressure is decreased. Elbel and Hrnjak (2004a) also used Kornhauser's approach to numerically investigate the effect of using an IHX on the performance of a transcritical R744 ejector system. They showed that the highest COPs can be achieved with ejector and IHX. Hernandez et al. (2004) carried out a global thermodynamic study of a hybrid compressor and ejector refrigeration system for a refrigeration application.

Arbel and Sokolov (2004) described a compression-enhanced ejector system capable of generating air conditioning by utilizing solar collectors as a heat source. All the components of the system are off-the-shelf items that are readily available in the air conditioning components marketplace. Like any other solar system, this one is not complete without some storage. Due to the low thermal efficiency of the system, it is wasteful to store large amounts of heat. Cold storage is therefore the only sensible alternative that may be achieved by phase changing materials, cold water, or ice. The results indicated that this system with R-142b is more efficient than the one operating with R-114. Thus using R-142b not only offers a "greener" system but also a better one. The efficiency of such system may be improved by raising the generator temperature. This may require a higher class of flat-plate collectors or even the usage of concentrating collectors. Although such an improvement might be appreciably high, it should nevertheless be realized that the maximum generator temperature is limited by the fact that the refrigerants tend to lose their chemical stability at elevated temperatures. A combined compression-enhanced ejector system, in which solar space-heating, air-conditioning, and hot water are produced, with moderate condensing temperatures, is a very feasible system. It can be used for more than one season, and the size of the collectors has been reduced, making this a very economical system. Li and Groll (2004, 2005) presented simulation results of transcritical R744 air conditioning systems with a two-phase ejector. Their analysis was also based on Kornhauser's approach. COP improvements of up to 16% were reported. No internal heat exchangers were included in the systems investigated. They offered a potential solution to a problem they foresaw in regard to controlling the system. A cycle was suggested in which part of the vapour coming from the vapour-liquid separator was injected to the liquid entering the evaporator. They claimed this was necessary to relax the existing constraint between the entrainment ratio and the vapour quality at the exit of the diffuser. Li and Groll (2005) proposed an ejector expansion transcritical CO_2 refrigeration cycle to improve the COP of the basic transcritical CO_2 cycle by reducing the expansion process losses. A constant pressure-mixing model for the ejector was established to perform the thermodynamic analysis of the ejector expansion transcritical CO_2 cycle. The effect of the entrainment ratio (0.50 to 0.70) and the pressure drop in the receiving section of the ejector (0.01 MPa to 0.05 MPa) on the relative performance of the ejector expansion transcritical CO_2 cycle was investigated for typical air conditioning operation conditions. The effect of different operating conditions (gas cooler temperature, 35°C to 49°C; gas cooler pressure, 8 to 13 MPa; and evaporator temperature, 0°C to 20°C) on the relative performance of the ejector expansion transcritical CO_2 cycle was also investigated using assumed values for the entrainment ratio and pressure drop in the receiving section of the ejector. They found that the COP of the ejector expansion transcritical CO_2 cycle could be improved by more than 16% over the basic transcritical CO_2 cycle for typical air conditioning operation conditions. Bartosiewicz et al. (2005) reported about multi-stage ejectors being used to simulate aerospace altitude testing of equipment by reducing test chamber pressures. They further mentioned the use of ejectors in aircraft propulsion systems for thrust augmentation purposes and to reduce the thermal signature of the exhaust gases. In 2003, the Denso corporation from Japan introduced a hot water heater using a transcritical R744 heat pump with two-phase ejector to the Japanese market.

Alexis (2005) studied in detail the irreversibilities in the 100 kW steam-ejector refrigeration system using frist and second law analysis. According to him a better quality of the ejector has more effect on the system performance than the better quality of other components, because the ejector at first and the condenser at second have the greater exergy loss of the system. The analysis showed that the exergy loss of the system is 53.2 kW and also the ejector and condenser had the greater exergy loss than other components, 54 and 27% of the total exergy loss of the system. In a subsequent publication by Li (2006), experimental R744 ejector data were presented as well, although the emphasis of the work was on modeling a transcritical R744 two-phase ejector system and to study the effects of different geometries and operational conditions. Interestingly, it was concluded that for ambient temperatures of more than 49°C, the ejector would not be capable of improving the performance in comparison to that of the baseline system with expansion valve. Li and Groll (2006) developed and validated with experimental results of a two-phase flow ejector expansion device model. They found that the motive nozzle expansion process had an isentropic efficiency of 95% but the suction nozzle had a very low isentropic efficiency of 26%. They also performed parametric studies to investigate the effects of ejector design parameters on the system performance. They concluded that when both systems have the same gas cooler, evaporator and compressor, the ejector expansion system could have 11% higher COP and a 9.50% higher cooling capacity over the basic system for the given design of a U.S. Military ECU (Electronic Control Unit) and operating conditions. Li (2006) reasoned that at these high ambient temperatures the entrainment ratio would drop significantly. This finding appears somewhat counter-intuitive, because the expansion work recovery potential should actually increase with increasing ambient temperature for otherwise unchanged conditions. Elbel and Hrnjak (2006b) used temperature-specific entropy diagrams to visualize the interference between expansion work recovery and internal heat exchange.

Selvaraju and Mani (2006) described experimental investigations of the performance of a 0.5 kW R134a vapour ejector refrigeration system. The operating conditions are chosen accordingly as, generator temperature between 338 K and 363 K, condenser temperature between 299 K and 310.5 K, and evaporator temperature between 275 K and 285.5 K. Six configurations of ejectors of different geometrical dimensions were selected for the parametric study. It was seen that the ejector with higher area ratio exhibited better performance. Sankarlal and Mani (2007) evaluated experimentally variation of performance of the ammonia ejector refrigeration by varying the operational parameters. The entrainment ratio increases with an increase in expansion and area ratios and it increases with decreases in the compression ratio. COP increases with increase in expansion ratio and area ratio and it also increases with decrease in compression ratio. Though diameters of nozzle and mixing chambers are different, performance of the ejector refrigeration system depends on area ratio alone. Elbel and Hrnjak (2007) presented the first high-side pressure control equation used to maximize the COP of a transcritical R744 two-phase ejector system. Furthermore, they identified the existence of mixing shock waves which they detected through static wall pressure distributions along the axis of the ejector. Nehdi *et al.* (2007) numerically investigated the performance of a vapour compression system using a two-phase ejector instead of an expansion valve. Among the fluids considered, R141b, yielded the highest COP improvements, 22%, over a comparable baseline system with expansion valve. They also studied the effect of the geometry of the ejector section ratio and the fluid nature in a new refrigeration cycle that combines an ejector cycle and compression cycle. They found that the geometric parameters of the ejector design have considerable effects on the system's performance. The maximum COP is obtained for optimum area ratio whose value is around 10. For the considered refrigerants, it had been observed that the best coefficient of performances are obtained with R141b and R408a. The COP values are 4.90 and 4.60 respectively at optimum area ratio for these refrigerants at given operating conditions. The COP of the improved cycle was found to be about 22% higher than the standard cycle. Deng *et al.* (2007) described a theoretical analysis of a transcritical CO_2 ejector expansion refrigeration cycle that uses an ejector as the main expansion device instead of an expansion valve. They found that system performance is strongly coupled to the ejector entrainment ratio that must produce the proper CO_2 quality at the ejector exit. Zha *et al.* (2007) presented a design and parametric investigation on ejector for R744 transcritical system. They found that the key point in the nozzle design is the parameters at throat of nozzle. They set the throat pressure to be the spinodal pressure to calculate the sonic velocity, which is also the CO_2 fluid velocity at the throat of nozzle. Yari and Sirousazar (2007) considered a new configuration of the ejector-vapour compression refrigeration cycle, which used an internal heat exchanger and intercooler to enhance the performance of the cycle. They found that the COP and second law efficiency values of the new ejector-vapour compression refrigeration cycle were on average 8.60 and 8.15 per cent higher than that of the conventional ejector-vapour compression refrigeration cycle with R125. They showed that the COP of the new ejector-vapour compression cycle is 21% higher than that of the conventional vapour compression cycle. Later, Elbel and Hrnjak (2006a) and Elbel and Hrnjak (2008) were able to verify their numerical predictions by experimental results. Their prototype was equipped with a needle extending into the throat of the motive nozzle, allowing for high-side pressure control. In comparison to the expansion valve system, their ejector setup simultaneously improved the COP and cooling capacity by 7% and 8%, respectively. Yu *et al.* (2008) proposed the integration of an ejector in the autocascade refrigeration cycle to improve the cycle performance. Theoretical computation model based on the constant pressure mixing model for the ejector was used to perform a thermodynamic cycle analysis for this novel autocascade refrigeration cycle with the refrigerant mixture of R23/R134a. For this novel autocascade refrigeration cycle operated at the condenser outlet temperature of 40°C, the evaporator inlet temperature of –40.3°C, and the mass fraction of R23 is 0.15, the pressure ratio of the ejector reaches to 1.35, the pressure ratio of compressor is reduced by 25.8% and the COP is improved by 19.1% over the conventional autocascade refrigeration cycle. Yapici *et al.* (2008) studied experimentally the performance of the R-123 ejector refrigeration system using six configurations of ejector with an axially movable primary nozzle and cylindrical mixing chamber at operating conditions with choking in the mixing chamber. The condenser pressure was chosen so that the secondary flow choking can occur even in the ejector with the smallest area ratio. The study was performed over a range of the ejector area ratio from 6.5 to 11.5 at the compression ratio 2.47. In the studied range, the experimental coefficient of performance of the system rises from 0.29 to 0.41, as the optimum generator temperature increases from 83° to 103°C. Similar results were also found in the parametric study when the efficiencies of the nozzle and diffuser are taken as 0.90. Chaiwongsa and Wongwises (2008) experimentally investigated the performance of the R-134a refrigeration cycle using a two-phase ejector as an expansion device. Motive nozzles having different outlet diameter (2.0, 2.5 and 3.0 mm) with inlet diameter, inlet length, convergent length, throat diameter and divergent length as 6, 32, 6, 0.9, 20 mm, respectively were taken for study. The variation of the heat source temperature has no significant effect on the primary mass flow rate. The results showed that the primary mass flow rate, secondary mass flow rate, recirculation ratio, compressor pressure ratio, and discharge temperature varyies directly with the heat sink temperature. On the other hand, the cooling capacity varies inversely with the heat sink temperature. Grazzini and Rocchetti (2008) investigated six objective functions as optimisation criteria of the design data of a steam ejector cycle, keeping the same boundary conditions (same cooling power, fluid, materials and source temperatures) and convergence limits and using a numerical optimisation of the cycle published by the authors. The six objective functions were COP, the ratio of COP of the cycle to the heat transfer area or the volume size of the heat exchangers, the second law or the exergy efficiency assuming condenser temperature as the reference state, Life cycle analysis (that considers the balance between the exergy output and the exergy used over the entire useful life of the system, including the "embodied energy" focused, in this case, on the exergy costs of the materials used for heat exchangers), and the entropy production of the thermodynamic universe. The comparison shows that the choice of the objective function decisively influences the robustness of the numerical code results and the convergence performances of the code. Using a direct search method like the Complex method, the objective function that shows better optimisation performance also allows better convergence efficiency. Angelino and Invernizzi (2008) theoretically investigated a cold generation system featuring a Rankine cycle powered refrigeration cycle actuated by a supersonic ejector in view of the thermo-fluid-dynamic optimization of the working fluid characteristics. A reference system was considered in which a Rankine cycle at moderate top temperature delivers its expansion power by means of an ideal turbine to an ideal compressor of a refrigeration cycle. Two main optimizing variables were ascertained: the fluid critical temperature and the complexity of the fluid molecule. The best performance of such reference cycle is around 80% of that of an ideal fully reversible, Carnot cycle based, system (COP of 2.0 for $t_{E,PC}$ = 150°C, $t_{E,RC}$ = 5°C, and t_C = 35°C). He et al. (2009) proposed a new predication approach using the multivariate grey model combined with grey relational analysis for the performance of ejector refrigeration systems as an alternate tool to a wide range of mathematical models developed by other researchers based on complex shock, choking and mixing phenomena happening in the ejector. This new tool is based on limited number of experimental results simple program and takes a short time. Results indicate that the proposed combined method provides significant improvement over the traditional methods in model simplification, time requirement and universal application. Wang et al. (2009) and Dai et al. (2009) proposed a new combined power and refrigeration cycle for the cogeneration, which combines the Rankine cycle and the ejector refrigeration cycle by adding an extraction turbine between heat recovery vapour generator and ejector. This combined cycle could be driven by the flue gas from gas turbine or engine, solar energy, geothermal energy and industrial waste heats. Parametric analysis and exergy analysis showed that the condenser temperature, the evaporator temperature, the turbine inlet pressure, the turbine extraction pressure and extraction ratio have significant effects on the turbine power output, refrigeration output, exergy efficiency and exergy destruction in each component in the combined cycle. They concluded that the biggest exergy destruction occurs in the heat recovery vapour generator, followed by the ejector and turbine. Therefore, it is significant to employ methods for reducing exergy destructions of these components, such as increasing the area of heat transfer and the coefficient of heat transfer in the heat recovery vapour generator, optimization design and manufacture in the ejector and turbine. They found that the combined cycle has a maximum exergy efficiency of 27.10% when turbine inlet pressure, turbine inlet temperature and turbine back pressure are 0.7852 MPa, 118.9°C and 0.1462 MPa, respectively. Varga et al. (2009a) simulated performance of a steam ejector using CFD for operating temperatures (90°C and 100°C) that would be suitable to run an air conditioner (10°C evaporator temperature) using solar thermal energy. It was found that at constant evaporator and generator pressures, by increasing area ratio between nozzle throat and constant area chamber cross section (range of 14–25), the entrainment ratio can be improved considerably but will result in a smaller critical back pressure. Therefore a single optimum can be identified as the area ratio that would bring the ejector to operate at critical mode for a given condenser temperature (25°C to 40°C). Obviously, this would require different ejectors for different operating conditions that occur due to variation in solar radiation and ambient temperature. In order to overcome this problem, a new feature – a spindle – was implemented and tested using CFD. It was found that by changing the spindle position, the effective nozzle area can be adjusted and an optimal area ratio between nozzle throat and constant area chamber cross section can be adjusted with a single ejector. Nozzle exit position affected both the critical back pressure and entrainment ratio. The optimum location for the nozzle exit was found to be 6 cm from the converging section inlet plane resulting in a 5% and 12% increase in the entrainment ratio and critical back pressure, respectively. The constant area section length did not influence the entrainment ratio, within the range considered in this work. Increasing constant area section up to 155 mm increased the critical back pressure. A further increase did not affect the performance indicators. Boumaraf and Lallemand (2009) theoretically evaluated the performance and the characteristics of the operating cycle of an ejector refrigerating system with the working fluids R142b and R600a. The simulation program included the a correlation of the ejector entrainment ratio established in different operating conditions at critical point from the conservation equations of the available 1-D model and considered the temperatures of the three thermal sources and local heat transfer coefficients for the boiler, the condenser and the evaporator. All the components of the system were dimensioned for a refrigerating power of 10 kW, the hot source temperature is equal to 120 and 130°C, whereas those of the intermediate and the cold sources are fixed at 35 and 10°C, respectively (dimensioning conditions). Then, the system performance is investigated in dimensioning conditions and in off-dimensioning conditions using the simulation program. The results showed that at fixed cold source temperature, the intermediate temperature corresponding to the critical mode with ($\Delta P_C \leq P^*_C$), the system COP decreases when the hot source temperature is higher than that of its dimensioning. Consequently, it is recommended to dimension the system components at the highest possible temperature in order to guarantee better performance in the case of an operating at lower temperature of the hot source. Also, it was noted that R142b leads to better performance of the system in all cases. This can be due to the fact that R142b is a heavier fluid than R600a. Guo and Shen (2009) applied the lumped method combined with dynamic model for performance prediction of solar-driven ejector refrigeration system for providing air conditioning to office buildings. The results of the mathematical simulation have demonstrated that the solar-driven ejector refrigeration system can be designed to meet the cooling requirements of air conditioning for office buildings. 'The results had shown that the condenser temperature influences more on the performance of the system than the generator temperature.' During the office working-time, i.e., from 9:00 a.m.to 17:00 p.m, the average COP and the average solar fraction of the system were 0.48 and 0.82 respectively when the operating conditions were: generator temperature (85°C), evaporator temperature (8°C) and condenser temperature varying with ambient temperature. The COP during most of the daytime remains steady between 0.43–0.53, except at 17:00, when it drops as low as 0.29. Compared with traditional compressor based air conditioner, the solar-driven ejector refrigeration system conserves more than 75% of electric energy when it is used to supply air conditioning during daytime for office buildings. Zhu and Li (2009) developed a theoretical ejector model for the performance evaluation on ejectors with both dry and wet vapour working fluids. Results had shown that the model fits the experimental data fairly well for both dry and wet vapour fluids. Varga *et al.* (2009b) estimated the different ejector efficiencies, through a most fundamental approach, using the results of a CFD model. Within the range of conditions considered in the work, the data indicated that nozzle efficiency can be considered independent of the operating conditions, and its value was only slightly affected by nozzle diameter. Suction efficiency was also constant, although its value dropped when pc was beyond the critical value. Zhu *et al.* (2009) employed CFD technique to investigate the effects of two important ejector geometry parameters: the primary Nozzle Exit Position (NXP) and the mixing section converging angle (λ), on its performance. Yari (2009) performed first and second laws of thermodynamics analysis of two-stage configuration of ejector-expansion transcritical CO_2 (TRCC) refrigeration cycle which uses an internal heat exchanger and intercooler to enhance the performance of the cycle. Based on the simulation results, it is found that, compared with the conventional two-stage transcritical CO_2 cycle, the COP and second law efficiency of the new two-stage cycle are about 12.5–21% higher than that of conventional two-stage cycle. 'It was also concluded that, the performance of the new two-stage transcritical CO_2 refrigeration can be significantly improved based on the presented new two-stage cycle.' Meyer *et al.* (2009) designed and fabricated a small scale steam jet ejector experimental setup that was consisted of an open loop configuration and the boiler operate in the temperature range of 85–140°C. The typical evaporator liquid temperatures ranged from 5°C to 10°C while the typical water-cooled condenser pressure ranged from 1.70 kPa to 5.63 kPa (condenser temperature of 15–35°C). The boiler was powered by two 4 kW electric elements while a 3 kW electric element simulates the cooling load in the evaporator. The electric elements are controlled by means of variacs. Primary nozzles with throat diameters of 2.5 mm, 3.0 mm and 3.5 mm were tested while the secondary ejector throat diameter remained unchanged at 18 mm. These primary nozzles allowed the boiler to operate in the temperature range of 85–110°C. When the nozzle throat diameter is increased, the minimum boiler temperature decreases. A primary nozzle with a 3.5 mm throat diameter was tested at a boiler temperature of 95°C, an evaporator temperature of 10°C and a critical condenser pressure of 2.67 kPa (22.6°C). The system's COP is 0.253. Chong *et al.* (2009) introduced the supersonic ejector into boosting the production of low pressure natural gas wells. The energy of high pressure gas wells, which was usually wasted through choke valves, was used as its power supply to boost the low gas production. The operating performance of natural gas ejectors was determined not only by the operating parameters but also by the structural parameters. The field experiment was carried out with the motive pressure ranging from 8 MPa to 13 MPa, induced pressure ranging from 1.0 MPa to 5.0 MPa and discharged pressure being 5.2 MPa. According to Nakagawa *et al.* (2009) CO_2 is an environment-friendly, safe and more suitable to ejector refrigeration cycle than to vapour compression cycle. They presented analyses of the decompression boiling phenomena and the decompression curves of CO_2 in the diverging sections of the nozzles tested in relation to divergence angle. Supersonic two-phase flow of CO_2 in the diverging sections of rectangular converging–diverging nozzles was investigated. The divergence angles with significant variation of decompression were 0.076°, 0.153°, 0.306° and 0.612°. This paper presents experimental decompression phenomena which can be used in designing nozzles and an assessment of Isentropic Homogeneous Equilibrium (IHE). Inlet conditions around 6–9 MPa, 20–37°C were used to resemble ejector nozzles of coolers and heat pumps. For inlet temperature around 37°C, throat decompression boiling from the saturated liquid line, supersonic decompression and IHE solution were obtained for the two large divergence angles. For divergence angles larger than 0.306°, decompression curves for inlet temperature above 35°C approached IHE curves. For divergence angles smaller than 0.306° or for nozzles with inlet temperature below 35°C, IHE had no solution. Kasperski (2009) presented the self-regulating behaviour of gravitational ejector refrigerator. He applied mathematical model that included only the phenomena at steady state. The accumulation of heat and mass flow rate of refrigerant had not been taken into consideration. According to him the temperatures and levels of refrigerant in the heat exchangers of a gravitational ejector refrigerant make a combined system of numerous mutual couplings. It has some significant properties of self-regulation. There are three kinds of reaction on the self-regulation system of the refrigerator: thermal-type when it is stimulated by changing the temperature of exchangers surrounding, ejector-type stimulated by changing the ejector's geometry or refrigerant kind, volume-type stimulated by changing the amount of refrigerant. The prediction of a specific reaction of a system balance is difficult but it may be obtained by a mathematical model and calculating experiments. Kairouani *et al.* (2009) introduced an ejector into the conventional multi-evaporators compression system to enhance the COP utilizing vapour pre-compression by ejector and proposed a novel multi-evaporators compression system. This new configuration increases the suction pressure. In fact, in the diffuser, the kinetic energy of the mixture is converted into pressure energy. The specific work of the compressor is reduced and then the COP of the system is improved as compared to the conventional refrigeration system and to the indirect refrigeration system. A one-dimensional mathematical model was developed using the equations governing the flow and thermodynamics based on the constant-area ejector flow model. The model included effects of friction at the constant area mixing chamber. The model was used to simulate the new system using some refrigerants known as "natural" (R290, R152a, R717 and R600a) and with some transitory fluids R141b which is a good working fluid for an ejector and R134a (having a high contribution to the greenhouse effect but always used in refrigeration). The theoretical results show that the COP of the novel cycle is better than the conventional system. From the results, it can be concluded that the novel cycle, using an ejector to the recompression of the vapour, can increase the cycle COP. For the same operating temperatures of the ejector refrigeration systems, R141b gives the most advantageous relative coefficient of performance. The new system is even more advantageous than the indirect refrigeration system since *this last* gives an improvement of 15% (California Energy Commission, 2004; Horton and Groll, 2003) relative to the conventional cycle.

5.3 EJECTOR

5.3.1 Types of Ejector and its Applications in Refrigeration and Air conditioning

Ejectors have vast applications in the chemical, petrochemical, pulp and paper, food, power, steel, and allied industries in connection with such operations as filtration, distillation, absorption, mixing, vacuum packaging, freeze drying, flash cooling, deaerating, dehydrating, degassing, mud handling and refrigeration and air conditioning, etc. The ejectors may also be integrated in the existing system/s to enhance the energy savings. They will handle both condensable and non-condensable gases and vapours as well as mixtures of the two and also liquids, powders and semi solids, etc. Depending on the purpose and field of application of ejector, it is also known as injector, eductor, diffusion pump, aspirator, venturi- pump and jet pump. The main difference in the case of injector and ejector is the discharge pressure at the diffuser exit. The diffuser exit pressure of the ejector is closer to that of the suction flow (driven fluid/material) while injector it is sometimes used for applications in which the diffuser discharge pressure can actually reach the pressure of the driving/motive fluid. The fluid/material flow exiting the ejector may consist of a single or multi-component fluid/s or material/s. The most commonly encountered ejector flow in refrigeration and air conditioning constitute a single component fluid, however, two/multi component fluids flow may also be sought. Table 5.1 shows the most commonly encountered fluid flows in refrigeration and air conditioning along with their application/s in this field.

5.3.2 Parts of Ejector

An ejector, irrespective of its type, consists of a motive or primary nozzle, a suction nozzle or secondary nozzle, a mixer, and a diffuser. The complete length of the ejector is divided into three sections, viz., suction section/chamber, mixing section/chamber and diffusing section/diffuser (Fig. 5.1). The primary/ motive nozzle and secondary/suction nozzle are part of suction chamber/section. The mixing section may be designed according to the type of mixing, viz., constant pressure mixing and constant area/velocity mixing of two fluid streams (driving/motive and driven/suction fluid).

5.3.3 Working Principle of Ejector

All ejectors operate on a common principle. A high velocity jet of propelling steam, air or other fluid (a gas, vapour or liquid) or even finely divided solids (powders) will be created by passing through the motive/primary nozzle of ejector. This high velocity jet of fluid/powder can entrain the gas, vapour, liquid or powders of same or different material than the motive material through the opening/inlet of suction chamber/nozzle and cause to flow at high velocity along with the motive stream. As per Bernoulli's principle for an inviscid flow, an increase in the velocity of the fluid there is a simultaneous decrease in pressure (and vice versa), thus due to high velocity jet, a pressure difference will be created between motive and suction fluid which leads to entrainment. The mixing of the motive and suction fluid occurs in the mixing section either at constant velocity/area or constant pressure. Then, the combined stream is directed into the diffuser section of an ejector which recovers some pressure by converting velocity into pressure. In effect, the high-velocity combined stream pushes against the discharge pressure of the ejector and maintains a pressure difference between the suction inlet and the discharge of the ejector. Figure 5.1 also shows a typical velocity

Table 5.1 Types of Commonly Encountered Single- and Multi-component Fluid Flow Ejector in Refrigeration and Air Conditioning

Type of ejector	Driving/Motive Flow	Driven/ Suction Flow	Exit Flow	Application/s Refrigeration and Air Conditioning	Comments
Gas/Vapour Jet Ejector	Gas/Vapour	Gas/Vapour	Gas/Vapour	Ejector refrigeration and air conditioning; transcritical carbon dioxide refrigeration with ejector as expansion device	Shock waves possible; two-phase flow can occur
Liquid Jet Ejector	Liquid	Liquid	Liquid	Vapour absorption refrigeration with ejector integration between generator and absorber	Multi-component fluid with single phase (liquid) flow without shock waves
Condensing Ejector	Vapour	Liquid	Liquid	Vapour absorption refrigeration with ejector integration between generator and condenser with using high temperature boiler and pump	Two-phase flow with condensation of driving vapour; strong shock waves
Two-Phase Ejector	Liquid	Vapour	Two-phase (Vapour and Liquid)	Vapour compression refrigeration with ejector as expansion device	Two-phase flow; shock waves possible

and pressure variations along the length of ejector. The conversion of pressure to velocity in the primary nozzle of the ejector and the conversion of velocity into pressure in the diffuser occurs.

The flow at the exit of the motive nozzle may be two-phase, depending on the state of the primary fluid. The flashing of the primary fluid inside the nozzle may be delayed due to thermodynamic and hydrodynamic non-equilibrium effects. The high-speed (mainly supersonic) jet of primary fluid exiting from primary nozzle starts interacting with the secondary fluid inside the suction chamber. Acceleration of the secondary fluid results due to momentum transfer from the primary fluid. An additional suction nozzle is used to pre-accelerate the relatively stagnant suction fluid. This helps to reduce excessive shearing losses caused by large velocity differences between the two fluid streams. Both the supersonic primary flow and the secondary flow might be choked inside the ejector depending on the operating conditions. Due to static pressure differences it is possible for the primary fluid core to fan out and to create a fictitious throat in which the secondary fluid reaches sonic condition before both streams thoroughly mix in the subsequent mixing section (Fig. 5.1). The mixing section is a segment having a constant cross-sectional area but often has a tapered inlet

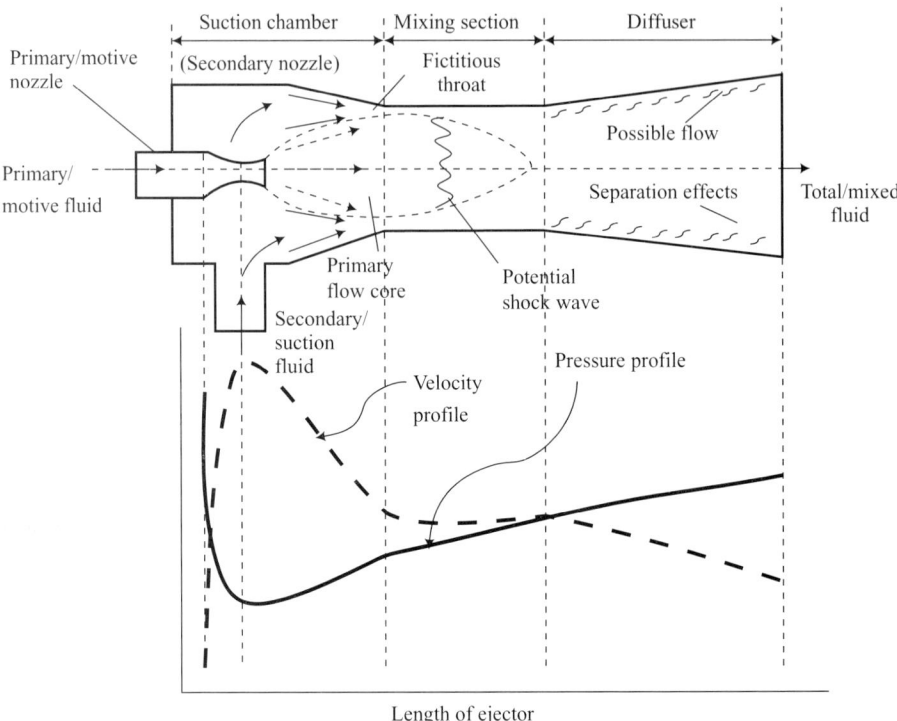

Fig. 5.1 Typical Constant Area/Velocity Mixing Ejector

section. The mixing of the two fluid streams occurs either at constant area/velocity associated with pressure changes or at constant pressure as a result of changes in cross-sectional area/velocity of the mixing section. In case of only gas/vapour or two phase fluids, the mixing process is frequently accompanied by shock wave phenomena resulting in a considerable pressure rise. After the mixing section, a diffuser is used to recover the remainder of the kinetic energy associated with the high velocity total/ combined fluid at the exit of the mixing section to convert it into potential energy, thereby increasing the static pressure. Typically, the total/combined fluid exiting the diffuser has a pressure in between that of the primary and the secondary streams entering the ejector. Therefore, mostly the ejector acts as a motive-flow driven fluid pump used to elevate the pressure of the entrained fluid.

The two major characteristics which can be used to determine the performance of an ejector are the suction pressure ratio and the mass entrainment ratio. The suction pressure ratio is defined as the ratio of diffuser exit pressure to the pressure of the suction flow entering the ejector. The mass entrainment ratio is defined as the ratio of suction mass flow rate to motive mass flow rate. A well-designed ejector is able to provide large suction pressure ratios and large mass entrainment ratios at the same time.

5.3.4 Ejector Efficiencies (Varga et al., 2009)

(i) Nozzle and Diffuser Efficiencies

In order to include irreversibilities associated with the primary nozzle, suction nozzle and diffuser, friction losses were introduced by applying isentropic efficiencies to the primary nozzle, suction nozzle and diffuser. However, there is some diversity in the literature on how losses in the mixing chamber are taken into account. The actual enthalpy change between the inlet and outlet of the nozzle is smaller than that of an isentropic process. Therefore, nozzle efficiency can be defined as the ratio of actual decrease in enthalpy of fluid to isentropic or ideal decrease in the enthalpy of the fluid passing through the nozzle. On the other hand, in the case of diffuser, the actual change in outlet and inlet enthalpy of fluid is greater than the isentropic process. Thus, diffuser efficiency can be defined as the ratio of isentropic or ideal increase in the enthalpy of the fluid to the actual increase in enthalpy of fluid passing through the diffuser. Therefore,

$$\eta_{nozzle} = \frac{h_{inlet} - h_{outlet}}{h_{inlet} - h_{outlet,isen}} \tag{5.1}$$

and

$$\eta_{diffuser} = \frac{h_{outlet,isen} - h_{inlet}}{h_{outlet} - h_{inlet}} \tag{5.2}$$

(ii) Entrainment Efficiency

Tyagi and Murty (1985) defined entrainment efficiency in the mixing chamber as the fraction of the kinetic energy in the motive fluid transmitted to the mixture:

$$\eta_{entr} = \frac{(m_{motive} + m_{suction})(h_{outlet,ejector} - h_{mix})}{m_{motive}(h_{inlet,motive-nozzle} - h_{outlet,motive-nozzle})} \tag{5.3}$$

(iii) Mixing Efficiency

Eames et al. (1995) and Huang and Huang et al. (1999) defined mixing efficiency as momentum transfer efficiency:

$$\eta_{mix} = \frac{(m_{motive} + m_{suction})v_{mix}}{(m_{motive}v_{outlet,motive-nozzle} + m_{suction}v_{outlet,suction-nozzle})} \tag{5.4}$$

Aly et al. (1999) and Korres et al. (2002) defined η_{mix} similarly to equation (5.4). However, it was assumed that the velocity of the secondary fluid at the primary nozzle exit plane was approximately zero. Yu et al. (2006) considered that losses in the mixing chamber can be written as:

$$\eta_{mix} = \left[\frac{(1+\lambda)v_{mix}}{v_{outlet,motive-nozzle}}\right]^2 \tag{5.5}$$

Cizungu et al. (2001) and Selvaraju and Mani (2004) defined mixing losses as a friction factor (f) in the form of the well known Darcy-Weisbach equation. If the velocities and pressures at the nozzle exit plane and at the mixing section are known, then **f** can be calculated as:

$$f = 2\frac{d_m}{L_m}\left[\frac{v_{nozzle,exit} - \lambda v_{sec.nozzle,exit}}{(1+\lambda)v_{mix}} + (P_{nozzle,exit} - P_{mix})\frac{A_{mix}}{(m_g + m_e)} - 1\right] \quad (5.6)$$

Cizungu et al. (2001) considered f_{mix} a constant value taken from the literature. Selvaraju and Mani (2004) calculated its value according to an empirical correlation for smooth walls. This approach is inherently incorrect, since most irreversible losses along the mixing process are due to the viscous shear layer between primary and secondary flow and not to wall friction. Zhu et al. (2007) defined an isentropic expansion efficiency due to frictional losses in the suction (mixing) chamber as:

$$\eta_{exp} = \left(\frac{d'_{pr,ch}}{d_{pr,ch}}\right)^2 \quad (5.7)$$

In this equation, $d'_{pr,ch}$ is the diameter of the primary flow at the cross section where the secondary fluid gets chocked and $d_{pr,ch}$ is the value considering an ideal case:

$$d'_{pr,ch} = \left[\frac{2+(\gamma-1)Ma^2_{pr,ch}}{2+(\gamma-1)Ma^2_{nozzle,exit}}\right]^{\frac{\gamma+1}{4(\gamma-1)}}\sqrt{\frac{Ma_{nozzle,exit}}{Ma_{pr,ch}}}d_{nozzle,exit} \quad (5.8)$$

Typical efficiency values published in the literature are shown in Table 5.2.

5.3.5 Modes of Ejector Operation

There are three different operational modes of the ejector: critical mode, subcritical mode and back flow mode, as shown in Fig. 5.2 (Munday and Bagster, 1977; Huang et al., 1999). The back pressure has distinct influences on the entrainment ratio in different modes:
  (i) No secondary flow is entrained into the ejector at a high back pressure.
  (ii) As the back pressure decreases to the subcritical region, the secondary flow is entrained proportionally and is sensible to the back pressure.
  (iii) In the critical mode, the entrainment ratio remains unchanged and is independent of the back pressure.

The thermodynamic cycle of the ejector refrigeration system with different working fluids: one dry vapour fluid (R141b) and two wet vapour fluids (R11 and steam) are qualitatively given on pressure–enthalpy chart in Fig. 5.3. It can be observed that for the wet vapour fluids there is phase change in the ejector as the slope of the isentropic line is greater than that of the saturated vapour line. Compared to R11, steam has a lower dryness fraction in the ejector.

Table 5.2 Ejector Efficiencies Considered in Literature

Reference	η_{nozz}	η_s	η_{entr}	η_{mix}	f_{mix}	η_{exp}	η_{diff}
Tyagi and Murti, 1985	0.9	-	0.8	-	-	-	0.9
Eames et al., 1995	0.85	-	-	0.95	-	-	0.85
Sun, 1996	0.85	-	-	-	-	-	0.85
Grazzini and Mariani, 1998	0.9	-	-	1	-	-	0.85
Aly et al., 1999	0.9	-	-	0.95	-	-	0.9
Huang and Chang, 1999	-	0.85	-	-	-	-	-
Huang et al., 1999	0.95	0.85	0.8 – 0.84	-	-	-	-
Sun, 1999	0.85	-	-	-	-	-	0.85
Rogdakis and Alexis, 2000	0.8	-	-	0.8	-	-	-
Cizungu et al., 2001	0.95	0.95	-	-	0.03	-	0.85
El-Dessouky et al., 2002	1	-	-	1	-	-	1
Selvaraju and Mani, 2004	0.95	0.95	-	-	f(Re)	-	0.85
Yarpici and Ersoy, 2005	0.85	-	-	-	-	-	0.9
Yu et al., 2006	0.85	-	-	0.95	-	-	0.85
Godefroy et al., 2007	0.8	0.95	-	0.935	-	-	0.8
Zhu et al., 2007	0.9 – 0.95	0.85	-	-	-	0.765 – 0.8075	-
Yu et al., 2008	0.9	-	-	0.85	-	-	0.85

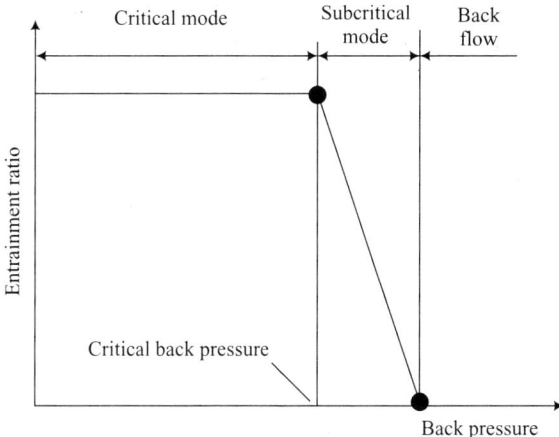

Fig. 5.2 Operational Modes of an Ejector

5.4 EJECTOR REFRIGERATION SYSTEM

5.4.1 Working Principle and Operation of Ejector Refrigeration System

If liquid (may be water) is sprayed into a chamber where a low pressure is maintained, a part of the liquid will evaporate. The enthalpy of evaporation will cool the remaining liquid to its saturation

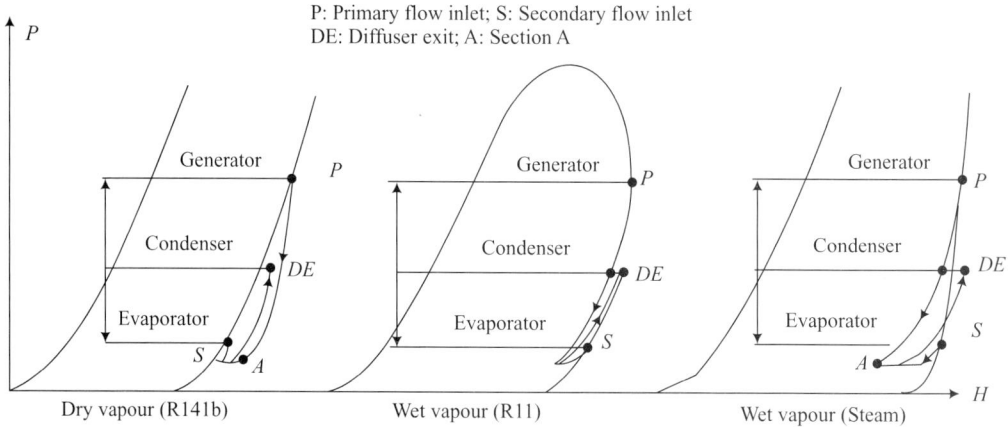

Fig. 5.3 Pressure–Enthalpy Charts of Ejector Refrigeration Cycles with Different Working Fluids

temperature at the pressure in the chamber. Obviously, lower temperature will require lower pressure. Water freezes at 0°C hence temperature lower than 4°C cannot be obtained with water. In this system, high velocity fluid (may be steam) is used to entrain the evaporating vapours. High-pressure motive fluid passes through either convergent or convergent-divergent nozzle where it acquires either sonic or supersonic velocity and low pressure (of the order of 0.009 kPa in case of steam jet refrigeration) corresponding to an evaporator temperature (of 4°C in case of steam jet refrigeration). The high momentum of motive fluid entrains or carries along with it the vapour evaporating from the evaporator (flash chamber). Because of its high velocity it moves the vapours against the pressure gradient up to the condenser where the pressure is (5.6-7.4 kPa in case of steam jet refrigeration) corresponding to condenser temperature of 35-45°C. The motive vapour and the evaporated vapour both are condensed and recycled.

Figure 5.4 shows a schematic of the ejector refrigeration system. It can be seen that this system requires a good vacuum to be maintained in case of steam jet refrigeration. Sometimes, booster ejector is used for this purpose. This system can be driven by low-grade energy that may be process steam in chemical plants or a boiler.

5.5 PERFORMANCE IMPROVEMENT OF EJECTOR REFRIGERATION SYSTEM

The major limitation of the ejector refrigeration system is its low COP because of low ejector entrainment ratio. Various methods have been suggested by several researchers for improving the performance of ejector refrigeration system. The two methods which are proposed are introduction of (i) heat exchangers and (ii) mechanical compressor.

5.5.1 Ejector Refrigeration System with Heat exchangers (Heat Regenerators)

Figure 5.5 shows basic ejector refrigeration system with two heat exchangers. In this case the refrigerant temperature is slightly increased in the preheater (heat exchanger installed between the ejector and condenser) and decreased in the sub-cooler (heat exchanger installed between evaporator and ejector) before entering the boiler and the evaporator, respectively. Consequently, the required

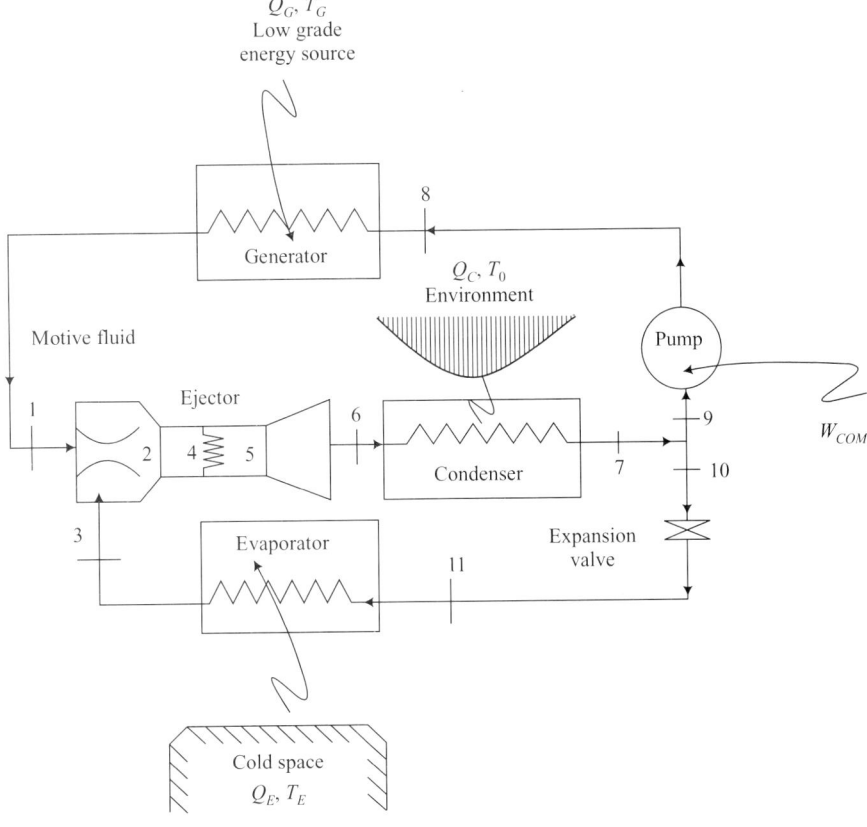

Fig. 5.4 Ejector Refrigeration System

heat input and the cooling load of the system are reduced. In the study of Dorantes and Lallemand (1995), it was shown that the use of heat recovery systems can improve the system performance (COP) by about 14% with R141b, 22% with R123 and 32% with RC318.

5.5.2 Ejector Refrigeration System with Mechanical Compressor

A) Booster Assisted Ejector Refrigeration System
The system of booster assisted ejector refrigeration is shown in Fig. 5.6. A low pressure ratio mechanical compressor is placed between the evaporator outlet and the ejector suction line. Therefore the suction pressure of the ejector is increased, thus increasing its performance.

B) Hybrid Compressor Ejector Refrigeration System
Figures 5.7 and 5.8 show the schematic diagram and the corresponding cycle thermodynamic states on a p–h plot, respectively of hybrid compressor ejector refrigeration system (Hernández et al., 2004). The interface for both the subsystems is the intercooler, whose cold streams are located

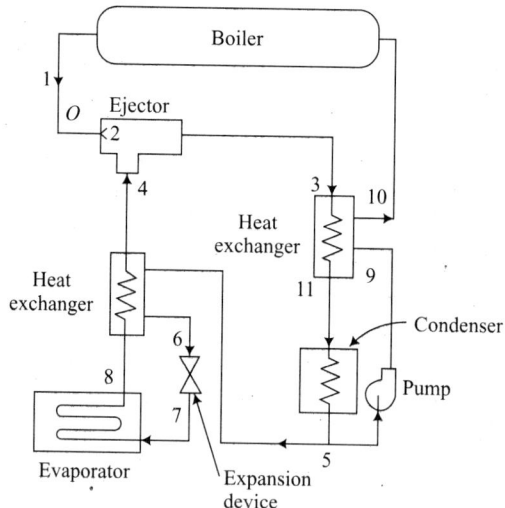

Fig. 5.5 Schematic Diagram of an Ejector Refrigeration System with Heat Exchangers (Heat Regenerators)

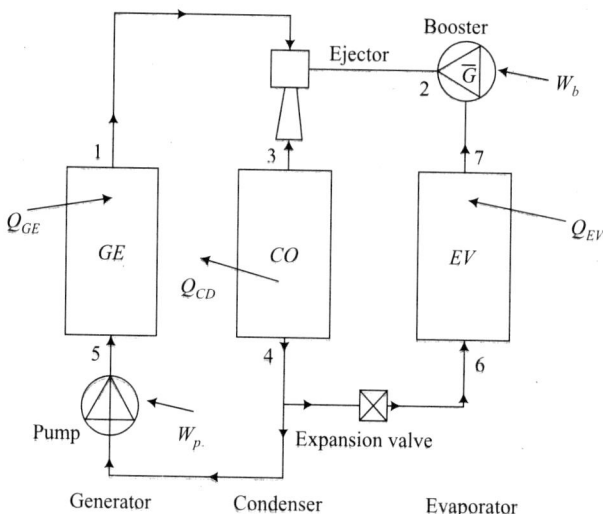

Fig. 5.6 Booster Assisted Ejector Refrigeration System

on the left side, the "evaporator" (at state points 6 and 8) for the ejector subsystem, while the hot streams are found on the right side, the "condenser" for the compression subsystem (state points 2 and 10). The two streams interact directly at a constant pressure and internal heat transfer causes the booster exit vapour to desuperheat until the saturation state is achieved.

The ejector subsystem comprises the generator, the condenser, the intercooler cold streams, the ejector, the pump and an expansion valve. The mechanical compression subsystem includes the intercooler hot streams, an expansion valve, the evaporator and the booster.

Use of Ejector in Refrigeration and Air Conditioning 205

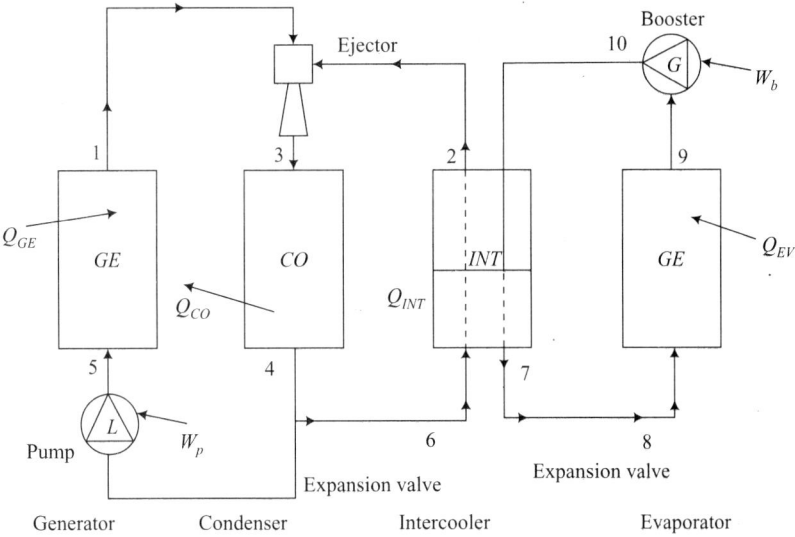

Fig. 5.7 Hybrid Compressor Ejector Refrigeration System

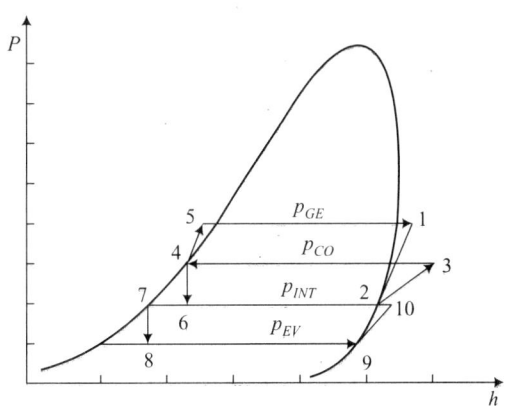

Fig. 5.8 Hybrid Compressor Ejector Refrigeration Cycle on P-h Diagram

At the ejector subsystem, the superheated vapour coming from the generator (at state 1) and the saturated vapour leaving the intercooler (at state 2), enter the ejector and mixing causes the superheated vapour emerge at state 3. The vapour enters the condenser, loses heat Q_C at constant condenser pressure P_C due to which a phase change occurs and the saturation state 4 is achieved. The saturated liquid is further divided into two streams with one going into the pump and the other going to the expansion valve. The pumping increases the pressure of the stream and the subcooled state 5 is reached. In the generator heat input Q_G is added which assists in evaporation of the subcooled liquid at constant generator pressure P_G until it reaches the superheated state 1. The other part of the condensed liquid flows through the expansion valve and a liquid-vapour saturated mixture at state 6 is obtained by means of an adiabatic expansion. This saturated mixture enters the intercooler where the saturated liquid is converted to saturated vapour state 2 by means of heat transfer (Q_{INT}) at constant intercooler pressure (P_{INT}). So, the ejector subsystem cycle is completed. At the mechanical compression subsystem, the superheated vapour coming from the booster enters the intercooler where it loses its sensible and latent heats (Q_{INT}) at constant intercooler pressure (P_{INT}), its phase change occurs and saturation state 7 is reached. The saturated liquid leaving the intercooler flows through the expansion valve resulting in a liquid-vapour saturated mixture state 8. The saturated mixture at state 8 enters the evaporator, where the saturated liquid refrigerant is evaporated at constant evaporator pressure P_e by absorbing latent heat Q_e to reach saturated vapour state 9. The booster compressor compresses the saturated vapour until the superheated state 10 is achieved. In this way, the compression subsystem and the system cycle are completed.

5.6 EJECTOR INTEGRATED VAPOUR COMPRESSION REFRIGERATION SYSTEM

The ejector is integrated in vapour compression system to minimize throttling losses (irreversibilities). The use of ejector increases suction pressure of the refrigerant at the inlet to the compressor. The ejector used in this system is a two-phase ejector instead of a single-phase ejector. Along with basic system components, viz., condenser, evaporator, compressor and throttling device the cycle includes an ejector and separator (refer Fig. 5.9). The main parts of the ejector are (a) the motive (primary) nozzle, (b) suction nozzle for secondary fluid, (c) mixing section, and (d) diffuser. High-pressure refrigerant, leaving the condenser (state point 6) is made to expand in the motive nozzle (state point 1). The drop in pressure causes acceleration of the liquid refrigerant in the motive nozzle. The high velocity refrigerant liquid causes the entrainment of the refrigerant vapour from evaporator through the suction nozzle (state point 9). The entrained and motive refrigerants mix in the mixing section

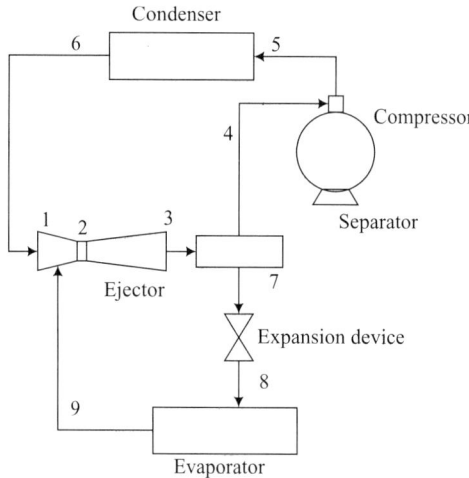

Fig. 5.9 Ejector Integrated Vapour Compression Refrigeration System

(state point 2). In the diffuser the mixture is decelerated causing the pressure of the mixture to increase the evaporator pressure. (state point 3). The two-phase mixture is separated into a saturated vapour (state point 4) and saturated liquid (state point 7) in the separator. The saturated liquid is expanded in an expansion device in which the refrigerant liquid pressure drops to evaporator pressure (state point 8) while the saturated vapour (state point 4) enters the compressor. Since the saturated vapour leaving the separator is at higher pressure than the evaporator pressure, therefore pressure ratio across the compressor reduces, thereby reducing work of compression. Since the pressure drop across the expansion device is reduced therefore the flashing of the refrigerant vapour decreases.(i.e., dryness fraction at state point 8 is reduced).

The various processes occurring in the cycle are shown in the Fig. 5.10 on P-h diagram and their detail are given below:

Process 6-1	: Isentropic expansion of refrigerant in the motive nozzle
Process 9-1	: Entertainment of vapour refrigerant (secondary fluid) from the evaporator due to suction created by motive steam at constant pressure
Process 1-2	: Mixing of liquid refrigerant (motive fluid) with the entertained refrigerant (secondary fluid) in mixing tube
Process 2-3	: Isentropic diffusion of mixed fluid (two-phase)
Process 3-4-7	: Separation of vapour and liquid refrigerant in the separator
Process 4-5	: Isentropic compression of the vapour refrigerant in the compressor
Process 5-6	: Condensation of superheated refrigerant in condenser
Process 7-8	: Throttling of liquid refrigerant coming from separator
Process 8-9	: Evaporation of liquid refrigerant

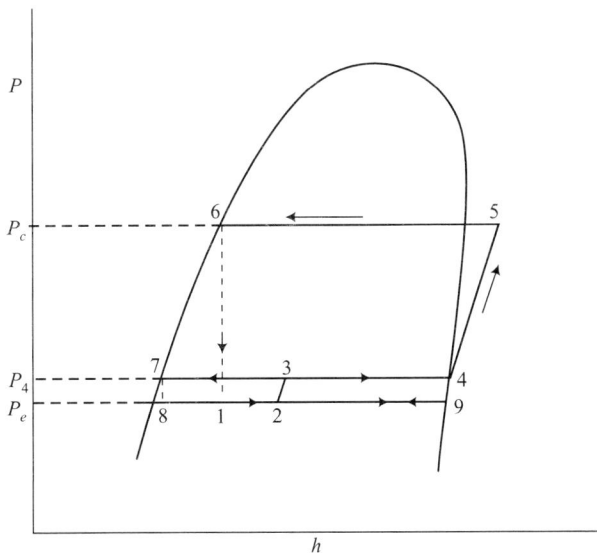

Fig. 5.10 P-h Diagram of Ejector Integrated Vapour Compression Cycle

5.6.1 Thermodynamic Modeling

This section deals with the thermodynamic modeling comprising the first law (energy) and second law (exergy) analysis of the ejector integrated vapour compression system. The thermodynamic modeling is based on the mass, energy and momentum principles and the assumptions given below to simplify the study.

Assumptions (Nehdi *et al.*, 2007):
1. The system operates under steady-state.
2. The pressure drop in various connecting pipelines and components is neglected.
3. The processes in the compressor, expansion valve, ejector are adiabatic.
4. The isentropic efficiency of motive nozzle is η_N.
5. The state of the refrigerant is assumed saturated at the exit from evaporator and condenser outlet.
6. One-dimensional flow model in the ejector.
7. Wall friction is neglected.

In the first section, the thermodynamic modeling of the ejector is discussed. Figure 5.11 shows the ejector details and pressure distribution along its length. The high pressure motive refrigerant (in saturated liquid state) leaving the condenser enters the ejector negligible velocity. The motive refrigerant expands, in the motive nozzle, to evaporator pressure (state point 1). Because of high velocity jet at low pressure created from motive fluid, the ejector entrains the refrigerant from the evaporator (secondary flow). Both fluids mix together in the mixing section. At this stage the fluid is in wet state. The wet refrigerant enters the diffuser section where its pressure increases due to diffusion

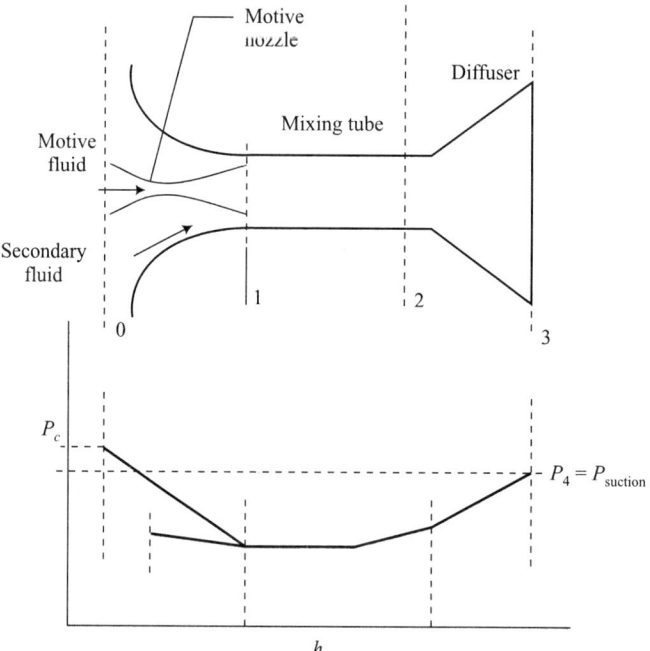

Fig. 5.11 Ejector Details and Pressure Distribution along its Length

of flow. The mixture leaves the ejector at state 3 at a pressure which is in between evaporator and condenser pressure and enters the separator where it is separated in saturated vapour and liquid phases.

A) Energy Analysis:

Motive Nozzle: The velocity, C_1, leaving the motive nozzle is determined

$$C_1 = \sqrt{2(h_6 - h_1)} = \sqrt{2\eta_N (h_6 - h_{1s})} \tag{5.9}$$

where h_1 is the enthalpy at the outlet of the motive nozzle.

Enthalpy, h_{1s}, can be calculated using the function relation given below:

$$h_{1s} = h(s_6, P_1) \tag{5.10}$$

The enthalpy, h_1, is calculated using the isentropic efficiency of the nozzle relation. Accordingly

$$h_1 = h_6 - \eta_N (h_6 - h_{1s}) \tag{5.11}$$

The density at the outlet of the motive nozzle is calculated from

$$\rho_1 = \rho(h_1, P_1) \tag{5.12}$$

The mass flow rate of motive fluid is given by equation (5.5.5)

$$\dot{m}_1 = \dot{m}_6 = \rho_1 A_1 C_1 \tag{5.13}$$

Mixing Tube: In this section, the governing equations in the mixing section are described. Applying continuity equation for mixing section, we get

$$\dot{m}_2 = \dot{m}_6 + \dot{m}_9 \tag{5.14}$$

$$\text{Also } \dot{m}_2 = \rho_2 A_2 C_2 \tag{5.15}$$

By neglecting frictional effect, secondary flow momentum, a momentum balance in the mixing section yields

$$(P_2 - P_1) A_2 = \dot{m}_1 C_1 - \dot{m}_2 C_2 \tag{5.16}$$

Combining equations 5.5.6-5.5.8, the pressure rise in the mixing section is given as:

$$\frac{(P_2 - P_1)}{\frac{1}{2}\rho_1 C_1^2} = 2(AR) - 2(1+r)^2 \left\{\frac{\rho_1}{\rho_2}\right\}(AR)^2 \tag{5.17}$$

where AR is area ratio $\dfrac{A_1}{A_2}$ and r represents entrainment ratio and is defined as

$$r = \frac{\dot{m}_9}{\dot{m}_6} \tag{5.18}$$

The density ratio can be approximated as given below (Chen, 1988):

$$\frac{\rho_2}{\rho_1} = \frac{r\rho_v}{(r+1)\rho_1} + \frac{1}{1+r} \qquad (5.19)$$

where ρ_v is the density of the refrigerant vapour leaving the evaporator. The velocity in the outlet of the mixing section is defined as

$$C_2 = \frac{1}{1+r}C_1 \qquad (5.20)$$

The conservation of energy at the mixing section yields

$$h_2 + 0.5C_2^2 = \frac{1}{1+r}h_6 + \frac{r}{1+r}h_9 \qquad (5.21)$$

where, secondary flow and condenser outlet flow kinetic energy were neglected.

The entropy at the outlet of the mixing section is calculated from

$$s_2 = s(P_2, h_2) \qquad (5.22)$$

Diffuser: This section describes the governing equation in the diffuser section. Applying principle of energy conservation to diffuser gives:

$$h_3 = h_2 + 0.5C_2^2 \qquad (5.23)$$

The velocity of the fluid leaving the diffuser is negligible, therefore pressure at the exit of the diffuser is calculated as

$$h_{3s} = h_2 + \eta_d \frac{C_2^2}{2} \qquad (5.24)$$

$$P_3 = P(s_2, h_{3s}) \qquad (5.25)$$

$$h_3 = h(P_3, x_3) \qquad (5.26)$$

Compressor: The power input to the compressor is calculated as

$$\eta_{sc} = \frac{h_{5s} - h_4}{h_5 - h_4} \qquad (5.27)$$

where $h_{5s} = h(P_5, s_4)$ \qquad (5.28)

W_{cs} is isentropic compressor work and W_{ca} is actual compressor work.

The entrainment ratio of the ejector, r, and the quality of the ejector outlet stream at state point 3, x_3, has to satisfy the following relation to meet the mass conservation constraint for steady-state operation of the cycle:

$$x_3 = \frac{1}{1+r} \qquad (5.29)$$

The isentropic efficiency of compressor can be determined from empirical relation proposed by (Brunin *et al.*, 1997)

$$\eta_{sc} = 0.874 - 0.0135\left(\frac{P_5}{P_4}\right) \quad (5.30)$$

$$w_c = h_5 - h_4 \quad (5.31)$$

The cooling capacity is defined as

$$q_e = r(h_9 - h_8) \quad (5.32)$$

The COP of the cycle is determined by the following relation:

$$\text{COP} = \frac{q_e}{w_c} \quad (5.33)$$

B) Exergy Analysis

Exergy analysis is usually aimed to determine the maximum performance of the system and identify the locations of exergy destruction and to show the direction for potential improvements.

Exergy destruction equations for condenser, compressor, expansion valve, evaporator and ejector are as follows:

$$\dot{ED}_{cond} = T_0(s_5 - s_4) \quad (5.34)$$

$$\dot{ED}_{comp} = (h_5 - h_6) - T_0(s_5 - s_6) \quad (5.35)$$

$$\dot{ED}_{ev} = rT_0(s_8 - s_7) \quad (5.36)$$

$$\dot{ED}_e = rT_0\left[(s_9 - s_8) + \frac{(h_8 - h_9)}{T_r}\right] \quad (5.37)$$

$$(5.38)$$

$$\dot{ED}_{ej} = T_0\left[(1+r)s_3 - s_6 + rs_9\right]$$

$$\dot{ED}_{total} = \dot{ED}_{comp} + \dot{ED}_{cond} + \dot{ED}_{ev} + \dot{ED}_e + \dot{ED}_{ej} \quad (5.39)$$

$$w_c = w_{rev} + \dot{ED}_{tatal} \quad (5.40)$$

$$\eta_{II} = \frac{rq_e\left|1 - \frac{T_0}{T_r}\right|}{w_c} = 1 - \frac{\dot{ED}_{total}}{w_c} \quad (5.41)$$

5.6.2 Results and Discussion

Figure 5.12 shows the effect of the ejector geometric area ratio on the COP and compressor pressure ratio (P_r) for the refrigerant R134a. It is observed that the COP increases until a maximum value is reached and then decreases with increasing ejector area ratio (Nehdi et al., 2007). The variation in COP is attributed to the change of pressure ratio in the compressor, when the compression suction pressure increases, the load on the compressor decreases, and conversely, also can be seen in Fig. 5.12. The compression-pressure ratio is minimum corresponding to maximum value of COP.

Figure 5.13 presents the variation of optimum ejector area ratio as a function of the evaporator and condenser temperatures. The optimum area ratio increases by increasing the evaporator and

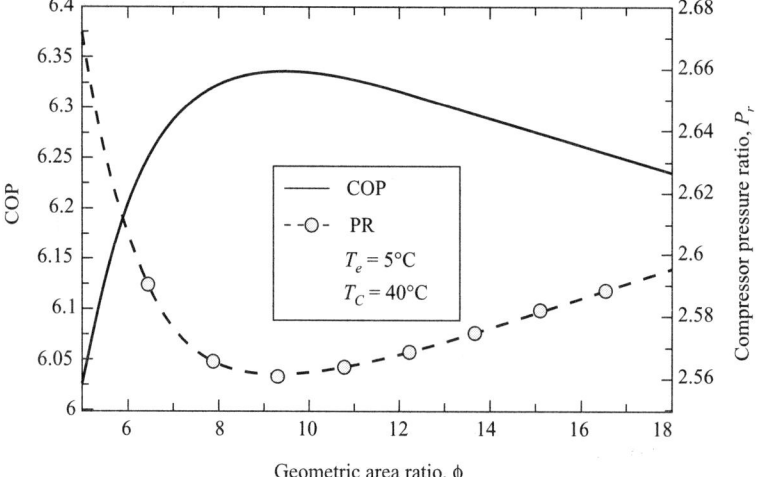

Fig. 5.12 COP and Compressor Pressure Ratio versus Ejector Geometric Area Ratio

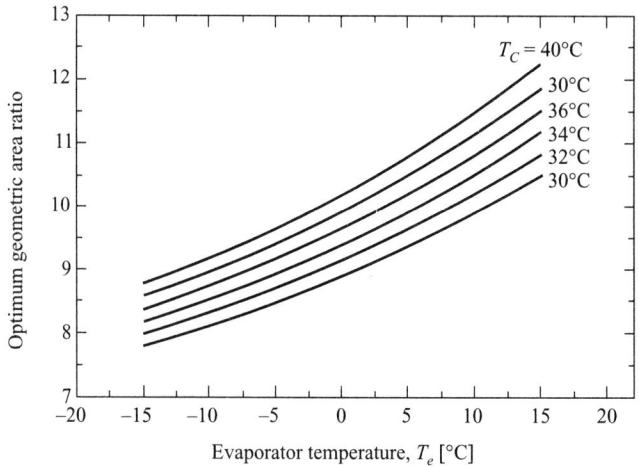

Fig. 5.13 Optimum Geometric Area Ratio vs. Evaporator Temperature Under Different Condenser Temperature

Use of Ejector in Refrigeration and Air Conditioning 213

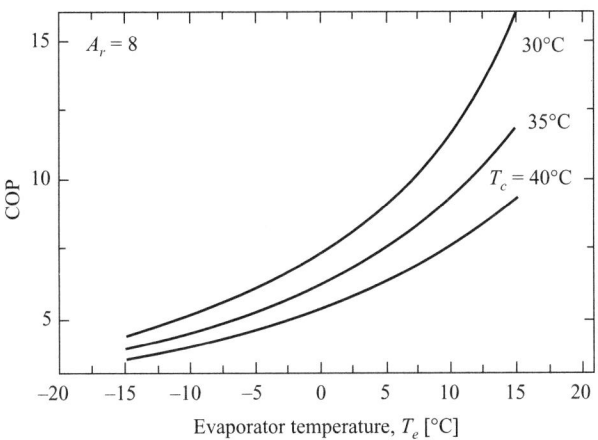

Fig. 5.14 COP vs. Evaporator Temperature Under Different Condenser Temperature

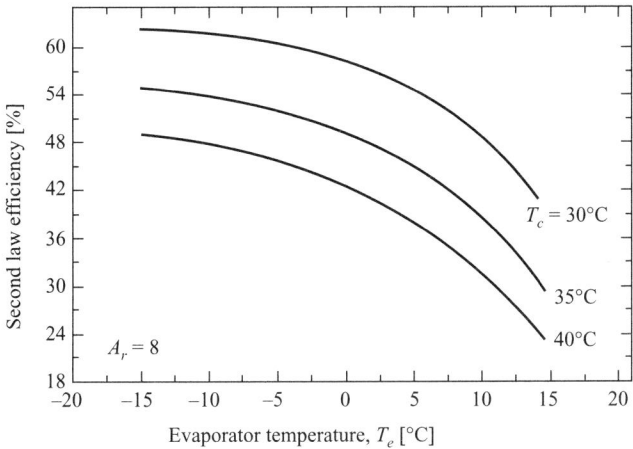

Fig. 5.15 Second Law Efficiency vs. Evaporator Temperature Under Different Condenser Temperature

condenser temperatures. The optimum ejector area ratio significantly depends on the operating conditions of the ejector integrated vapour compression cycle (Selvaraju and Mani, 2006).

Figure 5.14 shows that the coefficient of performance (COP) increases with the increasing evaporator temperature whereas exergetic efficiency shows reverse trend, i.e., it decreases with increasing evaporator temperature (Figure 5.15) and on decreasing condenser temperature both COP and exegetic efficiency increase.

Figure 5.16 represents that COP value of the ejector-compression cycle is higher by about 18% than that of the vapour compression cycle.

Figure 5.17 shows the variation of the second law efficiencies of the two cycles with change in evaporator temperature for a constant value of temperature of 40°C. It is observed that the

214 *Alternatives in Refrigeration and Air Conditioning*

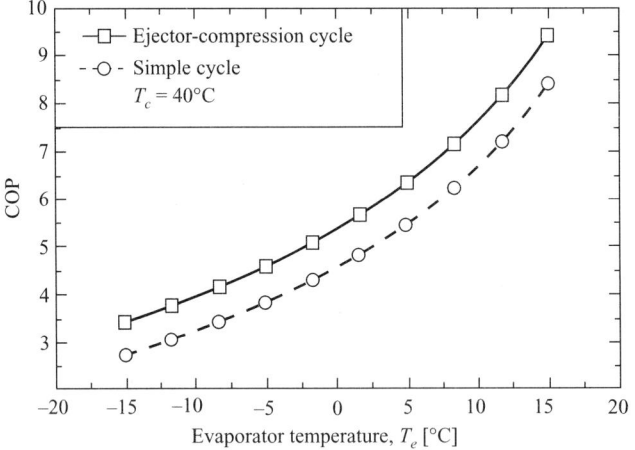

Fig. 5.16 COP Comparison Between the Vapour Compression and Ejector Compression Cycles

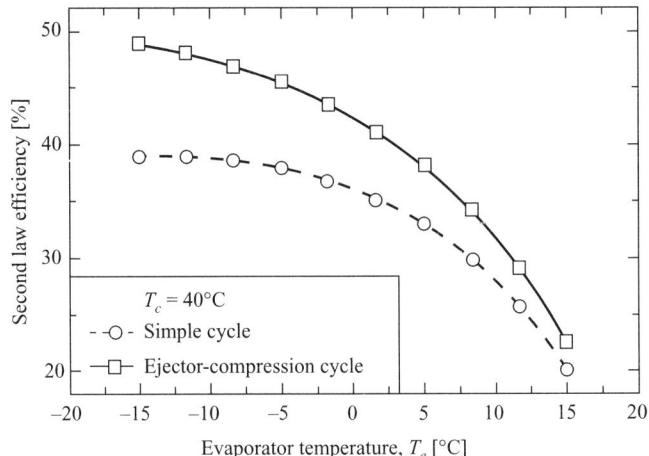

Fig. 5.17 Second Law Efficiency Comparison Between the Vapour Compression and Ejector Compression Cycles

second law efficiency of vapour compression cycle is on an average 14% lower than that of the ejector-compression cycle. It is therefore obvious that the integration of an ejector in place of expansion device effectively reduces the compression work and as a result greatly improves the system efficiency. In addition, it is found that the second law efficiencies of both the cycles decrease and converge with increase in evaporator temperature.'

5.7 TRANSCRITICAL CO_2 COMPRESSION REFRIGERATION SYSTEM WITH EJECTOR-EXPANSION DEVICE (TCCRSEJT)

Figure 5.18 represents the Transcritical CO_2 Compression Refrigeration System with Ejector-Expansion Device (TCCRSEJT), and the corresponding thermodynamic cycle is shown in Fig. 5.19.

This system consists of a compressor (COM), a gas cooler (GCO), an ejector (EJT), an evaporator (EVA) and an expansion valve (EXP). In this system an additional component, ejector (a thermo-compressor), has been employed as compared to the basic Transcritical CO_2 Refrigeration System (TCCRS) to recover some of the kinetic energy of the expansion process. Here, the compressor suction pressure is higher than it would be in TCCRS, resulting in less compression work and improved system performance. It is assumed that CO_2 exiting from the ejector (state 6) is in the gaseous/vapour phase only.

Carbon dioxide coming from the GCO, state 1, enters the motive nozzle of an ejector where its kinetic energy is increased and pressure reduced to state 2. Due to reduction in this pressure, the ejector sucks through the suction nozzle the vapours of CO_2 exiting from evaporator (state 3) and their state changes to 3a. These vapours at state 3a and the vapours emerging from nozzle are mixed together in the mixing section of the ejector, state 4, and finally come out of the ejector at state 6 after gaining pressure in the diffuser section of the ejector. The state of vapours of CO_2 at state 6, 7 and 9 is same. Depending upon the cooling load, some part of the vapours coming out of the ejector have been throttled from state 9 to 10 and the remaining vapours (used as motive gas in the ejector) have been compressed from state 7 to state 8. These high pressure and high temperature vapours cooled to state 1 in the gas cooler. The process 10-3 represents the evaporation of low temperature low pressure liquid CO_2 to saturated vapours.

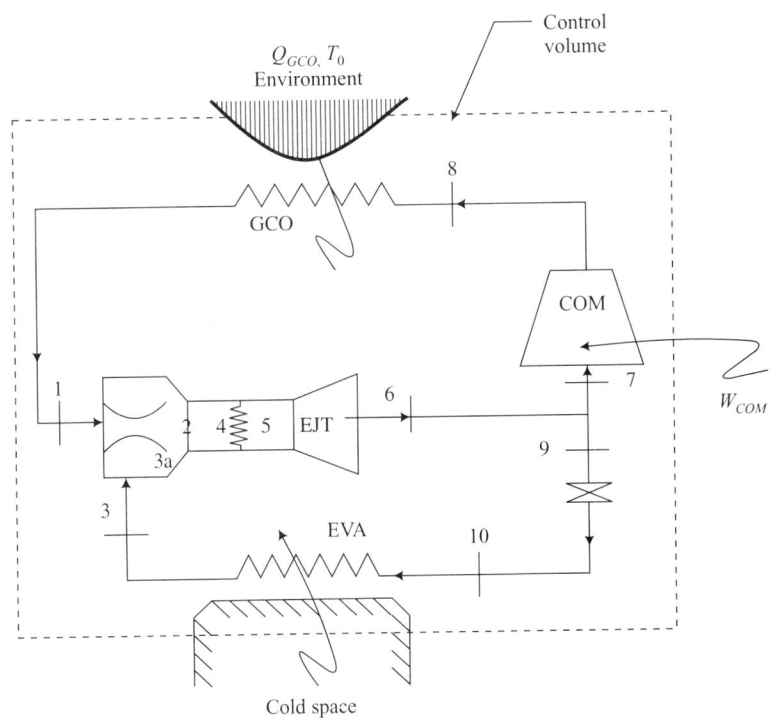

Fig. 5.18 Block Diagram of TCCRSEJT

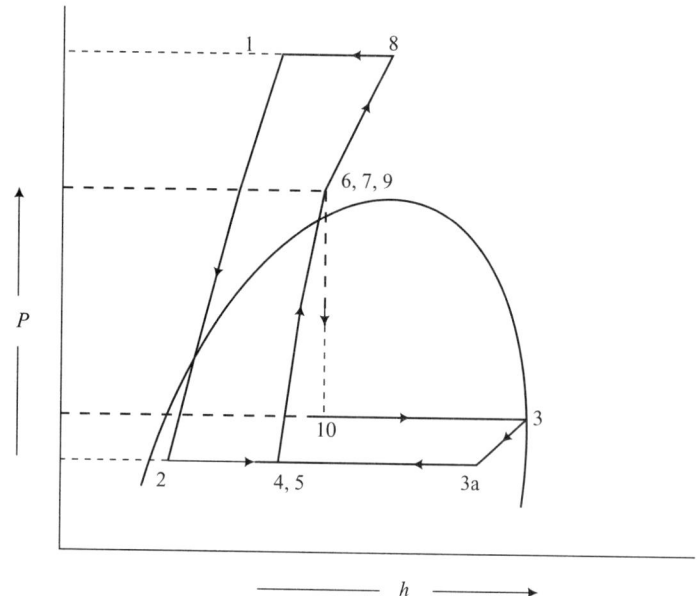

Fig. 5.19 Thermodynamic Cycle of TCCRSEJT

5.8 EJECTOR INTEGRATED ABSORPTION REFRIGERATION SYSTEMS

The interest in absorption refrigeration systems is growing because these systems can be operated by solar energy or low temperature waste heat available from various sources such as power plants (Aphornratana and Eames, 1995). The utilization of waste heat helps in reducing global warming and ozone depletion is also avoided because of use of ozone-friendly absorbent refrigerant pairs. The absorption refrigeration system with LiBr-H_2O particularly became more attractive because of the above reasons. However, COP of vapour absorption systems is significantly lower than those for vapour compression systems. This has restricted their wide application in commercial and industrial sectors. The performance of a single effect absorption system can be significantly improved by the incorporation of ejector in the system. Indeed, not only the capital investment cost is low because of its simplicity but also the incorporation of ejector in a conventional single effect system allows its COP achieve a value very close to that of COP of a typical double effect absorption system.

The performance of an absorption refrigeration system increases by increasing the absorber pressure or the generator temperature. Based on the above concept, various configurations of a hybrid ejector absorption refrigeration cycle have been proposed by several researchers to enhance the performance of the basic absorption refrigeration system. Kuhlenschmidt (1973) proposed an absorption system with two-stage generator, similar to that used in a double effect absorption system, with an aim to increase the absorber pressure at a level higher than the evaporator pressure and thus to reduce the solution concentration (refer Fig. 5.20).

In this case the ejector is placed between the evaporator and absorber. The refrigerant vapour in the second effect generator are at higher pressure than the refrigerant vapour leaving the evaporator and hence it is used as a motive fluid for entraining the refrigerant vapour from the evaporator. The

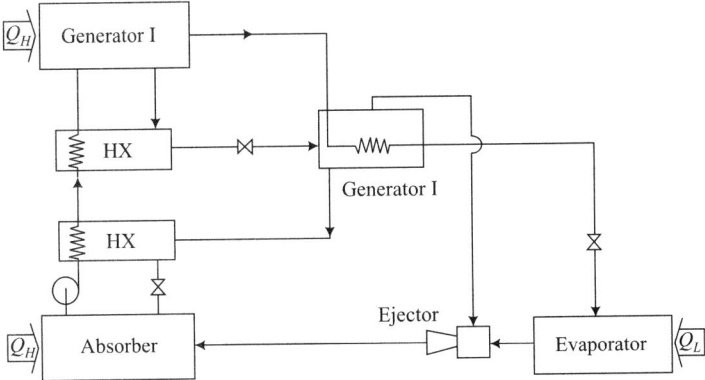

Fig. 5.20 Modified Double Effect Combined Ejector-Absorption Refrigeration Cycle (Without Condenser)

inclusion of ejector causes the absorber to operate at higher pressure than before, since the ejector discharge is at higher pressure than the evaporator pressure. Higher pressure in absorber makes possible to reduce the concentration of the solution lower than the concentration at which crystallization occurs in the system. This helps to operate the system at reduced evaporator temperature or at increased absorber temperature (such as an air-cooled unit).

The major drawback of this system is that a major part of the refrigerant vapour is directly discharged into the absorber without producing any cooling effect. This warrants lower COP of the system. To produce the same cooling capacity as before, the absorber needs to be oversized. No theoretical or experimental results of this system are available.

Chung *et al.* (1984) and Chen (1988) used an ejector in the solution circuit as represented in Fig. 5.21. It helps to maintain the absorber at a pressure higher than the evaporator pressure. In this configuration high pressure strong solution leaving the generator is used as the motive fluid for the ejector for entraining refrigerant vapour from evaporator. DMETEG/HCFC22 and DMETEG/R21 pairs have been experimentally investigated in this system. The results show that the pressure ratio of the order of 1.2 can be obtained between the absorber and evaporator. The improved pressure ratio causes the solution circulation ratio to reduce thereby reducing the generator heat duty and hence increase in COP of the system is anticipated for same operating parameters.

Aphornratana and Eames (1998) proposed another configuration as shown in Fig. 5.22. In this configuration, the ejector is positioned after the generator and before the condenser of a single effect LiBr/water absorption system. The ejector entrains the low pressure water vapour from the evaporator using steam leaving the generator as the motive fluid. The use of ejector helps in operating the generator at a pressure higher than the condenser. Thus, crystallization of the salt is avoided since the temperature of the solution is increased. The higher generator pressure generates more refrigerant hence, cooling effect is increased. The theoretical and experimental investigations show significant improvement in COP (0.86 to 1.04) in comparison to basic single effect absorption refrigeration system (0.7 to 0.8). However, with ejector the heat source temperature should be in the range of 190 to 210°C. The higher generator temperature requires better construction materials to avoid corrosion.

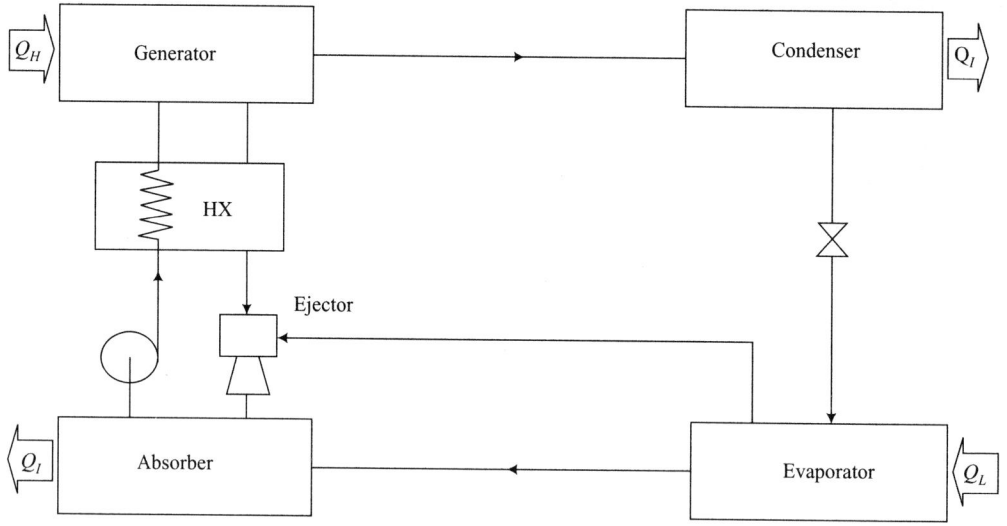

Fig. 5.21 Combined Ejector/Absorption System using DMETEG/HCFC22 and DMETEG/R21 as Working Fluids

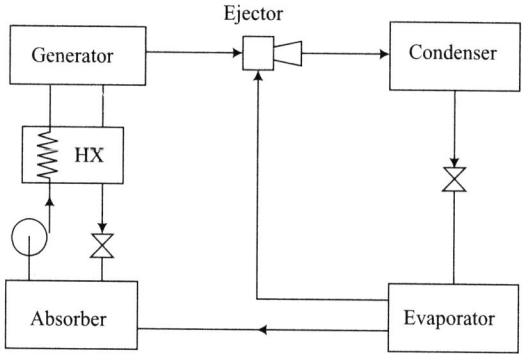

Fig. 5.22 Combined Ejector/Absorption Proposed by Aphornratana and Eames (1998).

5.9 CONCLUSION

In this chapter, ejector and its use in refrigeration and air conditioning is discussed. Firstly, the evolution of ejector and its diverse application areas are discussed. The different types of ejectors, parts of ejectors and their functions are described. The basic principle of ejector refrigeration system is explained. Further, in order to overcome its limitation of low COP, performance improvement methods like introduction of heat exchangers and mechanical compressors are studied. Ejectors can also integrated with different cooling technologies like vapour compression refrigeration systems and vapour absorption refrigeration systems. Thermodynamic modeling, energy and exergy analysis of ejector integrated vapour compression refrigeration system is carried out in detail. The study on transcritical carbon dioxide compression refrigeration system with ejector expansion device is also presented in brief. The ejector integrated absorption refrigeration systems are also given in the end.

NOMENCLATURE

d, d''	Diameter
f	Friction factor
h	Specific enthalpy
M	Mass
P	Pressure
q	Heat absorbed/rejected
r	Entrainment ratio
s	Specific entropy
v	Velocity
x	Dryness fraction
w	Specific work
A	Area
AR	Area ratio
C, V	Velocity
COP	Coefficient of performance
$\dot{E}D$	Exergy destruction
L	Length
Ma	Mach Number
P	Pressure
Q	Heat transfer
W	Work
η	Efficiency
λ	Converging angle
ρ	Density
γ	Index of compression or expansion

Subscripts

c	Condenser
ca	Actual compression
cs	Isentropic compression
d, diff	Diffuser,
e	Evaporator
ej	Ejector
exp	Expansion
Entr	Entrainment
g	Generator
m	Mass

mix	Mixing section
motive	Motive fluid
ne	Primary nozzle exit
nt	Primary nozzle throat
pr, ch	Primary flow at cross section where the secondary fluid gets choked
rev	Reversible
0	Dead state temperature
s, isen	Isentropic
sec	Secondary
sc	Compressor (isentropic)
1, 2, 3	State points
C	Compressor, condensation
E	Evaporator
N	Nozzle
V	Vapour
PC	Power cycle
RC	Refrigerating cycle
COM, comp	Compressor
II	Second law or Exergetic

REFERENCES

Addy, A.L., Dutton, J.C., 1981. Supersonic Ejector-diffuser Theory and Experiments. *Report UILU-ENG. University of Illinois at Urbana-Champaign, Urbana, IL, US,.* 82-4001.

Aly, N.H., Karmeldin, A., Shamloul, M.M., 1999. Modelling and simulation of steam jet ejectors, *Desalination 123*, 1-8.

Angelino G. and Invernizzi C., 2008. Thermodynamic optimization of ejector actuated refrigerating cycles. *Int. J. of Refrig.3 1*, 453-463.

Aphornratana, S., Chungpaibulpatana, S., Srikhirin, P., 2001. Experimental investigation of an ejector refrigerator: Effect of mixing chamber geometry on system performance. *Int. J. Energy Rese. 25*, 397-411.

Aphornratana, S., Eames, I. W., 1997. A small capacity steamejector refrigerator: experimental investigation of a system using ejector with movable primary nozzle. *Int. J. Refrigeration 20*, 352 – 358.

Aphornratana, S., Eames, I.W., 1995. Thermodynamic analysis of absorption refrigeration cycles using the second law of thermodynamics method. *International Journal of Refrigeration, Volume 18, issue 4*, 244-252.

Aphornratana, S., Eames, I.W., 1998. Experimental investigation of a combined ejector-absorption refrigerator. *Int J of Energy Res 22*, 195-207.

Arbel A. and Sokolov M., 2004. Revisiting solar-powered ejector air conditioner—the greener the better. *Solar Energy 77*, 57-66.

Bartosiewicz, Y., Aidoun, Z., Desevaux, P., Mercadier, Y., 2005. Numerical and experimental investigations on supersonic ejectors. *Int. J. Heat Fluid Fl 26*, 56-70.

Beithou, N., Aybar, H.S., 2000. A mathematical model for steamdriven jet pump. *Int. J. Multiphase Flow 26*, 1609-1619.

Boumaraf, L., Lallemand, A., 2009. Modelling of an ejector refrigerating system operating in dimensioning and off-dimensioning conditions with the working fluids R142b and R609a. *Appl Therm Eng 29*, 265-74.

Brunin, O., Feidt, M., Hivet, B., 1997. Comparison of the working domains of some compression heat pumps and a compression-absorption heat pump. *International Journal of Refrigeration 20(5)*, 308-318.

Butrymowicz, D., 2003. Improvement of compressor refrigeration cycle by means of two-phase ejector. *21st IIR International Congress of Refrigeration, Paper ICR0310, Washington DC, USA*.

California Energy Commission, 2004. Final Report – Investigation of Secondary Loop Supermarket Refrigeration Systems. Contract Number 500-98-039.

Chaiwongsa, P., Wongwises, S., 2008. Experimental study on R-134a refrigeration system using a two-phase ejector as an expansion device. *Applied Thermal Engineering vol. 28 issue 5-6*, 467-477.

Chang, Y. J. and Chen, Y. M., 2000. Enhancement of a Steam-Jet Refrigerator Using a Novel Application of the Petal Nozzle. *Experimental Thermal and Fluid Science 22*, 203-211.

Chen, L.T., 1988. A new ejector-absorber cycle to improve the COP of an absorption refrigeration system. *Applied Energy 30*, 37-51.

Chen, S.L., Yen, J.Y., Huang, M.C., 1998. An experimental investigation of ejector performance based upon different refrigerants, *ASHRAE Trans. 104*, 153-160.

Chen, Y.M., Sun, C.Y., 1997. Experimental study of the performance characteristics of a steam-ejector refrigeration system, *Exper. Therm. Fluid Sci. 15*, 384-394.

Chong, D.T., Yan, J.J., Wu, G.S., Liu, J.P., 2009. Structural optimization and experimental investigation of supersonic ejectors for boosting low pressure natural gas, *Appl. Therm. Eng. 29*, 2799 - 2807.

Chunannond, K., Aphornratana, S., (2004a). Ejectors: applications in refrigeration technology. *Renew. Sustain. Energ. Rev. 8*, 129-155.

Chung, H., Huor, M.H., Prevost, M., Bugarel, R., 1984. Domestic Heating Application of an Absorption Heat Pump, Directly Fired Heat Pumps. *Procs. Int. Conf., Uni. of Bristol, paper 2.2*.

Chunnanond, K., Aphornratana, S. (2004b). An experimental investigation of a steam ejector refrigerator: the analysis of the pressure profile along the ejector. *Appl Therm Eng 27*, 311-22.

Cizungu, K., Mani, A., Groll, M., 2001. Performance comparison of vapour jet refrigeration system with environment friendly working fluids. *Appl. Therm. Eng. 21 (5)*, 585-598.

Dai, Y.P., Wang, J.F., Gao, L., 2009. Exergy analysis, parametric analysis and optimization for a novel combined power and ejector refrigeration cycle. *Appl Therm Eng 29(10)*, 1983-90.

Dai, Y.P., Wang, J.F., Gao, L., 2009. Parametric optimization and comparative study of organic Rankine cycle (ORC) for low grade waste heat recovery. Energy *Conversion and Management 50*, 576-582.

Deng, J.-q. Jiang, P.-x. Lu, T. Lu, W., 2007. Particular characteristics of transcritical CO2 refrigeration cycle with an ejector. *Appl. Therm. Eng. 27*, 381-388.

Domanski, P.A., 1995. Theoretical evaluation of the vapour compression cycle with a liquid-line/suction-line heat exchanger, economizer, and ejector. *NISTIR 5606. NIST*.

Dorantes, R.J. and Lallemand, A. 1995. Prediction of performance of a jet cooling system operating with pure refrigerants or non azeotropic mixtures. *International J. Refrigeration, vol 18(1)*, 21-30.

Eames I.W., Aphornratana S., Haider H., 1995. A Theoretical and Experimental Study of a Small Scale Steam Jet Refrigerator. *Int. J. Refrig.*, 18(6), 378-386.

Eames, I.W., 2002. A new prescription for the design of supersonic jet-pumps: the constant rate of momentum change method, *Applied Thermal Engineering 22(2)*, 121-131.

Eames, I.W., Wu, S., 1999. A Theoretical and Experimental Study of Jet-pumps Powered by Low Grade Heat. *Recent Developments in Heat, Mass and Momentum Transfer, Research Sign Post, vol. 2 (), ISBN 81 86481 93 1*, 199-215.

Elbel S., Hrnjak P., 2008. Experimental Validation of a Prototype Ejector Designed to Reduce Throttling Losses Encountered in Transcritical R744 System Operation. *Int. J. of Ref.*, 31, 411-422.

Elbel, S., Hrnjak, P., 2004b. Flash gas bypass for improving the performance of transcritical R744 systems that use microchannel evaporators. *Int. J. Refrigeration 27*, 724-735.

Elbel, S., Hrnjak, P., 2006a. Experimental validation and design study of a transcritical CO2 prototype ejector system. *Proceedings of the 7th IIR-Gustav Lorentzen Conference on Natural Working Fluids, Trondheim, Norway.*

Elbel, S., Hrnjak, P., 2006b. A thermodynamic property chart as a visual aid to illustrate the interference between expansion work recovery and internal heat exchange. *11th International Refrigeration and Air Conditioning Conference at Purdue,Paper R165, West Lafayette, IN, USA.*

Elbel, S., Hrnjak, P., 2007. Experimental investigation of transcritical CO2 ejector system performance. *22nd IIR International Congress of Refrigeration, Paper ICR07-E1-72, Beijing, China.*

Elbel, S.W., Hrnjak, P.S., 2004a. Effect of internal heat exchanger on performance of transcritical CO2 systems with ejector. *10th International Refrigeration and Air Conditioning Conference at Purdue, Paper R166, West Lafayette, IN, USA.*

El-Dessouky, H., Ettouney, H., Alatiqi, I., Al-Nuwaibit, G., 2002. Evaluation of steam jet ejectors, *Chemical Engineering and Processing 41*, 551-561.

Elgozali, A., Linek, V., Fialov´a , M., Wein, O., Zahradný´k, J., 2002. Influence of viscosity and surface tension on performance of gas-liquid contactors with ejector type gas distributor. *Chem.E ng. Sci. 57*, 2987-2994.

Garris, C.A., Hong, W.J., Mavriplis, C., Shipman, J., 1998. A new thermally driven refrigeration system with environmental benefits. *Proceedings of the 33rd Intersociety Energy Conversion Engineering Conference, Paper IECEC-98-I088, Colorado Springs, CO, USA.*

Godefroy, J., Boukhanouf, R., Riffat, S., 2007. Design, testing and mathematical modelling of a small-scale CHP and cooling system (small CHP-ejector trigeneration). *Applied Thermal Engineering 27*, 68-77.

Grazzini, G., Mariani, A., 1998. A simple program to design a multi-stage jet-pump for refrigeration cycles, *Energy Convers. Manag. 39*, 1827-1834.

Grazzini, G., Rocchetti, A., 2008. Influence of the objective function on the optimisation of a steam ejector cycle. *Int. J. of Refrig., 31*, 510-515.

Guo, J., Shen, H.G., 2009.Modeling solar-driven ejector refrigeration system offering air conditioning for office buildings. Energy and Buildings 41, 175-181.

Harrell, G.S., Kornhauser, A. A., 1995. Performance tests of a two-phase ejector. *IECEC PAPER NO. CT-69 ASME*, 49-53.

He, S., Li, Y., Wang, R.Z., 2009. A new approach to performance analysis of ejector refrigeration system using grey system theory. *Applied Thermal Engineering 29*, 1592-1597.

Hernández, J.I., Dorantes, R.J., Best , R., Estrada, C.A., 2004. The behaviour of a hybrid compressor and ejector refrigeration system with refrigerants 134a and 142b. *Applied Thermal Engineering 24*, 1765-1783.

Horton, W.T., Groll, E.A., 2003. Secondary Loop Refrigeration in Supermarket Applications – A case study. *ICR0345 Proc 21st IIF/IIR Int Cong. Refrig, Washington DC, USA.*

Huang B.J., Chang J.M., Wang C.P., Petrenko V.A., 1999. A 1-D Analysis of Ejector Performance. *Int. J. of Refrig., 22*, 354-364.

Huang, B.J., Chang, J.M., 1999. Empirical correlation for ejector design. *Int. J. Refrigeration 22*, 379-388.

Huang, B.J., Jiang, C.B., Hu, F.L., 1985. Ejector performance characteristics and design analysis of jet refrigeration system. *J. Eng. Gas Turb. Power 107*, 792-802.

Huang,B.J.,Chang, J.M.,Petrenko,V.A., Zhuk,K.B., 1998.Asolar ejector cooling system using refrigerant R141b. *Sol. Energy 64*, 223-226.

Kairouani, L., Elakhdar, M., Nehdi, E., Bouaziz, N., 2009. Use of ejectors in a multi-evaporator refrigeration system for performance enhancement. *Int. J. Refrigeration 32*, 1173-1185.

Kasperski, J., 2009. Two kinds of gravitational ejector refrigerator stimulation. *Applied Thermal Engineering 29*, 3380-3385.

Kaushik S.C., 1989. Solar refrigeration and Space Conditioning. *Divyajyoti Prakashan, Jodhpur.*

Keenan J.H., Neumann E.P., 1942. A Simple Air Ejector. *ASME J. Appl. Mech.*, A75-A81.

Kemper, C.A., Harper, G.F., Brown, G.A., 1966. Multiple-phase ejector refrigerating system. *U.S. Patent 3,277,660*.

Kornhauser, A.A., 1990. The use of an ejector as a refrigerant expander. *Proceedings of the 1990 USNC/IIR-Purdue Refrigeration Conference, West Lafayette, IN, USA*.

Korres, C.J., Papaioannou, A.T., Lygerou, V., Koumoutsos, N.G., 2002. Solar cooling by thermal compression The dependence of the jet thermal compressor efficiency on the compression ratio. *Energy* 27, 795-805.

Kozinski, R.C., 1996. Vehicle air conditioning system utilizing refrigerant recirculation within the evaporator/accumulator circuit. *U.S. Patent 5,493,875*.

Kuhlenschmidt, D. 1973. Absorption Refrigeration System with Multiple Generator Stages. *US Patent No. 3717007*.

Li, D., 2006. Investigation of an ejector-expansion device in a transcritical carbon dioxide cycle for military ECU applications. *Purdue University, Ph.D. Dissertation, West Lafayette, IN, USA*.

Li, D., Groll, E. A., 2005. Transcritical CO_2 Refrigeration Cycle with Ejector-Expansion Device. *Int. J. of Refrig.*, 28, 766-773.

Li, D., Groll, E.A., 2006. Analysis of an Ejector Expansion Device in a Transcritical CO_2 Air Conditioning System. *7th IIR Gustav Lorentzen Conference on Natural Working Fluids, Trondheim, Norway*, May 28-31.

Li, D.Q., Groll, E.A., 2004. Transcritical CO2 Refrigeration Cycle with Ejector-Expansion Device. *International Refrigeration and Air Conditioning Conference.Paper 707, Purdue University*.

Lorentzen, G., 1983. Throttling, the internal haemorrhage of the refrigeration process. *Proceedings of the Institute of Refrigeration* 80, 39-47.

Menegay, P. and Kornhauser, A. A., 1996. Improvements to the ejector expansion refrigeration cycle. *Proceedings of the 31st Intersociety Energy Conversion Engineering Conference, Washington DC*, 702-706.

Menegay, P., 1991. Experimental investigation of an ejector as a refrigerant expansion engine. *Virginia Polytechnic Institute and State University, M.Sc. Thesis, Blacksburg VA, USA*.

Meyer, A.J., Harms, T.M., Dobson, R.T., 2009. Steam jet ejector cooling powered by waste or solar heat. *Renewable Energy* 34, 297-306.

Munday, J. T., Bagster, D. F., 1977. A new ejector theory applied to steam jet refrigeration. *Ind. Eng. Chem., Process Res. Dev.* 16(4), 442-9.

Nakagawa, M., Berana, M.S., Kishinec, A., 2009. Supersonic two-phase flow of CO_2 through convergingdiverging nozzles for the ejector refrigeration cycle. *Int. J. Refrig.* 32, 1195-1202.

Nakagawa, M., Takeuchi, H. 1998. Performance of two-phase ejector in refrigeration cycle. *Proceedings of the third International Conference on Multiphase Flow, Lyon, France. June 8–12*, 1-8.

Nehdi E., Kairouani L., Bouzaina M., 2007. Performance Analysis of the Vapour Compression Cycle Using Ejector as an Expander. *Int. J. Energy Res.*, 31, 364-375.

Ouzzane, M., Aidoun, Z., 2003. Model development and numerical procedure for detailed ejector analysis and design. *Appl. Therm. Eng.* 23 (18), 2337-2351.

Ozaki, Y., Takeuchi, H., Hirata, T., 2004. Regeneration of expansion energy by ejector in CO_2 cycle. *Proceedings of the 6th IIR-Gustav Lorentzen Conference on Natural Working Fluids, Glasgow, UK*.

Power, R.B., 1993. Steam Jet Ejectors for the Process Industries. *McGraw-Hill, New York, NY, USA*.

Riffat, S.B., Holt, A. 1998. A novel heat pipe/ejector cooler. *Applied Thermal Engineering* 18(3 - 4), 93-101.

Rogdakis, E.D., Alexis, G.K., 2000. Design and parametric investigation of an ejector in an air-conditioning system. *Appl.Therm. Eng.* 20 (2), 213-226.

Sankarlal, T., Mani, A., 2007. Experimental investigations on ejector refrigeration system with ammonia. *Renew Energy* 32, 1403-13.

Selvaraju, A., Mani, A., 2004. Analysis of a vapour ejector refrigeration system with environment friendly refrigerants. *Int. J. Therm. Sci.* 43 (9), 915-921.

Selvaraju, A., Mani, A., 2006. Experimental investigation on R134a vapour ejector refrigeration system. *Int. J. Refrigeration 29 (7)*, 1160-1166.

Sherif, S. A., Goswami, D. Y., Mathur, G. D., Iyer, S. V, Davanagere, B. S., Natarajan, S., Colacino, F., 1998. A Feasibility Study of Steam-Jet Refrigeration. *Int. J. Energy Res., 22*, 1323-1336.

Sokolov M., Hershgal D., 1990. Enhanced Ejector Refrigeration Cycles Powered by low Gade Heat. Part 2. Design Procedures.*Rev. Int. Froid.*, 13, 357-363.

Sokolov, M., Hershgal, D., 1991. Enhanced ejector refrigeration cycles powered by low grade heat. Part 3. Experimental results. *Int. J. Refrig. 14*, 24-31.

Sokolov, M., Hershgal, D., 1990a. Enhanced ejector refrigeration cycles powered by low grade heat. Part 1. Systems characterization, *Int. J. Refrig. 13*, 351-356.

Sun, D., 1999. Comparative study of the performance of an ejector refrigeration cycle operating with various refrigerants. *Energy Convers. Manage. 40 (8)*, 873-884.

Sun, D.W., 1996. Variable geometry ejectors and their applications in ejector refrigeration systems. *Energy 21 (10)*, 919-929.

Sun, D.W., 1998. Evaluation of A Combined Ejector-Vapour-Compression Refrigeration System. *Int. J. Energy Res., 22*, 333-342.

Sun, D.W., 1998a. Evaluation of a combined ejector-vapourcompression refrigeration system. *Int. J. Energ. Res. 22*, 333-342.

Sun, D.W., 1998b. Evaluation of a solar combined ejector-vapour compression refrigeration system. *International Journal of Energy Research 2*, 333-342.

Takeuchi, H., Nishijima, H., Ikemoto, T., 2004. World's first high efficiency refrigeration cycle with two-phase ejector: "ejector cycle". *SAE World Congress, Paper 2004-01-0916, Detroit, MI, USA*.

Tomasek, M.-L., Radermacher, R., 1995. Analysis of a domestic refrigerator cycle with an ejector. *ASHRAE Trans. 101*, 1431-1438.

Tyagi, K., Murty, K., 1985. Ejector-compression systems for cooling: utilising low grade waste. *Heat Recovery Syst 5*, 545-550.

Tyagi, K.P., Murty, K.N., 1985. Ejector-compression systems for cooling: Utilising low grade waste heat. *Heat Recovery Systems 5(6)*, 545-50.

Varga, S., Oliveira, A.C., Diaconu, B., 2009. Influence of geometrical factors on steam ejector performance – A numerical assessment. *Int J Refrig 32*, 1694-701.

Varga, S., Oliveira, A.C., Diaconu, B., 2009. Numerical assessment of steam ejector efficiencies using CFD. *Int. J. Refrig. 32*, 1203-1211.

Wang, J.H., Wu, J.H., Hu, S.S., Huang, B.J., 2009. Performance of ejector cooling system with thermal pumping effect using R141b and R365mfc. *Appl Therm Eng 29*, 1904-12.

Yapici, R., 2008. Experimental determination of the optimum performance of ejector refrigeration system depending on ejector area ratio. *Int J Refrig 31*, 1183-9.

Yapici, R., Ersoy, H.K., 2005. Performance characteristics of the ejector refrigeration system based on the constant area ejector flow model. *Energy Conversion and Management 46*, 3117-3135.

Yari, M., 2009. Performance analysis and optimization of a new two-stage ejectorexpansion transcritical CO_2 refrigeration cycle. *Int J Therm Sci 48(10)*, 1997-2005.

Yari, M., Sirousazar, M., 2007. Performance Analysis of the Ejector-Vapour Compression Refrigeration Cycle. *Proc. IMechE*, 221, Part A, *J. Power and Energy*, 1089-1098.

Yu, J., Zhao, H., Li, Y., 2008. Application of an ejector in autocascade refrigeration cycle for the performance improvement. *International Journal of Refrigeration 31*, 279-286.

Yu, J.L., Chen, H., Ren, Y.f., Li, Y.Z., 2006. A new ejector refrigeration system with an additional jet pump. *Appl Therm Eng*; 26, 312-9.

Zha S., Jakobsen A., Hafner A., Neksa P., 2007. Design and Parametric Investigation on Ejector for R-744 Transcritical System. International Congress of Refrigeration, *Beijing*, B1-743, 1-8.

Zhu, Y.H., Cai, W.J., Wen, C.Y., Li, Y.Z., 2009. Numerical investigation of geometry parameters for design of high performance ejectors. *Appl Therm Eng 29*, 898-905.

Zhu, Y.H., Li, Y.Z., 2009. Novel ejector model for performance evaluation on both dry and wet vapors ejectors. *Int. J. Refrigeration 32*, 21-31.

CHAPTER 6

Vortex Tube Refrigeration Systems

6.1 INTRODUCTION

The vortex tube or Ranque-Hilsch tube or Ranque tube or Hilsch tube or Maxwell's Demon or vortex refrigerator is a simple device which produces hot and cold gas streams simultaneously from a same source of compressed gas. This device has no moving parts and is very simple in construction and operation.

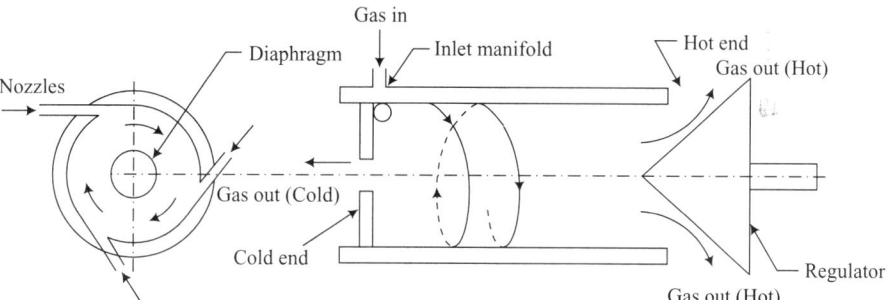

Fig. 6.1   Counter Flow Vortex Tube

The pressurized inlet gas is introduced to a supply manifold, Fig. 6.1, where it enters the simple tube through one or more tangential nozzles. This imparts a swirling or vortex motion to the inlet gas which subsequently spirals down the tube to right (hot end) of the inlet nozzles. A conical valve (regulator) at the end of this hot end of tube confines the exiting fluid to regions near the outer wall and restricts the gas in central portion of the hot end of tube from making a direct exit. A diaphragm with a suitable sized hole in the centre is placed immediately to the left (cold end) of the inlet nozzles and allows a certain portion of the fluid near the longitudinal exits to escape through the cold end of tube. This forms a counter flow vortex tube. If the cold stream of fluid is made to flow in the same direction (Fig. 6.2) as that of hot stream of fluid, then this forms a uniflow vortex tube. The temperature setting of cold gas stream in a vortex tube can be easily accomplished by simply inserting a thermometer in the cold gas exhaust and by adjusting the valve provided at the hot end.

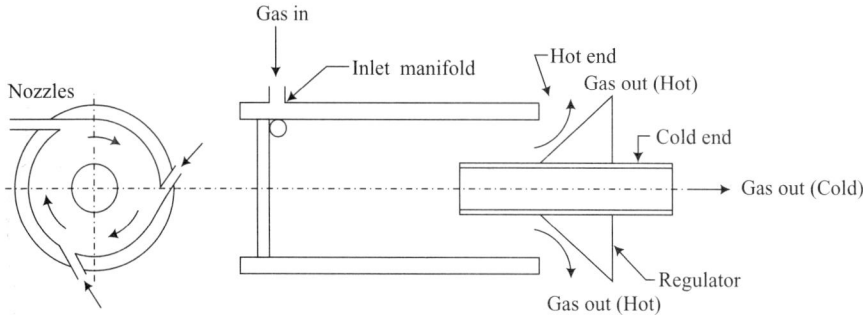

Fig. 6.2   Uniflow Vortex Tube

Most of the research on vortex tube was carried out to determine the mechanism of energy separation in it and its use in different fields besides refrigeration and air conditioning, or where it is considered an alternative to the existing RAC systems. Very few researchers explored its use in RAC systems to enhance their performance.

6.2   HISTORICAL BACKGROUND

Ranque (1933), a French metallurgist associated with a steel company, invented the vortex tube in around 1930. He observed that the temperatures of the central and peripheral layers of air in cyclone dust separators are different, the core having a lower temperature. This led him to develop the vortex tube. He obtained the French patent in 1932 and U.S. patent in 1934 (Ranque, 1934). He started a small firm to exploit the commercial potential for this strange device that produced hot and cold air with no moving parts. As the performance of vortex tube was very poor, so it was not popular. This led to failure of Ranque's firm. There was no further development in the vortex tube until 1945. A team of American scientists sent to study wartime German research in 1945, found that Rudolf Hilsch of Erlangen University, a German physicist, had performed detailed and systematic experiments on vortex tube. His comprehensive article on vortex tube was translated to English and published in the U.S.A. in 1947 (Hilsch, 1947; Scheper, 1951; Parulekar, 1961).

6.3   THEORY OF RANQUE-HILSCH OR VORTEX EFFECT

Till date, there is not a single theory available which will completely describe the vortex effect. The scientists on one hand worked to explain the energy separation phenomenon in the vortex tube, on the other hand tried to optimize the parameters to enhance its performance and explored its areas of new applications. Eiamsa-ard and Promvonge (2008) presented a detailed review on this effect and Yilmaz *et al.* (2009) published a review on design criteria for vortex tubes.

Different mathematical models based on mass, momentum and energy conservation principle have been proposed by researchers from time to time to explain the vortex effect. Some of the models are based on experimental results and others require the knowledge of computational fluid dynamics (CFD). In general, one can classify them as follows:

(a) Momentum exchange models (Ranque, 1934; Hilsch, 1947; Webster, 1950; Fulton, 1950; Deissler and Perlmutter, 1960; Reynolds, 1961)

(b) Heat exchanger models (Scheper, 1951; Lewins and Bejan, 1999; Cao et al., 2003)
(c) Acoustic streaming model (Kurosaka, 1982)
(d) Heat pump or refrigerator models (Ahlborn et al., 1998; Ahlborn and Gordon, 2000)
(e) Experiments based models (Stephan et al.,1983; Shannak, 2004)
(f) CFD/Numerical Simulation models (Frohlingsdorf and Unger, 1999; Aljuwayhel et al., 2005; Behera et al., 2005; Skye et al., 2006; Eiamsa-ard and Promvonge, 2007).

6.3.1 Momentum Exchange Models

According to Ranque (1934), the compressed external layers of air within the vortex tube have a low velocity while the expanded central layers have the greatest part of their energy in kinetic form and rotate at a very high angular velocity. In compressible fluids, the angular velocity of each layer is inversely proportional to the square of its diameter. Thus, such a distribution of velocities gives rise to considerable friction between the layers. If the layers are long enough, equilibrium will tend to be established, in which all the layers acquire the same angular velocity. Therefore, there is a centrifugal migration of energy. The central layers give their velocity to the external layers. He further explained the phenomenon and said that the rotating gas spreads out in a thick sheet on the wall of the tube and the inner layers of this sheet press upon the outer layers by centrifugal force and compresses them, thus heating them. At the same time the inner layers expand and grow cold. Friction between the layers is to be minimized, to which end the uniflow design is considered advantageous.

Hilsch (1947) described that an air passing through the orifice has been expanded in the centrifugal field from the region of high pressure at the wall of the tube to low pressure near the axis. During this expansion, it gives considerable part of its kinetic energy to the peripheral layers through internal friction. The peripheral layers then flow away with increased temperature. If there were no internal friction, the velocity of air would increase to supersonic value in the expansion from the circumference to the axis, sufficient pressure ratio being available. The internal friction is particularly effective in this range of velocities. It causes a flow of energy from the axis to the circumference by trying to establish a uniflow angular velocity across the entire cross section of the tube.

Webster (1950) hypothesized that the energy transfer outward from any given point in the whirling mass occurs as a sort of recoil reaction to the inward expansion of the gas at that point.

Fulton (1950), Van Deemter (1952) and Parulekar (1961) expressed that fresh gas succeeds in forming an almost free vortex in which angular velocity is low at the periphery and very high toward the centre. But friction between the layers of gas undertakes to reduce all the gas to the same angular velocity, amounting to the flow of work from centre to the outside of the vortex. At the same time, because the centre of the vortex is much colder than the outside, heat flows towards the centre, but not rapidly as the work flows. The inner gas is originally cooled by its expansion; it stays partly cold by giving away its kinetic energy to the outer gas by friction without receiving as much heat energy in return. The outer gas, in turn, receives more kinetic energy than it loses heat energy; this kinetic energy eventually becomes converted into internal energy through friction in the hot end of the tube. The flow is highly turbulent and the free vortex is supersonic. Van Deemter pointed out that the rotational velocity of the free vortex at the periphery decreases gradually from the plane of nozzle to the plane of the valve; therefore, there is a relative sliding between the two adjacent air planes, which are moving towards the valve. The result of this is a continuous transfer of energy from plane of the nozzle to that of the valve. This gives the explanation why the heating of air takes place as it proceeds towards the valve. Parulekar said that the turbulent mixing in the centrifugal field resulted in pumping of energy from the low-pressure region at the axis to the high-pressure region at the periphery. This is then transferred towards the valve in the form of momentum. This radial outflow energy due to turbulent mixing is much more than that of the inward flow of energy due to the formation of vortex. There is net transfer of energy radially towards the valve. Thus, a peripheral layer emerges a hot stream while axial layer emerges a cold stream.

Lay (1959) mathematically superposed the free vortex motion of the gas upon entrance to the tube to a compressible sink to give a spiral flow in the plane. The solution in the space is obtained by addition of a uniform axial velocity to the spiral flow. When viscosity effects were considered, the free vortex was shown to change into forced vortex.

Reynolds (1961) considered the dynamics of the turbulent flow of a compressible fluid using an order-of-magnitude analysis which also included energy fluxes in vortex flows to give an explanation of the energy separation in the Ranque tube. The simplified forms of the energy equation suggest that the significant energy movements within the vortex are due to four energy fluxes (i) The heat flux produced by turbulent mixing through the radial temperature and pressure gradients. (ii) The flux of total energy associated with the 'Archimedean' correlation which was discussed briefly in connection with vortex dynamics. (iii) The work fluxes associated with the two important Reynolds stresses. All these fluxes will commonly be outwards and will tend to cool the vortex core. He used experimental results to estimate the relative magnitudes of the contributions. The Archimedean effect seems to be the least important.

Sibulkin (1962) presented the qualitative picture of energy transfer processes in the vortex tube based on calculations using two concepts:
   (i) That the flow in the inlet plane may be divided into an annulus and an initially quiescent core, and that the subsequent velocity and temperature distributions may be calculated by an equivalent unsteady flow analysis; and
   (ii) That the temperature separation between the hot and the cold streams leaving the vortex tube may be found by calculating the change in energy of a fluid element travelling along the centre line of the vortex tube from its hot to cold end.

Balmer (1988) examined that compressibility need not be the governing mechanism in the temperature separation phenomena in vortex tube. The totally incompressible fluids can also produce the vortex effect. Experiments with liquid water in a commercial counter flow Ranque-Hilsch tube (designed for use with air) verify that significant temperature separation does in fact occur when a sufficiently high inlet pressure is used.

Arbuzov et al. (1997) observed the existence of large-scale structures in the form of a vertical double helix in a swirling Ranque flow. The structure of the vertical double helix is visualized in real time by the method of Hilbert bichromatic filtering. The experimental results had been interpreted on the basis that the most probable physical mechanism for Ranque effect is viscous heating of the gas in a thin boundary layer at the walls of the vortex chamber and the adiabatic cooling at the centre owing to the formation of an intense vortex braid near the axis.

Gutsol and Bakken (1998) had given a new explanation of the vortex effect. They pointed that the initial isothermal gas stream entering through the tangential inlet with a non-uniform velocity distribution becomes much more turbulent during the interaction with the cylindrical wall of the tube and with the main vortex flow. Therefore, inside this turbulent rotating bulk flow, micro-volumes with different circumferential velocities, but with equal temperature, appear. It is well known that, in a centrifugal field, elements with low tangential velocity move to the axis and elements with high tangential velocity move to the periphery. In the coordinate system connected with the bulk rotating flow the resulting force accelerates these elements in the opposite radial directions. In this way, a radial separation of elements with different kinetic energies takes place. The subsequent adiabatic expansion of the central elements with low kinetic energy in the radial pressure field of the vortex flow produces the low-temperature flow. The deceleration of the peripheral high-kinetic-energy elements due to friction produces the high-temperature flow.

Trofimov (2000) applied the model of a continuous medium with distributed internal angular momenta (angular-momentum medium) to the description of real flow in Ranque tube. He explained the inhomogeneity of temperature by the properties of a complex (three-parameter) thermodynamic system. Its increase is caused by the strengthening of the angular momentum field under the action of angular velocity in the vortex tube, while its decrease is due to the destruction of this field because of a rapid decrease in vorticity in the near-axial region of the tube. Both these processes provide additional entropy production. The same mechanism is operative in incompressible flows.

6.3.2 Heat Exchanger Models

Scheper (1951) said that heat transfer occurs radially from the inner core outward by virtue of static temperature gradient in that direction. The heat sink is provided by the outer layers, which are at lower static temperature due to the nozzle expansion. This heat transfer raises the stagnation temperature of the outer gas, which produces the heat flow, while at the same time the stagnation temperature of the core is lowered producing a cold flow through the orifice. He considered a vortex tube as a counter flow heat exchanger operative between hot air, cold air and air exiting from nozzles to determine the temperatures of temperature of hot and cold air streams.

Lewins and Bejan (1999) in their vortex tube optimization theory presented a model for temperature separation in a vortex tube by considering it as a counterflow heat exchanger between inlet gas, hot gas, cold gas and gas exiting from nozzles. They supposed that the gas expands in the vortex and produced a maximum internal temperature difference. They considered there would be heat exchange between the two streams formed at different temperatures; this exchange will be either by thermal diffusion or by its equivalent in turbulent exchange.

Cao *et al.* (2003) have formulated a correlation of cold mass flow fraction to the temperature separation effect based on making an analogy between a vortex tube and a heat exchanger. They imagined that there is a zero thickness tube wall in a vortex tube of a diameter approximately equal to that of the cold orifice, which is rotating at every axial position at the same angular velocity as the vortex. The heat transfer across this imaginary tube surface is proportional to the bulk static temperature difference and to the overall film coefficient of heat transfer occasioned by the relative axial velocity of the oppositely moving streams. A difference from Scheper's theory is that they supposed the gas expands in the vortex and produces an internal temperature difference. There will be a heat exchange between the two streams formed at different temperatures. This exchange will be by thermal diffusion so that one would hope that the vortex tube is a poor heat exchanger.

Bilga and Kaushik (2008) proposed a new analytical model of temperature separation in a Ranque Hilsch vortex tube based on ramming effect and heat exchanger. This new model closely conforms to the previous model of Ahlborn et al. (1998) and available experimental results. The major advantage of this new model is that it is sensitive to both thermodynamic and geometric parameters while Ahlborn et al.'s model does not take into account the geometry of the vortex tube. They found that the percentage deviations of temperature difference between inlet and cold gas, and that of hot and inlet gas for the case of air from available experimental results using new model at high input pressure (5 atm.), were about 0.5% to 31%, while with Ahlborn et al.'s model results were approximately 8% to 37%. For the case of CH_4 and CO_2 also this new model provides better result, higher pressure and high velocity of cold fraction. This new model for CO_2 shows a maximum 6.9% deviation from the available experimental results for temperature separation at 5 atm. input pressure.

6.3.3 Acoustic Streaming Model

Kurosaka (1982) demonstrated through an analysis and experiments that the acoustic streaming caused by the vortex whistle, a spinning wave with discrete frequency in swirling flows, is in great part responsible for the Ranque-Hilsch effect in the vortex tube of uniflow type: near the periphery of the tube, the induced 'd.c.' tangential velocity adds to the steady swirl, transforming the latter to a forced vortex and leading to the total temperature separation in the radial direction. To the extent that the temperature separation arises owing to the unsteadiness in flow, which is the converse of the phenomenon of Rijke tubes and thermally driven acoustic oscillations of liquid helium where the difference in temperature gives rise to unsteady disturbances. In the broader context, to the extent that he seeks in the present mechanism an organized origin distinct from the stochastic process, this falls under the same morphological group as the study of the large- scale structure in a mixing layer.

6.3.4 Heat Pump or Refrigerator Models

Ahlborn et al. (1998) identified the temperature splitting phenomenon of a Ranque-Hilsch vortex tube as a natural heat pump mechanism, which is enabled by a secondary circulation. They had developed an analytical model that quantitatively explained the experimental results. According to them, the primary vortex in the vortex tube sets up a secondary circulation that acts like the working fluid in a heat pump. The secondary stream absorbs heat near the axis at low pressure. Then an adiabatic compression and heating takes place as the gas is moved towards the periphery where the energy is transferred to very cold fluid elements which are just emerging out of the injection nozzles. This process together with the conversion of kinetic energy into the hot stream, accounts fully for the observed Ranque-Hilsch effect. But their model does not predict the effect of geometric parameters of vortex tube on energy separation.

Ahlborn and Gordon (2000) showed that the thermal and fluid dynamics of the vortex tube bear the signature of a classic cooling cycle, and quantified its performance as a thermodynamic machine. In the process, they developed simple analytic formulae for the temperature and pressure profiles within the tube.

Gao (2005) formulated a modified Ahlborn et al.'s model, which takes into account the normalized pressure ratio which depends on the geometric parameters of vortex tube. In this model, he showed that to improve the performance, the pressure drop over the inlet nozzle should be minimum and it is necessary to increase the exhaust velocity at the exit of the inlet nozzle.

6.3.5 Experiments Based Models

Shannak (2004) reported a simple analytical model based on mass and energy balances as well as on experimental results, where the friction losses and the heat losses are considered. The geometry of the vortex tube is in a first approach simplified to that of tee junction system. Some coupling of both branches of tee junction comes about through the pressure field and furthermore through the definition of the internal energy in the vortex tube.

6.3.6 CFD/Numerical Simulation Models

Frohlingsdorf and Unger (1999) simulated numerically by using the code system CFX the compressible flow and energy separation phenomena in the vortex tube. He expressed that in order to calculate successfully the energy separation the numerical model had to be extended by integrating relevant terms for the shear stress induced mechanical work.

Aljuwayhel *et al.* (2005) has developed a two-dimensional axi-symmetric CFD model that exhibits the general behaviour expected from a vortex tube. The CFD model is subsequently used to investigate the internal thermal fluid processes that are responsible for the vortex tube's temperature separation behaviour. The model shows that the vortex tube flow field can be divided into three regions that correspond to: flow that will eventually leave through the hot exit (hot flow region), flow that will eventually leave through the cold exit (cold flow region), and flow that is entrained within the device (recirculating region). They studied the underlying physical processes by calculating the heat and work transfers through control surfaces defined by the streamlines that separate these regions. It was found that the energy separation exhibited by the vortex tube can be primarily explained by a work transfer caused by a torque produced by viscous shear acting on a rotating control surface that separates the cold flow region and the hot flow region. This work transfer is from the cold region to the hot region whereas the net heat transfer flows in the opposite direction and therefore tends to reduce the temperature separation effect.

Behera *et al.* (2005) evolved a new approach in optimising the design of vortex tube through CFD analysis using k–ε turbulence model of the Star-CD code. The accuracy of the CFD simulations was also validated experimentally.

Skye *et al.* (2006) created a CFD model of a commercial vortex tube for use as a design tool in optimizing vortex tube performance. The model was developed using a two-dimensional (2D) steady axisymmetric model (with swirl) that utilized the standard k-ε turbulence equations. The RNG k-ε turbulence model was investigated. The comparison between the CFD model and the measured experimental data yielded promising results.

Eiamsa-ard and Promvonge (2007) used a staggered finite volume approach with the standard k–ε turbulence model and an algebraic stress model (ASM) to carry out all the computations in order to provide an understanding of the physical behaviour of the flow, pressure and temperature in a vortex tube. The results predicted by both turbulence models generally are in good agreement with measurements but the ASM performs better agreement between the numerical results and experimental data.

6.4 VARIOUS PHENOMENA IN A VORTEX TUBE

Considering the steady state operation of the vortex tube and taking the control volume as shown in Fig. 6.3, apply a mass balance and energy balance.

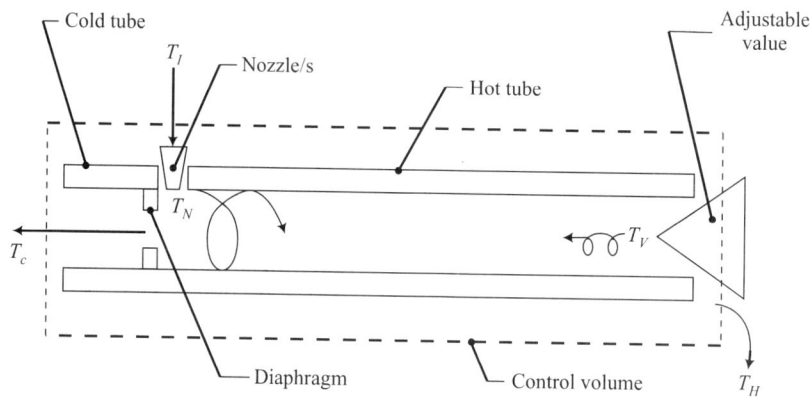

Fig. 6.3 Block Diagram of Vortex Tube

Mass Balance

$$m_C + m_H - m_I = 0 \tag{6.1}$$

Let
$$\mu = \frac{m_C}{m_I} \tag{6.2}$$

Using equations (6.1) and (6.2)

$$\frac{m_H}{m_I} = (1 - \mu) \tag{6.3}$$

Energy Balance

Steady state steady flow energy equation for any control volume can be written as,

$$\Sigma m \left(h + 1/2\, u^2 + gz \right) + \Sigma Q + \Sigma W = 0 \tag{6.4}$$

Here, the incoming mass, incoming heat and outgoing work from the control volume are taken as positive.

Consider the vortex tube is completely adiabatic, i.e. no heat interaction of the control volume with the surrounding. Also there are no work interactions because the vortex tube is not used for that purpose.

$$\Rightarrow \quad \Sigma Q = 0 \text{ and } \Sigma W = 0$$

So, equation (6.3.4) for the case of vortex tube becomes,

$$\Sigma m \left(h + 1/2\, u^2 + gz \right) = 0$$

$$\Rightarrow m_C \left(h_C + \frac{1}{2} u_C^2 + gz_C \right) + m_H \left(h_H + \frac{1}{2} u_H^2 + gz_H \right) - m_I \left(h_I + \frac{1}{2} u_I^2 + gz_I \right) = 0 \tag{6.5}$$

For ideal gas, the mass flow rate and ramming or stagnation temperature can be written as,

$$m = PAM \sqrt{\frac{\gamma}{RT}} \quad \text{(Cengel and Boles, 2006)} \tag{6.6}$$

$$h = C_p T = \frac{\gamma}{\gamma-1} RT = \frac{a^2}{\gamma-1} \quad \text{(Ahlborn et al., 1994)} \tag{6.7}$$

$$T_{RAM} = T\left[1 + \left(\frac{\gamma-1}{2}\right)M^2\right] \ldots \text{(Cengel and Boles, 2006)} \tag{6.8}$$

Where $M = \dfrac{u}{\sqrt{\gamma RT}}$ \hfill (6.9)

From equation (6.2), (6.3) and (6.6),

$$\mu = \frac{(PAM)_C}{(PAM)_I} \sqrt{\frac{T_I}{T_C}} \tag{6.10}$$

$$(1-\mu) = \frac{(PAM)_H}{(PAM)_I} \sqrt{\frac{T_I}{T_H}} \tag{6.11}$$

Using equation (6.2) to (6.11) one gets a generalized energy equation for a vortex tube as,

$$T_{RAM_I} = \mu T_{RAM_C} + (1-\mu)T_{RAM_H} + \left(\frac{\gamma-1}{2}\right)\left(\frac{g}{\gamma R}\right)[\mu z_C + (1-\mu)Z_H - Z_I] \tag{6.12}$$

This equation clearly illustrates that ramming effect is predominant within a vortex tube.

OPTION: I
When potential energy differences are neglected, which is the usual case as elevation difference between the inlet gas stream and cold and hot exit gas streams are very less. From equation (6.12),

$$T_{RAM_I} = \mu T_{RAM_C} + (1-\mu)T_{RAM_H} \tag{6.13}$$

OPTION: II
When kinetic energy differences associated with different streams are negligible then from equation (6.3.12),

$$T_I = \mu T_C + (1-\mu)T_H + \left(\frac{\gamma-1}{2}\right)\left(\frac{g}{\gamma R}\right)[\mu z_C + (1-\mu)z_H - z_I] \tag{6.14}$$

OPTION: III
When both the kinetic and potential energies differences associated with different streams are neglected, then from equation (6.12),

$$T_I = \mu T_C + (1-\mu) T_H \tag{6.15}$$

234 Alternatives in Refrigeration and Air Conditioning

This equation is most widely used as energy balance equation in all the research papers and other associated literature on vortex tube and can be presented in different ways which gives the presence of different phenomena existing within the vortex tube.

Case I: *Heat exchanger phenomena within the vortex tube*

Equation (6.15) can be written in the following form:

$$\frac{(T_H - T_I)}{(T_H - T_N)} = \frac{\mu(T_H - T_C)}{(T_H - T_N)} \tag{6.16}$$

Out of many possibilities, consider one possible heat exchanger similar to taken by Scheper (1951) which may be operative between inlet, hot and cold fluids (Fig. 6.4).

Now,
$$\varepsilon = \frac{\text{Actual Heat Transfer}}{\text{Maximum Possible Heat Transfer}}$$

$$\varepsilon = \frac{(T_H - T_I)}{(T_H - T_N)} \tag{6.17}$$

Here,
$$(T_H - T_I) \le T_I \left(\frac{\gamma - 1}{2}\right) M_I^2 \quad \text{(Ahlborn et al., 1994)} \tag{6.18}$$

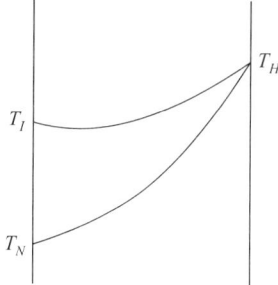

Fig. 6.4 Heat Exchanger

Comparing equations (6.16) and (6.17), one can say that there exists a heat exchanger within the vortex tube.

The effectiveness of this heat exchanger may be calculated assuming a very thin cylindrical surface between cold fluid and hot fluid as done by Lewins and Bejan (1999) and Cao et al. (2003).

Case II: *Heat engines, heat pumps and refrigerators phenomena within the vortex tube*

Equation (6.15) can be presented in the following different ways which shows the presence of heat engines, heat pumps and refrigerators within the vortex tube:

$$\frac{\left(1 - \dfrac{T_I}{T_H}\right)}{\left(1 - \dfrac{T_C}{T_H}\right)} = \mu \quad \Rightarrow \quad \frac{\eta_{CHE(H-I)}}{\eta_{CHE(H-C)}} = \mu \tag{6.19) [Ref. Fig. 6.5]}$$

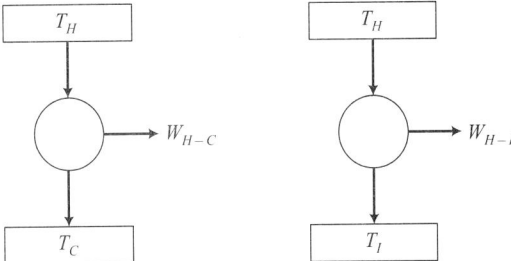

Fig. 6.5 Carnot Heat Engine

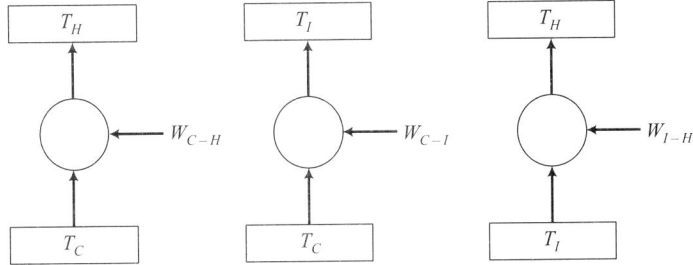

Fig. 6.6 Carnot Heat Pump/Refrigerator

$$\frac{\left(\dfrac{T_H}{T_H - T_C}\right)}{\left(\dfrac{T_H}{T_H - T_I}\right)} = \mu \quad \Rightarrow \quad \frac{(COP)_{CHP(H-C)}}{(COP)_{CHP(H-I)}} = \mu \qquad (6.20) \text{ [Ref. Fig. 6.6]}$$

$$\frac{\left(\dfrac{T_C}{T_H - T_C}\right)}{\left(\dfrac{T_C}{T_I - T_C}\right)} = (1 - \mu) \quad \Rightarrow \quad \frac{(COP)_{CR(C-H)}}{(COP)_{CR(C-I)}} = \mu \quad \ldots \quad (6.21) \text{ [Ref. Fig. 6.6]}$$

$$\left(1 - \frac{T_I}{T_H}\right)\left(\frac{T_H}{T_H - T_C}\right) = \mu \quad \Rightarrow \quad \eta_{CHE(H-I)}(COP)_{CHE(H-I)} = \mu \qquad (6.22)$$

Likewise, variety of equations can be derived. From the above analysis it is obvious that one cannot model the vortex tube as a single heat engine, heat pump or refrigerator, because these equations always show the presence of more than one device. Thus, combinations of these devices have to be taken into account for analysis.

Case III: *Acoustic phenomena within the vortex tube*

Using equations (6.7) and (6.15) one can write

$$a_I^2 = \mu a_C^2 + (1-\mu) a_H^2 \tag{6.23}$$

Thus, this equation which gives the relationship among the sonic velocities in different fluid streams clearly shows the presence of acoustic phenomena within the vortex tube which had also been modeled by Kurosaka (1982).

6.5 MODELING OF TEMPERATURE SEPARATION IN A VORTEX TUBE

6.5.1 Mathematical Model

Consider the flow of compressible gas through the Ranque Hilsch Tube (Fig. 6.3). Applying the steady state steady flow energy equation to nozzle/s of the vortex tube, one can get the following relations (Cengel and Boles, 2006):

$$\frac{T_I}{T_N} = 1 + \left(\frac{\gamma-1}{2}\right) M_N^2 \tag{6.24}$$

$$\frac{P_I}{P_N} = \left[1 + \left(\frac{\gamma-1}{2}\right) M_N^2\right]^{\frac{\gamma}{\gamma-1}} \tag{6.25}$$

$$\frac{\rho_I}{\rho_N} = \left[1 + \left(\frac{\gamma-1}{2}\right) M_N^2\right]^{\frac{1}{\gamma-1}} \tag{6.26}$$

Now applying the mass balance and the steady state steady flow energy equation to the whole control volume of the vortex tube, and assuming same value of specific heat for all gas streams, one gets,

$$T_I = \mu T_C + (1-\mu) T_H \tag{6.27}$$

To know the pressure at the outlet of a nozzle, i.e., P_N from the known values of P_I, P_C and T_I radial dynamics in the entrance plane has to be investigated.

The generalized 3-D momentum equation in the radial direction can be written as (Ogawa, 1993),

$$\frac{\partial U_r}{\partial \tau} + U_r \frac{\partial U_r}{\partial r} + U_\theta \frac{\partial U_r}{r \partial \theta} + U_z \frac{\partial U_r}{\partial z} = -\frac{1}{\rho}\frac{\partial P}{\partial r} + \frac{U_\theta^2}{r} + F_r \tag{6.28}$$

The tangentially mounted inlet nozzle/s of cross-sectional area (A_N) causes strong vortex motion in the entrance plane. Neglecting the U_r and its variations in comparison to U_θ and ignoring body forces F_r (Ogawa, 1993; Ahlborn et al., 1994), for steady state conditions at the entrance plane, the equation (6.28) becomes,

$$\frac{1}{\rho}\frac{\partial P}{\partial r} = \frac{U_\theta^2}{r} \tag{6.29}$$

Equation (6.29) can be integrated to know the pressures in the entrance plane provided the velocity profile is known. In the present model the velocity profile given by Ogawa Combined Vortex Model (Ogawa, 1993) has been considered, Fig. 6.7, which is as follows:

Quasi-forced vortex

$$U_\theta = Kr(1 - A_r) \quad (6.30)$$

$$\text{for} \quad 0 \leq r \leq r_t$$

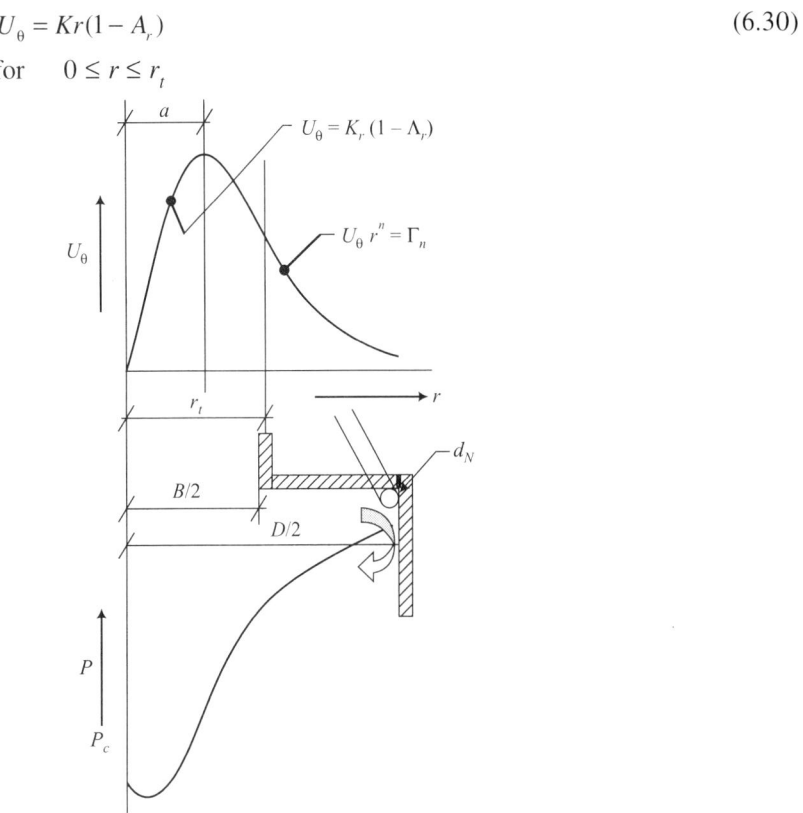

Fig. 6.7 Velocity and Static Pressure Distribution of Ogawa Combined Vortex Model

Quasi-free vortex

$$U_\theta r^n = \Gamma_n \quad (6.31)$$

$$\text{for } r_t \leq r \leq D/2$$

where Γ_n = Vortex circulation

A = Vortex vorticity

Using equations (6.28), (6.29) and (6.30) the static pressure distribution at the entrance plane can be determined, which is as follows (Ogawa, 1993):

For quasi-forced vortex,

$$P - P_C = 2\rho U_{\theta M}^2 \left(\frac{r}{a}\right)^2 \left[1 - \frac{2}{3}\frac{r}{a} + \frac{1}{8}\left(\frac{r}{a}\right)^2\right] \quad (6.32)$$

For quasi-free vortex,

$$P - P_{r_t} = 2\rho U_{\theta M}^2 \left(\frac{r}{a}\right)^2 \frac{(1-A_{r_t})^2}{n}\left[1-\left(\frac{r_t}{r}\right)^{2n}\right] \qquad (6.33)$$

where,

$$\Gamma_n = K r_t^{(1+n)}(1-Ar_t) \qquad (6.34)$$

$$K = \frac{2U_{\theta M}}{a} \qquad (6.35)$$

$$Ar_t = \frac{1+n}{2+n} \qquad (6.36)$$

$$a \Lambda = \frac{1}{2} \qquad (637)$$

$$\frac{U_{\theta M}}{U_N} = \frac{1}{4}\left(\frac{2+n}{1+n}\right)^2 \left(\frac{PERI_D - PERI_N}{PERI_{r_t}}\right)^n \qquad (6.38)$$

$$PERI_D = \pi D \qquad (6.39)$$

$$PERI_N = N\pi d_N \ ...\ \text{for circular cross-section nozzles} \qquad (6.40a)$$

$$PERI_N = 2N(b+w) \ ...\ \text{for rectangular cross-section nozzles} \qquad (6.40b)$$

$$PERI_{r_t} = 2\pi r_t \qquad (6.41)$$

Now, assuming $B = 1.40\ r_t$ (i.e., radius of diaphragm $= 0.70\ r_t$) and $n = 0.50$, for circular cross-sectional nozzle/s, the static pressure difference $(P_N - P_C)$ can be easily found using equations (6.32) to (6.41), which is as follows:

$$\left(\frac{P_N - P_C}{P_N}\right) = 0.4444 \gamma M_N^2 \left(\frac{D - Nd_N}{B}\right)\left\{1.531 - \frac{B}{D}\right\} \qquad (6.42)$$

Further, using equations (6.4.2) and (6.4.19), one can get,

$$\frac{P_I}{P_C} = \frac{\left[1+\left(\frac{\gamma-1}{2}\right)M_N^2\right]^{\frac{y}{y-1}}}{\left[1 - 0.4444\gamma M_N^2 \left(\frac{D - Nd_N}{B}\right)\left\{1.531 - \frac{B}{D}\right\}\right]} \qquad (6.43)$$

So, from the known values of P_I, T_I, P_C, γ (i.e., thermodynamic parameters) and D, B, L, N, d_N (i.e., geometric parameters) one can determine the value of M_N and then using equation (6.24) T_N can be determined.

Vortex Tube Refrigeration Systems

When the flowing compressible fluid is made or becomes stagnant then there is a rise in temperature and pressure of the fluid, this phenomenon is known as **Ramming Effect**.

Assuming that the gas exiting from the nozzle/s of the vortex tube at temperature T_N and pressure P_N is rammed twice, once in the secondary flow (Ahlborn and Groves, 1997) and then at the adjustable valve. Thus, the temperature of the gas at the surface of adjustable valve becomes,

$$T_V = T_N \left[1 + \left(\frac{\gamma - 1}{2} \right) M_N^2 \right]^2 \qquad (6.44)$$

This T_V can be viewed as a hypothetical highest temperature of the gas at the surface of the adjustable valve. In actual situations, when a stream of gas strikes any surface in the flow passage then all the particles of the gas do not ram/strike completely with the concerned surface. Some particles of the gas may remain at the same speed or their velocity partially reduces and bounces back or changes its direction. As there does not exist any concept of partial ramming or stagnation in gas dynamics, so complete ramming of all the particles of the gas has been considered at the two surfaces, one at secondary circulating gas surface and other at the surface of adjustable valve. It is assumed that the particles that are moved towards the adjustable valve will attain the temperature equal to the stagnation temperature after ramming at the secondary circulating gas surface. These particles are further rammed at the adjustable valve surface and attain the temperature equal to T_V.

Consider a counter flow heat exchanger as shown in Fig. 6.8 and applying the energy balance and assuming different streams of gas has same value of specific heat, one can get,

$$\frac{T_V - T_H}{T_V - T_N} = \mu - \mu \left(\frac{T_H - T_C}{T_V - T_N} \right) \qquad (6.45)$$

The effectiveness of this heat exchanger can be expressed as follows:

$$\varepsilon = \frac{(1 - \mu) m_I c_P (T_V - T_H)}{(mc_P)_{\min} (T_V - T_N)} \qquad (6.46)$$

For $\mu < 1/2$

$$\mu m_I c_P < (1 - \mu) m_I c_P$$

$$\therefore \quad (mc_P)_{\min} = \mu m_I c_P \qquad (6.47)$$

Using equations (6.46) and (6.47),

$$\varepsilon_1 = \left(\frac{1 - \mu}{\mu} \right) \left(\frac{T_V - T_H}{T_V - T_N} \right) \qquad (6.48)$$

From equations (6.45) and (6.48) one gets,

$$(T_H - T_C) = (T_V - T_N) \left(1 - \frac{\varepsilon_1}{(1 - \mu)} \right) \qquad (6.49)$$

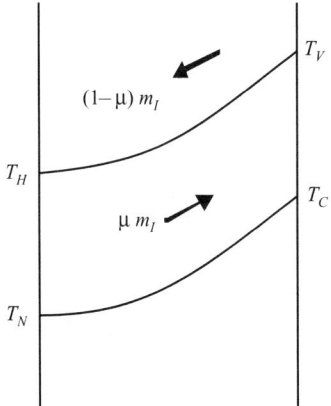

Fig. 6.8 Counter Flow Heat Exchanger

For $\mu > 1/2$

$$(1-\mu)m_I c_P < \mu\, m_I c_P$$

$$\therefore (mc_P)_{min} = (1-\mu)m_I c_P \tag{6.50}$$

Using equations (6.45) and (6.49),

$$\varepsilon_2 = \left(\frac{T_V - T_H}{T_V - T_N}\right) \tag{6.51}$$

From equations (6.44) and (6.50) one gets,

$$(T_H - T_C) = (T_V - T_N)\left(1 - \frac{\varepsilon_2}{\mu}\right) \tag{6.52}$$

For $\mu = 1/2$

The relationship for $(T_H - T_C)$ given by equations (6.49) or (6.52) can be used.

Further, the relationship developed by Lewins and Bejan (1999) for effectiveness of this type of counter flow heat exchanger by assuming the two gas streams separated by a very thin cylindrical surface of gas itself, has been considered which are as follows:

For $\mu < 1/2$

$$\varepsilon_1 = \frac{1 - \exp\left[-\dfrac{N_o TU}{\mu}\left(\dfrac{1-2\mu}{1-\mu}\right)\right]}{1 - \dfrac{\mu}{(1-\mu)}\exp\left[-\dfrac{N_o TU}{\mu}\left(\dfrac{1-2\mu}{1-\mu}\right)\right]} \tag{6.53a}$$

$$NTU = \frac{N_o TU}{\mu} \tag{6.53b}$$

Here, *NTU* represents the non-dimensional heat exchanger size based on the stream with smaller heat capacity and $N_o TU$ is known as modified non-dimensional heat exchanger size.

For μ > 1/2

$$\varepsilon_2 = \frac{1 - \exp\left[-\frac{N_o TU}{(1-\mu)}\left(\frac{2\mu-1}{1-\mu}\right)\right]}{1 - \frac{(1-\mu)}{\mu}\exp\left[-\frac{N_o TU}{(1-\mu)}\left(\frac{2\mu-1}{1-\mu}\right)\right]} \tag{6.54}$$

For μ = 1/2

$$NTU = \frac{N_o TU}{(1-\mu)} \tag{6.55}$$

$$\varepsilon = \frac{2 N_o TU}{1 + 2 N_o TU} \tag{6.56a}$$

$$NTU = 2 N_o TU \tag{6.56b}$$

where,
$$N_o TU = \frac{2\pi k L}{m_I c_P} \tag{6.57}$$

$$m_I = \rho_N A_N U_N = m_N \tag{6.58}$$

$$U_N = a_N M_N \tag{6.59}$$

$$a_N = \sqrt{\gamma R T_N} \tag{6.60}$$

The equations (6.49) and (6.52) developed here are of the same form as given by Cao *et al.* (2003). Instead of T_V, they used T_H in their relations.

6.5.2 Results and Discussion

The new model based on ramming effect and heat exchanger phenomena in a Ranque Hilsch vortex tube presented in Section 6.4.1 can be simulated using any available software, viz., MATLAB, EES, Microsoft Excel, etc., or using any programming language such as C^{++}, C, and FORTRAN, etc. In the present research, EES software (Klein and Alvaradro, 2005) is used and results are compared with the available analytical model of Ahlborn *et al.* (1998), the experimental results of Martynovskii and Alekseev (1956) and experimental results of Promvonge and Eiamsa-ard (2005).

Table 6.1 Input Parameters for Theoretical Model of Vortex Effect

Input Parameter	Martynovskii and Alekseev Experiments	Promvonge and Eiamsa-ard Experiments
Gas	Air; CH_4; CO_2	Air
P_I	3 atm. and 5 atm.	3.50 bar
P_C	1.28 atm. (assumed)	1.30 bar (assumed)
T_I	Air : 20°C at both pressures CH_4: 21°C at 3 atm. and 22°C at 5 atm. CO_2: 1.50°C at 3 atm. and -0.50°C at 5 atm.	29°C
D	9 mm	16 mm
L	450 mm	45 D
d_N	2.30 mm	$D/9$
B	4 mm	0.50 D
N	1	1, 2

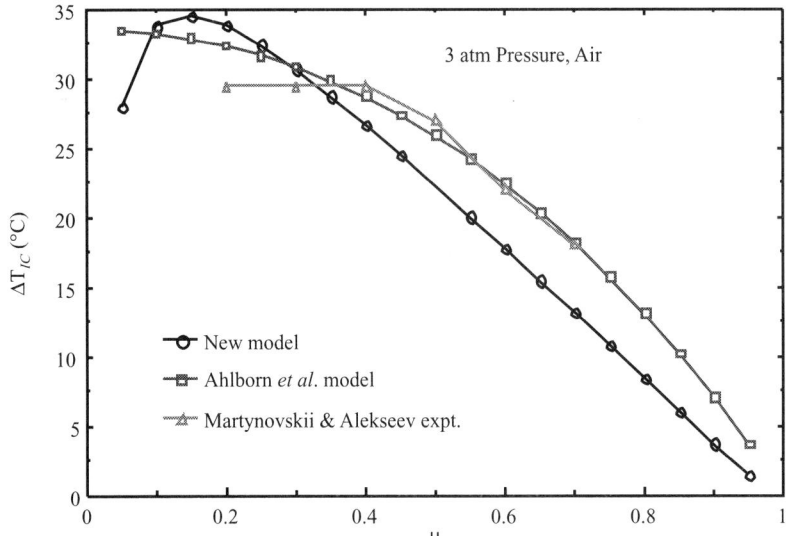

Fig. 6.9 Comparison of New Model with Experimental Results of Martynovskii and Alekseev (1956), and Ahlborn *et al*. (1998) Model for ΔT_{IC} of Air at 3 atm

The input parameters for the comparison of new model with that of Ahlborn *et al*. (1998) and available experimental results have been given in Table 6.1. The results of comparison of the new model with available analytical model of Ahlborn *et al*. (1998) and the experimental results of Martynovskii and Alekseev (1956) have been presented in Figs. 6.9 to 6.12, 6.13 to 6.16 and 6.17 to 6.20 for Air, CH_4 and CO_2 respectively. In the case of air at low value of pressure (3 atm.) the value of ΔT_{IC} and ΔT_{HI} for various values of μ given by Ahlborn *et al*. (1998) model are much close to the experimental results of Martynovskii and Alekseev (1956). The new model shows large deviations from the experimental results than Ahlborn *et al.* (1998) model at low pressure. But at higher pressure (5 atm.) the results of ΔT_{IC} and ΔT_{HI} for µ greater than 0.35 provided by the new model has been much better than the Ahlborn *et al.* model.

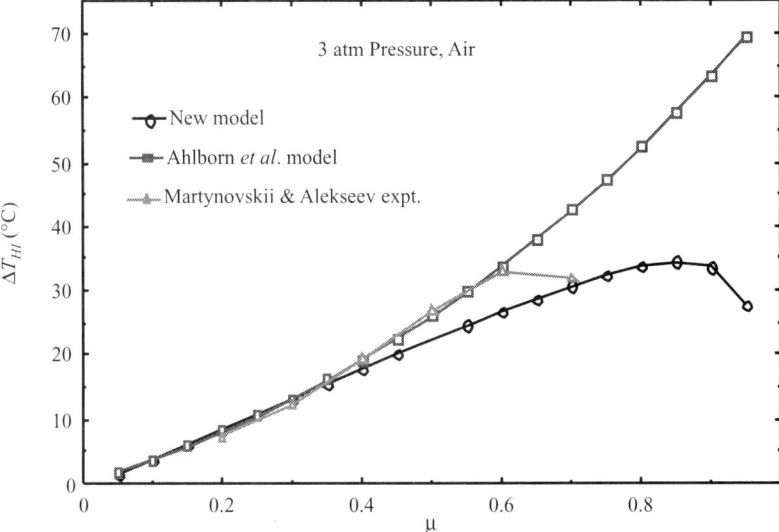

Fig. 6.10 Comparison of New Model with Experimental Results of Martynovskii and Alekseev (1956), and Ahlborn *et al.* (1998) Model for ΔT_{HI} of Air at 3 atm

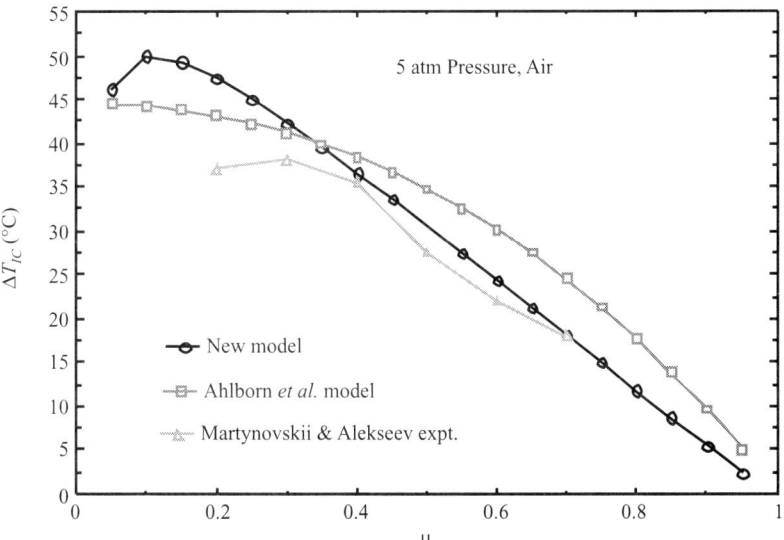

Fig. 6.11 Comparison of New Model with Experimental Results of Martynovskii and Alekseev (1956), and Ahlborn *et al.* (1998) Model for ΔT_{IC} of Air at 5 atm

The results of ΔT_{IC} and ΔT_{HI} for the flow of CH_4 through the vortex tube show almost the similar trend as for the case of Air, i.e., Ahlborn *et al.* (1998) model give more closer results to the experimental results of Martynovskii and Alekseev (1956) than the new model at low value of inlet gas pressure (3 atm.) and for higher pressure (5 atm.) corresponding to μ below 0.5. However, at higher pressure (5 atm.) and μ above 0.5 the new model gives more accurate results than Ahlborn *et al.* (1998) model.

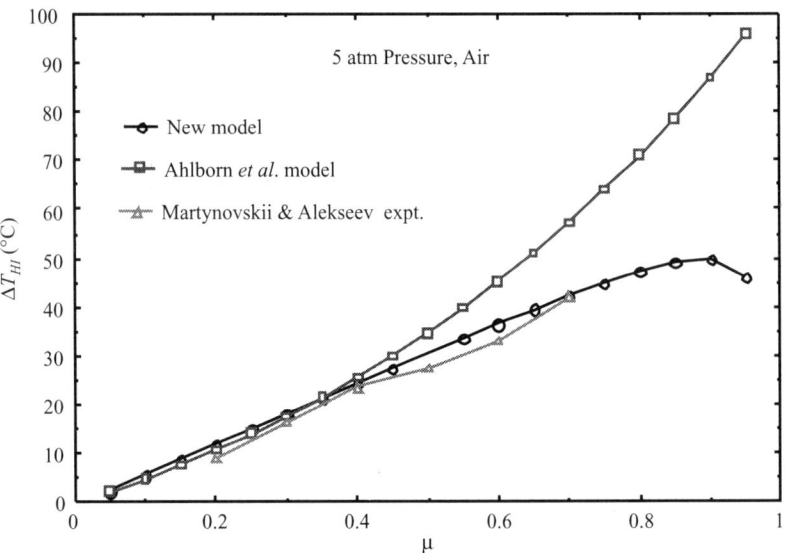

Fig. 6.12 Comparison of New Model with Experimental Results of Martynovskii and Alekseev (1956), and Ahlborn *et al.* (1998) Model for ΔT_{HI} of Air at 5 atm

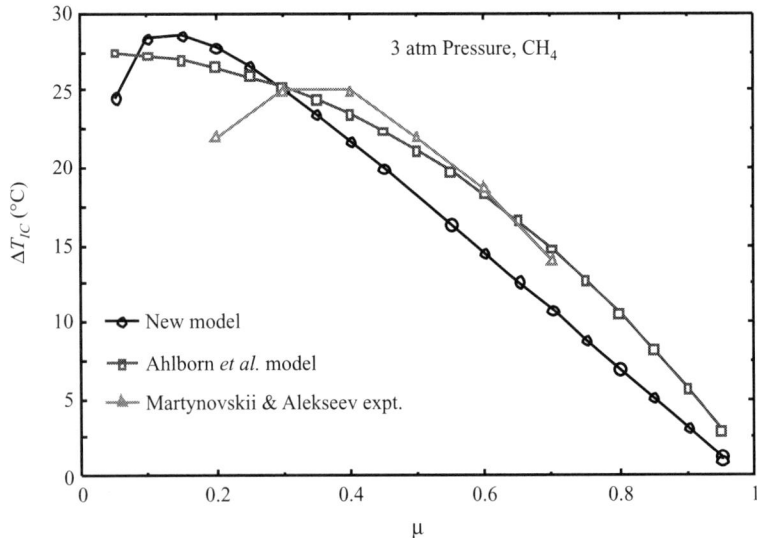

Fig. 6.13 Comparison of New Model with Experimental Results of Martynovskii and Alekseev (1956), and Ahlborn *et al.* (1998) Model for ΔT_{IC} of CH_4 at 3 atm

Similarly, the values of ΔT_{IC} and ΔT_{HI} for CO_2 gas at low inlet gas pressure (3 atm.) Ahlborn *et al.* (1998) model is more accurate than the new model. But at higher value of the inlet gas pressure (5 atm.) and µ above 0.30 the new model is more nearer to the experimental results.

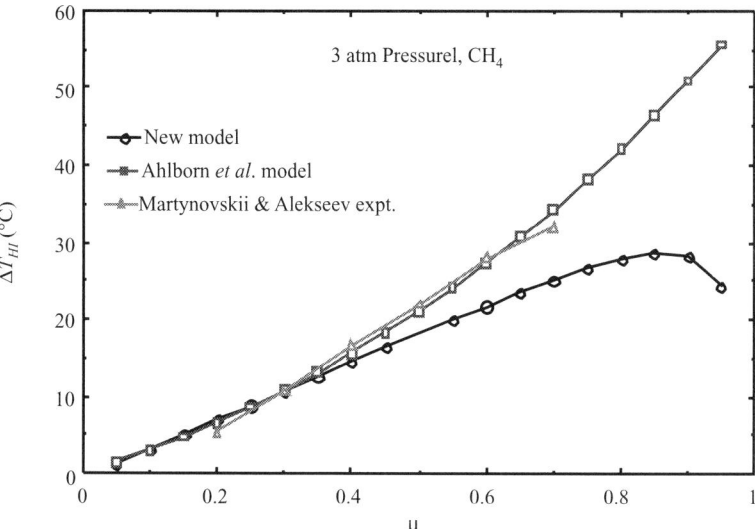

Fig. 6.14 Comparison of New Model with Experimental Results of Martynovskii and Alekseev (1956), and Ahlborn *et al.* (1998) Model for ΔT_{HI} of CH_4 at 3 atm

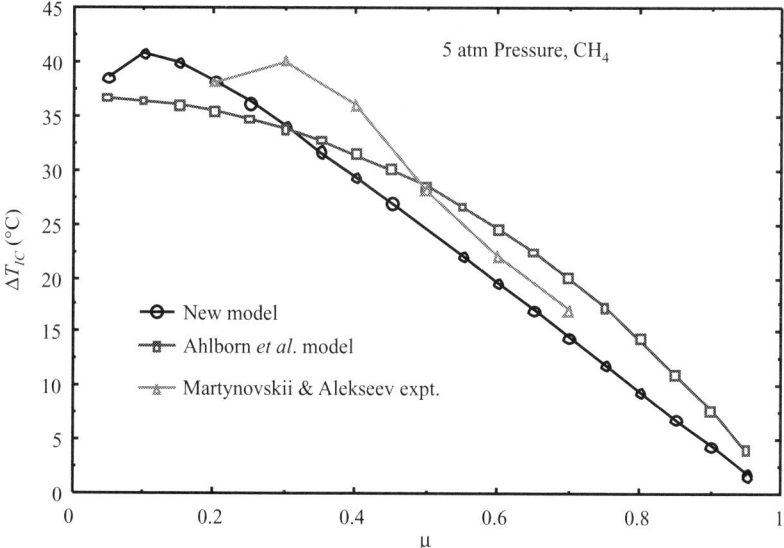

Fig. 6.15 Comparison of New Model with Experimental Results of Martynovskii and Alekseev (1956), and Ahlborn *et al.* (1998) Model for ΔT_{IC} of CH_4 at 5 atm

The percentage deviations of ΔT_{IC} and ΔT_{HI} for all the three gases are presented in Table 6.2 which clearly show the superiority of new model than the Ahlborn *et al.* (1998) model at high pressures and higher values of µ. The main advantage of the new model is that it also gives the information about the geometric parameters of the vortex tube. But Ahlborn *et al.* (1998) model is insensitive to the geometric parameters.

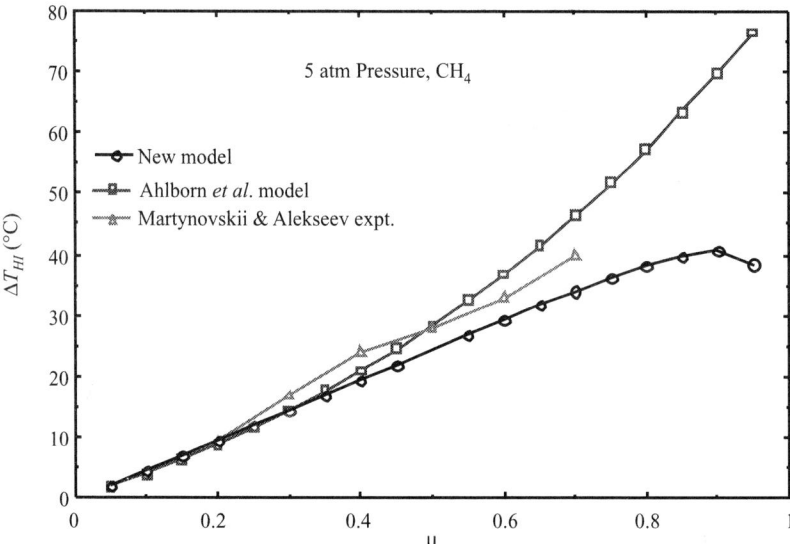

Fig. 6.16 Comparison of New Model with Experimental Results of Martynovskii and Alekseev (1956), and Ahlborn *et al.* (1998) Model for ΔT_{HI} of CH_4 at 5 atm

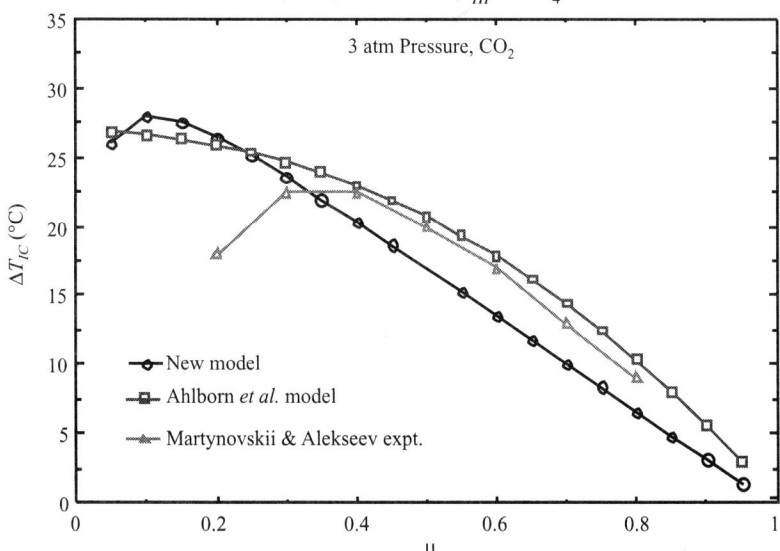

Fig. 6.17 Comparison of New Model with Experimental Results of Martynovskii and Alekseev (1956), and Ahlborn *et al.* (1998) Model for ΔT_{IC} of CO_2 at 3 atm

Vortex Tube Refrigeration Systems 247

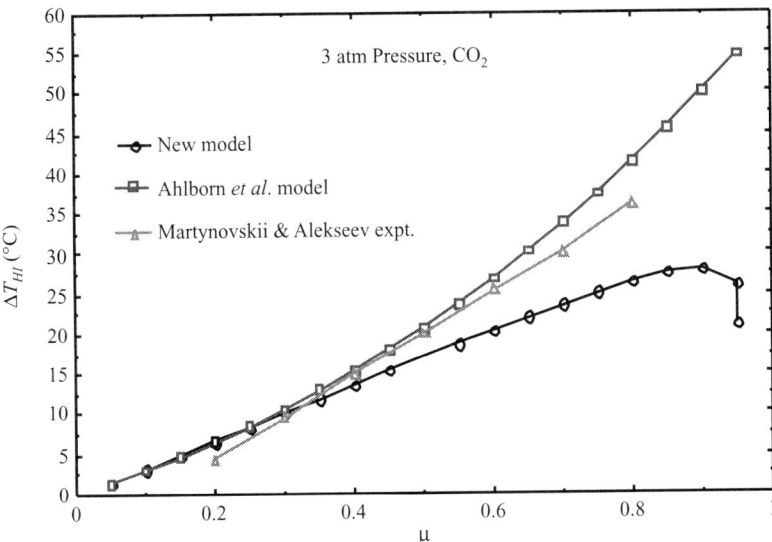

Fig. 6.18 Comparison of New Model with Experimental Results of Martynovskii and Alekseev (1956), and Ahlborn *et al.* (1998) Model for ΔT_{HI} of CO_2 at 3 atm

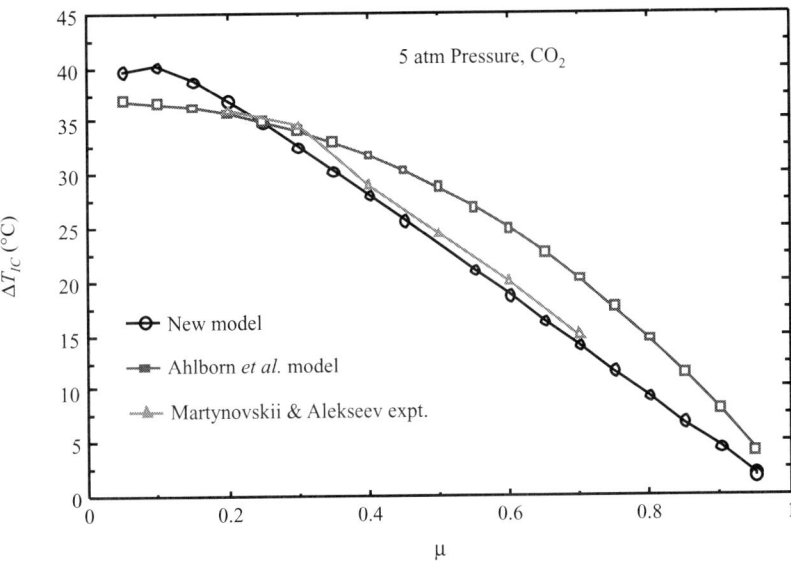

Fig. 6.19 Comparison of New Model with Experimental Results of Martynovskii and Alekseev (1956), and Ahlborn *et al.* (1998) Model for ΔT_{IC} of CO_2 at 5 atm

248 *Alternatives in Refrigeration and Air Conditioning*

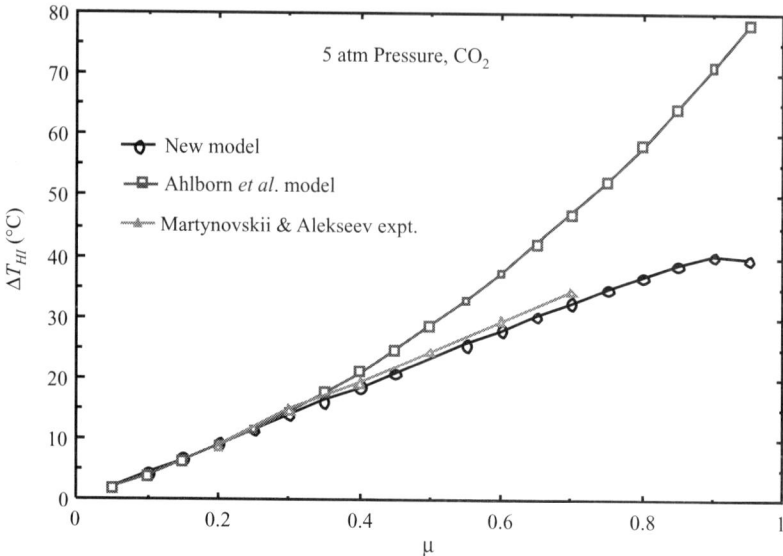

Fig. 6.20 Comparison of New Model with Experimental Results of Martynovskii and Alekseev (1956), and Ahlborn *et al.* (1998) Model for ΔT_{HI} of CO_2 at 5 atm

The results of temperature difference of hot and inlet air for the new model are quite close to experimental results of Promvonge and Eiamsa-ard (2005) (Fig. 6.21). Large deviations from the experimental results have been observed for temperature difference of inlet and cold air (Fig. 6.22) but still this new model provides better results than the Ahlborn *et al.* (1998) model. It is clear from Figs. 6.21 and 6.22 that the values of the ΔT_{HI} and ΔT_{IC} for $N = 1$ and $N = 2$ for Ahlborn *et al.* (1998) model are same. But for the case of new model the aforesaid results are different, thus represents that new model is sensitive to the change in the value of N, i.e., a geometric parameter of the vortex tube.

Table 6.2 Percentage Deviation of ΔT_{IC} and ΔT_{HI}

Gas P_I = 5 atm		New Model		Ahlborn *et al.* (1998) Model	
		Min. % Deviation	Max. % Deviation	Min. % Deviation	Max. % Deviation
AIR	For ΔT_{IC}	(+) 0.56	(+) 28.00	(+) 8.03	(+)37.09
	ΔT_{HI}	(+) 0.57	(+) 31.56	(+) 8.04	(+)37.09
CH_4	For ΔT_{IC}	(+) 0.53	(−) 18.56	(−) 6.92	(+)17.12
	ΔT_{HI}	(+) 0.52	(−) 18.58	(−) 6.94	(+)16.13
CO_2	For ΔT_{IC}	(+) 2.50	(−) 6.87	(−) 0.78	(+)34.80
	ΔT_{HI}	(+) 2.50	(−) 5.61	(−) 0.79	(+)37.12

(+) means more value than experimental
(−) means less value than experimental

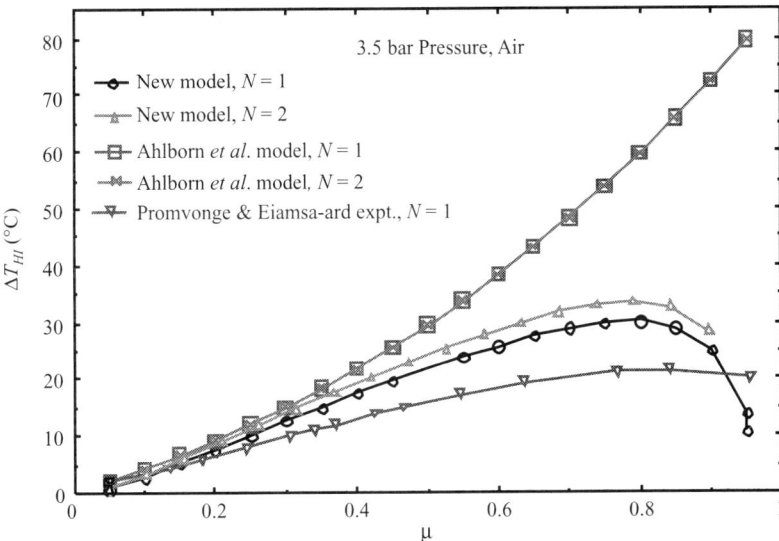

Fig. 6.21 Comparison of New Model with Experimental Results of Promvonge and Eiamsa-ard (2005), and Ahlborn *et al.* (1998) Model for ΔT_{HI} of Air

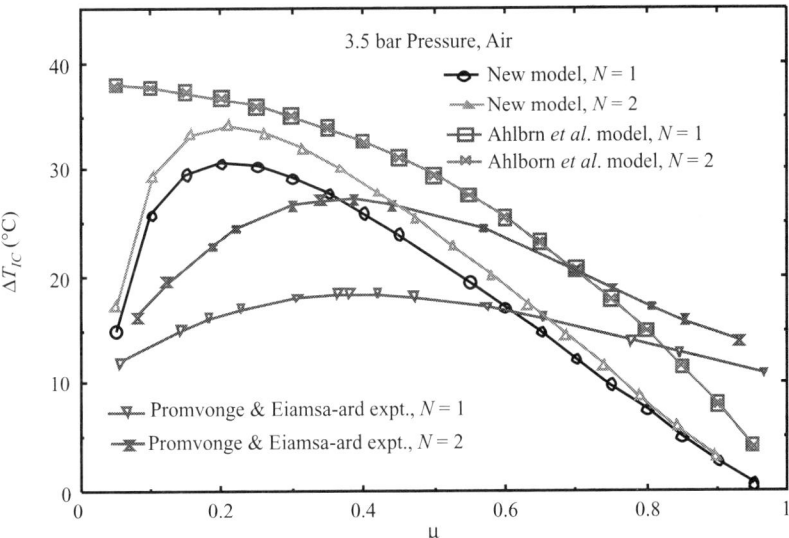

Fig. 6.22 Comparison of New Model with Experimental Results of Promvonge and Eiamsa-ard (2005), and Ahlborn *et al.* (1998) Model for ΔT_{IC} of Air

250 *Alternatives in Refrigeration and Air Conditioning*

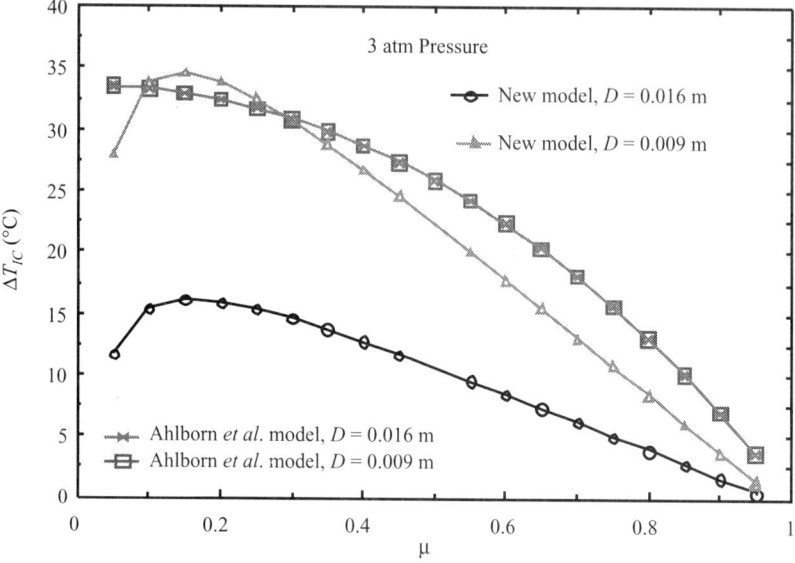

Fig. 6.23 Effect of Diameter of Hot Tube on ΔT_{IC} for Air

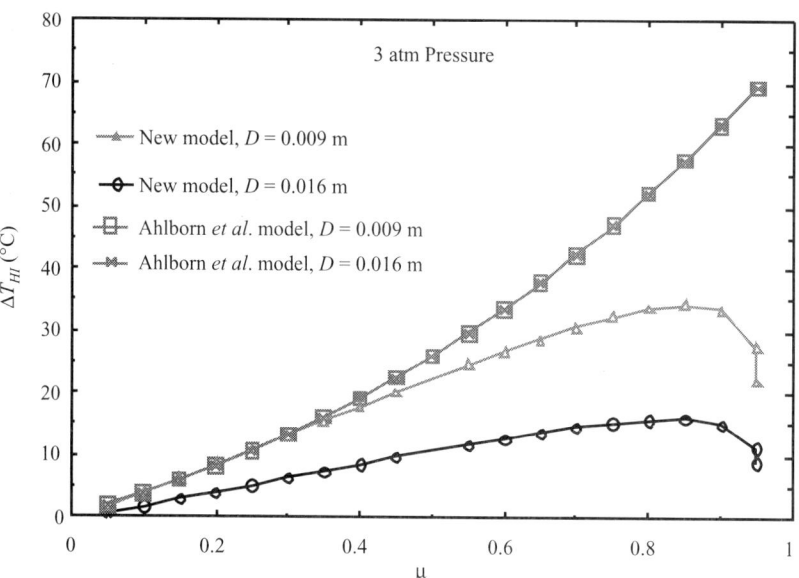

Fig. 6.24 Effect of Diameter of Hot Tube on ΔT_{HI} for Air

Figures 6.23-6.24 and 6.25-6.26 shows the effect of variation of hot tube diameter and number of nozzles respectively keeping other parameters constant as that of Martynovskii and Alekseev (1956) experiments for air. The model of Ahlborn *et al.* (1998) does not show any effect of geometric parameters, e.g., D and N. The bigger diameter tube gives larger energy separation. Similarly, the higher number of nozzles provides increased values of ΔT_{IC} and ΔT_{HI}. This trend is same as the available experimental results (Figs. 6.21 and 6.22). Thus, the major advantage of this new model is that it is sensitive to both thermodynamic parameters and geometric parameters. However, this new model has not shown expected results for all the parameters. This is because, the vortex effect is such a complex phenomenon that even the CFD models developed by other researchers do not conform to the performance so accurately to all the types and geometries of the vortex tubes tested experimentally so far. A generalized model can only be developed when complete phenomenon is clear. Thus, this new model cannot be used to design optimized vortex tube, but can only be used to subdesign the dimensions of the vortex tube required for preliminary investigations of vortex tube integrated systems.

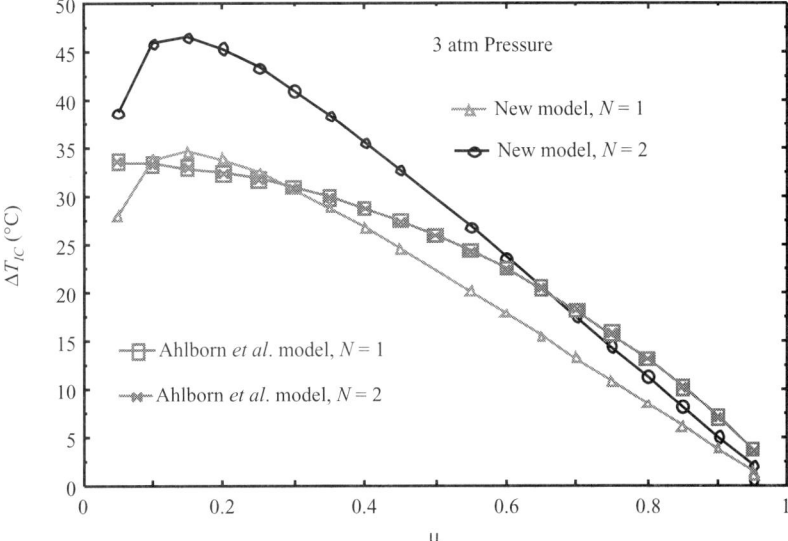

Fig. 6.25 Effect of Number of Nozzles on ΔT_{IC} for Air

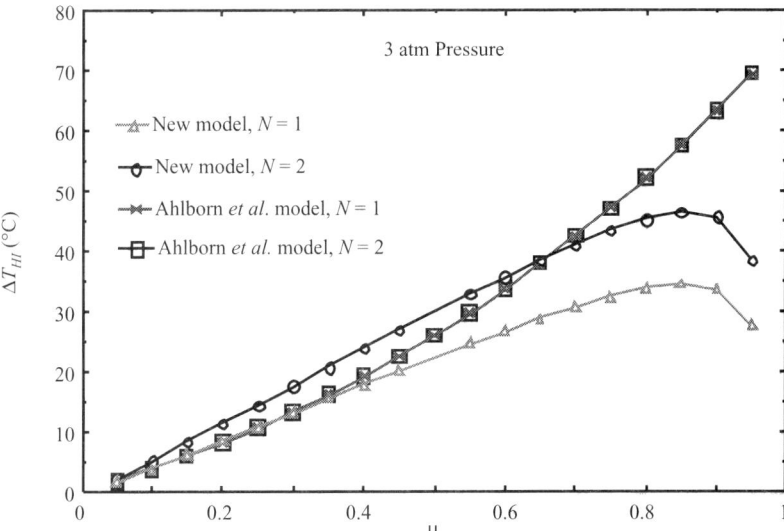

Fig. 6.26 Effect of Number of Nozzles on ΔT_{HI} for Air

6.6 EXERGY ANALYSIS OF THE VORTEX TUBE

6.6.1 Exergy Analysis

The exergy is the maximum work which can be obtained from a given form of energy using environmental parameters as the reference state (Kotas, 1985). The exergy destruction (ED) or entropy generation or irreversibility represents the deviations of the systems from the ideal systems which have ED equal to zero. Any system may be 100% efficient from energy point of view but that may not be an ideal system. The exergy analysis is a very effective thermodynamic tool to quantify the irreversibilities existing within the system.

The basic exergy balance equation applied to the control volume (CV) of any open type thermodynamic system under steady state steady flow conditions is given as below:

(Rate of Exergy incoming to CV) – (Rate of Exergy outgoing from CV) = Rate of ED or Rate of Entropy Generation or Irreversibility Rate within the CV

Consider the operation of vortex tube under steady state steady flow conditions. Further, neglect any kind of heat interactions of the vortex tube with the surroundings (i.e., vortex tube is insulated) and assume negligible potential energy and kinetic energy differences are associated with gas streams. Apply the exergy balance to the CV of vortex tube (Fig. 6.3) one gets,

$$\frac{ED}{m_I} = e_I - \mu e_C - (1-\mu) e_H \tag{6.61}$$

where,

$$e = (h - h_0) - T_0 (s - s_0) \tag{6.62}$$

For ideal gas (compressible fluid) (Balmer, 1988; Ogawa, 1993)

$$(h - h_0) = c_p (T - T_0) \tag{6.63}$$

$$(s - s_0) = c_p \ln \frac{T}{T_0} - R \ln \frac{P}{P_0} \tag{6.64}$$

So using equations (6.61) to (6.64), we get by assuming c_p as constant and same for all the fluid streams within the working temperature and pressure range, and $P_C \sim P_H$,

$$\frac{ED}{m_I} = T_0 \left\{ c_p \ln \left[\frac{\left(\frac{T_H}{T_C}\right)^{(1-\mu)}}{1 + (1-\mu)\left(\frac{T_H}{T_C} - 1\right)} \right] + R \ln \frac{P_I}{P_C} \right\} \tag{6.65}$$

For ideal conditions (reversible process), $ED = 0$

Thus, from equation (6.65),

$$\frac{\left(\dfrac{T_{HREV}}{T_{CREV}}\right)^{(1-\mu)}}{1+(1-\mu)\left(\dfrac{T_{HREV}}{T_{CREV}}-1\right)} = \left(\frac{P_C}{P_I}\right)^{\frac{\gamma-1}{\gamma}} \quad (6.66)$$

From equation (6.15) for ideal conditions,

$$T_I = \mu\, T_{CREV} + (1-\mu)\, T_{HREV} \quad (6.67)$$

Knowing the values of μ, P_C, P_I, T_I and γ of a particular gas and using equations (6.66) & (6.67) one can determine the values of T_{CREV} and T_{HREV} for ideal conditions within the vortex tube.

In almost all literature available on the vortex tube so far, equation (6.15) is used to describe the energetic balance of the phenomena. Generally, the value of T_I and μ are known, thus the values of T_C and T_H cannot be determined using this single equation. So other equations are required which describe the relation between T_C and T_H based on some parameters of the phenomena. As mentioned in the introduction that different types of models are available which describe this relationship. In the present research, a new model based on the ramming effect and heat exchanger in the Ranque Hilsch vortex tube has been used in the simulation of the vortex effect (Section 6.4).

The ramming of the gas is considered at two places within the vortex tube, viz., firstly at secondary circulation (Ahlborn and Groves, 1997) and secondly at adjustable valve (Fig. 6.3) of vortex tube. Further, a counter flow type heat exchanger is considered operating between hot gas, ram gas and cold gas, gas exiting from the nozzle/s of the vortex tube. The velocity profile within the vortex tube has been assumed to be given by Ogawa Combined Vortex Model (Ogawa, 1993). The outcome of this is a set of mathematical expressions, which describe the relationship between T_C and T_H. Therefore, knowing the thermodynamic and geometric parameters of the vortex tube one can easily determine T_H and T_C.

6.6.2 Results and Discussion

The energy analysis of the vortex tube presented in section 6.3 shows the existence of various types of phenomenon in a vortex tube, viz., heat engine, heat pump, refrigerator, ramming effect, heat exchanger and acoustics. So, one can use different ways to model the temperature separation in the vortex tube. Further, the temperature separation and exergy destruction calculations have performed using the various equations described in Section 6.5.1 using Engineering Equation Solver Software (Klein and Alvaradro, 2005). The in-built property data in this software for different gases have been used. The simulated results using new model for temperature separation based on ramming effect and heat exchanger within vortex tube, and experimental data of Martynovskii and Alekseev (1956) have been compared with results for temperature separation for reversible process within the vortex tube. These results have been presented graphically in Figs. 6.27 to 6.30 and in Table 6.3. In Figs. 6.27 to 6.30 the logarithmic scale is used for y-axis in order to have more clarity of presenting large data in a single figure. The input parameters for new model have been presented in Table 6.3. These parameters are same as that of Martynovskii and Alekseev (1956) experiments.

Figures 6.27 to 6.30 show that the values of ΔT_{IC} and ΔT_{HI} for ideal conditions in the vortex tube are in the increasing order for different gases as follows:

$$CO_2 < CH_4 < Air$$

This order is same as discussed by the Martynovskii and Alekseev (1956). The experimental results for 5 atm inlet pressure deviate from this order. At this pressure the values of ΔT_{IC} and ΔT_{HI} are in the increasing order as $CO_2 < Air < CH_4$. This might be due to the presence of water vapours in the air which decreases the values of ΔT_{IC} and ΔT_{HI} due to transfer of heat to the cold air by condensation of water vapours at high pressure. The simulated results for ΔT_{IC} and ΔT_{HI} are in the same order as that of ideal conditions. This might be because of dry gas property data has been used in the calculations.

Table 6.3 Input Data for Analysis of Vortex Tube

Parameter	Details
Gas	Air/ CH_4/ CO_2
P_3	5 atm
P_1	1 atm
T_3	20°C for Air 21°C for CH_4 -0.50°C for CO_2
D	9 mm
L	450 mm
B	4 mm
N	1
d_N	2.30 mm

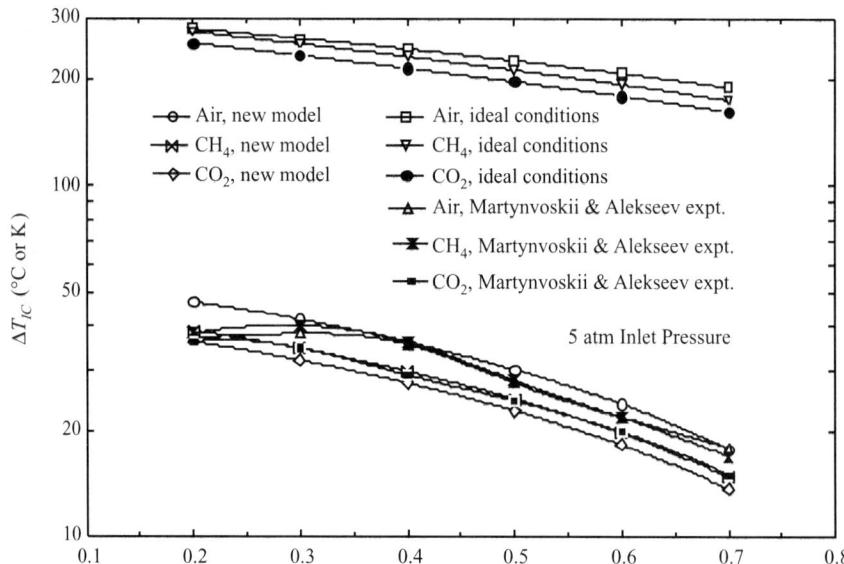

Fig. 6.27 Comparison of Simulated and Experimental Data for ΔT_{IC} of 5 atm Air, CH_4 & CO_2 with Results at Ideal Conditions

Vortex Tube Refrigeration Systems 255

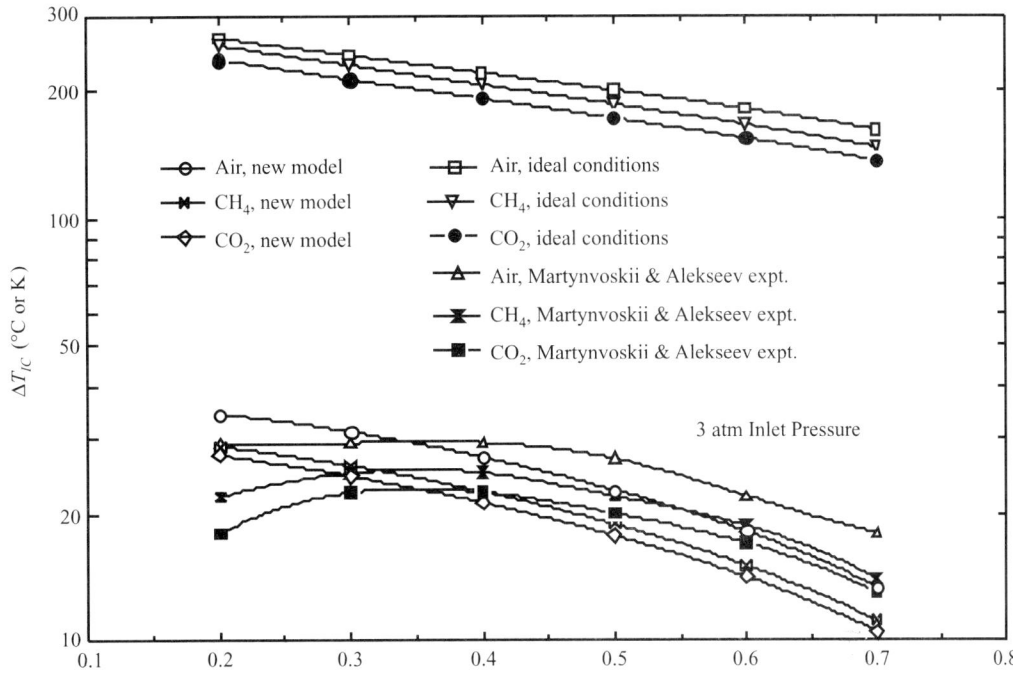

Fig. 6.28 Comparison of Simulated and Experimental Data for ΔT_{IC} of 3 atm Air, CH_4 & CO_2 with Results at Ideal Conditions

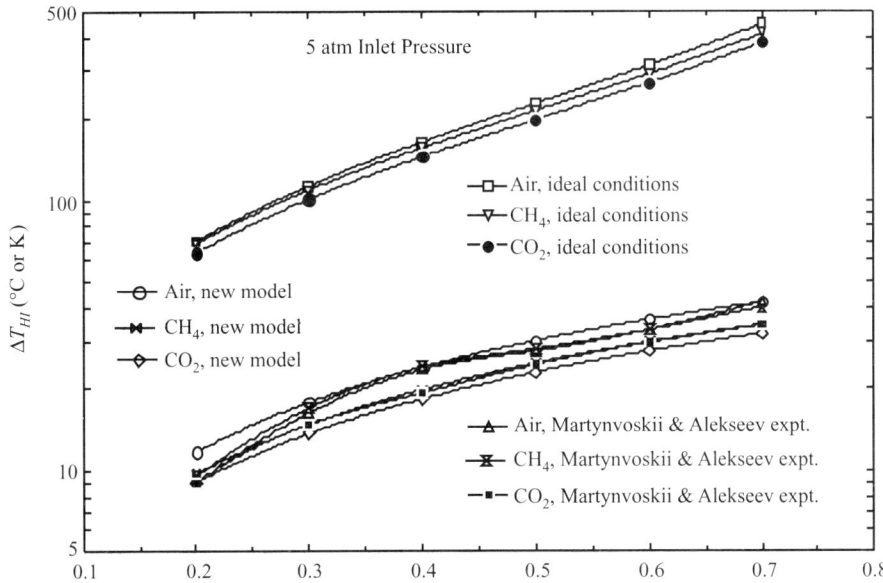

Fig. 6.29 Comparison of Simulated and Experimental Data for ΔT_{HI} of 5 atm Air, CH_4 & CO_2 with Results at Ideal Conditions

256   *Alternatives in Refrigeration and Air Conditioning*

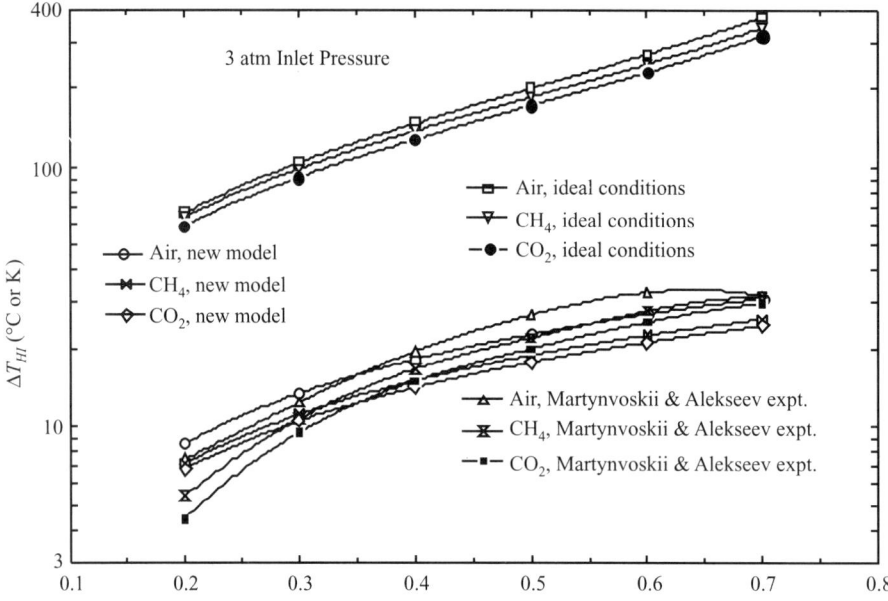

Fig. 6.30 Comparison of Simulated and Experimental Data for ΔT_{HI} of 3 atm Air, CH_4 & CO_2 with Results at Ideal Conditions

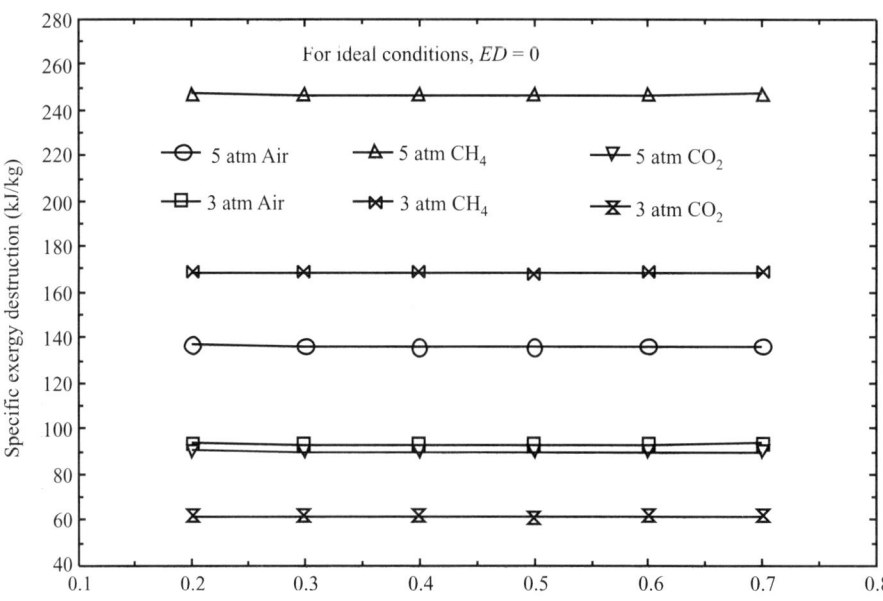

Fig. 6.31 Effect of Pressure and Type of Gas on Simulated ED having Input Temperature Different for Each Gas as per Table 6.3

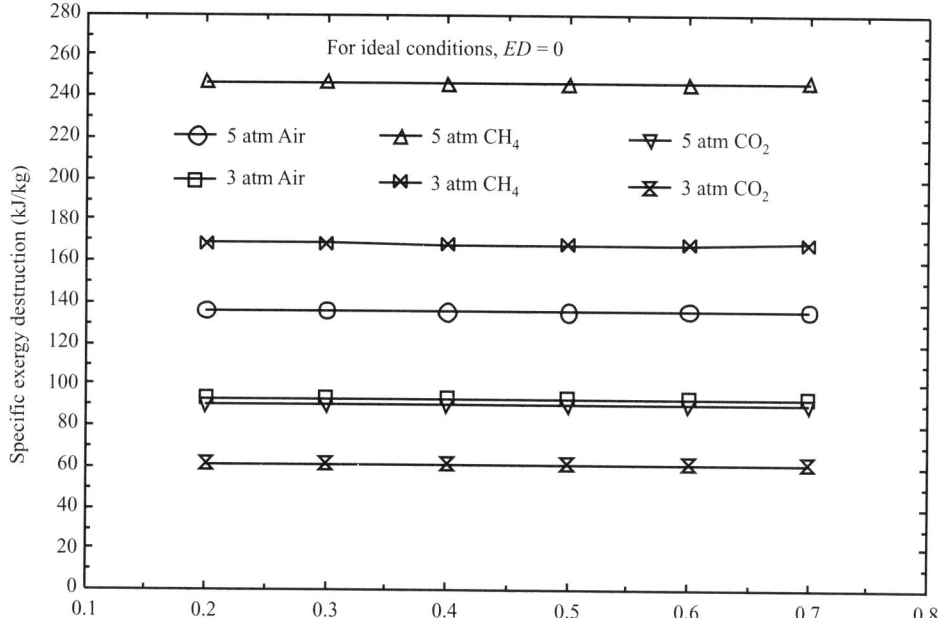

Fig. 6.32 Effect of Pressure and Type of Gas on Simulated *ED* having Input Temperature Same (20°C) for Each Gas

The order of exergy destruction within the vortex tube is different from the temperature difference and is as follows in the decreasing order: CH_4, air and CO_2 (Fig. 6.31). This means that the gas with large molecular weight leads to lower exergy destruction. This might be due to the presence of more centrifugal forces which contribute to effective energy separation. Further, Fig. 6.31 shows that the exergy destruction is higher at higher pressures for all the three gases. The simulated results for ED within the vortex tube for these three gases having same inlet temperature of 20°C have been shown in Fig. 6.32. These results also have the same trend as for the previous results shown in Fig. 6.31. The exegetic analysis has quantified the departure of the actual vortex tube and theoretical model of vortex tube from the ideal conditions (ref. Figs. 6.27 to 6.30). The inference can be drawn from these graphs that there exists a wide scope to improve the actual vortex tube considered here to reach near the ideal vortex tube.

Table 6.4 Percentage Deviations of Simulated Exergy Destruction

S. No.	Gas	M	ED_{expt} (kJ/kg)	ED (kJ/kg)	% Deviation of *ED*
1.	5 atm Air	0.20	137.02	136.70	− 0.23
2.		0.30	136.51	136.36	− 0.11
3.		0.40	136.14	136.18	0.03
4.		0.50	136.33	136.16	− 0.12
5.		0.60	136.41	136.27	− 0.11
6.		0.70	136.39	136.50	0.08

Contd...

7.	3 atm Air	0.20	93.56	93.43	−0.14
8.		0.30	93.29	93.22	−0.08
9.		0.40	92.93	93.11	0.19
10.		0.50	92.68	93.09	0.43
11.		0.60	92.72	93.14	0.46
12.		0.70	92.71	93.28	0.62
13.	5 atm CH_4	0.20	246.95	247.14	0.08
14.		0.30	245.66	246.64	0.40
15.		0.40	245.01	246.36	0.55
16.		0.50	245.41	246.30	0.36
17.		0.60	245.71	246.44	0.30
18.		0.70	245.94	246.78	0.34
19.	3 atm CH_4	0.20	169.10	168.84	−0.16
20.		0.30	168.52	168.53	0.01
21.		0.40	167.94	168.35	0.24
22.		0.50	167.71	168.31	0.36
23.		0.60	167.60	168.39	0.47
24.		0.70	167.93	168.59	0.39
25.	5 atm CO_2	0.20	90.03	90.12	0.09
26.		0.30	89.72	89.94	0.25
27.		0.40	89.66	89.85	0.20
28.		0.50	89.62	89.82	0.23
29.		0.60	89.66	89.87	0.23
30.		0.70	89.80	89.98	0.20
31.	3 atm CO_2	0.20	61.74	61.55	−0.31
32.		0.30	61.51	61.44	−0.12
33.		0.40	61.31	61.37	0.10
34.		0.50	61.22	61.36	0.23
35.		0.60	61.18	61.39	0.33
36.		0.70	61.27	61.46	0.31

The percentage deviations of exergy destruction calculated using new model from the exergy destruction determined for experimental data using equation (6.65) has been presented in Table 6.4. Only (+) 0.62% maximum deviation from the experimental results has been observed for simulated ED for 3 atm air. For other cases these deviations are quite less than this value. The ED firstly decreases with increase in the value of cold mass flow fraction (μ) and reaches a minimum level and then it starts increasing with further increase in the value of μ. The minima point occurs for μ between 0.40 and 0.50 for all the gases considered in this study. Although the maxima point for ΔT_{IC} occurs between 0.20 and 0.40 for actual as well as modelled vortex tube. Likewise, one can determine the behaviour of other parameters of the vortex tube using the energetic and exergetic analysis.

6.7 VORTEX TUBE INTEGRATED REFRIGERATION AND AIR CONDITIONING SYSTEMS

Bartlett (1960) patented a novel vapour cycle refrigeration system which utilizes a vortex tube as a heat exchanger in place of the conventional condenser for dissipating heat to the surrounding atmosphere. He proposed a vortex tube of the type having a closed end hot tube so that none of the refrigerant escapes to the atmosphere. This type of vortex tube operates with the temperature of the fluid flowing in the hot pipe above the temperature admitted to the tube. This temperature difference between the inlet temperature of the gas and the temperature of the gas in the hot tube results in a greater temperature differential between the temperature of the gas and the surrounding coolant, thus permitting a decrease in the size of heat exchange surface required.

Kawashima and Araki (1987) described in their patent that integrating a vortex tube between the compressor and the condenser in a reversed Rankine cycle will intensify heat recovery and improve the performance of the system.

Christensen et al. (2001) carried out experimental and theoretical investigations on the possibility of using a vortex tube in refrigeration systems as expansion device. The theoretical analysis demonstrates that it should be possible to improve the COP up to 20% compared with a traditional refrigeration system. CFD calculations had been carried out considering CO_2 as an ideal gas resulting in a large temperature separation (difference between warm and cold side of the vortex tube). They repeated the calculations considering CO_2 as real gas and found these did not result in a large temperature separation. They produced a prototype vortex tube with the same geometry as used in the CFD calculations. The vortex tube was mounted on a CO_2 test rig. The measurements were carried out with the same conditions as during the CFD calculations and demonstrated the same result. The vortex tube was tested in different pressure and temperature areas, however, without resulting in the desired temperature separation. In addition, another vortex tube was produced which had to expand into the two-phase area. The vortex tube was produced in such a way that the liquid occurring during the expansion was extracted from the flow and for that reason, should not be able to influence the Ranque-Hilsch effect. The measurements demonstrated that the warm and the cold outlet of the vortex tube had the saturation temperature and thus there was no heating of CO_2. They concluded that separation of liquid and gas phase and subsequent separation of the gas phase into a cold and a warm part followed by a cooling of the warm stream resulted in theoretical improvements in COP of conventional transcritical CO_2 compression refrigeration. However, they did not hold in reality with a vortex tube used with refrigerants, i.e., fluids not behaving as ideal gases.

Hebecker and Bittrich (2006) concluded that vortex tube is a failure in improving the efficiency of other cyclic processes such as heat pump or steam turbine, etc. They are hopeful of its use in electrostatic precipitation.

Fedorov *et al.* (2010) got U.S. patent for vortex tube refrigeration systems and methods. They disclosed that the vortex tube transcritical CO_2 compression refrigeration system with two compressors could have COP about 4.50 and that for conventional transcritical CO_2 compression refrigeration system about 2.80 only under typical conditions. They also presented the transcritical CO_2 compression refrigeration cycle on P-h diagram and comparative graph of performance of transcritical CO_2 compression refrigeration cycle with one and two compressors and conventional transcritical CO_2 compression refrigeration cycle.

6.7.1 Vortex Tube Integrated Transcritical CO_2 Compression Refrigeration and Air Conditioning Systems

The energy and exergy analysis of conventional TCCRS (transcritical CO_2 compression refrigeration system). TCCRS with internal heat exchanger (IHX) and vortex tube integrated transcritical CO_2 compression refrigeration and air conditioning systems has been presented in this section. Two types of configurations (i) with single compressor (ii) with double compressors for vortex tube integrated transcritical CO_2 compression refrigeration system (VTITCCRS) have been analyzed. The comparative study of all these systems and parametric study of VTITCCRS with double compressor has also been presented.

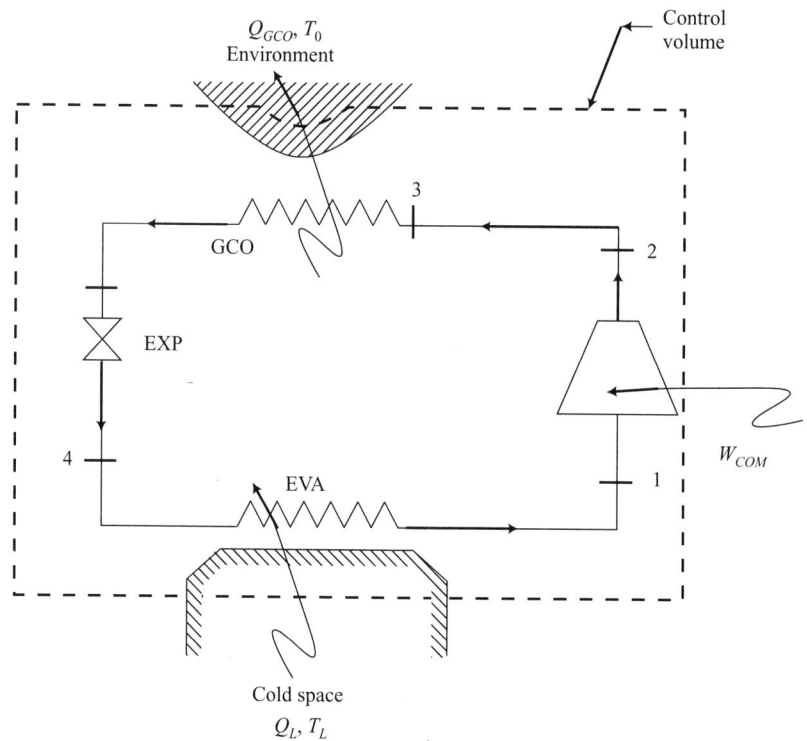

Fig. 6.33 Conventional Transcritical CO_2 Compression Refrigeration System

In a conventional transcritical CO_2 compression refrigeration system (Fig. 6.33) the refrigerant vapours (CO_2) coming out of the evaporator (EVA) are compressed above the critical conditions (P_{cr} = 73.77 bar and T_{cr} = 31.10°C). The compressed hot refrigerant is cooled upto slightly above the cooling medium temperature by using a gas cooler (GCO) instead of a condenser. Then, this high temperature and high pressure CO_2 is throttled using expansion device (EXP) to evaporator pressure and temperature. The conventional transcritical CO_2 compression refrigeration cycle (1-2-3-4) has been represented on P-h diagram in Fig. 6.34.

The use of internal heat exchanger (IHX) in TCCRS (ref. Fig. 6.35) leads to enhancement of its performance. Thermodynamic cycle 1a-2a-3a-4a is shown in Fig. 6.34 is for TCCRS with IHX.

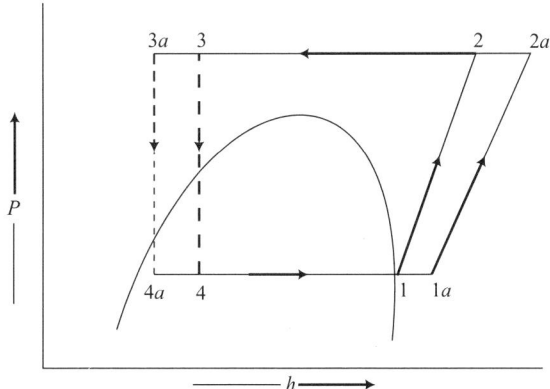

Fig. 6.34 Thermodynamic Cycle for Conventional TCCRS (cycle 1-2-3-4-1) and TCCRS with IHX (cycle 1a-2a-3a-4a-1a)

A vortex tube integrated transcritical CO_2 vapour compression refrigeration system with single compressor (COM) has been shown in Fig. 6.36 and its corresponding thermodynamic cycle has been represented on P-h diagram in Fig. 6.37. In this system a vortex tube has been integrated between the gas cooler and expansion device and the high pressure high temperature pressure CO_2 coming out of gas cooler has been passed through the vortex tube and separated into two different temperature CO_2 streams. The cold temperature stream has been throttled or expanded to evaporator pressure and temperature, and hot stream's pressure has been reduced using a back pressure valve (BPV). The efflux of vapours from evaporator has been mixed with refrigerant exiting from BPV. These mixed vapours have been compressed using a single compressor to the gas cooler pressure.

A vortex tube integrated transcritical CO_2 compression refrigeration system with double compressors has been sketched in Fig. 6.38 and its corresponding thermodynamic cycle in Fig. 6.39. There is no BPV in this system instead an additional second compressor is used to compress the intermediate pressure vapours of CO_2. The discharge from the first compressor (COM I) has been mixed with hot CO_2 vapours coming out of vortex tube. Then, these mixed CO_2 vapours have been compressed to gas cooler pressure using a second compressor (COM II). Rest of the system components and thermodynamic cycle are same as that for VTITCCRS with single compressor.

262 *Alternatives in Refrigeration and Air Conditioning*

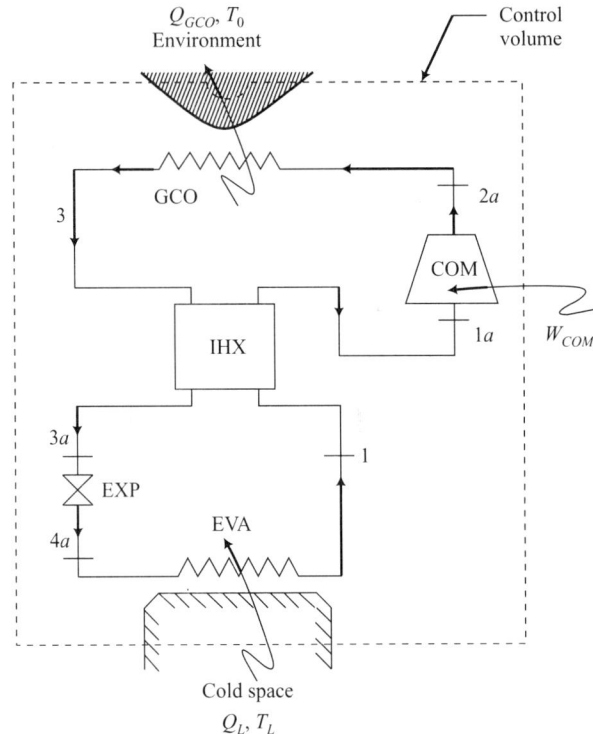

Fig. 6.35 Transcritical CO_2 Compression Refrigeration System with IHX

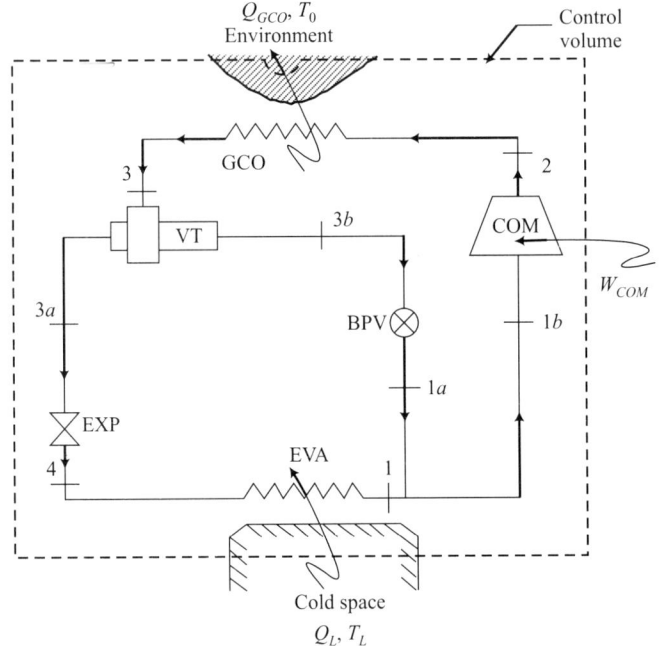

Fig. 6.36 Vortex Tube Integrated Transcritical CO_2 Compression Refrigeration System with Single Compressor

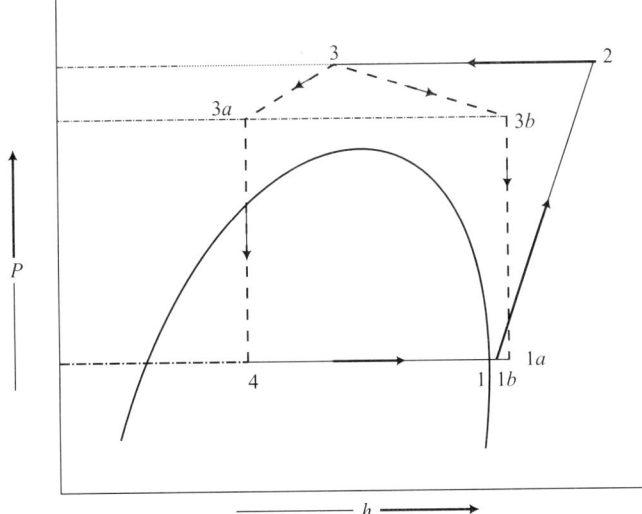

Fig. 6.37 Thermodynamic Cycle of VTITCCRS with Single Compressor

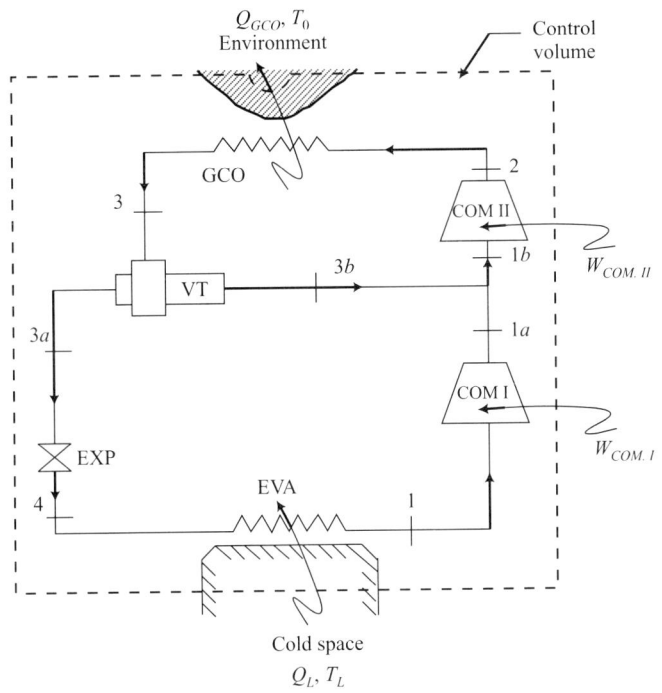

Fig. 6.38 Vortex Tube Integrated Transcritical CO_2 Compression Refrigeration System with Double Compressors

264 *Alternatives in Refrigeration and Air Conditioning*

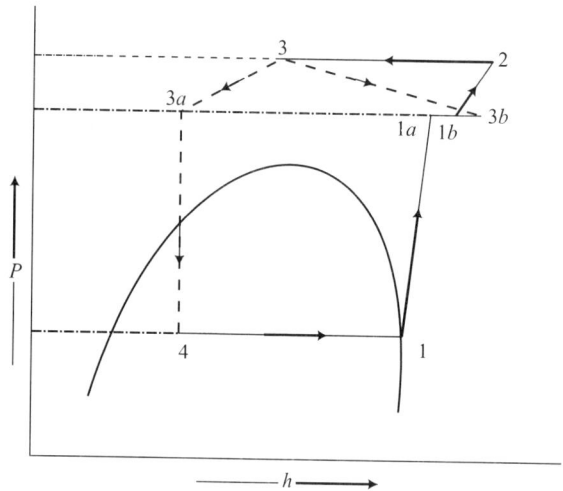

Fig. 6.39 Thermodynamic Cycle of VTITCCRS with Double Compressors

6.7.2 Energy and Exergy Analysis of Vortex Tube Integrated Transcritical CO_2 Compression Refrigeration and Air Conditioning Systems

An energy and exergy analysis of the different systems described in Section 6.6.1 has been performed using Engineering Equation Solver software (Klein and Alvaradro, 2005) and its inbuilt property data for carbon dioxide. The results of energetic performance of these four systems have been validated with published data available in literature (Fedorov *et al.*, 2010) and presented in the next section.

The elementary model for VTITCCRS with double compressors (Figs. 6.38 and 6.39) consists of the following major parts:

1. Compressor I
2. Mixer
3. Compressor II
4. Gas cooler
5. Vortex tube
6. Expansion device
7. Evaporator

The following equations have been used for energy and exergy analysis of the VTITCCRS with double compressors to know the energetic and exergetic performance as well as the exergy destruction in its individual components:

$$m_{1a} = m_1 \qquad (6.68)$$

$$m_2 = m_{1b} \qquad (6.69)$$

$$m_{1b} = m_{1a} + m_{3b} \qquad (6.70)$$

$$m_{3b} = (1 - \mu) m_3 \qquad (6.71)$$

$$m_3 = m_2 \qquad (6.72)$$

$$m_{3a} = \mu m_3 \qquad (6.73)$$

$$m_4 = m_{3a} \tag{6.74}$$

$$m_1 = m_{3a} \tag{6.75}$$

$$P_1 = \text{saturation pressure at } T_1 \tag{6.76}$$

$$h_1 = f(T_1, x_1); \text{ for saturated conditions, } x_1 = 1 \tag{6.77}$$

$$s_1 = f(T_1, x_1) \tag{6.78}$$

$$s_{1a} = s_1 \tag{6.79}$$

$$h_{1a} = f(P_{1a}, s_{1a}) \tag{6.80}$$

$$h_3 = f(P_3, T_3) \tag{6.81}$$

$$s_3 = f(P_3, T_3) \tag{6.82}$$

$$P_{1a} = P_{1b} = P_{3a} = P_{3b} \tag{6.83}$$

$$P_3 = P_2 \tag{6.84}$$

T_{3a} and T_{3b} can be obtained using the mathematical model for energy separation in a vortex tube based on ramming effect and heat exchanger within it (ref. Section 6.3).

$$h_{3a} = f(P_{3a}, T_{3a}) \tag{6.85}$$

$$s_{3a} = f(P_{3a}, T_{3a}) \tag{6.86}$$

$$h_{3b} = f(P_{3b}, T_{3b}) \tag{6.87}$$

$$s_{3b} = f(P_{3b}, T_{3b}) \tag{6.88}$$

$$m_{1b} h_{1b} = m_{1a} h_{1a} + m_{3b} h_{3b} \tag{6.89}$$

$$s_{1b} = f(h_{1b}, P_{1b}) \tag{6.90}$$

$$s_2 = s_{1b} \tag{6.91}$$

$$h_2 = f(P_2, s_2) \tag{6.92}$$

$$h_4 = h_{3a} \tag{6.93}$$

$$s_4 = f(h_4, P_4) \tag{6.94}$$

$$m_1 = \frac{Q_L}{(h_1 - h_4)} \tag{6.95}$$

$$W_{COM\,I} = m_{1a} h_{1a} - m_1 h_1 \tag{6.96}$$

$$W_{COM\,II} = m_2 h_2 - m_{1b} h_{1b} \tag{6.97}$$

$$W_{TOT} = W_{COM\,I} + W_{COM\,II} \tag{6.98}$$

$$\text{COP} = \frac{Q_L}{W_{TOT}} \tag{6.99}$$

In general,

$$e = (h - h_0) - T_0(s - s_0) \tag{6.100}$$

Therefore,

$$ED_{COM\ I} = m_1 e_1 + E_{COM\ I} - m_{1a} e_{1a} \quad (6.101)$$

$$E_{COM\ I} = W_{COM\ I} \quad (6.102)$$

$$ED_{COM\ II} = m_{1b} e_{1b} + E_{COM\ II} - m_2 e_2 \quad (6.103)$$

$$E_{COM\ II} = W_{COM\ II} \quad (6.104)$$

$$ED_{GCO} = m_2 e_2 + E_{GCO} - m_3 e_3 \quad (6.105)$$

$$E_{GCO} = Q_{GCO}\left(1 - \frac{T_0}{T_R}\right) \quad (6.106)$$

Here,

$$T_R = T_0 \quad (6.107)$$

$$ED_{VT} = m_3 e_3 - m_{3a} e_{3a} - m_{3b} e_{3b} \quad (6.108)$$

$$ED_{EXP} = m_{3a} e_{3a} - m_4 e_4 \quad (6.109)$$

$$ED_{EVA} = m_4 e_4 + E_{EVA} - m_1 e_1 \quad (6.110)$$

$$E_{EVA} = Q_L\left(1 - \frac{T_0}{T_L}\right) \quad (6.111)$$

$$T_L = T_1 + 10 \quad (6.112)$$

$$ED_{MIX} = m_{1a} e_{1a} + m_{3b} e_{3b} - m_{1b} e_{1b} \quad (6.113)$$

$$ED_{TOT} = ED_{COM\ I} + ED_{COM\ II} + ED_{GCO} + ED_{VT} + ED_{EXP} + ED_{EVA} + ED_{MIX} \quad (6.114)$$

$$\varepsilon = 1 - \frac{ED_{TOT}}{W_{TOT}} \quad (6.115)$$

$$T_{MAX} = T_2 \quad (6.116)$$

Similarly, the mass, energy and exergy balance can be applied to individual components or overall system of other three cases (section 6.6.1), viz., conventional TCCRS, TCCRS with IHX and VTITCCRS with single compressor, to determine the energetic and exergetic performance as well as exergy destruction.

6.7.2.1 Results and Discussion

The details of input data considered in the present investigations are given in Table 6.5. Figure 6.40 represents the comparison of results of the theoretical models of all four different systems with that of published work (Fedorov et al., 2010). The percentage deviations for the case of VTITCCRS with double compressors from the published results are presented in Table 6.6. The results of COP using theoretical model with input data of Table 6.5 fairly match with the published results of COP

(Fedorov *et al.*, 2010) having maximum deviation = – 7.07%. The deviations for theoretical models of other systems from the published results (Fedorov *et al.*, 2010) are also quite less than this value (Fig. 6.40).

Table 6.5 Input Data for Analysis of TCCRS and VTITCCRS

S. No.	Input Parameter	For Validation of Results with Published Work	For Parametric Study
1.	T_L	—	– 10°C to 20°C
2.	T_{EVA}	10°C	$(T_L – 10)$°C
3.	T_{GCO}^{o}	40°C	20°C to 60°C
4.	P_{GCO}	135 bar	90 bar to 200 bar
5.	P_{VT}^{o}	45 bar to 100 bar	45 bar to 100 bar
6.	μ	0.50	0.45 to 0.95
7.	T_0		– 25°C
8.	P_0		– 1.01 bar
9.	Cooling Capacity	10 TR	10 TR
10.	Refrigerant	Carbon Dioxide	Carbon Dioxide

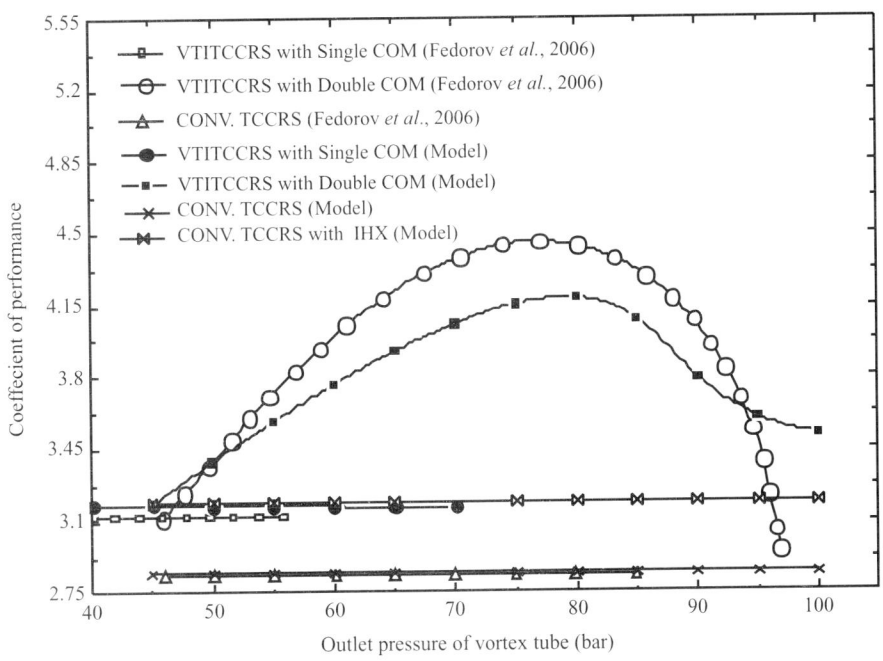

Fig. 6.40 Comparison of Theoretical Models of Different Systems with Published Work (Fedorov *et al.*, 2010)

Table 6.6 Deviation of Theoretical Model of VTITCCRS with Double Compressors

S. No.	Outlet pressure of vortex tube (bar)	Coefficient of performance (Fedorov et al., 2010)	Coefficient of performance (model)	Deviation %
1.	50	3.39	3.39	0.04
2.	55	3.71	3.58	− 3.46
3.	60	3.97	3.76	− 5.25
4.	65	4.21	3.93	− 6.75
5.	70	4.37	4.06	− 7.07
6.	75	4.44	4.15	− 6.41
7.	80	4.44	4.18	− 5.78
8.	85	4.30	4.08	− 5.04
9.	90	4.04	3.79	− 6.01
10.	95	3.43	3.60	5.03

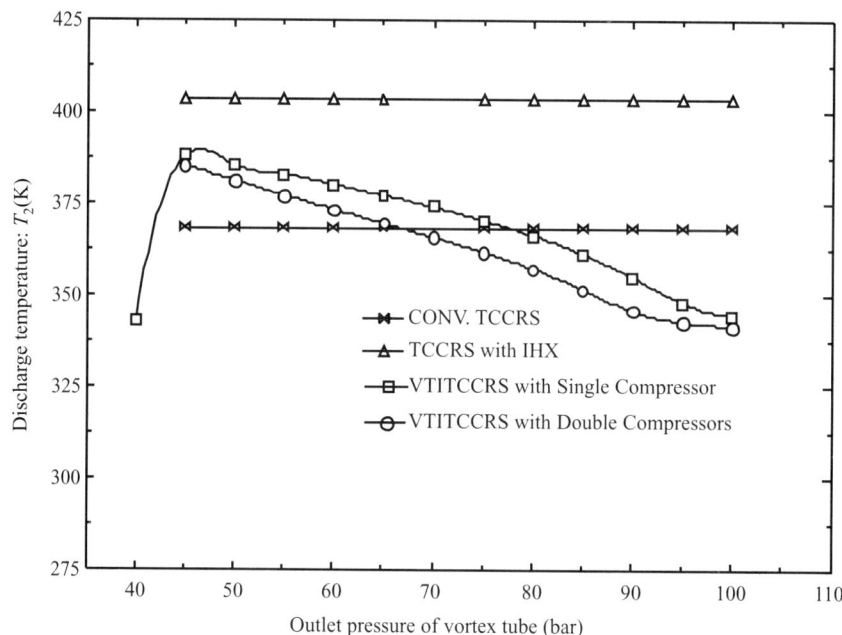

Fig. 6.41 Variation of Discharge Temperature of High Pressure Stage Compressor

Here the compression process for conventional TCCRS has been considered with 90% isentropic efficiency and for other systems isentropic compression process has been taken. It has been observed that the COP of the VTITCCRS with double compressors is higher than the other systems. A maximum COP ~ 4.50 has been reported at 76 bar outlet pressure of vortex tube and the theoretical model give maximum COP ~ 4.20 at approx. 79 bar outlet pressure of vortex tube. The COP values reported for VTITCCRS with single compressor and conventional TCCRS are 3.12 and 2.84 respectively, and the corresponding values with theoretical model are 3.17 and 2.85. This means about 50% enhancements in COP can be achieved by VTITCCRS having double compressors. The value of COP remains almost constant with increase in outlet pressure of vortex tube for VTITCCRS with single compressor and conventional TCCRS (no VT present). The COP increases with increase in outlet pressure of vortex tube, attains maximum value at certain pressure and then starts decreasing with further increase in outlet pressure of vortex tube for the case of VTITCCRS.

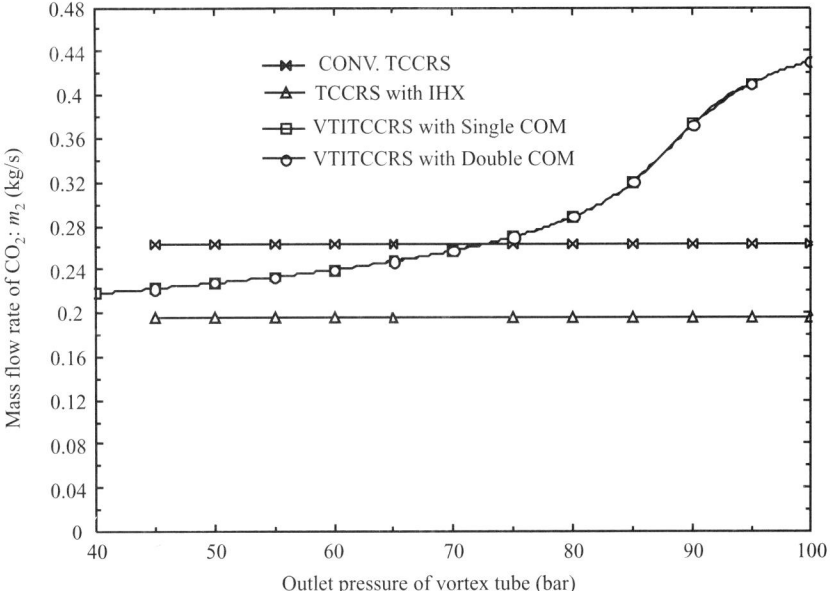

Fig. 6.42 Mass Flow Rate of CO_2 Through High Pressure Stage Compressor

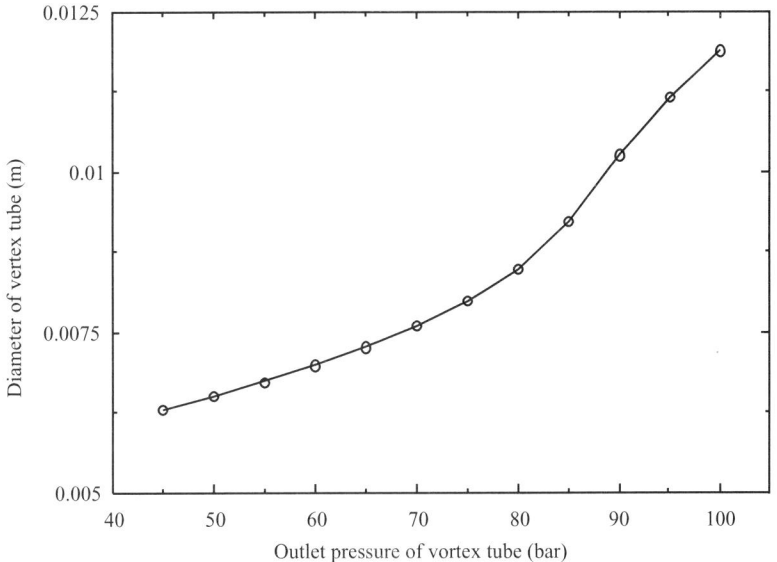

Fig. 6.43 Diameter of Vortex Tube Required at Various Outlet Pressures of Vortex Tube

This behaviour may be due to decrease in the discharge temperature of the high-pressure stage compressor is more than the rise in the mass flow rate of this compressor at lower pressures. However, at higher pressures the rise in mass flow rate dominates the decrease in the temperature that results in higher work input and hence less COP (ref Figs. 6.41 & 6.42). The higher value of COP of VTITCCRS with double compressors than the other systems is due to the less discharge temperature of second compressor, which leads to less work input required.

The diameter required at different outlet pressure of the vortex tube can be noted from Fig. 6.43. There is requirement of bigger diameter vortex tube at higher outlet pressure of vortex tube to compensate the increased mass flow rate of refrigerant through the evaporator.

The effect of cold space temperature on COP, exergetic efficiency and exergy destruction in each component of VTITCCRS with double compressors has been presented in Figs. 6.44 and 6.45. The COP increases with the rise in the cold space temperature due to low discharge temperature from compressors at higher values of evaporator temperature, subsequently less work input.

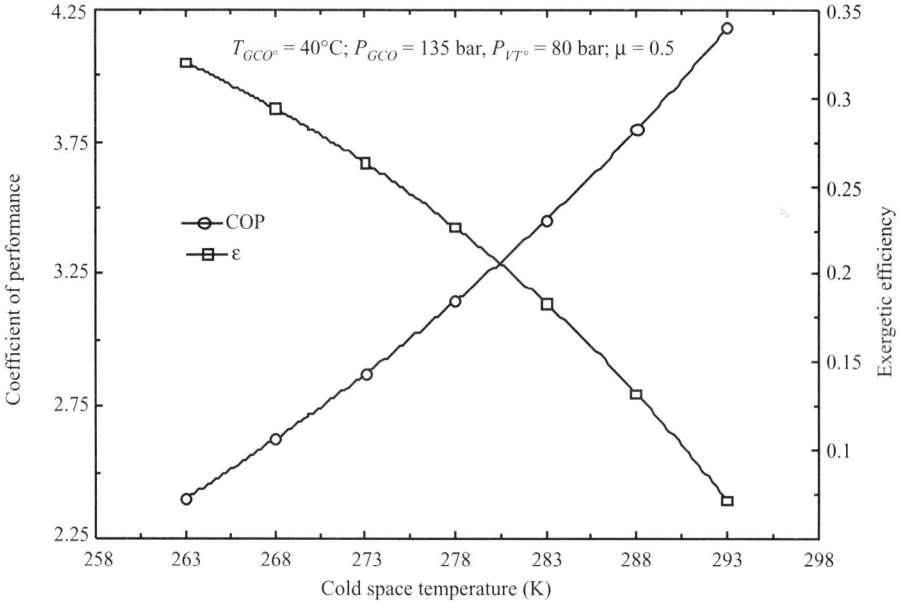

Fig. 6.44 Effect of Cold Space Temperature on COP and Exergetic Efficiency

The results for exergetic efficiency represents opposite behaviour than that of COP. An exergetic efficiency decreases with increase in cold space temperature. This may be mainly because decrease in total exergy destruction is less than the decrease in the work input with increase in the cold space temperature requirement as is seen in Fig. 6.45. The curve for total work input has more slope than the total exergy destruction in the system. Thus, this system gives lower exergetic efficiency for air conditioning applications than if it is used for domestic refrigeration or display cabinets. It may be mentioned here that COP is the first law of efficiency and exergetic efficiency is the second law of efficiency (which is always less than 100%). The opposing trends of COP and exergetic efficiency in Fig. 6.44 implies that at higher values of cold space temperature despite having higher values of

COP, the system is far from ideal as its exergetic efficiency is quite less. Thus, there exists large scope for improvements within the system. On the other hand, at lower temperatures of cold space, the system exhibits poor COP but higher exergetic efficiency, and thus, scope for further improvements w.r.t. ideal case is comparatively less.

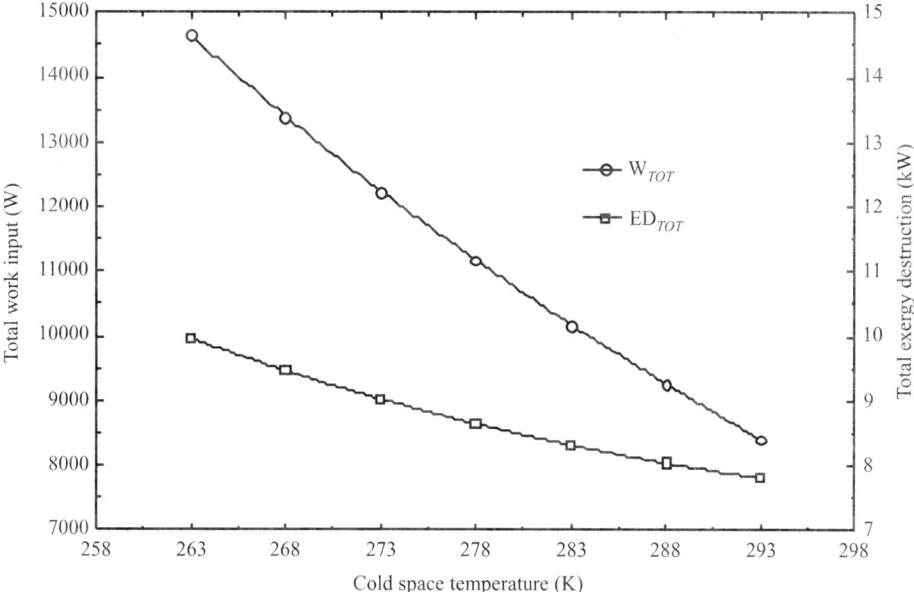

Fig. 6.45 Effect of Cold Space Temperature on Total Work Input and Total Exergy Destruction

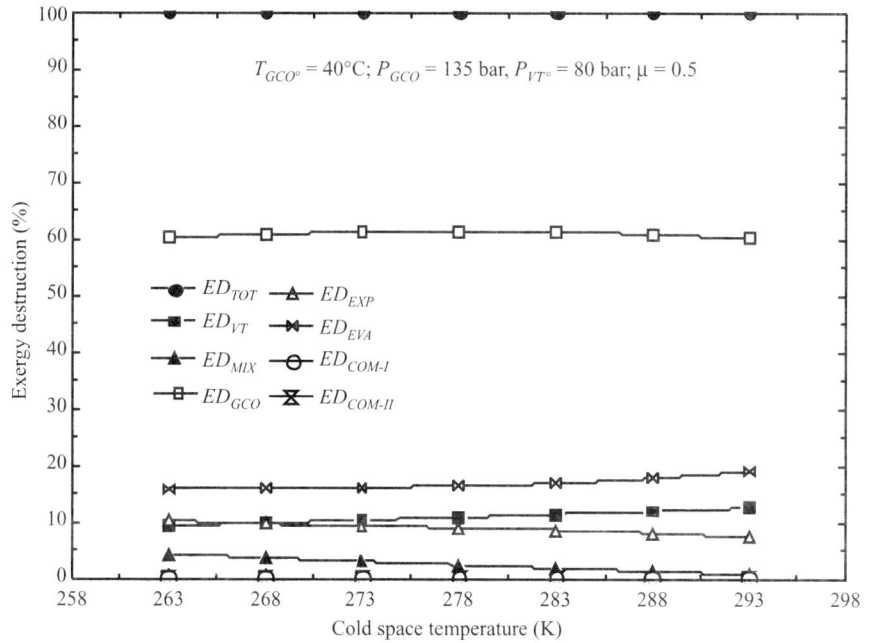

Fig. 6.46 Effect of Cold Space Temperature on Exergy Destruction

272 *Alternatives in Refrigeration and Air Conditioning*

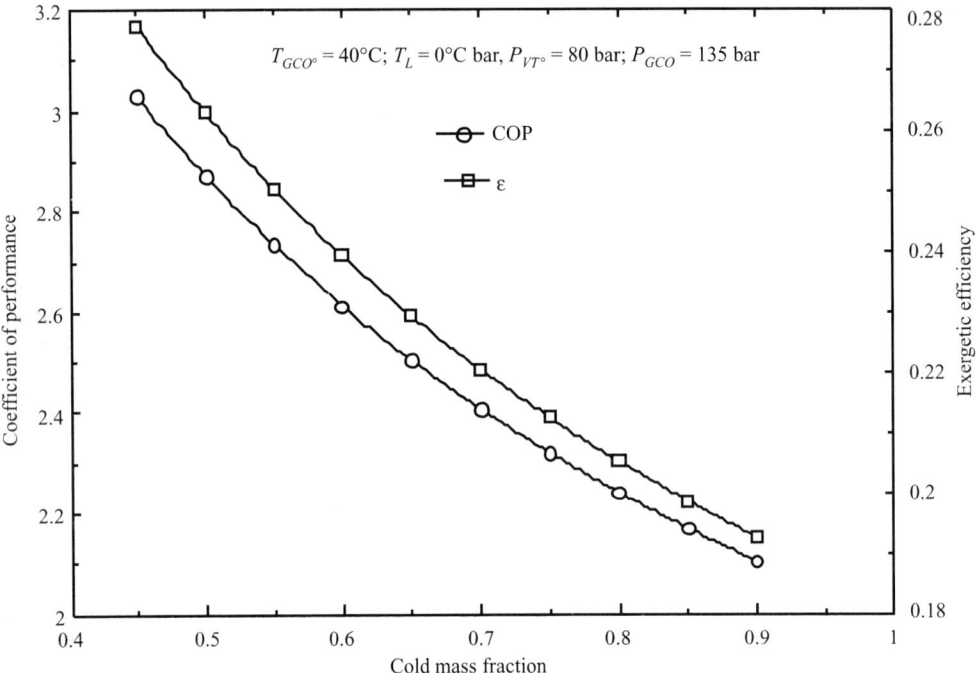

Fig. 6.47 Effect of Cold Mass Fraction on COP and Exergetic Efficiency

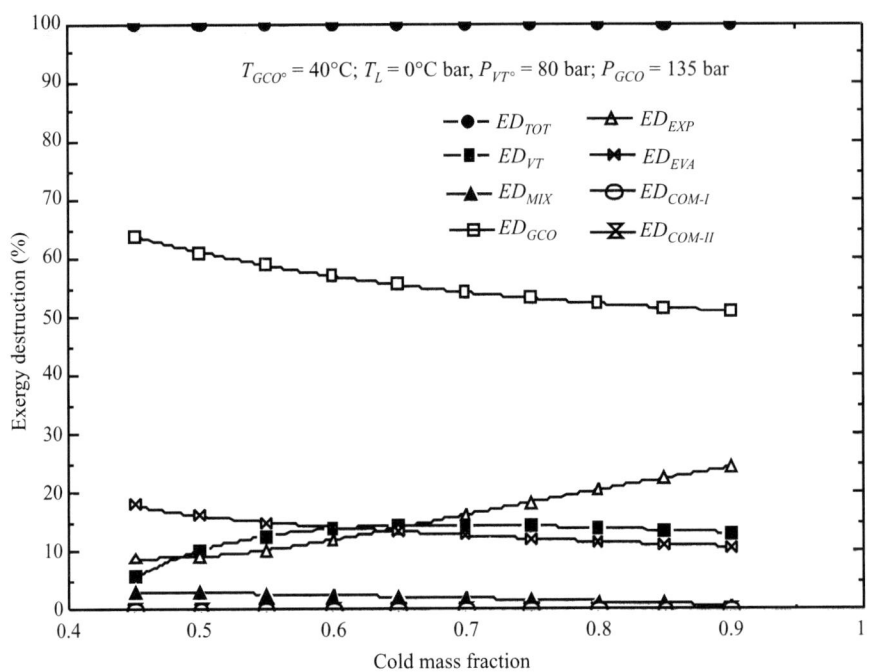

Fig. 6.48 Effect of Cold Mass Fraction on Percentage Exergy Destruction

The percentage of exergy destruction in each component of VTITCCRS with double compressors has been shown in Fig. 6.46. The gas cooler exhibits more exergy destruction (about 60%) than the other components, which have less than 20% exergy destruction. The order of exergy destruction in other components is as $ED_{EVA} > ED_{VT} > ED_{EXP} > ED_{MIX}$. As the compression process is assumed ideal, the exergy destruction in the compressors is nil. With increase in the cold space temperature, the percentage exergy destruction in evaporator and vortex tube increases, and in expansion device and mixer, it decreases.

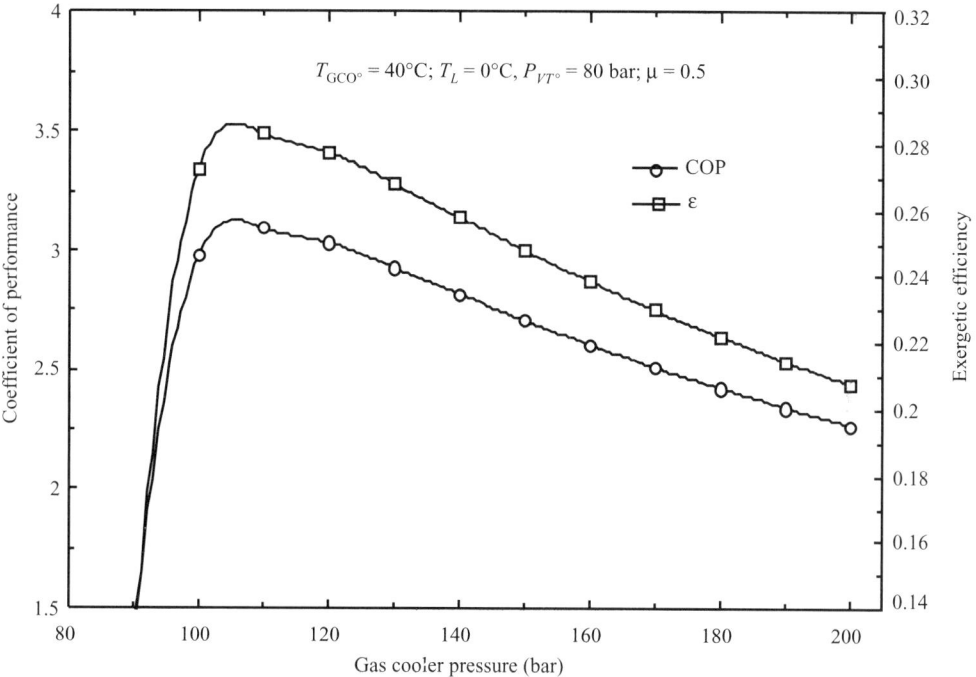

Fig. 6.49 Effect of Gas Cooler Pressure on COP and Exergetic Efficiency

Both the COP and exergetic efficiency decreases with increase in the cold mass fraction (Fig. 6.47). Figure 6.48 shows that the percentage exergy destruction in the expansion device increases and for gas cooler, evaporator and mixer it decreases with increase in the value of cold mass fraction. In the case of vortex tube the percentage exergy destruction first increases, reaches maximum value and then starts decreasing. The total exergy destruction, total work input, mass flow rate through the first as well as second compressor and discharge temperature of high stage compressor increases with increase in the cold mass fraction.

With increase in the gas cooler pressure from 90 bars to 200 bars, the COP and exergetic efficiency first increases, reaches maximum at approx. 105 bars and then starts decreasing with further increase in the gas cooler pressure (ref. Fig. 6.49). The effect of gas cooler pressure on the percentage exergy destruction in the individual components has been presented in Fig. 6.50.

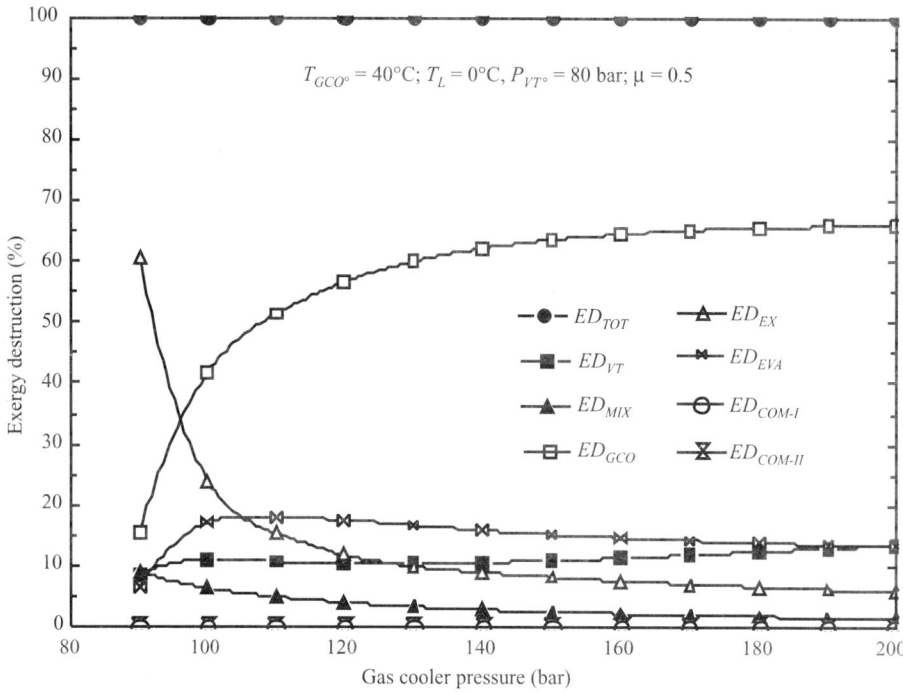

Fig. 6.50 Effect of Gas Cooler Pressure on Percentage Exergy Destruction

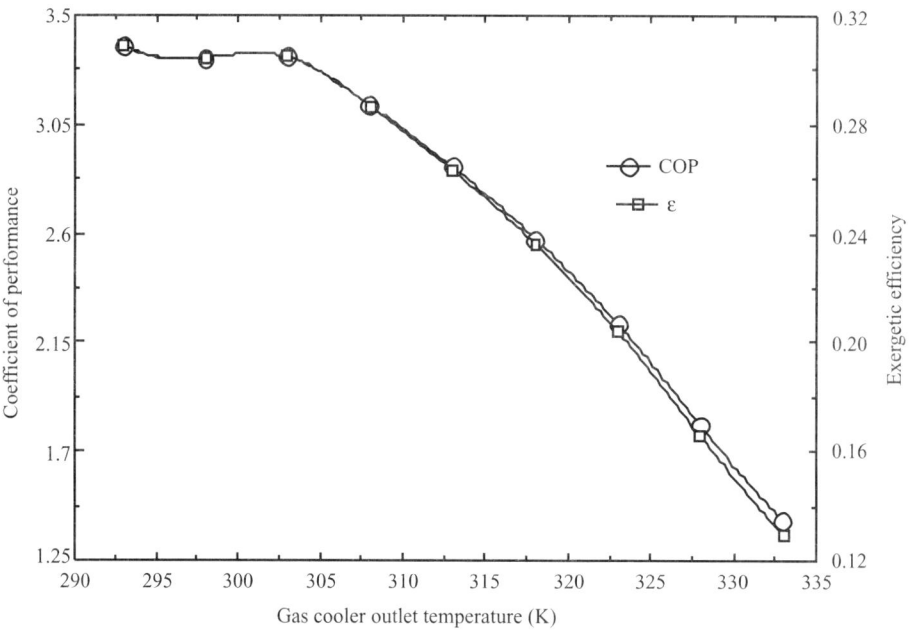

Fig. 6.51 Effect of Gas Cooler Outlet Temperature on COP and Exergetic Efficiency

The COP and exergetic efficiency decreases when the temperature of the gas exiting from the gas cooler increases (Fig. 6.51). This implies that the water cooled gas cooler provides better performance than the air cooled one. Figure 6.52 represents the effect of gas cooler outlet temperature on the percentage exergy destruction in each component of the system.

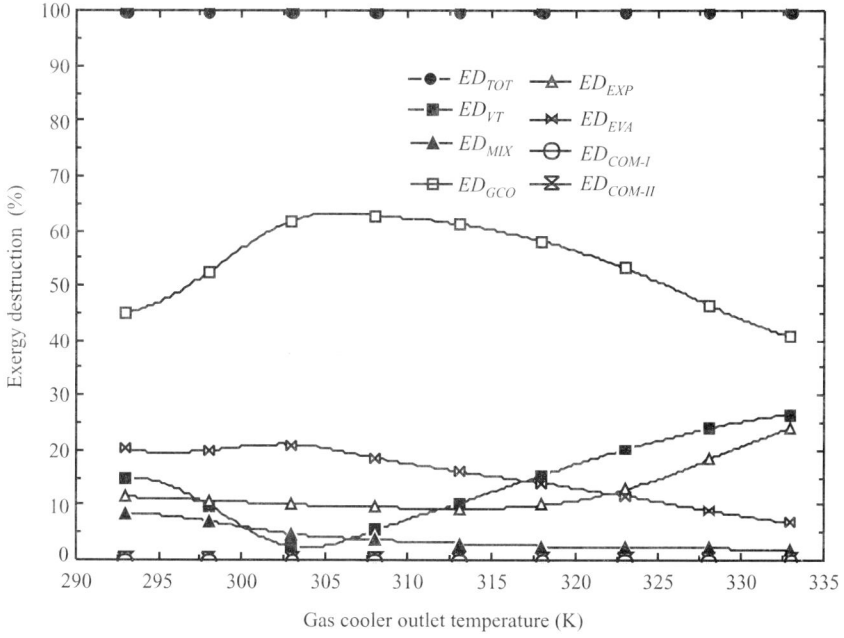

Fig. 6.52 Effect of Gas Cooler Outlet Temperature on Percentage Exergy Destruction

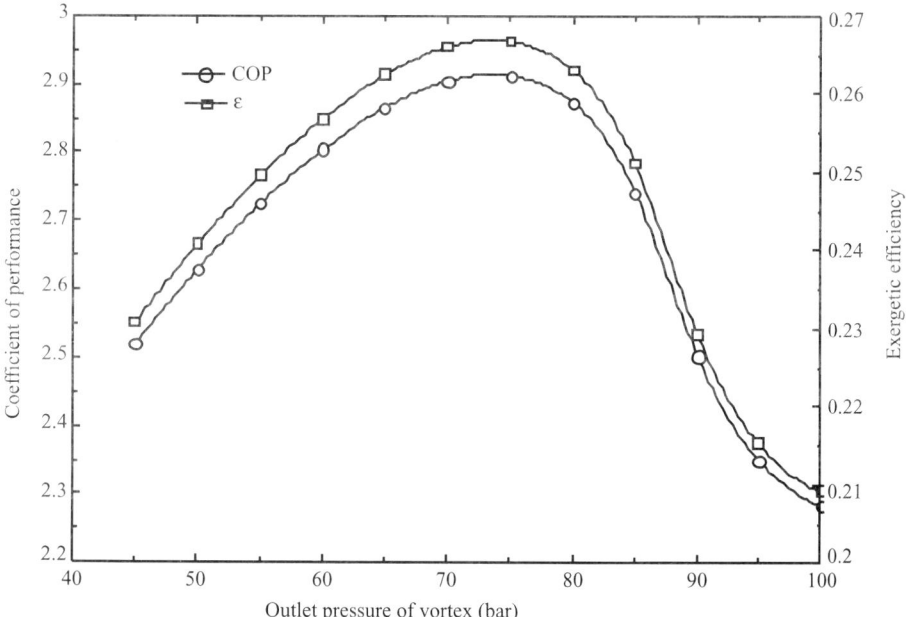

Fig. 6.53 Effect of Outlet Pressure of Vortex Tube on COP and Exergetic Efficiency

276 *Alternatives in Refrigeration and Air Conditioning*

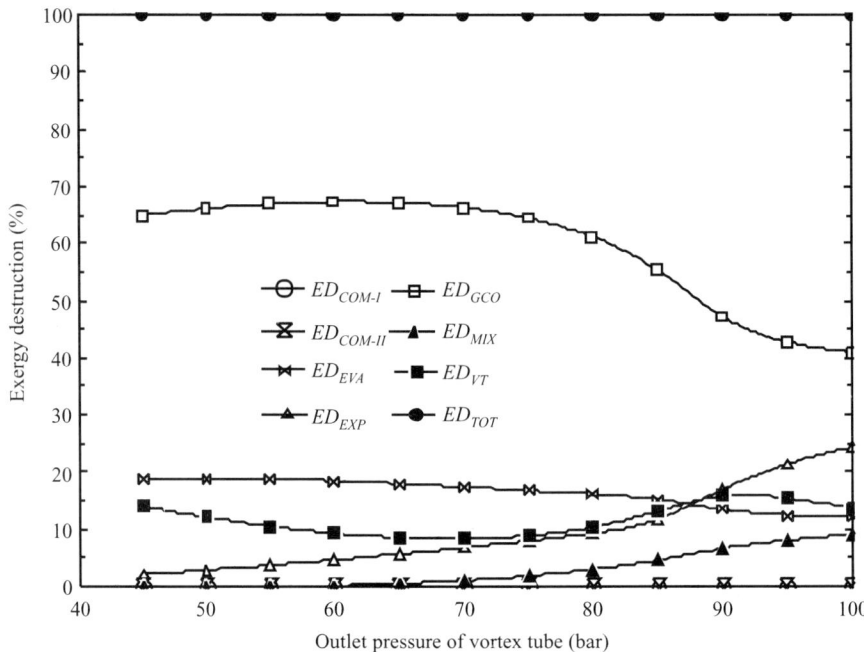

Fig. 6.54 Effect of Outlet Pressure of Vortex Tube on Percentage Exergy Destruction

The variation of the COP and exergetic efficiency with increase in the outlet pressure of the vortex tube has been shown in Fig. 6.53. The maximum COP and exergetic efficiency occur at 75 bars outlet pressure of the vortex tube. Figure 6.54 represents the effect outlet pressure of the vortex tube on the percentage of exergy destruction. The order of percentage exergy destruction for outlet pressure of vortex tube below 85 bars is as $ED_{GCO} > ED_{EVA} > ED_{VT} > ED_{EXP} > ED_{MIX}$ and for outlet pressure of vortex tube above 85 bars it is as $ED_{GCO} > ED_{EXP} > ED_{VT} > ED_{EVA} > ED_{MIX}$.

6.7.3 Enhancement of Quality of Waste Heat Recovery from Vapour Compression Refrigeration System Using Vortex Tube Integration

6.7.3.1 *Thermodynamic Model of Vortex Tube Integrated Vapour Compression Refrigeration System (VTIVCRS)*

The block diagram of the vortex tube integrated vapour compression refrigeration system having waste heat reclaim option has been shown in Fig. 6.55. It consists of a compressor (COM), vortex tube (VT), two desuperheaters (DS-I and DS-II) to reclaim waste heat, condenser (CON), expansion valve/device (EXP) and an evaporator (EVA). In the present system, the condenser is assumed to be of air cooled type. The waste heat in the form of latent heat at a temperature close to the environment has been rejected to the environment through this condenser. Only superheat of the refrigerant vapours has been reclaimed through the two desuperheaters.

The basic thermodynamic cycle for the aforesaid processes has been shown on pressure-enthalpy (P-h) coordinates in Fig. 6.56. The high temperature and high pressure refrigerant vapours

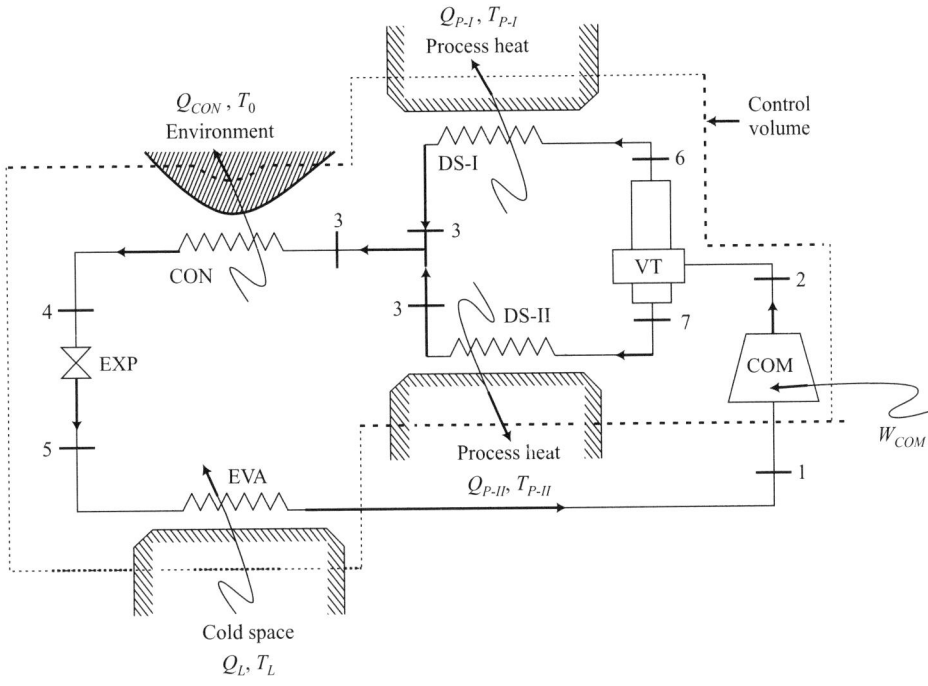

Fig. 6.55 Block Diagram of Vortex Tube Integrated Vapour Compression System

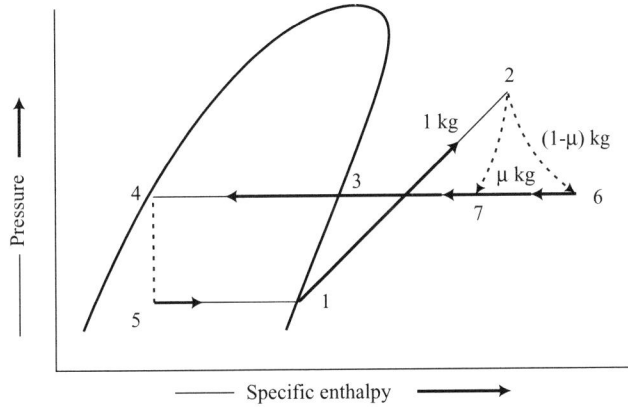

Fig. 6.56 Representation of VTIVCRS Processes on P-h Diagram

coming from the compressor discharge (state 2) are passed through the vortex tube where further enhancement of temperature of some fraction of total refrigerant vapours occurs. These vapours come out from the hot end of the vortex tube (state 6). The remaining vapours that are comparatively at low temperature discharge through the cold end of vortex tube (state 7). The high temperature and low temperature superheat available in these refrigerant vapours has been considered to be completely utilized in the form of a process heat through DS-I and DS-II respectively.

It has been assumed that the pressure of the vapours coming from hot end and cold end of VT is same and equal to the condenser pressure. The temperature and pressure of the refrigerant vapours effluxes from DS-I and DS-II also considered same (state 3). These vapours have been mixed together before entry to condenser. After condensation saturated liquid refrigerant (state 4) has been throttled in the EXP to evaporator pressure and temperature (state 5). The evaporation of the low temperature and low pressure liquid refrigerant occurs after absorbing heat from a space/substance to be cooled. Finally, the saturated vapours coming from the evaporator (state 1) has been compressed isentropically to the high pressure (state 2).

Following are the major assumptions considered in the elementary model of VTIVCRS:
1. The pressure of the refrigerant vapours effluxes from cold end and hot end of VT are same and equal to condenser pressure.
2. The compression is isentropic and expansion of refrigerant is isenthalpic.
3. The saturated vapours and saturated liquid refrigerant enters the compressor and expansion device respectively.
4. The pressure drops in all heat exchangers (desuperheaters, condenser and evaporator) and pipelines are negligible.
5. No heat losses (or gains) to (or from) ambient from (or to) the system components.
6. The superheat of the refrigerant vapours exiting from vortex tube has been completely utilized as process heat.
7. The temperature of the respective process heat is equal to the average temperature of the respective refrigerant vapours coming out of VT.
8. The refrigerant vapours behave as a perfect gas.
9. VTIVCRS has been compared with simple VCRS (with and without heat reclaim options).

The elementary model of VTIVCRS consists of the following major parts:
- Compressor
- Vortex tube
- Desuperheater-I
- Desuperheater-II
- Condenser
- Expansion valve/device
- Evaporator

Compressor

A single-stage compressor having isentropic compression and 100% volumetric efficiency has been considered. The work required and exergy destruction within the compressor can be calculated using the following equations:

$$s_2 = s_1 \tag{6.117}$$

The temperature T_2 of the refrigerant vapours can be found corresponding to the value of s_2.

$$W_{COM} = mR(h_2 - h_1) \tag{6.118}$$

$$ED_{COM} = mR(e_1 - e_2) + E_{COM} \tag{6.119}$$

$$E_{COM} = W_{COM} \tag{6.120}$$

Obviously, exergy destruction comes out to be zero as the compression process is assumed ideal.

Vortex Tube

A new model based on the ramming effect and heat exchanger in the Ranque Hilsch vortex tube has been used to determine the temperature separation in the vortex tube (ref. section 3.3). In this model, the ramming of the gas is considered at two places within the vortex tube, viz., firstly at secondary circulation and secondly at adjustable valve of vortex tube. Further, a counter flow type heat exchanger is considered operating between hot air, ram air and cold air, air exiting from the nozzle/s of the vortex tube. The velocity profile within the vortex tube has been assumed to be given by Ogawa Combined Vortex Model (Ogawa, 1993). One can also use the analytical model of Ahlborn et al. (1998) or any other available model for determining the temperature separation in the vortex tube.

$$h_2 = \mu h_7 + (1 - \mu) h_6 \quad (6.121)$$

$$EDVT = mR (e_2 - \mu e_7 - (1 - \mu) e_6) \quad (6.122)$$

$$T_{MAX} = T_6 \quad (6.123)$$

Desuperheaters

The desuperheaters are basically heat exchangers used to reclaim the waste heat which can be used as process heat in some applications. The average temperature of the waste heat within desuperheaters can be obtained using the following equations:

$$T_{SUP-I} = \frac{m_R \mu (h_6 - h_3)}{(s_6 - s_3)} \quad (6.124)$$

$$T_{SUP-II} = \frac{m_R (1 - \mu)(h_7 - h_3)}{(s_7 - s_3)} \quad (6.125)$$

$$Q_{SUP-I} = mR (1 - \mu) (h_6 - h_3) \quad (6.126)$$

$$Q_{SUP-II} = mR \mu (h_7 - h_3) \quad (6.127)$$

$$ED_{SUP-I} = mR (1 - \mu) (e_6 - e_3) - E_{SUP-I} \quad (6.128)$$

$$E_{SUP-I} = Q_{SUP-I} \left(1 - \frac{T_0}{T_{r-I}}\right) \quad (6.129)$$

$$T_{r-I} = T_0 \quad (6.130)$$

$$ED_{SUP-II} = mR \mu (e_7 - e_3) - E_{SUP-II} \quad (6.131)$$

$$E_{SUP-II} = Q_{SUP-II} \left(1 - \frac{T_0}{T_{r-II}}\right) \quad (6.132)$$

$$T_{r-II} = T_0 \quad (6.133)$$

$$T_{P-I} = T_{SUP-I} \qquad (6.134)$$

$$T_{P-II} = T_{SUP-II} \qquad (6.135)$$

$$Q_{P-I} = Q_{SUP-I} \qquad (6.136)$$

$$Q_{P-II} = Q_{SUP-II} \qquad (6.137)$$

$$ED_{P-I} = mR\,(1-\mu)\,(e_6 - e_3) - E_{P-I} \qquad (6.138)$$

$$E_{P-I} = Q_{P-I}\left(1 - \frac{T_0}{T_{P-I}}\right) \qquad (6.139)$$

$$ED_{P-II} = mR\,\mu\,(e_7 - e_3) - E_{P-II} \qquad (6.140)$$

$$E_{P-II} = Q_{P-II}\left(1 - \frac{T_0}{T_{P-II}}\right) \qquad (6.141)$$

If the average temperature of the waste heat is quite low (i.e., nearly T_0) then its utility is quite less. In this case the respective exergy destruction can be calculated using equation (6.128) or (6.131) otherwise expressions (6.138) and (6.140) are used.

Condenser

The air-cooled condenser has been considered in the present study. The waste heat in the form of only latent heat is assumed to be rejected from the refrigerant in the condenser. So, although the amount of waste heat is quite large but its quality is not sufficient to be effectively utilized for some applications. This is because designers always try to keep the condenser temperature low to achieve higher performance.

The quantity of heat to be rejected and exergy destruction in the condenser can be obtained from the following expressions:

$$Q_{CON} = mR\,(h_3 - h_4) \qquad (6.142)$$

$$ED_{CON} = mR\,(e_3 - e_4) - E_{CON} \qquad (6.143)$$

$$E_{CON} = Q_{CON}\left(1 - \frac{T_0}{T_{r-C}}\right) \qquad (6.144)$$

$$T_{r-C} = T_0 \qquad (6.145)$$

Expansion Device

As the thermodynamic process considered here is isenthalpic, without any loss or gain of heat from the environment, so,

$$h_5 = h_4 \qquad (6.146)$$

$$ED_{EX} = mR\,(e_4 - e_5) \qquad (6.147)$$

Evaporator

It is assumed that only two-phase conditions exist within the evaporator and no superheating of the refrigerant vapours occurs. The mass flow rate of the refrigerant required to compensate the cooling load and exergy destruction in evaporator can be determined as follows:

$$m_R = \frac{Q_L}{(h_1 - h_5)} \tag{6.148}$$

$$ED_{EVA} = mR(e_5 - e_1) - E_{EVA} \tag{6.149}$$

$$E_{EVA} = Q_L \left(1 - \frac{T_0}{T_L}\right), \text{ where } T_L < T_0 \tag{6.150}$$

VTIVCR overall System

The energy and exergy balance can also be employed directly to the control volume of VTIVCRS (ref. Fig. 6.55) in order to determine the overall energetic and exergetic performance of the system. Following are the final expressions obtained for coefficient of performance (COP) and exergetic efficiency of the system:

$$COP_{TOT} = \frac{Q_L + Q_{P-I} + Q_{P-II}}{W_{COM}} \tag{6.151}$$

$$\varepsilon = \frac{-E_{EVA} + E_{P-I} + E_{P-II}}{E_{COM}} \tag{6.152}$$

The overall exergy destruction within the system can be determined by simply adding the exergy destruction of individual components of the system.

$$ED_{TOT} = ED_{COM} + ED_{VT} + ED_{P-I} + ED_{P-II} + ED_{CON}$$
$$+ ED_{EX} + ED_{EVA} \tag{6.153}$$

Alternatively, the exergetic efficiency can be obtained as follows:

$$\varepsilon = 1 - \frac{ED_{TOT}}{E_{COM}} \tag{6.154}$$

$$ED_{TOT} = (E_{EVA} + E_{COM}) - (E_{CON} + E_{P-I} + E_{P-II}) \tag{6.155}$$

6.7.3.2 Simulation

The simulation of the following three different cases has been performed using the input data presented in Table 6.7:

- **Case A:** Simple VCRS with and without heat reclaim option (Figs. 6.57 and 6.58)
- **Case B:** Simple VCRS at increased discharged pressure (PMAX) (Figs. 6.57 and 6.58)
- **Case C:** VTIVCRS (Figs. 6.55 and 6.56)

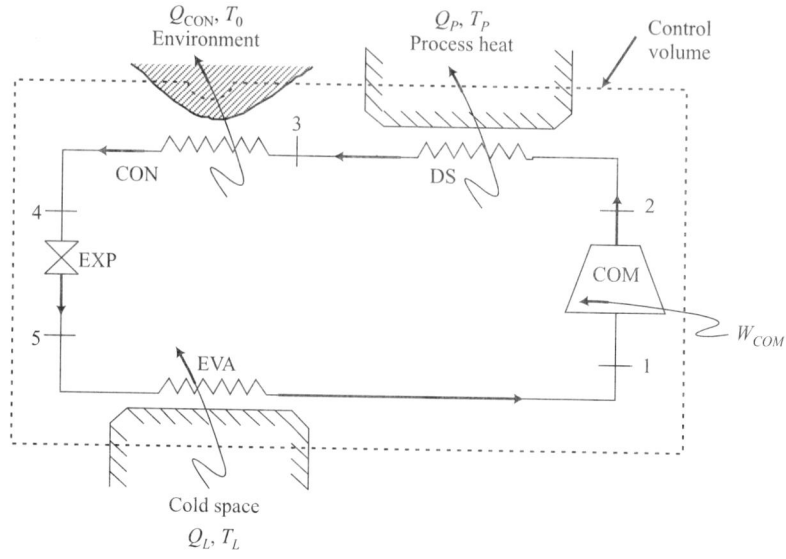

Fig. 6.57 Block Diagram of Simple VCRS with Heat Reclaim Option

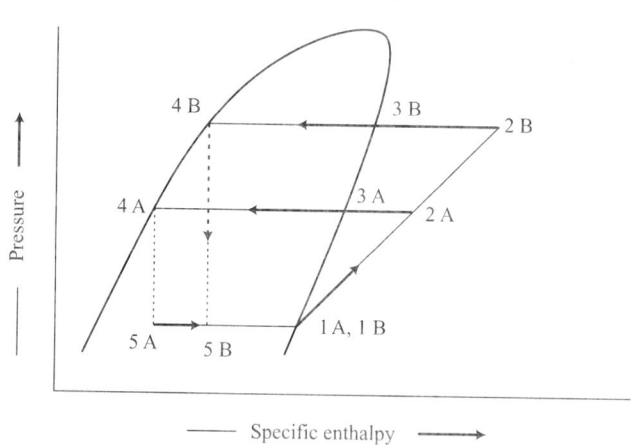

Fig. 6.58 Thermodynamic Cycle of Case-A (Cycle 1A-2A-3A-4A-5A-1A) and Case-B (Cycle 1B-2B-3B-4B-5B-1B)

The thermodynamic simulation has been carried out writing a computer program using the Engineering Equation Solver software (Klein and Alvaradro, 2005). The property data of R152a built in the software has been used.

Table 6.7 Input Data for Analysis of VCRS and VTIVCRS having Waste Heat Recovery Option

S. No.	Parameter	Details
1.	Refrigerant	R152a
2.	Q_L	1 T_R
3.	T_L	20°C (Air conditioner)
4.	T_{AMB}	25°C, 30°C, 35°C, 40°C, 45°C
5.	T_{CON}	(T_{AMB} + 10°C)
6.	T_{EVA}	−20°C to 10°C
7.	T_0	T_{AMB}
8.	P_0	1.01 bar
9.	P_{MAX}	(P_{CON} + 4) bar
10.	L	15 D
11.	d_N	D/9
12.	B	D/2
13.	N	3
14.	µ	0.80

The following terms have been defined for the comparison of these three cases:

(a) Fraction of waste heat reclaim (FR)

It is defined as the ratio of an amount of waste heat reclaimed in desuperheater/s to the total waste heat from the system.

$$FR = \frac{Q_P}{Q_{CON} + Q_{SUP}} \ldots \text{(Kaushik and Singh, 1995)}\ldots\ldots \quad (6.156)$$

This provides the information regarding the percentage of total waste heat reclaimed as process heat.

(b) Energy Savings (ES)

This term can be used to compare the waste heat reclaim option of the different cases w.r.t. the simple VCRS (CASE-A) without heat reclaim option. It is defined as the difference between the amount of the waste heat reclaimed and the additional work spent then the work required for simple VCRS without heat reclaim option

$$ES = Q_P - (W_{COM} - W_{COM-A}) \quad (6.157)$$

(c) Waste Heat Reclaim Advantage (WHRA)

It is defined as the ratio of the additional waste heat reclaimed to the additional work than simple VCRS (CASE-A) with waste heat reclaim option spent.

284 *Alternatives in Refrigeration and Air Conditioning*

$$\text{WHRA} = \frac{Q_P - Q_{P\text{-}A}}{W_{COM} - W_{COM\text{-}A}} \qquad (6.158)$$

This term is useful to determine the benefit of using particular case w.r.t. Case-A.

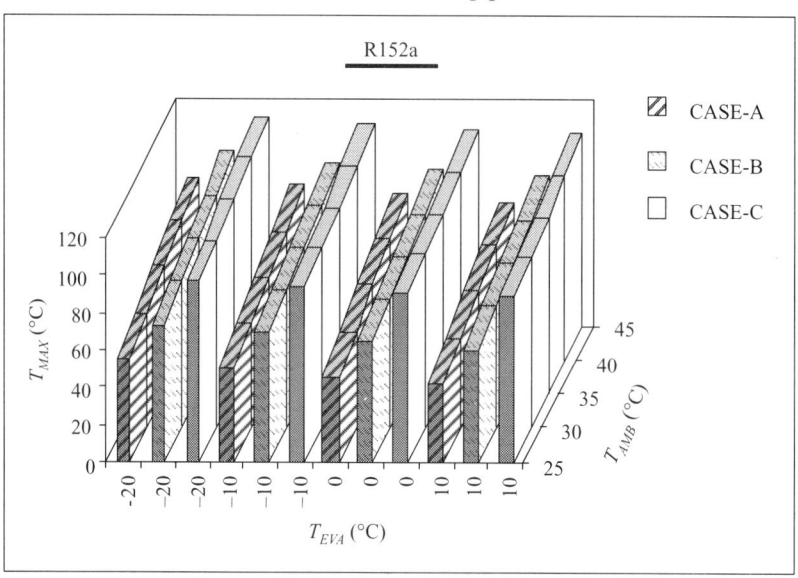

Fig. 6.59 Comparison of Maximum Obtainable Temperature

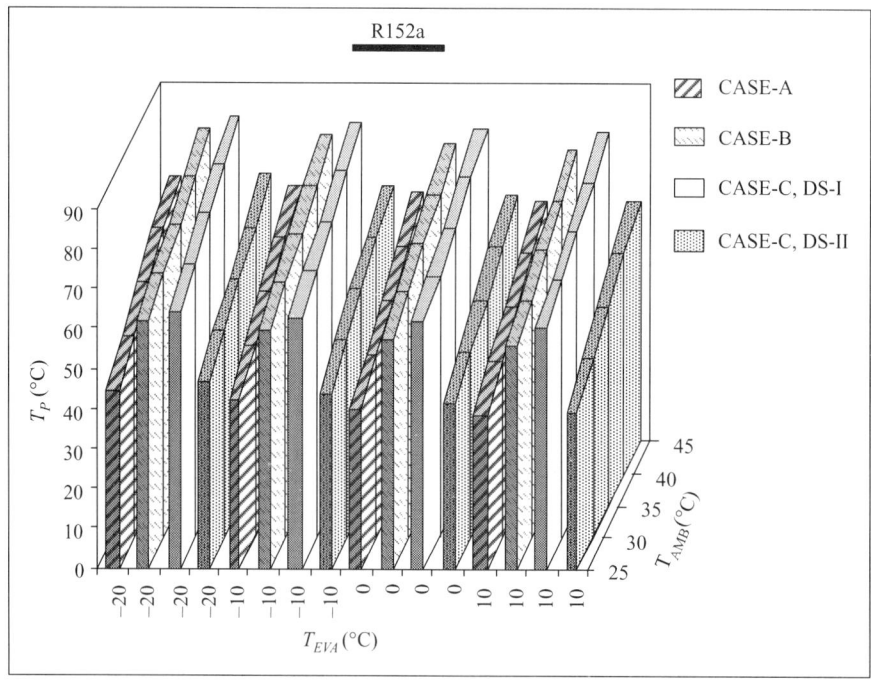

Fig. 6.60 Comparison of Average Temperature of Waste Heat Reclaimed as Process Heat

6.7.3.3 Results and Discussion

Figure 6.59 shows the comparison of the maximum temperature obtainable for the three cases of study. The value of T_{MAX} varies for Case-C between 86°C and 112°C, for Case-B between 60°C and 93°C and for Case-A between 41°C and 79°C in the range of T_{EVA} and T_{AMB} taken for this study. This clearly demonstrates the enhancement of quality of waste heat by integrating the vortex tube in the vapour compression refrigeration system. The value of T_{MAX} decreases with increase in the evaporator temperature but increases with increase in the ambient temperature. Same trend is observed for T_p (see Fig. 6.60). Although the fraction of high grade process heat available is less, but the temperature of the other fraction is still higher than that of Case-A. So, two different types of applications can be achieved by these available process heats. One possible application may be to get hot water and the other may be hot box to keep the products hot.

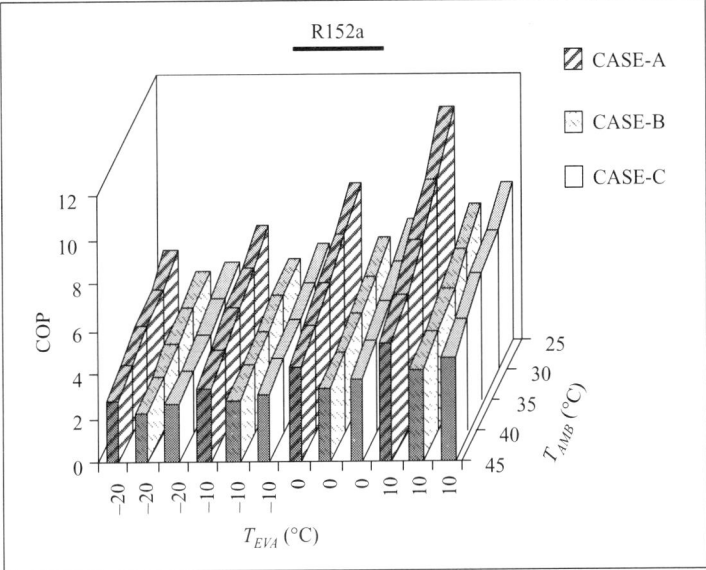

Fig. 6.61 Comparison of Overall Coefficient of Performance

Figures 6.61 and 6.62 represent the comparative overall energetic performance and exergetic performance respectively. Both the performances increase with increase in the evaporator and ambient temperature. Although the coefficient of performance for Case-C is less than that of Case-A, but their exergetic efficiencies are almost same at all evaporator and ambient temperatures. This clearly shows that Case-C can be successfully replaced with case-A, as two types of process heats of higher quality than the Case-A may be obtained from it. The Case-B shows poor overall energetic and exergetic performance than the other cases.

286 *Alternatives in Refrigeration and Air Conditioning*

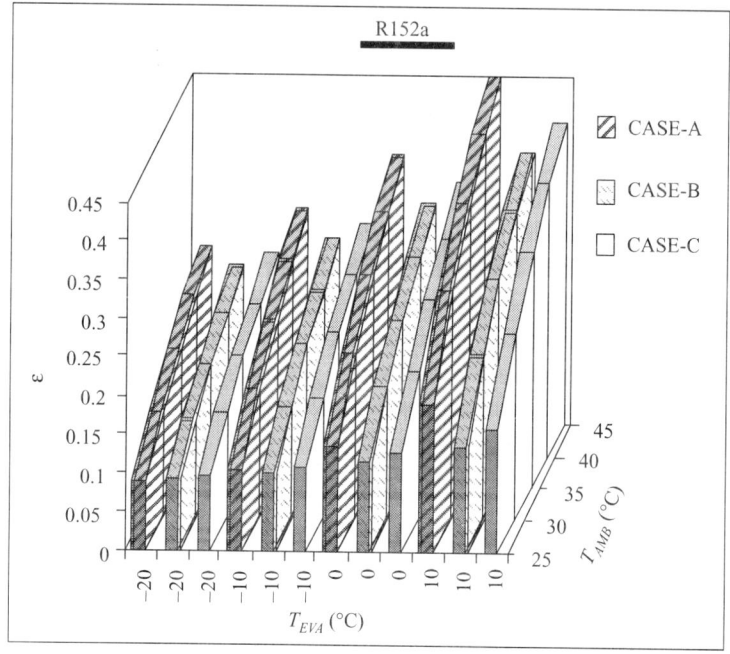

Fig. 6.62 Comparison of Exergetic Efficiency

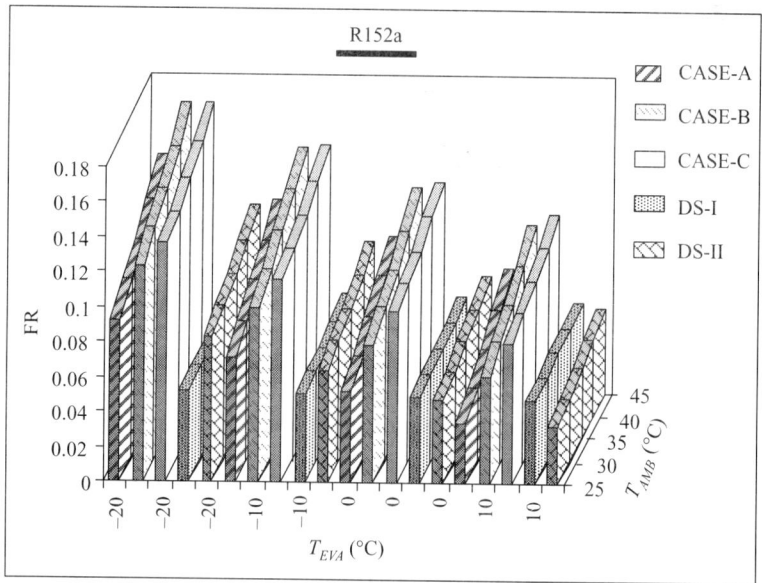

Fig. 6.63 Comparison of Fraction of Waste Heat Reclaim

The value of FR for Case-C is higher than the others (see Fig. 6.63) despite of higher mass flow rate of refrigerant for Case-B (ref. Fig. 6.64). This is because of the thermodynamic properties of refrigerant. Although the waste heat reclaim in Case-B is slightly higher than Case-C, but the total heat to be rejected for Case-B is much more than that of Case-C, which leads to lower value of FR for Case-B. The value of FR for Case-A is much lower than the others. FR decreases with increase in evaporator temperature and increases with increase in the ambient temperature for all the cases.

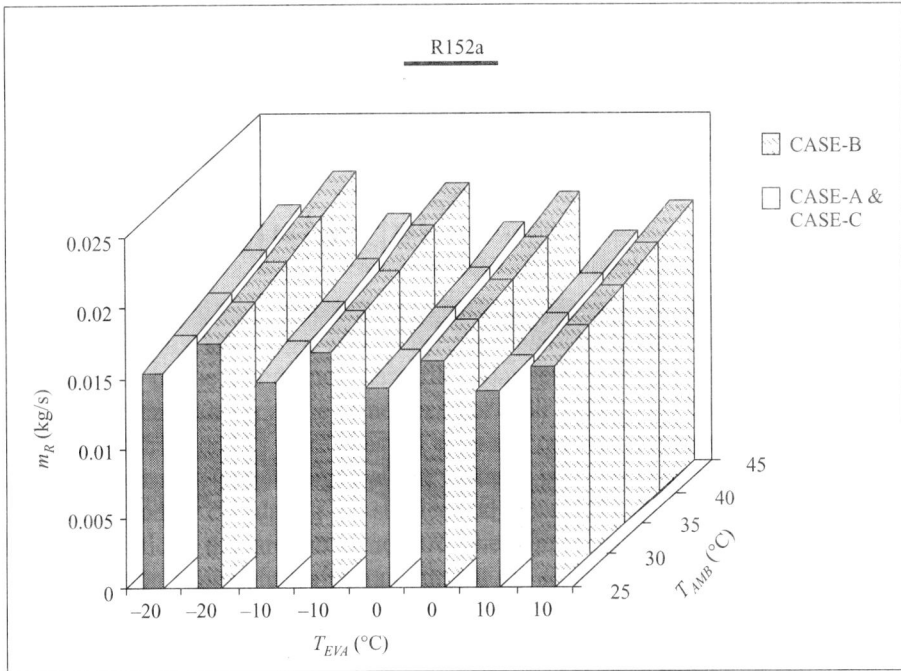

Fig. 6.64 Comparison of Refrigerant Flow Rate

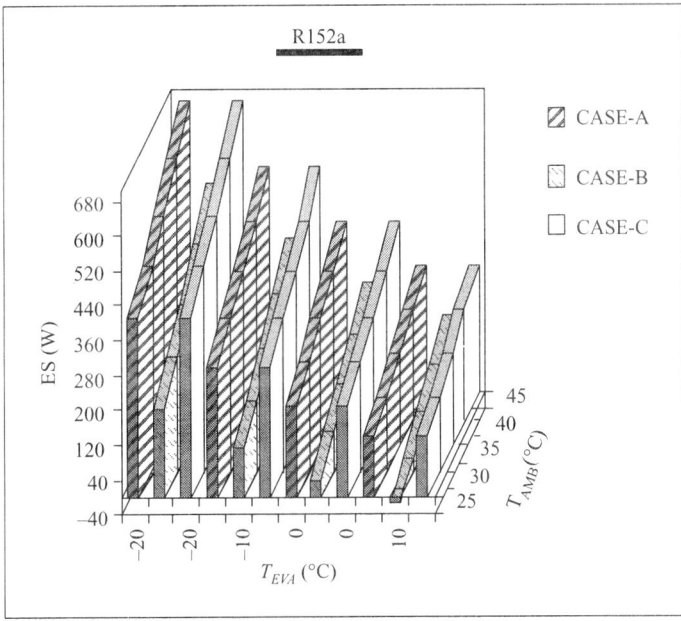

Fig. 6.65 Comparison of Energy Saving

The energy savings (ES) are same for Case-A and Case-C and are higher than that of Case-B (see Fig. 6.65). Its value decreases with increase in the evaporator temperature and increases with increase in ambient temperature. The waste heat reclaim advantage (WHRA) is 1 for VTIVCRS and about 0.43–0.54 for Case- B (ref Fig. 6.66). The value of WHRA remains same for Case-C at all the values of evaporator and ambient temperature, but for Case-B its value slightly increases with increase in the ambient temperature and almost same with increase in evaporator temperature.

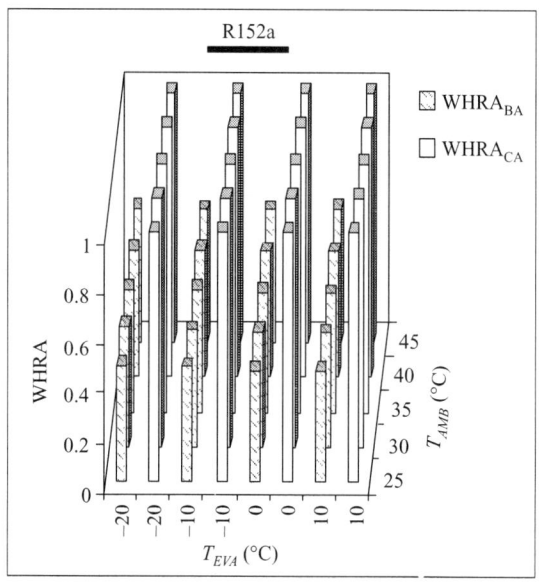

Fig. 6.66 Comparison of Waste Heat Reclaim Advantage

The required compression work has been presented in Fig. 6.67. The work of compression for Case-A is least than others, but quality of waste heat reclaim is also least.

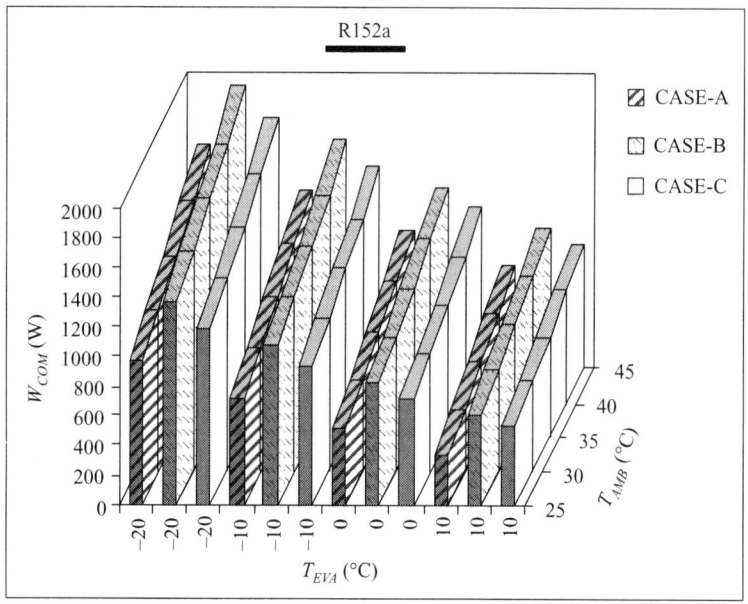

Fig. 6.67 Comparison of Required Compression Work

Although the W_{COM} for Case-C is higher than Case-A and less than that of Case-B, but it has WHRA equal to 1. This means additional work input has been reclaimed in the form of equal amount of process heat. On the other hand, Case-B consumes excessive energy and its recovery in the form of process heat is also poor (WHRA = 0.43 – 0.54). Thus, Case-C is a better option than that of other cases as it provides higher quality process heat. It also provides better opportunities for the use of process heat in two different levels of quality with less destruction of exergy. But additional initial investment on one more compressor (or replacement with high capacity compressor) and a vortex tube has to be born to save exergy destruction and to use the energy in the more efficient way.

The diameter of the vortex tube required for Case-C can be noted from Fig. 6.68 and other design dimensions of the VT can be calculated using Table 6.5. The value of D varies between 14 to 18 mm in the selected range of temperatures. At lower evaporator and lower ambient temperature higher diameter of the vortex tube is required.

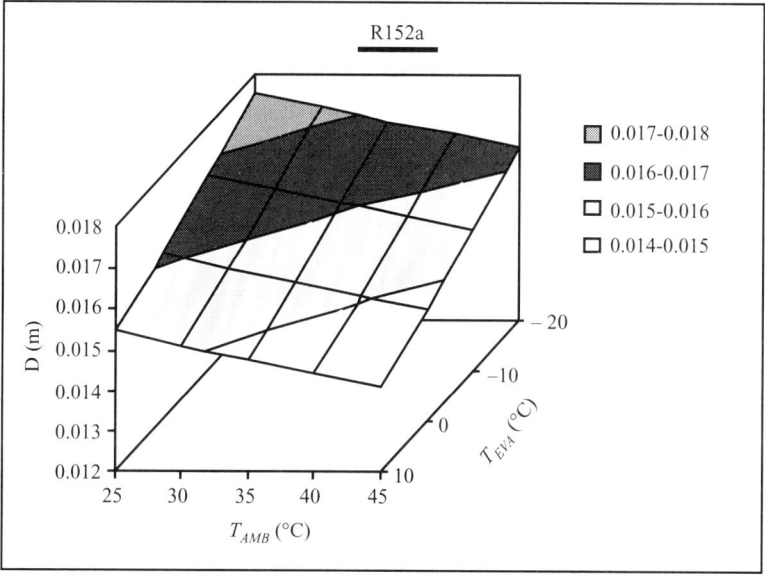

Fig. 6.68 Variation of Diameter of Vortex Tube

Due to lack of availability of the real data temperature separation of refrigerants when passed through VT, the results presented in this section seem to be conservative. The model of VT used in this simulation shows acceptable validation for carbon dioxide (ref. Section 3.3). Kawashima and Araki (1987) given a small data for R22 vortex tube integrated system, which is quite encouraging. Therefore, the results presented here can open a new window for the research activities in this direction.

290   *Alternatives in Refrigeration and Air Conditioning*

6.8 CONCLUSION

In this chapter, the concept of vortex tube and its application in refrigeration and air conditioning systems is discussed comprehensively. The review of various theories and models which explain the vortex effect is carried out. A new mathematical model of vortex tube based on ramming effect and heat exchange process is developed and then energy and exergy analysis is also carried out. The integration of vortex tube with refrigeration and air conditioning systems is done. Subsequently, the thermodynamic analysis of vortex tube integrated refrigeration and air conditioning system is carried out. At the end of the chapter, the use of vortex tube for the enhancement of quality of waste heat recovery from vapour compression refrigeration system is proposed and analysed.

Nomenclature

Abbreviations

BPV	:	Back pressure valve
COM	:	Compressor
COM I	:	Compressor I
COM II	:	Compressor II
CON	:	Condenser
DS	:	Desuperheater
DS-I	:	Desuperheater I
DS-II	:	Desuperheater II
EVA	:	Evaporator
EXP	:	Expansion valve/device
GCO	:	Gas cooler
IHX	:	Internal heat exchanger
MIX	:	Mixer
SEP	:	Separator
TR	:	Tons of refrigeration
VT	:	Vortex tube
VT-I	:	Vortex tube I
VT-II	:	Vortex tube II

Symbols

A	:	Cross-sectional area
B	:	Diameter of diaphragm
COP	:	Coefficient of performance
D	:	Diameter of hot tube or vortex tube
E	:	Rate of thermal exergy destruction

ED	:	Rate of exergy destruction
ES	:	Energy savings
F	:	External body force
FR	:	Fraction of total waste heat reclaimed
K	:	Constant
L	:	Length of hot tube; length of heat exchanger
M	:	Mach number
N	:	Number of nozzles
NTU	:	Number of transfer units
$NoTU$	:	Modified number of transfer units
P	:	Pressure
PR	:	Pressure ratio of compressor
$PERI$	:	Perimeter
Q	:	Rate of heat exchange; heat interactions
R	:	Gas constant
T	:	Temperature
U	:	Velocity
V	:	Velocity
W	:	Work; work interactions
$WHRA$	:	Waste heat reclaim advantage
a	:	Sonic velocity
b	:	Height of rectangle for rectangular cross-sectional nozzle
c	:	Specific heat
d	:	Diameter; diameter of nozzle
e	:	Specific exergy
g	:	Acceleration due to gravity
h	:	Specific enthalpy
k	:	Coefficient of thermal conductivity
m	:	Mass flow rate
n	:	Index for velocity profile
r	:	Radius
s	:	Specific entropy
t	:	Time
u, v	:	Velocity of fluid
w	:	Width of rectangle for rectangular cross-sectional nozzle

x	:	Dryness fraction
z	:	Elevation
γ	:	Ratio of specific heats
ε	:	Effectiveness; exergetic efficiency
η	:	Efficiency of heat engine
μ	:	Cold mass fraction
ρ	:	Density
τ	:	Time
Γ	:	Vortex circulation
Λ	:	Vortex vorticity
Δ	:	Difference/drop
Σ	:	Summation

Subscripts

AMB	:	Ambient
C	:	Cold fluid
CHE	:	Carnot heat engine
CHP	:	Carnot heat pump
COM	:	Compressor
$COM\ I$	:	Compressor I
$COM\ II$	:	Compressor II
CON	:	Condenser
CR	:	Carnot refrigerator
$CREV$	:	Cold fluid at ideal conditions
D	:	Vortex tube
$DS\text{-}I$	:	Desuperheater I
$DS\text{-}II$	:	Desuperheater II
EJT	:	Ejector
EVA	:	Evaporator
EXP	:	Expansion valve/device
GCO	:	Gas cooler
$GCO°$	:	Outlet of gas cooler
H	:	Hot fluid
HC	:	Between hot and cold fluid
HI	:	Between hot and inlet fluid
$HREV$	:	Hot fluid at ideal conditions

I	:	Inlet fluid
IC	:	Between inlet and cold fluid
L	:	Cold space/substance
MAX	:	Maximum
MIX	:	Mixer
N	:	Nozzle/s
P	:	Pressure; process heat
P-I	:	Process heat from DS-I
P-II	:	Process heat from DS-II
R	:	Refrigerant
RAM	:	Ramming
SEP	:	Separator
SUP-I	:	Superheat from DS-I
SUP-II	:	Superheat from DS-II
TOT	:	Total/overall
V	:	At adjustable valve
VT	:	Vortex tube
VT-I	:	Vortex tube I
VT-II	:	Vortex tube II
VT°	:	Outlet of vortex tube
min	:	Minimum
n	:	Index for velocity profile
r	:	Radius; radial direction
rt	:	Radius at which velocity of quasi-forced and quasi-free vortex are equal
t	:	Time
z	:	Axial direction
θ	:	Circumferential direction
θ *M*	:	Maximum in circumferential direction
0	:	Ambient
1; 2; 3;…	:	Thermodynamic state points
r-I	:	Reservoir for DS-I
r-II	:	Reservoir for DS-II
r-C	:	Reservoir for condenser
r-E	:	Reservoir for evaporator
r-G	:	Reservoir for gas cooler

REFERENCES

Ahlborn B.K., Gordon J.M., 2000. The Vortex Tube as a Classic Thermodynamic Refrigeration Cycle. *J. of Applied Physics*, 88(6), 3645-3653.

Ahlborn B., Groves S., 1997. Secondary Flow in a Vortex Tube. *Fluid Dynamics Research*, 21, 73-86.

Ahlborn B.K., Keller J.U., Rebhan E., 1998. The Heat Pump in a Vortex Tube. *J. of Non-Equilibrium Thermodynamics*, 23(2), 159-65.

Ahlborn B.K., Keller J.U., Staudt R., Treitz G., Rebhan E., 1994. Limits of Temperature Separation in a Vortex Tube. *J. Phy. D: Appl. Phys.*, 27, 480-488.

Aljuwayhel N.F., Nellis G.F., Klein S.A., 2005. Parametric and Internal Study of the Vortex Tube Using a CFD Model. *Int. J. Refrig.*, 28(3), 442-450.

Arbuzov V.A., Dubnishchev Y. N., Lebedev A.V., Pravdina M. K., Yavorski N.I., 1997. Observation of Large-Scale Hydrodynamic Structures in a Vortex Tube and the Ranque Effect. *Technical Physics Letters*, 23(12), 938-940.

Balmer R.T., 1988. Pressure Driven Ranque Hilsch Temperature Separation in Liquids. *J. of Fluid Engineering*, 110, 161-164.

Bartlett J.L., 1960. Refrigeration System with Vortex Means. U.S. Patent No. 2, 920,457.

Behera U., Paul P.J., Kasthurirengan S., Karunanithi R., Ram S.N., Dinesh K., Jacob S., 2005. CFD Analysis and Experimental Investigations towards Optimizing the Parameters of Ranque-Hilsch Vortex Tube. *Int. J. of Heat and Mass Transfer*, 48, 1961-1973.

Bilga P.S., Kaushik S.C., 2007. Heat Exchanger, Heat Engine, Refrigerator, Heat Pump and Acoustics Phenomenon in a Vortex Tube. Proc. *Int. Conf. on Recent Trends in Mechanical Engineering ICRTME*, 4-6 Oct, Ujjain Engineering College, Ujjain, vol. 1, TH15-TH20.

Bilga P.S., Kaushik S.C., 2008. Modeling of Temperature Separation Based on Ramming Effect and Heat Exchanger in a Ranque Hilsch Vortex Tube. *Int. J. Green Energy*, 5(5), 373-387.

Cao Y., Qi Y.F., Luo E.C., Wu J.F., Gong M.Q., Chen G.M., 2003. Study of a Vortex Tube by Analogy with a Heat Exchanger. *Cryocoolers12*, Kluwer Academic/Plenum Publishers, 615-620.

Cengel T.A., Boles M.A., 2006. Thermodynamic: An Engineering Approach. Tata McGraw Hills, 307.

Christensen K.G., Heiredal M., Schneider P., 2001. Energy Savings in Refrigeration by Means of a New Expansion Device. *Energy, Refrigeration and Heat Pump Technology*, 1-53, Journal No. 1223/99-0006 (www.hfc-fri.dk).

Deissler R.G., Perlmutter M., 1960. Analysis of the Flow and Energy Separation in a Vortex Tube. *Int. J. Heat Mass Transfer*, 1, 173-191.

Eiamsa-ard S., Promvonge P., 2007. Numerical Investigation of the Thermal Separation in a Ranque-Hilsch Vortex Tube. *Int. J. Heat Mass Transfer*, 50(5-6), 821-832.

Eiamsa-ard S., Promvonge P., 2008. Review of Ranque-Hilsch Effects in Vortex Tubes. *Renew. Sustain Energy Rev.*, 12 (7), 1822-1842.

Fedorov A.G., Wadell R., Launay S., 2010.Vortex Tube Refrigeration Systems and Methods. U.S. Patent No. US 7,669,428 B2.

Fröhlingsdorf W., Unger H., 1999. Numerical Investigations of the Compressible Flow and the Energy Separation in the Ranque-Hilsch Vortex Tube. *Int. J. of Heat and Mass Transfer*, 42, 415-422.

Fulton C.D., 1950. Ranque's Tube. *J. of the ASRE, Refrigerating Engineering*, 473-479.

Gao C., 2005. Experimental Study on the Ranque - Hilsch Vortex Tube. Ph.D. Thesis, Technishe Universiteit Eindhoven, ISBN 90-386-2361-5.

Gutsol A., Bakken J. A., 1998. A New Vortex Method of Plasma Insulation and Explanation of the Ranque Effect. *J. Phys. D: Appl. Phys.* 31, 704-711.

Hebeccker D., Bittrich P., 2006. Energietransformation mit Hilfe des Wirbelrohres. Chemie Ingenieur Technik, 78(5), 535-542.

Hilsch R., 1947. The Use of the Expansion of Gases in a Centrifugal Field as a Cooling Process. *Review of Scientific Instruments*, 18(2), 108-113.

Kaushik S.C., Singh M., 1994. Feasibility and Design Studies for Heat Recovery from a Refrigeration System with a Canopus Heat Exchanger. *Heat Recovery Sys & CHP*, 15(7), 665-673.

Kawashima J., Araki N., 1987. Method of Intensifying Heat in Reversed Rankine Cycle and Revered Rankine Cycle Apparatus for Conducting the Same. U.S. Patent No. 4,646,524.

Klein A., Alvarado F., 2005. Engineering Equation Solver, Version 7.441, *F Chart Software*, Middleton, WI.

Kotas T.J., 1985. The Exergy Method of Thermal Plant Analysis. *Butterworths*, ISBN 0-408-01350-8.

Kurosaka M., 1982. Acoustic Streaming in Swirling Flow and the Ranque - Hilsch (Vortex Tube) Effect. *J. of Fluid Mechanics*, 124,139-172.

Lay J.E., 1959. An Experimental and Analytical Study of Vortex Flow Temperature Separation by Superposition of Spiral and Axial Flows Part 2. Transactions of ASME- *J. of Heat and Mass Transfer*, 213- 222.

Lewins J, Bejan A., 1999. Vortex Tube Optimization Theory. *Energy*, 24, 931-943.

Martynovskii, V.S., Alekseev, V.P., 1956. Investigation of the Vortex Thermal Separation Effect for Gases and Vapours. *Sov. Phys-Tech. Phys.*, 2233-2243.

Parulekar B.B., 1961. Short Vortex Tube. *J. of Refrig.*, 74-80.

Promvonge P., Eiamsa-ard S., 2005. Investigation on the Vortex Thermal Separation in a Vortex Tube Refrigerator. *ScienceAsia*, 31, 215-223.

Ogawa A., 1993. Vortex flow. CRC Press, 31 & 132-135.

Ranque G.J., 1933. Experiments on Expansion in a Vortex Tube with Simultaneous Exhaust of Hot Air and Cold Air. *Le de Physique et le Radium*, 4(7), 112-114.

Ranque G.J., 1934. Method and Apparatus for Obtaining from a Fluid under Pressure Two Currents of Fluids at Different Temperatures. U.S. Patent No. 1952281.

Reynolds A. J., 1961. Energy Flows in a Vortex Tube. *Zeitschrift fur Angewandte Mathematik Physik*, 12(4), 343-357.

Scheper G.W., 1951. The Vortex Tube - Internal Flow Data and Heat Transfer Theory. *J. of ASRE, Refrig. Engg.* 59, 985-989.

Shannak B.A., 2004. Temperature Separation and Friction Losses in Vortex Tube. *Heat and Mass Transfer*, 40, 779-785.

Sibulkin M., 1962. Unsteady, Viscous, Circular Flow, Part 3 Applications to Ranque- Hilsch Vortex Tube. *J. of Fluid Mechanics*, 12, 269-293.

Skye H.M., Nellis G.F., Klein S.A., 2006. Comparison of CFD Analysis to Empirical Data in a Commercial Vortex Tube. *Int. J. of Refrig.*, 29, 71-80.

Stephan K., Lin S., Durst M., Huang F., Seher D., 1983. An Investigation of Energy Separation in a Vortex Tube. *In. J. of Heat Mass Transfer*, 26(3), 341-348.

Trofimov V.M., 2000. Physical Effect in Ranque Vortex Tubes. *JETP Letters*, 72(5), 249-252.

Van Deemter J.J., 1952. On the Theory of Ranque-Hilsch Cooling Effect. *Applied Science Research-A*, 3, 174-196.

Webster D.S., 1950. An Analysis of the Hilsch Vortex Tube. *J. of the ASRE, Refrig. Engg.*, 163-171.

Yilmaz M., Kaya M., Karagoz, Erdogen S., 2009. A Review on Design Criteria for Vortex Tubes. *Heat Mass Transfer*, 45, 613-632.

CHAPTER 7

Thermoelectric Refrigeration

7.1 INTRODUCTION

In the preceding chapters, cooling devices have used either shaft work or thermal energy as the input energy option for producing refrigeration. Thermoelectric cooling is another potential option for refrigeration that uses electrical energy directly without converting it into mechanical form to obtain refrigerating effect. The process of thermoelectric cooling is based on the principles of reversible thermoelectric effects, i.e., a temperature difference and heat absorption/rejection may be established between two junctions of dissimilar conductor materials charged by a direct electric current. Since there is no mechanical energy transfer, no moving parts are present and hence the system is free from gear and wear problems as well as noise, etc. The major attractions for thermoelectric coolers are simplicity in design, light weight and static components as compared to other cooling devices.

Historically, Seebeck in 1822 reported that an electric current is produced in a closed circuit made of dissimilar conductors at different junction temperatures. Peltier in 1834 observed that if a current is passed through a junction of two dissimilar materials, absorption or rejection of thermal energy occurs at the junctions. In 1857, Thomson related these thermoelectric effects and also discovered that for a single material with a uniform temperature gradient along its length, an amount of heat is absorbed or rejected if a current is passing through the conductor.

In thermoelectric refrigeration systems absence of moving parts eliminates vibration problem as well as regular maintenance. Therefore, it can be best suited for systems where vibration is undesirable. In addition there is no wear due to rubbing as such the life is expected to be almost infinite as compared with other systems. The load can be easily controlled by means of adjusting the current to meet the requirement. These are generally very compact in size since even the system boundary may be used as the cooling surface as exhibited in Fig. 7.1. Therefore, the manufacturing cost of that wall is taken care of by the evaporator surface. It is lighter in weight for the same capacity of refrigeration. Since electric current passes through conductors, there is no problem of leakage of refrigerant, which is most undesirable in other refrigeration systems.

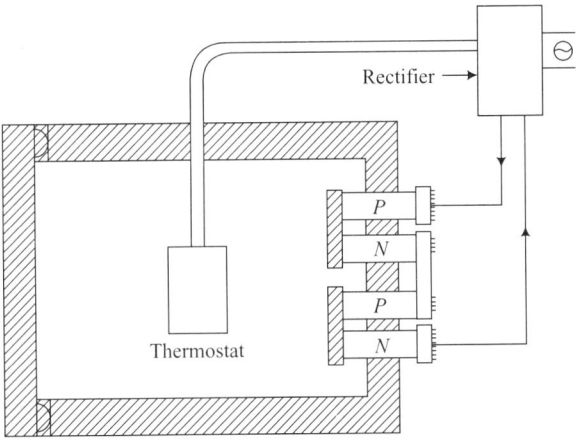

Fig. 7.1 Thermoelectric Cooling Module Device

Further, in a conventional VCR system, the leakage of refrigerant from the system causes the drastic decrease in the cooling capacity in addition to extra cost for the refrigerant and charging operations. While a thermoelectric refrigeration system operates at the same capacity for long and eliminates the cost of charging and extra materials. It is most easy to operate as a heat pump, just by reversing the terminals. Hence, a thermoelectric refrigeration system can be considered an year round air conditioner. Since no refrigerant is used, there is no question of toxication, etc. and can be used directly for air conditioning. Its design and manufacturing are rather much simpler than the other refrigeration systems. It is most suitable for the production of cooling suit. Main disadvantage of thermoelectric refrigeration system is the unavailability of suitable material of high figure of merit. Presently, the total cost of refrigeration system is a few times higher than the vapour-compression or other systems for a few ton capacity. In addition, the running cost is found to be much higher as compared to vapour-compression system. That is why the vapour-compression or other systems of refrigeration are most commonly employed. The overall COP of a thermoelectric refrigeration system is of the order of 0.1 to 0.2. It can be operated in any position in contrast with vapour-absorption, vapour compression or steam-jet ejector refrigeration systems.

The COP (energy efficiency) of thermoelectric refrigerators, based on currently available materials and technology, is still lower than its compressor counterparts. However, a marketable thermoelectric refrigerator can be made with an acceptable COP. Moreover, further improvement in the COP may be possible through improving module contact-electrical conductance, thermal interfaces and heat exchangers. With its environmental benefit, a thermoelectric refrigerator provides an alternative to consumers who are environmentally conscious and willing to spend a little bit more money to enjoy their quiet operation, and more precise and stable temperature control.

7.2 THERMOELECTRIC AND THERMOMAGNETIC EFFECTS

a) Seebeck Effect

The thermometric effects that underlie thermoelectric energy conversion can be conveniently discussed with reference to the schematic of a thermocouple shown in Fig. 7.2. It can be considered a circuit formed from two dissimilar conductors, *a* and *b* (referred as thermocouple legs, arms, thermoelements, or simply elements and sometimes as pellets by device manufacturers) which are connected electrically in series but thermally in parallel. If the junctions at A and B are maintained at different temperatures T_1 and T_2 and $T_1 > T_2$ an open circuit electromotive force (emf), E is developed between C and D and it can be given as

$$E = \alpha_{ab}(T_1 - T_2) \qquad (7.1)$$

where α_{ab} is the differential Seebeck coefficient between the elements *a* and *b* and also defined as $\alpha_{ab} = \dfrac{dE}{dT}$ known as thermoelectric power for the material combination *a* and *b*. For small temperature differences the relationship is linear. Although by convention α is the symbol for the Seebeck coefficient, which is also referred to as the thermo emf or thermoelectric power. The sign of α_{ab} is positive if the emf causes a current to flow in a clockwise direction around the circuit and is measured more often in μV/K.

b) Peltier Effect

If in Fig. 7.2 the reverse situation is considered with an external e.m.f. source applied across C and D and a current I flows in a clockwise sense around the circuit then a rate of heating ($\dot{Q}$) occurs at one Junction between *a* and *b* and a rate of cooling ($-\dot{Q}$) occurs at the other. The ratio of $\dot{Q}$ to I defines the Peltier coefficient given by $\pi_{ab} = \dfrac{\dot{Q}}{I}$, and it is positive if A is heated and B is cooled, and is measured in watts per ampere (W/A) or in volts (V).

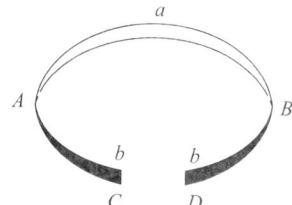

Fig. 7.2 Schematic of a Thermocouple

c) Thomson Effect

The last of the thermoelectric effects, the Thomson effect relates to the rate of generation of reversible heat $\dot{Q}$ which results from the passage of a current along a portion of a single conductor along which there is a temperature difference (dT), provided the temperature difference is small.

Thus, $$Q = \tau \cdot I \cdot \dfrac{dT}{dx}$$

where τ is the Thomson coefficient and x is the distance along the length of the conductor. Although the Thomson effect is not of primary importance in thermoelectric devices but it should not be neglected in detailed calculations.

Mathematically, the thermoelectric effects can be expressed in the following way:

The Seebeck voltage is given by equation 7.1

$$E = \alpha_{ab}(T_1 - T_2)$$

where T_1 and T_2 are the junction temperatures and **the Peltier heat** is the rate of energy absorbed or rejected and is proportional to the junction current (I), flowing through if, i.e.,

$$Q_p = \pi_{ab} I \qquad (7.2)$$

where π_{ab} is the Peltier coefficient. And **Thomson heat** absorbed or rejected along the length of the single conductor is given by

$$Q = \tau I \cdot \frac{dT}{dx} \qquad (7.3)$$

d) The Kelvin Relationships

It must be mentioned here that these thermoelectric coefficients are related with each other for any given pair of thermoelectric materials. Kelvin established the following relations:

$$\pi_{ab} = T \cdot \alpha_{ab} \qquad (7.4)$$

and

$$\tau_a - \tau_b = T \cdot \frac{d\alpha_{ab}}{dT} \qquad (7.5)$$

where τ_a and τ_b are the Thomson coefficients for the thermoelectric elements.

e) Nernst and Ettinghausen Effects

All the thermoelectric coefficients discussed above are in general dependent on magnetic field B with the principal parameters used in thermoelectrics, the Seebeck and Peltier effects having corresponding thermomagnetic coefficients, the Nernst and Ettinghausen effects, respectively. In the Nernst effect a transverse electric field E_y is produced by a longitudinal temperature gradient dT/dx in the presence of a magnetic field B_z and in the Ettinghausen effect a transverse temperature gradient is produced by a longitudinal electric current with the thermomagnetic coefficients as given by

$$|N| = \frac{\frac{E_y}{B_z}}{\frac{dT}{dx}} \qquad (7.6)$$

$$|P| = \frac{\frac{1}{I_x B_z}}{\frac{dT}{dx}} \qquad (7.7)$$

These thermoelectric effects are reversible in nature. Hence, if a current is made to pass through a circuit of two materials, there is absorption of thermal energy at the cold junction temperature and

300 *Alternatives in Refrigeration and Air Conditioning*

rejection of heat at the hot junction temperature. Thus, reversed thermoelectric effect, viz., Peltier effect is responsible for the operation of a thermoelectric refrigerator. In addition, with the flow of current (I) in a closed thermoelectric circuit, an irreversible generation of heat (I^2R) due to Joule effect will occur in the conductors where R is the total resistance in the circuit. Also when one junction of a thermocouple is at a higher temperature than the other junction, there would be conduction of heat from the warm end to the cold end. This irreversible conduction effect is expressed as $K(T_1 - T_2)$, where K is the overall thermal conductance and T_1, T_2 are the junction temperatures. Based on these thermoelectric effects, thermoelectric heat engine and thermoelectric refrigerator have been devised. It has also been proposed from time to time that thermocouples should be used in generating electricity from solar energy. Thermoelectric heat engines or generators have the advantages of being flexible in operating conditions, and can be used at moderate temperatures with quite small temperature differences between the source and the sink. The overall efficiency of these thermoelectric generator and refrigerator systems may be low but simplicity of the device is additional advantage over other alternatives.

In case of solar thermoelectric refrigeration a combination of thermoelectric heat engine and refrigerator is used. The thermoelectric generator draws its heat from solar heat source and converts it into electrical energy. This generator output supplies electrical power for the operation of a thermoelectric refrigerator. Thus, the combination of thermoelectric generator and thermoelectric refrigerator makes it possible to utilise solar energy for producing refrigeration.

7.3 ANALYSIS OF THERMOELECTRIC REFRIGERATOR

A conventional thermoelectric refrigerator consists of a thermoelectric circuit made of two conductor materials A and B (one n-type and another of p-type semiconductor) as shown in Fig. 7.3 where the two hot and cold junctions are at temperatures T_1 and T_2 respectively ($T_1 > T_2$). The cold end is attached to a metal plate or some other heat transfer surface exposed to the space to be cooled and is equivalent to the evaporating coil of a vapour compression refrigeration system. The hot end is attached to a warm metal plate which causes rejection of heat to the surroundings and performs the function of a condenser as in a vapour compression refrigerator. An external battery is used to provide a direct current circulation through the circuit and thus is equivalent to the compressor component of a vapour compression refrigeration process.

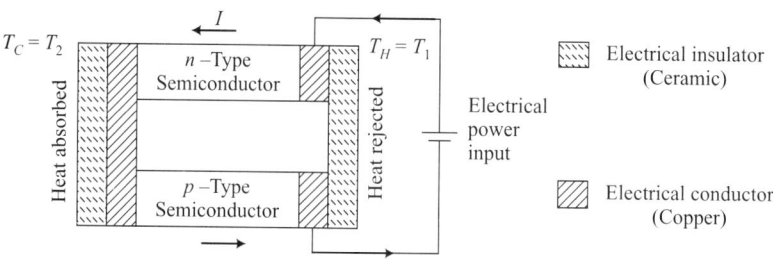

Fig. 7.3 Thermoelectric Cooling (Peltier Effect)

The thermoelectric circuit can be considered consisting of current source I, total electrical resistance R and thermal conductance K. There are three reversible processes, viz., Peltier, Thomson and Seebeck effects and two irreversible processes, viz., Joule heating and the heat conduction effects occurring simultaneously. It is logical to assume that one half of the Joulean heat (I^2R) is transferred to each junction and it has also been proved that the net effect of the combined Peltier and Thomson effects is to assume Peltier coefficients at the hot and cold junctions as given by $\alpha_m T_1$ and $\alpha_m T_2$ where α_m is the mean value of $\alpha_{ab}(T)$ between the temperatures T_1 and T_2. With these simplifications and assumptions of constant material properties, one may write the following heat balance equations for the hot and cold junctions:

$$Q_1 = T_1 \alpha_m I - K(T_1 - T_2) + \frac{I^2 R}{2} \tag{7.8}$$

$$Q_2 = T_2 \alpha_m I - K(T_1 - T_2) - \frac{I^2 R}{2} \tag{7.9}$$

The power input supplied by the battery is given by

$$W = Q_1 - Q_2 = E.I = (T_1 - T_2) \alpha_m I + I^2 R \tag{7.10}$$

and thus electromotive force $E = (T_1 - T_2) \alpha_m + IR$. \hfill (7.11)

The coefficient of performance of the thermoelectric refrigerator is given by

$$\text{COP} = \frac{Q_2}{W} = \frac{T_2 \alpha_m I - K(T_1 - T_2) - 1/2\, I^2 R}{(T_1 - T_2) \alpha_m I + I^2 R} \tag{7.12}$$

It is thus seen that COP of the thermoelectric device is also temperature controlled. The thermoelectric COP approaches Carnot COP $\equiv$ when K and R are zero. Both the COP and the refrigerating effect are functions of the current I. Thus, there is an optimum current (at which Q_2 is maximum) given by

$$I = I_0 = \frac{T_2 \alpha_m}{R} \tag{7.13}$$

and correspondingly

$$Q_{2max} = K \left[\frac{T_2^2}{2} Z_r - (T_1 - T_2) \right] \tag{7.14}$$

where $Z_r = \frac{\alpha_m^2}{KR}$ is called the figure of merit of the thermoelectric material combination for the refrigerator. This parameter expresses in quantitative form the requirements that the reversible thermoelectric effects should be large and irreversible effects of electrical resistance and thermal conductance should be small. Similarly, the optimum current for maximum COP is obtained to be

$$I = I_0 = \frac{\alpha_m (T_1 - T_2)}{R \left(\sqrt{1 + Z_r T_m} \right) - 1} \tag{7.15}$$

302 *Alternatives in Refrigeration and Air Conditioning*

where, $T_m = \dfrac{T_1 + T_2}{2}$ is the average junction temperature and maximum COP is given by the expression,

$$\text{COP}_{max} = \left(\dfrac{T_2}{T_1 - T_2}\right) \dfrac{\left(M_r - \dfrac{T_1}{T_2}\right)}{(M_r + 1)} \qquad (7.16)$$

where

$$M_r = \sqrt{1 + Z_r\, T_m} \qquad (7.17)$$

Similarly, the maximum thermal efficiency of a thermoelectric generator can be given by

$$\eta_{max} = \dfrac{T_1 - T_2}{T_1} \cdot \dfrac{(M_g - 1)}{\left(M_g + \dfrac{T_2}{T_1}\right)} \qquad (7.18)$$

where $M_g = \sqrt{1 + T_m\, Z_g}$, Z_g is the figure of merit for the generator. Thus, both for the thermoelectric generator and thermoelectric refrigerator, the performance depends on the junction temperatures at the hot and cold ends and the respective figure of merits defined in general,

$$Z = \dfrac{\alpha_m^2}{KR} \qquad (7.19)$$

Higher is the value of this parameter Z for each device, better is the performance. Figures 7.4 and 7.5 show the variations of COP_{max} of TEC and conversion efficiency of TEG with the temperature difference $\Delta T = T_1 - T_2$ for different values of Z. It is seen from Fig. 7.4, that for a fixed Z (material), COP_{max} decreases with increase in $(T_1 - T_2)$ while for fixed $(T_1 - T_2)$, the actual performance approaches Carnot performance as $Z \to \infty$. It is seen from Fig. 7.5, that the conversion efficiency increases with increase in hot side temperature of thermoelectric generator for fixed T_2 and as Z increases, conversion efficiency also increases. For given design conditions T_1 and T_2 are known and Z is fixed by the choice of the thermoelectric materials. For improved performance of the thermoelectric devices, the material should have high Seebeck coefficient, low thermal conductance and high electrical conductivity. For the parameter Z to be high, the product KR should be minimum and hence the performance of the thermoelectric device is dependent on the dimensions of the thermoelectric elements as discussed below (Chang, 1963).

For any combination of thermoelectric elements A and B, let

k_a, k_b = Thermal conductivities of A and B

ρ_a, ρ_b = Electrical resistivities of A and B

L_a, L_b = Lengths of the thermoelectric elements A and B

A_a, A_b = Cross-sectional areas of the thermoelements A and B.

Then, we have

$$K = \frac{A_a\, k_a}{L_a} + \frac{A_b\, k_b}{L_b} \tag{7.20}$$

$$R = \frac{L_a\, \rho_a}{A_a} + \frac{L_b\, \rho_b}{A_b} \tag{7.21}$$

and if we define $\gamma = \left(\dfrac{A_a\, L_a}{A_b\, L_b} \right)$,

$$\begin{aligned} KR &= (\gamma\, k_a + k_b) \left(\frac{\rho_a}{\gamma} + \rho_b \right) \\ &= \rho_a\, K_a + \rho_b\, k_b + \gamma\, k_{ab} + \frac{1}{\gamma} K_b\, \rho_a \end{aligned} \tag{7.22}$$

so for (KR) to be minimum, $\gamma = \sqrt{\dfrac{k_b\, \rho_a}{k_a\, \rho_b}}$ \hfill (7.23)

and hence, $KR = (KR)_0 = \left(\sqrt{\rho_a\, k_a} + \sqrt{\rho_b\, k_b} \right)^2$ \hfill (7.24)

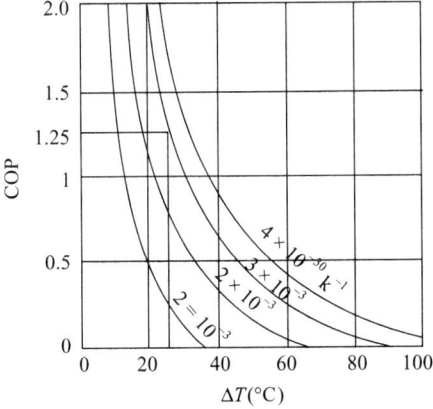

Fig. 7.4 Theoretical Coefficient of Performance of Thermoelectric Module Against Temperature Difference for Different Z Values

and therefore,
$$Z = \frac{\alpha_{ab}^2}{\left(\sqrt{\rho_a\, k_a} + \sqrt{\rho_b\, k_b} \right)^2} = \left[\frac{\alpha_{ab}}{\left(\sqrt{\rho_a\, k_a} + \sqrt{\rho_b\, k_b} \right)} \right]^2 \tag{7.25}$$

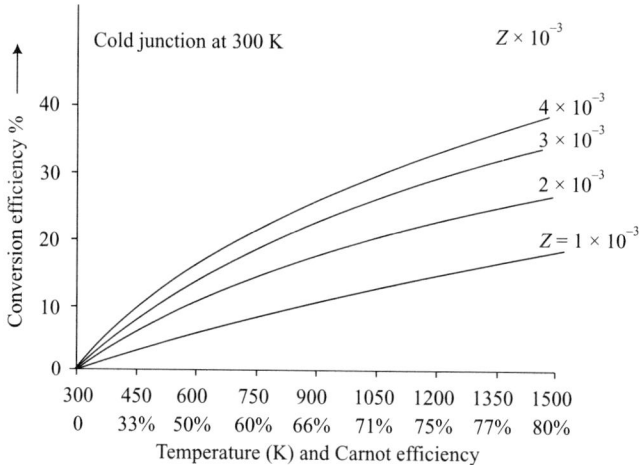

Fig. 7.5 Conversion Efficiency as a Function of Temperature and Figure of Merit of Thermocouple Material

Thus, as with any thermodynamic system, it is desirable to know the properties of the materials that are suitable for thermoelectric generators and refrigerators.

7.4 MULTISTAGE THERMOELECTRIC REFRIGERATION SYSTEMS

In order to improve the performance of a thermoelectric refrigeration system to achieve lower temperature, multistaging is used since maximum temperature difference achieved in a single-stage is not very high. Figure 7.6 shows a multistage system which can also be reduced to an equivalent thermal system. Then, for the first stage:

$$COP_1 = \frac{Q_c}{W} = \frac{Q_c}{Q_1 - Q_c} \qquad (7.26)$$

i.e.
$$Q_1 = Q_c\left(1 + \frac{1}{COP_1}\right) \qquad (7.27)$$

Similarly, for the second stage one can get:

$$COP_2 = \frac{Q_1}{Q_2 - Q_1} \qquad (7.28)$$

i.e.
$$Q_2 = Q_1\left(1 + \frac{1}{COP_2}\right) \qquad (7.29)$$

Substituting for Q_1 from the equation above

$$Q_2 = Q_c\left(1 + \frac{1}{COP_1}\right)\left(1 + \frac{1}{COP_2}\right) \qquad (7.30)$$

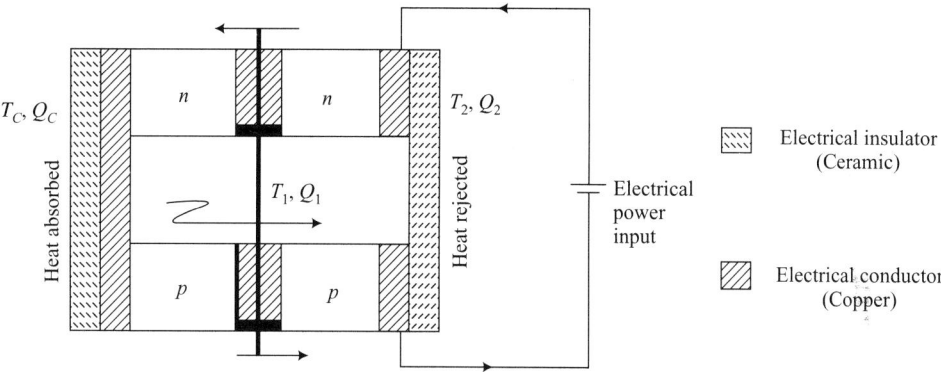

Fig. 7.6 Multistage (Two-stage) Thermoelectric Refrigeration System

Similarly, it can be proved that for n stages

$$Q_n = Q_c \left(1 + \frac{1}{COP_1}\right)\left(1 + \frac{1}{COP_2}\right)\cdots\left(1 + \frac{1}{COP_n}\right) = Q_c \prod_{i=1}^{n}\left(1 + \frac{1}{COP_i}\right) \quad (7.31)$$

COP of an n stage system can be written as:

$$COP_n = \frac{Q_c}{Q_n - Q_c} \quad (7.32)$$

or

$$Q_n = Q_c \left(1 + \frac{1}{COP_n}\right) \quad (7.33)$$

Thus, equations (7.31) and (7.33) yield:

$$COP_n = \frac{1}{\prod_{i=1}^{n}\left(1 + \frac{1}{COP_i}\right) - 1} \quad (7.34)$$

If we consider n-stage refrigeration system with the coefficient of performance of every stage to be equal and denoted by COP, one can obtain from equation (7.34) as:

$$COP_n = \frac{1}{\left(1 + \frac{1}{COP'}\right)^n - 1} \quad (7.35)$$

The coefficient of performance for each stage of an n-stage cascade system is given by an approximate expression:

$$COP' = n\,(COP_s + 1/2) - 1/2 \quad (7.36)$$

306 *Alternatives in Refrigeration and Air Conditioning*

Here it has been assumed that the temperature difference for each stage is equal to $(T_h - T_c)/n$ where T_h and T_c are the upper and lower temperature limits for the system and $T_m = \dfrac{T_h + T_c}{2}$. Then COP_s is given by:

$$COP_s = \frac{T_C}{T_h - T_C} \frac{[(1+T_m Z)^{0.5} - T_h/T_c]}{[(1+ZT_m)^{0.5} + 1]} \tag{7.37}$$

Using equation (7.37) into equation (7.36), it is found as:

$$COP_{n\text{-stage}} = \frac{1}{\left\{1 + 1/\left[n\left(COP_s + \dfrac{1}{2}\right) - \dfrac{1}{2}\right]\right\}^n - 1} \tag{7.38}$$

In case of infinite number of stages of a cascade system, the coefficient of performance can be given by:

$$COP_\infty = \frac{1}{\exp\left[1/\left(COP_s + \dfrac{1}{2}\right)\right] - 1} \tag{7.39}$$

However, the infinite number of cascade systems cannot be realized from the practical point of view. Hence, two- to three-stage systems can be practically a feasible choice.

Figure 7.7 shows the variation of COP for different number of stages with varying operating conditions. It is seen that as the number of stages increases, the coefficient of performance increases.

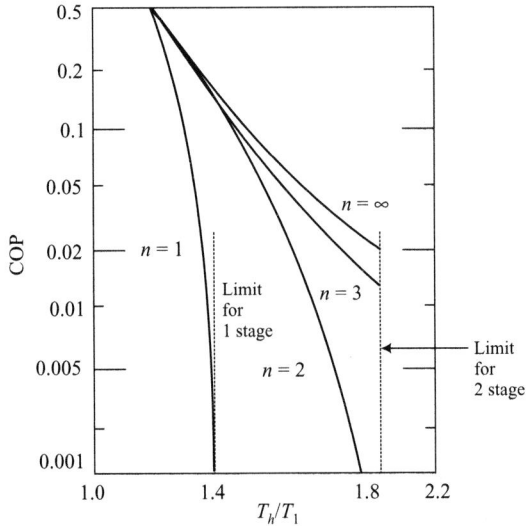

Fig. 7.7 Variation of COP for Different Number of Stages

7.4.1 Two-Stage Thermoelectric Cooler

Electrically Parallel Configuration

A brief outline of a two-stage electrically parallel thermoelectric cooling system is presented here with notations given in Fig. 7.8. The two-stage thermoelectric cooling system presented in Fig. 7.8 is called electrically parallel two-stage thermoelectric cooling system, because the electrical connection in each stage is made parallel with the power source. There is total 'M' 'number of thermocouples in the two-stage thermoelectric cooler. The number of thermocouples in the first stage is n and in second stage is m, and therefore, $M = n + m$.

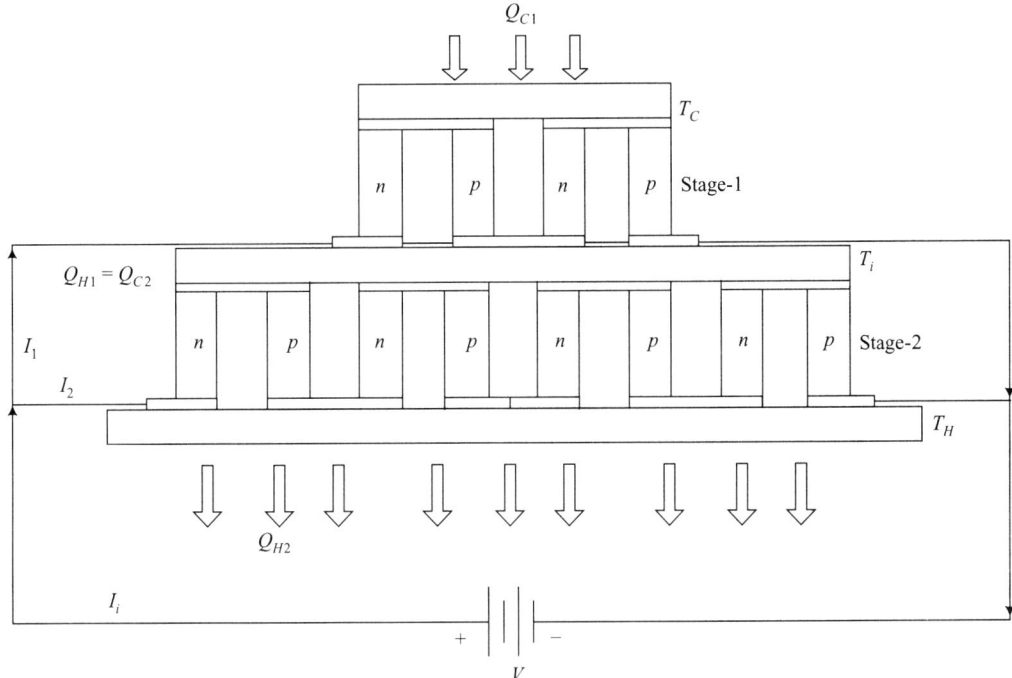

Fig. 7.8 Two-stage Thermoelectric Cooler (Electrically Parallel)

By the first law of thermodynamics, the energy balance at the hot and cold junctions of the electrically parallel two-stage thermoelectric cooler is given as follows:

$$Q_{C1} = \alpha_1 T_C I_1 - \frac{I_1^2 R_1}{2} - K_1(T_i - T_C) \tag{7.40}$$

$$Q_{H1} = \alpha_1 T_i I_1 + \frac{I_1^2 R_1}{2} - K_1(T_i - T_C) \tag{7.41}$$

$$Q_{C2} = \alpha_2 T_i I_2 - \frac{I_2^2 R_2}{2} - K_2(T_H - T_i) \tag{7.42}$$

$$Q_{H2} = \alpha_2 T_H I_2 + \frac{I_2^2 R_2}{2} - K_2(T_H - T_i) \tag{7.43}$$

where α_1, α_2 are $n\alpha_{pn}$, $m\alpha_{pn}$, and the thermal conductance and electrical resistance also follows the same notation.

Heat rejected from stage 1 is the heat absorbed in stage 2. Therefore, $Q_{H1} = Q_{C2}$, and by equating these equations one can find the inter-stage temperature T_i.

$$T_i = \frac{\left[\frac{I_1^2 R_1}{2} + K_1 T_C + \frac{I_2^2 R_2}{2} + K_2 T_H\right]}{[\alpha_2 I_2 - \alpha_1 I_1 + K_2 + K_1]} \tag{7.44}$$

Since the configuration is electrically parallel, the voltage drop in the two stages of the thermoelectric cooler will be equal and it can be given as

$$\alpha_1(T_i - T_C) + I_1 R_1 = \alpha_2(T_H - T_i) + I_2 R_2 \tag{7.45}$$

According to Kirchhoff's current law, the total current at any junction is balanced, i.e.,

$$I_i = I_1 + I_2 \tag{7.46}$$

Solving equations 7.45 and 7.46, I_1 and I_2 are found as:

$$I_1 = \left[\frac{I_i R_2}{R_1 + R_2}\right] + \left[\frac{\alpha_2 \Delta T_2 - \alpha_1 \Delta T_1}{R_1 + R_2}\right] \tag{7.47}$$

$$I_2 = \left[\frac{I_i R_1}{R_1 + R_2}\right] - \left[\frac{\alpha_2 \Delta T_2 - \alpha_1 \Delta T_1}{R_1 + R_2}\right] \tag{7.48}$$

where $\Delta T_1 = T_i - T_C$ and $\Delta T_2 = T_H - T_i$

Eliminating I_1 and I_2 from $Q_{H1} = Q_{C2}$, one gets a quadratic equation in I_i as follows:

$$I_i^2 \left[\frac{R_1 R_2}{2(R_1 + R_2)}\right] - I_i \left[\frac{T_i(\alpha_1 R_1 - \alpha_2 R_2)}{R_1 + R_2} - \frac{2(R_1 R_2)(\alpha_2 \Delta T_2 - \alpha_1 \Delta T_1)}{R_1 + R_2}\right] +$$

$$\left[\frac{(\alpha_1 + \alpha_2)(\alpha_2 \Delta T_2 - \alpha_1 \Delta T_1)}{R_1 + R_2} + \frac{(\alpha_2 \Delta T_2 - \alpha_1 \Delta T_1)^2}{2(R_1 + R_2)} + (K_2 \Delta T_2 - K_1 \Delta T_1)\right] = 0 \tag{7.49}$$

After calculating I_i from the above quadratic equation and substituting the value of I_i in equation 7.40, the cooling power can be obtained.

The power input to the electrically parallel two-stage thermoelectric cooler can be given as

$$W_{in} = (Q_{H1} - Q_{C1}) + (Q_{H2} - Q_{C2}) \tag{7.50}$$

$$W_{in} = \alpha_1 I_1 \Delta T_1 + I_1^2 R_1 + \alpha_2 I_2 \Delta T_2 + I_2^2 R_2 \tag{7.51}$$

Then the COP of two-stage electrically parallel thermoelectric cooler can be obtained as the ratio of Q_{C1} to the W_{in}.

$$\text{COP}_{ep} = \frac{Q_{C1}}{W_{in}} = \frac{\alpha_1 T_C I_1 - \dfrac{I_1^2 R_1}{2} - K_1(T_i - T_C)}{\alpha_1 I_1 \Delta T_1 + I_1^2 R_1 + \alpha_2 I_2 \Delta T_2 + \Delta T_2 + I_2^2 R_2} \tag{7.52}$$

Electrically Series Configuration

The two-stage thermoelectric cooler with electrically series configuration is shown in Fig. 7.9. There is total 'M' number of thermocouples in the two-stage thermoelectric cooler. The number of thermocouples in the first stage is n and in second stage is m, and therefore, $M = n + m$.

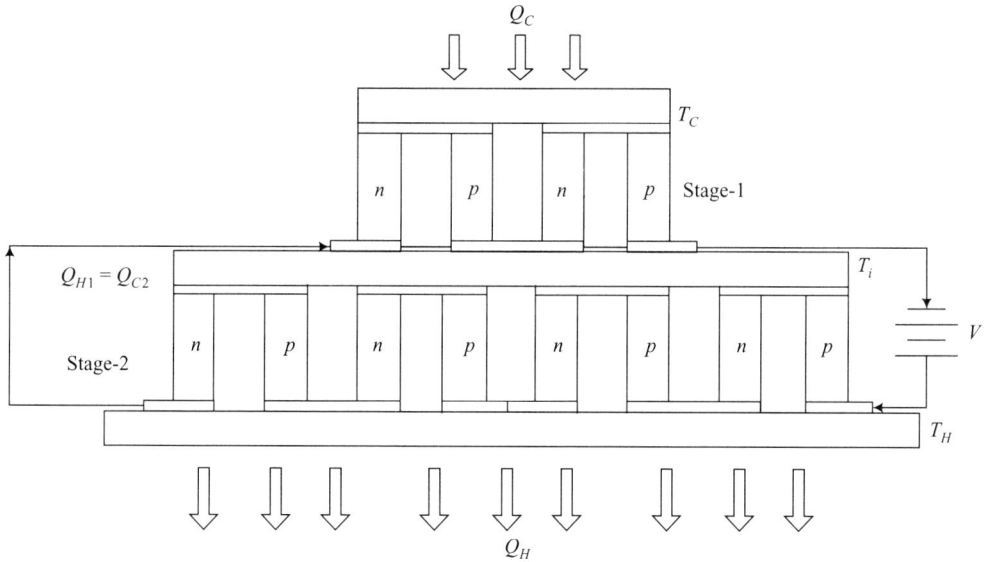

Fig. 7.9 Two-stage Thermoelectric Cooler (Electrically Series)

By the first law of thermodynamics, the energy balance at the hot and cold junctions of the electrically series-two-stage thermelectric cooler is given as follows:

$$Q_{C1} = \alpha_1 T_C I - \frac{I^2 R_1}{2} - K_1 (T_i - T_C) \tag{7.53}$$

$$Q_{H1} = \alpha_1 T_i I + \frac{I^2 R_1}{2} - K_1(T_i - T_C) \tag{7.54}$$

$$Q_{C2} = \alpha_2 T_i I - \frac{I^2 R_2}{2} - K_2(T_H - T_i) \tag{7.55}$$

$$Q_{H2} = \alpha_2 T_H I + \frac{I^2 R_2}{2} - K_2(T_H - T_i) \tag{7.56}$$

The inter-stage temperature T_i can be calculated by equating $Q_{H1} = Q_{C2}$, since the heat rejected in the first stage is the heat absorbed by the second stage, and the inter-stage temperature can be derived as follows:

$$T_i = \frac{\left[\dfrac{I^2(R_1 + R_2)}{2} + K_2 T_H + K_1 T_C\right]}{[I(\alpha_2 - \alpha_1) + K_1 + K_2]} \tag{7.57}$$

Since the configuration is electrically series, the current flowing through the two stages of the thermoelectric cooler will be equal and it can be given as

$$I = \left[\frac{V_{in}}{R_1 + R_2}\right] - \left[\frac{\alpha_1 \Delta T_1 + \alpha_2 \Delta T_2}{R_1 + R_2}\right] \tag{7.58}$$

where ΔT_1, ΔT_2 have the appropriate meaning, and V_{in} is the sum of voltages in the individual stages.

$$V_{in} = V_1 + V_2 = \alpha_1 \Delta T_1 + \alpha_2 \Delta T_2 + I(R_1 + R_2) \tag{7.59}$$

The power input to the electrically series two-stage thermoelectric cooler can be given as

$$W_{in} = I(\alpha_1 \Delta T_1 + \alpha_2 \Delta T_2) + I^2(R_1 + R_2) \tag{7.60}$$

Then the COP of two-stage electrically series thermoelectric cooler can be obtained as the ratio of Q_{C1} to the W_{in}.

$$\text{COP}_{es} = \frac{Q_{C1}}{W_{in}} = \frac{\alpha T_C I - \dfrac{I^2 R_1}{2} - K_1(T_i - T_C)}{I(\alpha_1 \Delta T_1 + \alpha_2 \Delta T_2) + I^2(R_1 + R_2)} \tag{7.61}$$

7.5 ACTUAL THERMOELECTRIC REFRIGERATION SYSTEM

The actual design consideration in the thermoelectric refrigeration system requires the knowledge of optimum size, optimum number of couples, etc., in a module of a given thermoelectric material.

Also, one should know the contact resistance between the thermoelectric elements and the copper plates to which these elements are soldered. Similarly, the increase in thermal resistances due to junction has to be incorporated. To achieve the desired performance high grade thermal insulation should be provided between the thermoelectric elements. Further, the thermocouples should be insulated from the outside circuit using appropriate electric insulation having low thermal resistance. To protect the life of insulation and thermoelectric elements a good quality moisture resisting element should be provided as shown in Fig. 7.10.

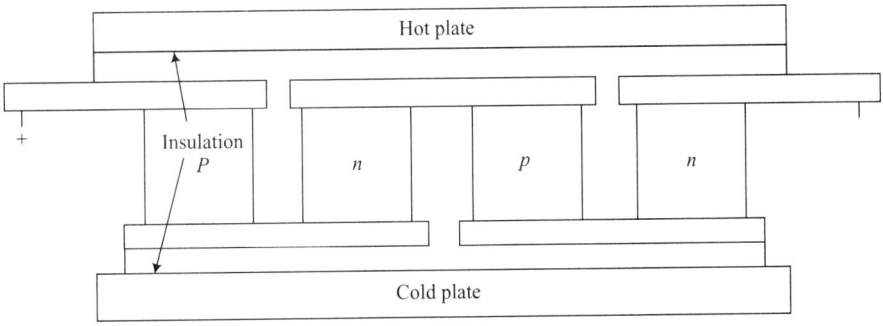

Fig. 7.10 Actual Thermoelectric Refrigeration System

7.5.1 Typical Thermoelectric Domestic Refrigerators (Riffat *et al.*, 2001)

The thermoelectric refrigeration has made little impact on the domestic refrigeration market, although Peltier modules have been employed in portable cool-boxes for medicine/serum transport and for picnic items storage, stabilise the temperature of solid-state lasers, to cool infrared detectors and charge-coupled devices, and to increase the operating speed and reduce unwanted noise of integrated circuits. The Peltier module (Fig. 7.11) is a unique cooling device, in which the electron gas serves as the working fluid. It is inherently noiseless, reliable and environment-friendly. In recent years, concerns of environmental pollution due to the use of CFCs in conventional domestic refrigerators have encouraged increasing activities in research and development of domestic refrigerators using Peltier modules. It consists of a number of n-type and p-type semiconductor thermoelements connected in series by copper strips and sandwiched between two electrically insulating, but thermally conducting ceramic plates. Heat is pumped from one side of the module to the other by charge carriers (electrons or holes) when a d.c. electric current is applied. The energy efficiency of a thermoelectric refrigerator is mainly determined by the COP of the Peltier module and the heat-transfer effectiveness of the heat exchangers. Although the basic structure of a thermoelectric refrigerator is essentially the same, their configurations may differ significantly depending on the heat exchangers employed. The following are the typical configurations of thermoelectric refrigerators:
  1. Fin type heat exchangers with forced convection on both sides (cold and hot, i.e., inside and outside of cabin), refer Fig. 7.12 (TER – 1)
  2. Fin type heat exchangers with forced convection on cold side and liquid circulation (or thermosyphon) at hot side, refer Fig. 7.13 (TER – 2)
  3. Liquid circulation (or thermosyphon) on both sides (TER – 3)

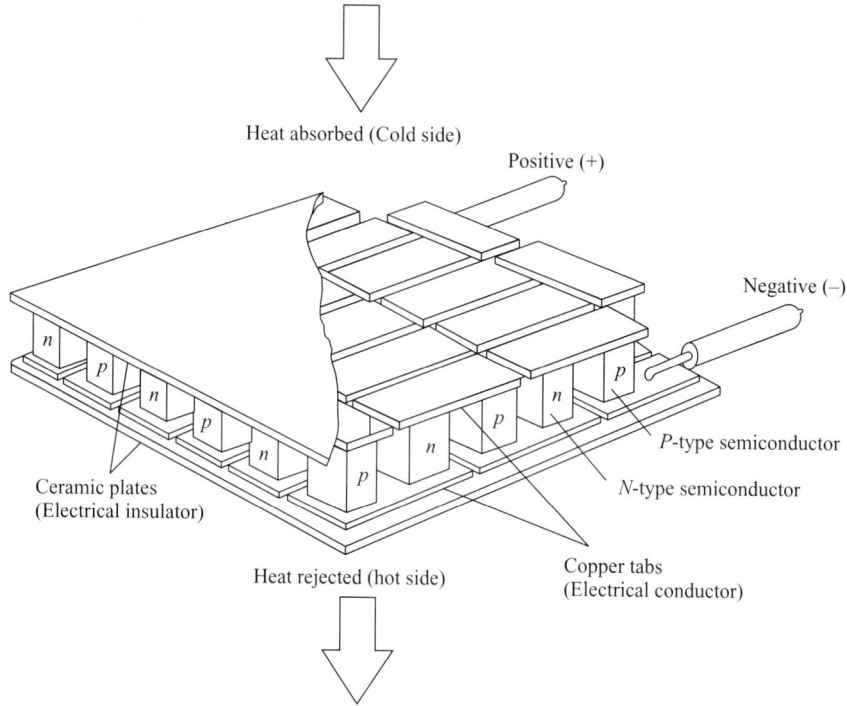

Fig. 7.11 Single-stage Peltier Module, Showing Details of Element Integration (Riffat *et al.*, 2001)

a) Technical Specifications:
TER - 1

Peltier modules	: CP2-127, MELCOR, USA, Nominal maximum pumping capacity of 120 W
Heat exchangers	
Material	: Aluminium with bonded fins
Overall dimensions for the inside	: 100 mm × 100 mm × 30 mm
No. of Fins for the inside	: 24 fins (100 mm × 20 mm × 1 mm)
Overall dimensions for the outside	: 200 mm × 150 mm × 70 mm
No. of Fins for the outside	: 33 fins (200 mm × 60 mm × 1 mm)
Fan inside	: 01; brushless d.c. (3W)
Fans outside	: 02; brushless d.c. (3W each)
Capacity of compartment	: 90 litres

TER - 2

Peltier module	: 40 mm × 40 mm × 5 mm
Heat dissipated from the hot side	: 50 to 150 W
Gives maximum heat flux	: around 10 W

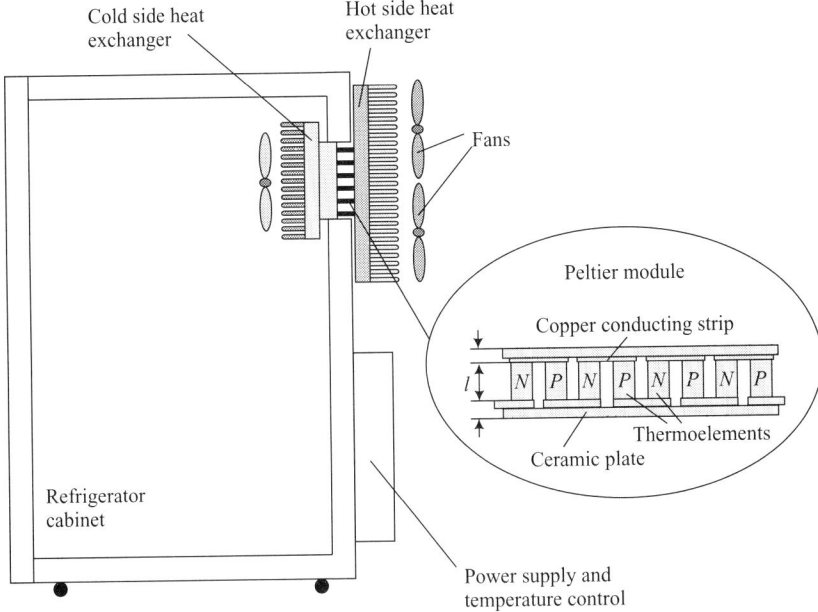

Fig. 7.12 Schematic of Thermoelectric Refrigerator having Fin-type Heat Exchanger on the Inside and Outside of the Refrigerated Cabinet (TER-1)

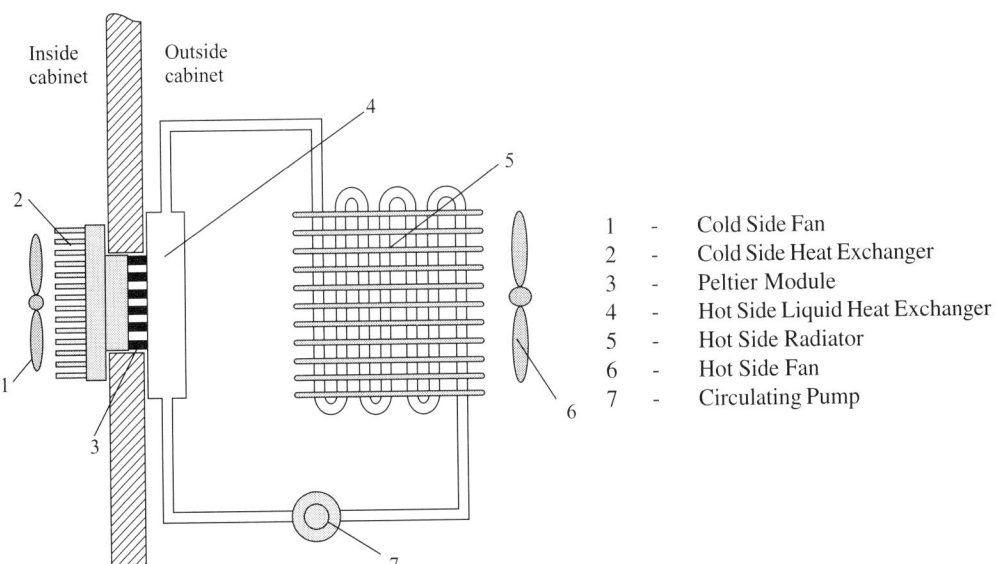

1 - Cold Side Fan
2 - Cold Side Heat Exchanger
3 - Peltier Module
4 - Hot Side Liquid Heat Exchanger
5 - Hot Side Radiator
6 - Hot Side Fan
7 - Circulating Pump

Fig. 7.13 Thermoelectric Refrigerator Consisting of a Fin-type Heat Exchanger on the Inside of the Refrigerated Cabinet and a Liquid Circulating Heat Exchanger on its Outside (TER - 2)

314 *Alternatives in Refrigeration and Air Conditioning*

TER - 3

Make	: Mitsubishi Electric Corporation, Japan
Capacity of cooling cabinet	: 40 litres.
Nominal power consumption	: 120 W (electric)

Conventional Compressor Refrigerator (CCR)

Make	: Lec Refrigeration Plc., UK
Capacity of a freezer compartment	: 25 litres
Capacity of cooling cabinet	: 90 litres

b) Comparative Study

Figures 7.14–7.16 represent the results of comparison of the aforesaid typical thermoelectric refrigerators.

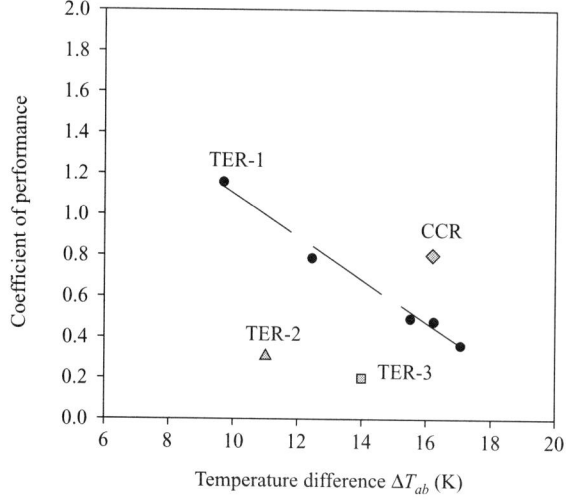

Fig. 7.14 Comparison of Coefficient of Performance

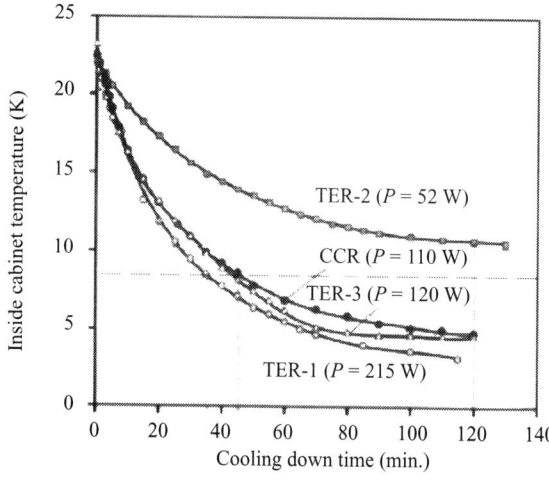

Fig. 7.15 Cooling Down Period for TER-1, TER-2, TER-3 and CCR

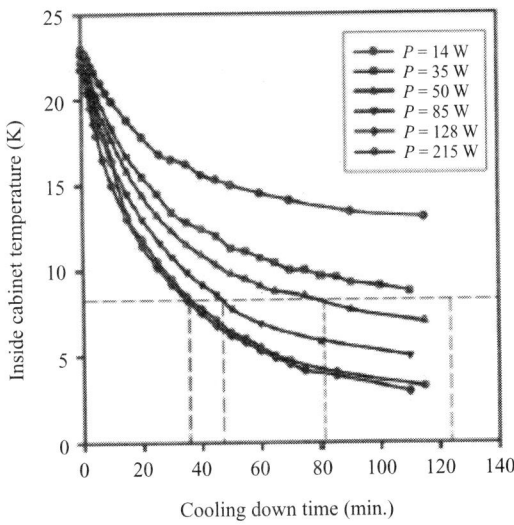

Fig. 7.16 Cooling Down Period for TER-1 for Different Input Powers

7.6 DESIGN ASPECTS OF SOLAR THERMOELECTRIC REFRIGERATOR

A solar thermoelectric refrigerator is based on the fact that a thermoelectric generator consisting of a small number of thermocouples generally produces a small e.m.f. but can easily be made to yield a large current. Thermoelectric generator has the advantage of being operated from a low grade heat source and is therefore an easy means of converting solar energy into electrical energy. This special feature is attractive if the electrical load is a thermoelectric refrigerator (also consisting of a small number of thermocouples requiring a large current and small e.m.f.). Thus, a combination of thermoelectric generator and thermoelectric refrigerator make a highly compatible combination for utilising solar energy for producing refrigeration (Vella et al., 1976 and Goldsmid et al., 1980).

The design of the thermoelectric module is an important aspect. The alternate p-type and n-type elements may be soldered to the heat transfer surfaces with the thermocouples connected in series electrically. The spaces separating the elements are filled with proper insulation. The choice of the insulating material is rather difficult because an electrical insulator may be good thermal conductor. Two parameters are important while designing a thermoelectric module: the current (I) and the ratio of element cross-sectional area to length (A/L). The former parameter needed for maximum coefficient of performance increases as the latter parameter increases. However, as A/L increases, the maximum refrigerating capacity increases but correspondingly higher currents are required. Thus, thermoelectric cooling is desirable for small cooling capacity requirements and is the most practical method if proper material with high figure of merit is made available.

In a solar thermoelectric refrigerator there is one thermocouple acting as a refrigerator and a number of identical couples acting as a generator. The hot junction of the generator is at T_1, the cold junction of the refrigerator is at T_2 and all the junctions in contact with the heat sink are at the same temperature T_0. The overall efficiency of the combined thermoelectric device is defined as the ratio of the heat extracted by the refrigerator to the heat entering the generator is the product of the maximum thermal efficiency of the generator and maximum COP of the refrigerator and thus given by following equation:

$$\varepsilon = \left(\frac{T_2}{T_O - T_2}\right)\left(\frac{T_1 - T_O}{T_1}\right)\left(\frac{M_g - 1}{M_r + 1}\right)\left(\frac{M_r - \frac{T_O}{T_2}}{M_g - \frac{T_O}{T_1}}\right) \qquad (7.62)$$

where M_g and M_r are parameters pertaining to the generator and the refrigerator respectively.

Vella et al. (1976) presented performance of a solar thermoelectric refrigerator and obtained optimum number of the couples in the generator and refrigerator subsystems. It is found that the number of the thermocouples in the generator must be selected according to the temperature differences across the two parts of the device. It is desirable and feasible to make both the generator and refrigerator from single thermocouples. However, it is rather difficult to obtain worth, while temperature depressions and cooling capacities using a single couple thermoelectric generator unless concentrated solar energy is used. It is however possible to reduce the number of couples in the generator from 4 to 2 if improved thermoelectric materials are used even when flat plate solar collectors are employed.

Goldsmid et al. (1980) investigated the possibility of using thermocouples made from Bi_2Te_3 alloys in thermoelectric generators using commercial thermoelectric modules and conventional flat plate solar collector as heat source. The use of an assymmetric stationary concentrator with Ge-Si alloy thermocouples is found to be more desirable in view of improved efficiency.

Some of the practical problems encountered in solar thermoelectric refrigerators are:
(i) There are substantial temperature differences between parts of the collector plate and the hot junction of the generator.
(ii) The design of the heat sink demands close contact between the refrigerator and the generator thereby necessitating a common sink temperature. However, since only a small rise of temperature of the sink for the refrigerator can be accepted for maintaining a reasonable COP, the sink for the generator part can also rise in temperature by a small amount. If separate heat sinks are provided for the generator and the refrigerator, this problem could he solved.
(iii) In the design and performance analysis of solar thermoelectric devices, it is assumed that junction temperatures are fixed while in practice this assumption is not valid and both the hot and cold junctions are of variable temperature.
(iv) The thermoelements are also assumed to have n- and p-type legs of identical thermoelectric properties and similar dimensions. This is rarely met in practice.

The overall poor efficiency of thermoelectric devices could partially be avoided by introducing suitable improvements in the heat economy of the system. The thermal energy loss due to high thermal conductivity can be recovered by using an auxiliary gas according to the regenerative principle. If a low temperature refrigeration is required it is necessary to connect the thermoelements in series. The two stages should be separated by a material which is a good thermal conductor and a poor electric conductor. When connected in series, each element must remove both the transferred heat and the Joule heat generated in the preceding elements. The COP of the multistage thermocouple will be higher than that of the single-stage thermocouple when the temperature difference is large. For N-stages of refrigeration, the overall COP is given by (Soo, 1968)

$$\frac{1}{\text{COP}} = \left[\sum_{i=1}^{N}\left(1+\frac{1}{\text{COP}_i}\right)-1\right] \tag{7.63}$$

Thus, cascading offers high promise for thermoelectric refrigerators. It is also desirable to obtain optimum intermediate temperatures in the cascaded system for maximising overall COP assuming that each stage is electrically insulated from adjacent stages in thermal contact. However, the analytical procedure is too cumbersome.

On the other hand, thermoelectric generators can never achieve the Carnot efficiency allowed by the second law of thermodynamics, unless one is operating within small ranges of temperature. Since the temperature difference $T = T_1 - T_2$ of a single-stage generator cannot be made arbitrarily large, thermoelectric generators can be cascaded with some improvement in efficiency (Harman, 1958). Consider two stages in series with increasing temperatures T_2, T_0 and T_1 at the respective junctions, we have

$$\eta_1 = \frac{W_1}{Q_1} = \frac{Q_1 - Q_0}{Q_1} \tag{7.64}$$

$$\eta_2 = \frac{W_2}{Q_0} = \frac{Q_0 - Q_2}{Q_0} \tag{7.65}$$

and thus, the overall efficiency is given by Soo, 1968

$$\eta = \frac{Q_1 - Q_2}{Q_1} = 1 - (1-\eta_1)(1-\eta_2) \tag{7.66}$$

Hence, for N-stages in series the overall efficiency is given by N

$$\eta = 1 - \prod_{i=1}^{N}(1-\eta_i) \tag{7.67}$$

Thus, cascading leads to higher output voltage in thermoelectric generators. Solar thermoelectric refrigerators are generally operated in varying climatic conditions and so subjected to a fluctuating heat flux as well as fluctuating hot junction temperature. Onyegegbu (1982) studied the performance of a solar thermoelectric cooler subjected to a modulated heat input approximated by a sinusoidal heat input superimposed over a steady heat input. The following conclusions were drawn from the modeling study of a solar thermoelectric cooler:

(i) When a solar thermoelectric cooler is subjected to a modulated heat input with a non-zero -mean, modulation has no effect on the mean value of the performance parameters.

(ii) As the ratio of the number of couples in the generator to the number of couples in the refrigerator increases, the current in the system increases and thus cooler with same number of couples in the generator and refrigerator is the most efficient arrangement.

(iii) While, the deviations from mean value heat input have no effect on mean value performance parameters, they significantly affect the corresponding fluctuations about any mean performance parameter.

(iv) Two basic design considerations for solar thermoelectric refrigerators are: (a) operation at maximum coefficient of performance and (b) operation at minimum cold space temperature. For the modulated heat input conditions, this corresponds to (a) operation at maximum

318 *Alternatives in Refrigeration and Air Conditioning*

oscillation COP and (b) operation at maximum cold space temperature oscillation per given solar heat oscillation. It is found that high frequency modulation with low N_g/N_r (ratio of the number of couples in the generator to that of the refrigerator) is best suited for the first design while low frequency modulation with high N_g/N_r ratio is most suitable for the second design (Onycgegbu, 1982).

7.7 CHOICE OF THE THERMOELECTRIC MATERIALS

The choice of the materials for thermoelectric devices is determined by their primary properties which are dependent on the concentration and mobility of charge carriers. Figure 7.17 shows the

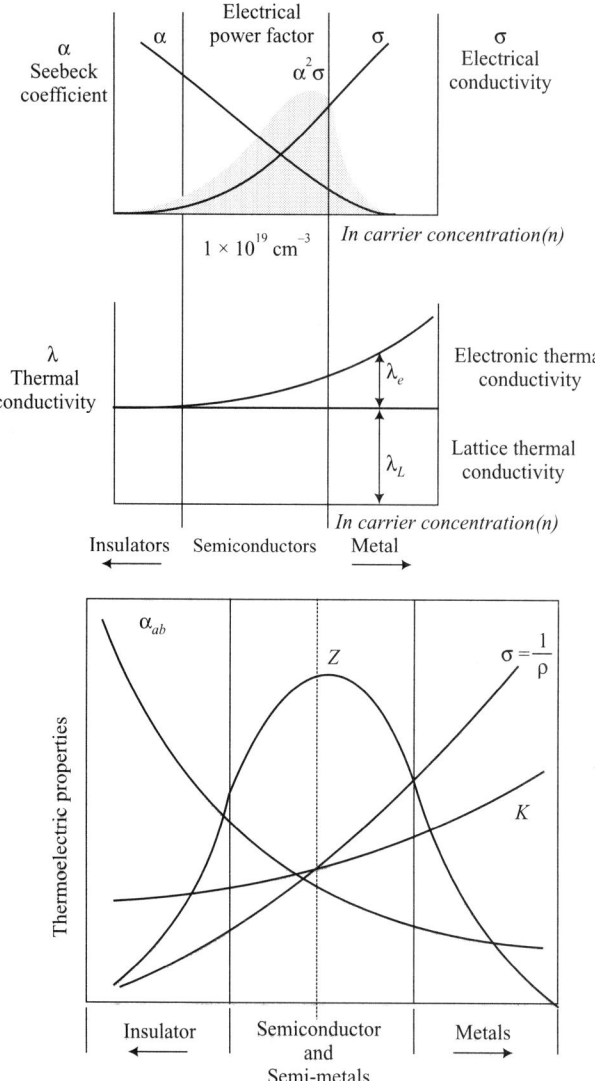

Fig. 7.17 Schematic Dependence of Electrical Conductivity, Seebeck Coefficient, Power Factor, and Thermal Conductivity on Concentration of Free Carriers

dependence of the thermoelectric properties on the electron density (carrier concentration) and explains the reason for suitability of semiconductors for thermoelectric devices.

Since the figure of merit parameter depends on thermal conductances, electrical resistivity and the Seebeck coefficient which are functions of the free charge carrier concentration in the material, metals have small values of Z because of small thermoelectric power and large thermal conductivity (Ioffe, 1957). Insulators are also poor thermoelectric materials because of their small electrical conductivity. The materials that have large Z are semi-metals and the semiconductors. Most commonly used semiconductors which have shown great promise are Lead Telluride (PbTe), Bismuth telluride (Bi_2Te_3) and Antimony Telluride (Sb_2Te_3) having Z value in the range 0.0014 to 0.0028 K^{-1}. Thus, assuming Z = 0.003, per Kelvin (K^{-1}) T_1 = 318 K, T_2 = 300 K, M = 1.389 and maximum COP is less than 30% of the COP of the Carnot device working between the same temperatures. Materials with higher Z value are being developed to make thermoelectric devices competitive with other alternatives. Electrical contact resistance plays an important role on the actual performance of thermoelectric devices and hence, proper thermoelectric modules are to be made. Typical parameters for thermoelectric elements are given in Table 7.1.

Table 7.1 Typical Parameters for Thermoelectric Element (ASHRAE Hand Book of Fundamentals, 1967)

Parameter	Typical Value
Thermoelectric power α_{ab}	0.00021 V/K
Electrical resistivity, ρ	0.0001 Ω cm
Thermal conductivity, K	0.015 W/cm K
Electrical contact resistance, r	0.00001 to 0.0001 Ω cm^2
Figure of merit, Z_{max}	0.003 K^{-1}

Thermoelectric phenomena are exhibited in almost all conducting materials (except for superconductors below T_c). Because the figure-of-merit varies with temperature, a more meaningful measure of performance is the dimensionless figure-of-merit ZT where T is absolute temperature. However, only those materials which possess a ZT > 0.5 are usually regarded as suitable thermoelectric materials.

Established thermoelectric materials (which are employed in commercial applications) can be conveniently divided into three groups with each dependent upon the temperature range of operation (Fig. 7.18). Alloys based on bismuth in combinations with antimony, tellurium, and selenium are referred to as low-temperature materials and can be used at temperatures up to around 450 K. These are the materials universally employed in thermoelectric refrigeration and have no serious contenders for applications over this temperature regime. The intermediate temperature range up to around 850 K is the regime of materials based on lead telluride while thermoelements employed at the highest temperatures are fabricated from silicon germanium alloys and operate up to 1300 K.

Although the above-mentioned materials remain the cornerstone for commercial applications in thermoelectric generation and refrigeration, significant advances have been made in synthesizing new materials and fabricating material structures with improved thermoelectric performance. Efforts focused on improving the figure-of-merit by reducing the lattice thermal conductivity. Two research avenues are currently being pursued. One is a search for so-called 'phonon glass-electronic crystals'

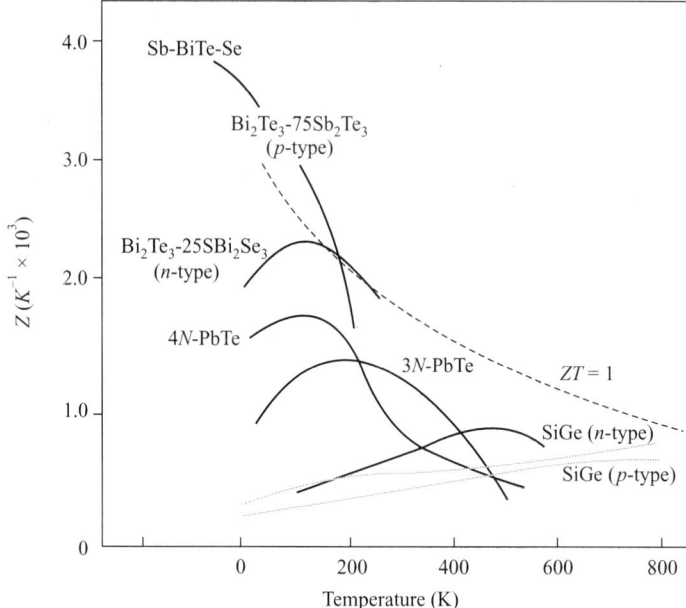

Fig. 7.18 Performance of the Established Thermoelectric Materials

in which it is proposed that crystal structures containing weakly bound atoms or molecules that "rattle" within an atomic cage should conduct heat like a glass, but conduct electricity like a crystal. Candidate materials receiving considerable attention are the filled skutterudites and the clathrates.

During the past decade material scientists have been optimistic in their belief that low-dimensional structures such as quantum wells (materials which are so thin as to be essentially of two dimensions (2D), quantum wires (extremely small cross section and considered to be of one-dimensional, and referred to as nanowires) quantum dots which are quantum confined in all directions and superlattices (a multiple-layered structure of quantum wells) will provide a route for achieving significantly improved thermoelectric figures-of-merit. The expectation is that the reduced dimensions of these structures will result in an increase in phonon interface scattering and a consequent reduction in lattice thermal conductivity.

Although low-dimensional structures would find immediate application in microelectronics, at present the technology is expensive and applying it to bulk devices is problematic. In some respects nanowires appear a more attractive proposition for thermoelectric applications than quantum well superlattices because the geometry of the current flow is more favorable and the fabrication process is more compatible with integrated technology.

Attempts are also being made to improve the competitiveness of thermoelectric material in directions other than the figure-of-merit. Efforts have also been focused on increasing the electrical power factor, decreasing cost, and developing environment-friendly materials. When the fuel cost is low or essentially free, as in waste heat recovery, then the cost per watt is mainly determined by the power per unit area and the operating period. The rare-earth compound $YbAl_3$, although possessing a relatively low figure-of-merit, with power factor almost three times that of bismuth telluride while MgSn has almost the same performance but costs less than a quarter of the price.

7.8 IMPROVEMENT IN THERMOELECTRIC FIGURE OF MERIT

Empirically, one can predict an upper bound by combining the best electronic properties of any known crystalline material with a phonon conductivity that is typical of glass. By this means, one arrives at a highest value of ZT equal to about four. Recently, however, it has been demonstrated that certain low-dimensional structures can have values of the figure-of-merit that exceed those obtained earlier.

FOM, denoted by Z is the combination of material properties, which are important in designing the thermoelectric (TE) systems. It has the unit of (1/K). To make it dimensionless, mean temperature T_m is multiplied with FOM.

$$ZT_m = \frac{\alpha_{12}^2 T_m}{KR} \quad \text{(dimensionless)} \tag{7.68}$$

where, $T_m = \frac{T_h + T_c}{2} = T$ (7.69)

By optimizing the geometry of the thermoelements, we can have ZT as follows:

$$ZT = \frac{\alpha_{12}^2 T}{\left(\sqrt{k_1 \rho_1} + \sqrt{k_2 \rho_2}\right)^2} \tag{7.70}$$

where k is the thermal conductivity and ρ is the resistivity of the thermoelement, subscripts 1 and 2 correspond to the leg1 and leg 2 of a thermocouple and α_{12} correspond to the mean seebeck coefficient of the thermocouple. It would be useful to have a figure of merit that could be assigned to a particular material, so that the materials could be compared individually. If a thermocouple is made up of p-type and n-type modifications of the same material, then $\alpha_1 = \alpha_2$, $\rho_1 = \rho_2$ and $k_1 = k_2$, then the above equation reduces to

$$ZT = \frac{\sigma \alpha^2 T}{k} = \frac{\sigma \alpha^2 T}{k_e + k_l} \tag{7.71}$$

In the above equation, $\sigma \alpha^2$ is called the power factor, where σ is electrical conductivity. The thermal conductivity can be contributed by electrons and phonons, $= k_e + k_l$, where k_l is the contribution by phonons and can be said as lattice thermal conductivity. The seebeck coefficient α for one level device can be given as follows:

$$\alpha = \frac{E - \mu}{qT} \tag{7.72}$$

where E is the energy levels and μ is the chemical potential or fermi energy level and q is the electron charge. To get higher ZT, the power factor has to be higher or the thermal conductivity has to be lower as possible. From the above equation, to get higher seebeck coefficient, $(E - \mu)$ have to be higher, but to have high electrical conductivity $(E - \mu)$ have to be minimum, e.g., in the case of metals. This is the reason for semiconductors materials are having relatively higher ZT values as compared to metals.

7.8.1 Search for New Materials

One can have high ZT by combining best electronic properties of any crystalline material with a material with less ZT. The best TE materials are solid solutions based on bismuth telluride, Bi_2Te_3, having ZT of 1 in 1950s and since then no significant improvements have been seen (Goldsmid, 1986). If the ZT value about 20 is reached, thermoelectric energy converters can have comparable efficiency with other energy or power converters and may have 50% efficiency as compared to Carnot engine.

A value of ZT > 2 was observed at high pressure for p-type $Sb_{1.5}Bi_{0.5}Te_3$ (Polvani et al., 2001). It was found that the dimensionless FOM at 9 GPa is equal to 2.2, but it is not advisable to operate at such high pressure. Efforts were made to have high ZT by suppressing k_l. Slack (1979) has set a lower bound for mean free path for the phonons of the order of lattice spacing, combining this with typical values of specific heat and the speed of sound, k_l should be no more than about 0.1–0.2 W/mK and by substituting this in k_l equation (7.49), one can have ZT of 4.

In the last 50 years, research was directed to find the new materials with improved properties. Ioffe (1957) showed that within a set of elements, k_l is decreased with increasing mean atomic weight. It was also suggested that the ratio of carrier mobility to lattice conductivity increases with the atomic weight (Ioffe and Ioffe,1954). This leads to Bi_2Te_3 as a choice of TE material (Goldsmid, 1954 and Cahill et al., 1992). Bi_2Te_3 has one of the highest value of power factor at ordinary temperature, and a low value of lattice thermal conductivity. It was found that compounds with empty spaces in their crystal structure can have low values for lattice thermal conductivity. These voids may be occupied by loosely bound atoms, such material is known as phonon glass-electron crystal (PGEC) having k_l as for amorphous material combined with typical electronic properties of a crystal. Skutterudites and Clathrates materials are some of examples of PGEC. These materials have seebeck coefficient of ±200 µV/K and having less k_l, they can have high ZT at higher temperature, e.g., Eu filled Skutterudite at 700 K (Lamberton et al., 2002) having ZT of 1.1 has been observed, similar values have been observed for other Skutterudites and some of the Clathrates (Nolas and Slack, 2001). Even though these materials have low k_l, they are inferior to Bi2Te3 because, these materials have lower power factor at room temperature.

Due to improvements in nano electronics, it has become easier to reduce the module size of a thermoelement. If we make the module width very small, but not less than the lattice constant, the electron will travel from one end to other end without any change in its direction. In case of such a ballistic transport, there may be reduction in the lattice thermal conductivity. Harman et al. (2002) have produced an n-type quantum dot structure based on PbSe-PbTe and have observed ZT of 1.6. A thermoelectric couple made from this material in conjunction with a metallic leg produced a cooling of about 44 K, when 700 mA current flows through it. This result is remarkable because PbTe-PbSe alloys in bulk form are inferior to Bi_2Te_3 alloys at ordinary temperature. Similar results were observed in p-type superlattice based Bi_2Te_3 and Sb_2Te_3 having ZT of 4, with a current flowing in the cross plane direction. Most of improvement is attributed to the reduction in lattice thermal conductivity to a value of $1/5^{th}$ of that in bulk samples.

Ghosal et al. (2002) made multiple contacts to the flat surface of p-type and n-type Bi2Te3 alloys. The radius of each contact was typically of 0.6 µm which shows ZT of 1.4 compared to 0.84 for commercial modules made from the same material. Lin et al. (2000) reported changes in the band structure as the diameter of the wires reduced, then they determined the seebeck coefficient as a function of carrier density and calculated ZT at 77 K, that yielded ZT of 6 for n-type Bi wires along the trigonal axis when carrier concentration is 10^{18} cm^{-3} and the diameter is 5nm. They stated that the seebeck coefficient of –400 µV/K, that is much greater than crystalline Bi. High power factor is because of electrical conductivity of about 3.4×10^6 Ω-1m-1, which is the result of enlarged density of states.

Cahill *et al.* (2003) gave the interface thermal conductance for a number of pair of materials. If suppose we are dealing with superlattice, that have a span of 5 nm, then there will be 2×10^8 interfaces/m. If the resistance is solely due to interfaces, the lattice conductivity could be as low as 0.15 W.mK.

If we make use of values of electrical conductivity and seebeck coefficient as given by Lin *et al.* (2000) Power factor at 77 K can be obtained as 41.9 W/mK and if k_l is 2.9 W/mK, then we get ZT of 6, if we change k_l to the minimum value of 0.1 W/mK ZT will become substantially greater. With this lower k_l if we reduce the carrier concentration to have high seebeck coefficient of –630 µV/K, we may have ZT of 20 as shown in Fig. 7.19. If we need high seebeck coefficient, we need substantial energy gap. Lin *et al.* (2000) have shown that energy gap is greater than 100 meV, when the diameter of the wire is less than 10 nm and it rapidly becomes larger as the wire becomes even narrower.

Lin *et al.* (2000) have reported that the mobility of the electron decreases and its value at 77 K is 17.5 times the value at 300 K (Abeles and Meiboom, 1956) and the density of states is greater by a factor of 7.7 at higher temperature and thus ZT at 300 K should be 1.7 times its value at 77 K.

Figure 7.19 shows the COP to that of a Carnot cycle as a function of ZT, the lower curve corresponds to T_h/T_c of 1.5 and the upper curve corresponds to T_h/T_c of 1. The COP would be as high as 35% of the ideal value for ZT of 4 and more than 60% for this value when ZT of 20. Once this ZT is achieved, thermoelectric devices can have the potential to replace all conventional refrigeration cycles available as of now.

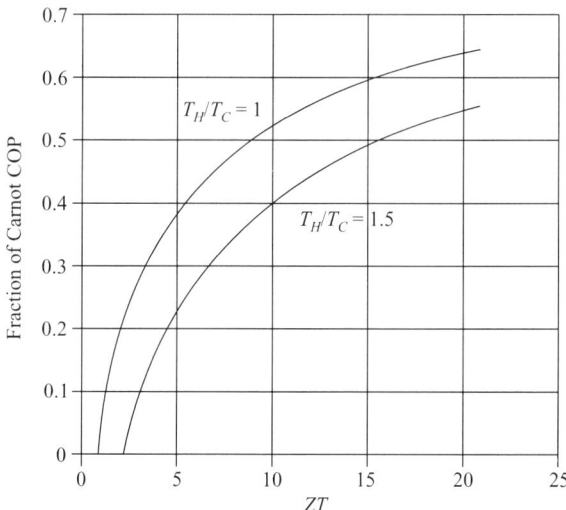

Fig. 7.19 Plot of COP of Thermoelectric Refrigerator (as a Fraction of Carnot COP) vs ZT

7.8.2 Solid State Thermionic Converters

In the nano structured configuration, ballistic transport is the predominating phenomenon. Vining and Mahan (1999) has demonstrated that thermionic devices might be superior if a reduction in the lattice conductivity is more than compensated for deterioration in the electronic properties. Logvinov and Gurevich (2003) found that the heat conduction by phonons is a limiting factor for typical solid state thermocouple. For size shorter than the thermal diffusion length (nm), the electron system can be regarded as isothermal (when the phonon system is adiabatic) having no energy interaction with phonons. The dimensionless FOM in this condition is given by:

$$ZT = \frac{\sigma \alpha^2 T}{\lambda_e} \tag{7.73}$$

Nevertheless as Logvinov and Gurevich (2003) have suggested, it will be very difficult to have a adiabatic situation for phonons. Figure 7.20 shows the ZT in the absence of phonon thermal conductivity. The heat conduction due to phonons is also absent in energy converter based on thermionic diode or vacuum thermoelement (Ioffe 1956). Thermionic energy conversion at room temperature needs work function (ϕ) of 0.3 eV, such small work function may not be possible. It has been shown (Nolas and Goldsmid 1999), that the COP of vacuum thermionic cooler with an emitter having a work function of 0.3eV, will be much better than that of thermoelectric cooler with

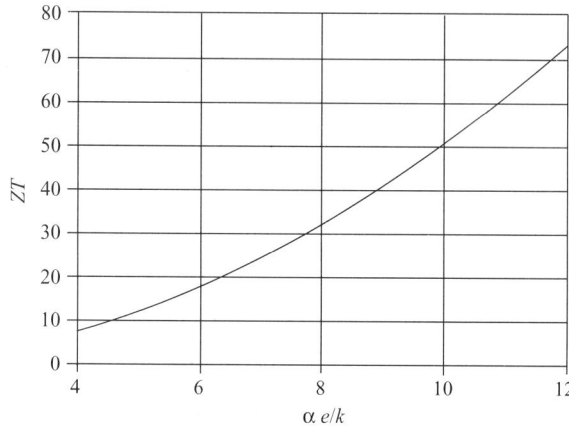

Fig. 7.20 Plot of ZT vs. Seebeck Coefficient in the Absence of Phonon Conductivity

ZT of 4, i.e., COP of vacuum thermoelement having work function of 0.3 eV is similar to thermoelectric cooler with ZT of 20. Although it is difficult to find an emitter having such low work function at present.

Thus, it is possible to have a refrigerator of this kind in near future, so the whole refrigeration and air conditioning industry may shift from conventional system to this novel technology.

7.9 CONCLUSION

In the last few decades thermoelectric refrigeration has gained importance in variety of applications like space applications, electronic cooling, automobiles, and some low-power consumer applications like water coolers of low capacity. TEC is highly reliable because of no moving parts, noise-free operation, it requires less maintenance and has simple operation, e.g., domestic cooling and heating operation becomes simple by reversing the polarity of current flow through the TEC. Thermoelectric generator can utilize low grade and high grade heat energy for its operation. This makes this kind of device suitable for solar energy conversion and automobile applications. However, its practical applications were limited because of its low conversion efficiency. If new materials were found with high ZT, thermoelectric energy conversion will play a significant role in refrigeration and air conditioning industry. A thermionic cooler has better prospects in future as compared to a thermoelectric cooler and hence a hybrid system is more desirable.

REFERENCES

Abeles, B., & Meiboom, S. (1956). Galvanomagnetic effects in bismuth. *Physical Review*, *101*(2), 544.

Cahill, D.G., Ford, W.K., Goodson, K.E., Mahan, G.D., Majumdar, A., Maris, H.J., Merlin, R., and Phillpot, S.R., Nanoscale thermal transport, J. Appl. Phys., 93, 793 – 818, 2003.

Cahill, D.G., Watson, S.K., and Pohl, R.O., Lower limit to the thermal conductivity of disordered crystals, Phys. Rev. B, 46, 6131 – 6139, 1992.

Cutler, P.H., Miskovsky, N.M., Kumar, N., and Chung, M.S., New results on microelectronic cooling using the inverse Nottingham effect. Low temperature operation and efficiency, Proc. Electrochem. Soc., 28, 99 – 111, 2000.

Foster, P.R. Two Stage Thermoelectric refrigeration. *ASHRAE Transactions*, 70, 312-318, 1964.

Ghoshal, U., Ghoshal, S., McDowell, C., Shi, L., Cordes, S., and Farinelli, M., Enhanced thermoelectric cooling at cold junction interfaces, Appl. Phys. Lett., 80, 3006 – 3008, 2002.

Goldsmid, H. J., J. E. Giutronich, and M. M. Kaila. "Solar thermoelectric generation using bismuth telluride alloys." *Solar Energy* 24.5. 435-440, 1980.

Goldsmid, H.J. and Douglas, R.W., The use of semiconductors in thermoelectric refrigeration, Br. J. Appl. Phys., 5, 386 – 390, 1954.

Goldsmid, H.J. and Sharp, J.W., Estimation of the thermal band gap of a semiconductor from Seebeck measurements, J. Electron. Mater., 28, 869 – 872, 1999.

Goldsmid, H.J., Electronic Refrigeration, Pion, London, 1986.

Goldsmid, H.J., Solid state and vacuum thermoelements, Proceedings of the 22nd International Conference on Thermoelectrics, La Grande Motte, France, pp. 433 – 438, 2003.

Harman, T.C. Multiple Stage Thermoelectric Generation of Power, J. Appl. Phys. 29, 1471, 1958.

Harman, T.C., Taylor, P.J., Walsh, M.P., and LaForge, B.E., Quantum dot superlattice thermoelectric materials, Science, 297, 2229 – 2232, 2002.

Heremans, J.P., Thrush, C.M., Morelli, D.T., and Wu, M.-C., Thermoelectric power of bismuth nanocomposites, Phys. Rev. Lett., 88, 68011 – 68014, 2002.

Ioffe, A.F. and Ioffe, A.V., Dokl. Akad. Nauk SSSR, 97, 821, 1954.

Ioffe, A.F., Semiconductor Thermoelements and Thermoelectric Cooling, Infosearch, London, 1957.

Lamberton, G.A., Bhattacharya, S., Littleton, R.T., Kaeser, M.A., Tedstrom, R.H., and Tritt, T.M., High figure-of-merit in Eu-filled $CoSb_3$-based skutterudites, Appl. Phys. Lett., 80, 598 – 600, 2002.

Lin, Y.-M. and Dresselhaus, M.S., Thermoelectric properties of superlattice nanowires, Phys. Rev. B, 68, 5304 – 5314, 2003.

Lin, Y.-M., Sun, X., and Dresselhaus, M.S., Theoretical investigation of thermoelectric transport properties of cylindrical Bi nanowires, Phys. Rev. B, 62, 4610 – 4623, 2000.

Logvinov, G. and Gurevich, Y., Upper value of thermoelectric figure-of-merit for isotropic semiconductors, Proceedings of the 22nd International Conference on Thermoelectrics, La Grande Motte, France, p. 23, 2003.

Miskovsky, N.M. and Cutler, P.H., Microelectronic cooling using the Nottingham effect and internal field emission in a diamond (wide-bandgap material) thin-film device, Appl. Phys. Lett., 75, 2147 – 2149, 1999.

Nolas, G.S. and Goldsmid, H.J., A comparison of projected thermoelectric and thermionic refrigerators, J. Appl. Phys., 85, 4066 – 4070, 1999.

Nolas, G.S. and Slack, G.A., Thermoelectric clathrates, Am. Sci., 89, 136 – 141, 2001.

Onyegegbu, S. O. "Performance of a modulated solar thermoelectric cooler." *Energy Conversion and Management* 22.1. 39-46, 1982.

Polvani, D.A., Meng, J.F., Chandra Shekar, N.V., Sharp, J., and Badding, J.V., Large improvement in thermoelectric properties in pressure-tuned p type $Sb1.5Bi0.5Te3$, Chem. Mater., 13, 2068 – 2071, 2001.

Riffat, S. B., S. A. Omer, and Xiaoli Ma. "A novel thermoelectric refrigeration system employing heat pipes and a phase change material: an experimental investigation." *Renewable Energy* 23.2: 313-323, 2001.

Slack, G.A., The thermal conductivity of nonmetallic crystals. In Solid State Physics, Vol. 34, H.Ehrenreich, F. Seitz and D. Turnbull, Eds., pp. 1 – 71. Academic Press, New York, 1979.

Soo, Shao-lee. "Direct energy conversion." (1968).

Vella, G. J., L. B. Harris, and H. J. Goldsmid. "A solar thermoelectric refrigerator." *Solar Energy* 18.4. 355-359, 1976.

Venkatasubramanian, R., Silvola, E., Colpitts, T., and O'Quinn, B., Thin-film thermoelectric devices with high room temperature figures-of-merit, Nature, 413, 597 – 602, 2001.

Vining, C.B. and Mahan, G.D., The B factor in multilayer thermionic refrigeration, Appl. Phys. Lett., 86, 6852 – 6853, 1999.

CHAPTER 8

Thermoacoustic Refrigeration

8.1 INTRODUCTION

Thermoacoustics is a combination of acoustics and thermodynamics branches of science and engineering which concerns with the utilization of sound energy transfer and its pressure oscillations in order to transfer heat on a macroscopic level. The acoustics deals with the properties of sound waves whereas thermodynamics focuses on the temperature variation and heat transfer. Bryon Higgins (1777) observed the acoustic oscillations, in a tube, at certain positions of a hydrogen flame [1]. Putnam and Dennis (1956) carry out a survey related to the phenomena of acoustic oscillations observed by Higgins [2]. Rijke (1859) discovered strong acoustic oscillations by placing a heated wire screen at the one-fourth length from the bottom end of an open ended pipe [3]. The theory of thermoacoustics was firstly created by Kirchhoff in 1868 and then Lord Rayleigh discussed the possibility of pumping heat by sound energy in 1887 [4]. There was a little research carried out in this area until Nikolaus Rott discussed thermoacoustic effects caused by heated surfaces maintaining large amplitude acoustic oscillations [5-7]. The oscillations caused by heat were well explored by several other researchers [8-10].

Gifford and Longsworth (1966) discovered the pumping of heat along the inner surface of a closed tube, known as pulse tube, by sustaining pressure pulses at low frequency [11]. Merkli and Thomann (1975) also observed refrigeration cooling effect in a cylindrical acoustic tube [12]. The first thermoacoustic heat pump was built by J.C. Wheatley, G.W. Swift and coworkers at Los Alamos National Laboratory in the early 1980s using loudspeaker as acoustic driver at one end of a closed tube having a stack of fibre-glass plates placed at other end [13]. This invention has drawn attention of researchers towards thermoacoustics refrigeration.Wheatley, Swift, and others [14-16] have improved the existing theory of thermoacoustics in a broader thermodynamical perspective.

Thermoacoustic heat pumps and refrigerators use an acoustic driver (e.g., loudspeaker), which drives an intense sound wave in a gas filled acoustical resonator, to pump heat from a source at low temperature to a sink at relatively high temperature. Thermoacoustic refrigerator uses high amplitude sound waves, generated using either thermal or mechanical energy, in a pressurized gas in order to generate a temperature difference between the hot and cold sides of a stationary element called the stack [17]. The main components of a thermoacoustic device are a sealed pressure vessel that may incorporate a Helmholtz Resonator to shorten the device and minimize losses; an acoustic driver

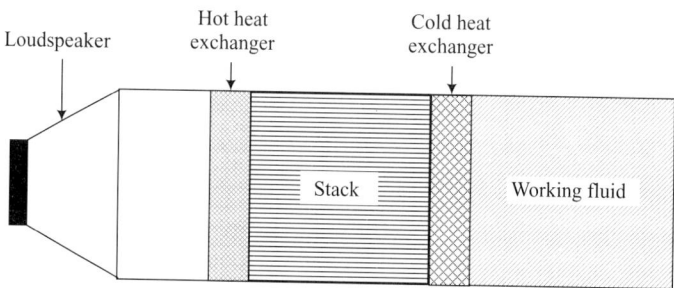

Fig. 8.1 Schematic Diagram of Thermoacoustic Refrigerator [17]

(e.g., a loudspeaker); a porous component known as stack or regenerator, and two heat exchangers with large area to volume ratio. The acoustic driver generates a high amplitude sound wave which makes the gas resonant, and hence large temperature and pressure oscillations into a resonator containing a stack or regenerator. The oscillation by the sound pressure produces adiabatic compression and expansion of the gas which creates a temperature gradient along the length of the stack. The temperature gradient is used to remove heat from the cold side and reject it at the hot side of the system. Thus, it has potential to offer an alternative refrigeration option with some technological improvements so as to increase COPs.

Although the concept of thermoacoustic refrigeration was discovered long back, the technology is still evolving and there is no commercial system available except for few examples of prototypes. A prototype thermoacoustic refrigerator made by G.W. Swift produced 3W of cooling at a temperature of –29°C (–20°F) and a sink temperature of 25°C (77°F) [15]. Some of the prototypes were designed for air conditioning applications with cooling capacity of 20W [19-20]. Tijani *et al.* (2002) attained a temperature of –65°C (–85°F) from thermoacousticrefrigerator by optimizing the size of different parts of the device [21]. A prototype with a cooling capacity of 119W was designed for an ice-cream freezer by Poese *et al.* (2004) to provide –24.6°C (–12.3°F) temperature with the COP of 0.81 [22]. In a recent study, Nsofor and Ali (2009) maximizes the theoretical cooling capacity of the device by optimizing the pressure for a given frequency [23]. Paek *et al.* (2007) simulated and carried out optimization analysis of standing wave thermoacoustic refrigerator and suggested that the second law efficiency increases with increasing temperature difference and reaches a maximum for 80°C (144°F) [24]. Zink *et al.* (2010) encouraged the thermoacoustic refrigeration for its low cost, high reliability and environmental benefits [25]. Currently, research towards the development of flow through designs (open systems) of thermoacoustic refrigerator is taking place which will reduce or eliminate the use of heat exchangers. It is also necessary to carry out extensive research so as to make it technically feasible as well as economically viable.

8.2 WORKING PRINCIPLE OF THERMOACOUSTIC CYCLE

The basic principle behind the thermoacoustics is that sound waves exert pressure on the molecules of the medium through which it is travelling. These sound waves or pressure waves propagate through the medium by means of molecular collisions. The molecular collisions cause a disruption in the medium in order to create constructive and destructive interference. The constructive

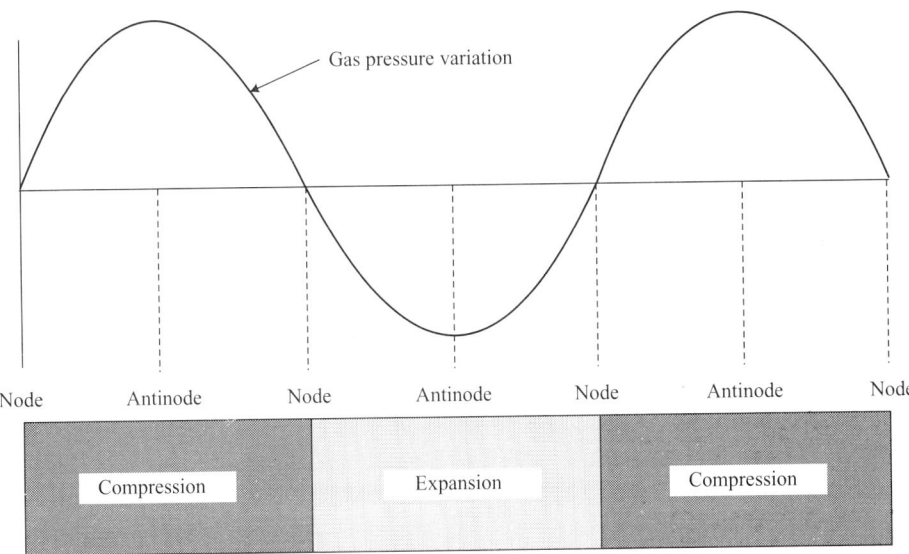

Fig. 8.2 Illustration of Pressure Wave Representing Nodes and Antinodes

interference occurs at crest produces compression in the molecules whereas destructive interference occurs at trough produce rarefaction and expand the molecules as shown in Fig. 8.2. This phenomenon of compression and expansion of the medium is utilized to produce refrigeration effect and is the basis of the thermoacoustic refrigerator.

The thermoacoustic device, such as shown in Fig. 8.3, generates refrigeration effect by pumping heat from the cold heat exchanger to the hot heat exchanger. Since the temperature of the working fluid (i.e., gas) is increased due to compression, the heat is dissipated from the working fluid through the hot heat exchanger placed near the maximum compression phase of stack. Whereas heat is absorbed by the working fluid from cold heat exchanger near the maximum expansion phase of stack. This process results in a net transfer of heat from cold side to the hot side of the stack. A closer look of Fig. 8.3 explains the mechanism of thermoacoustic refrigeration with the help of an oscillating gas parcel of the working fluid. Initially in process 1-2, the gas parcel is compressed adiabatically (dW) while moving towards pressure antinode (towards left of the stack). The compression increases the temperature of the gas parcel which results in the irreversible heat transfer from gas parcel to stack near hot heat exchanger (dQ in process 2-3). The gas parcel undergoes adiabatic expansion during its return to the initial position (in process 3-4). The expansion decreases its temperature and results in irreversible heat transfer (dQ) from stack to the gas parcel (in process 4-1). If the direction of the above processes (and also Fig. 8.3) are reversed, the thermoacoustic device will operate as a thermoacoustic engine or prime mover.

The four processes of thermoacoustic refrigeration cycle, as shown in Fig. 8.4, are:
1. Process 1–2: Working fluid is compressed adiabatically while being displaced toward the pressure antinode.
2. Process 2–3: Working fluid is further compressed while irreversible heat transfer occurs from working fluid to the stack.

330 *Alternatives in Refrigeration and Air Conditioning*

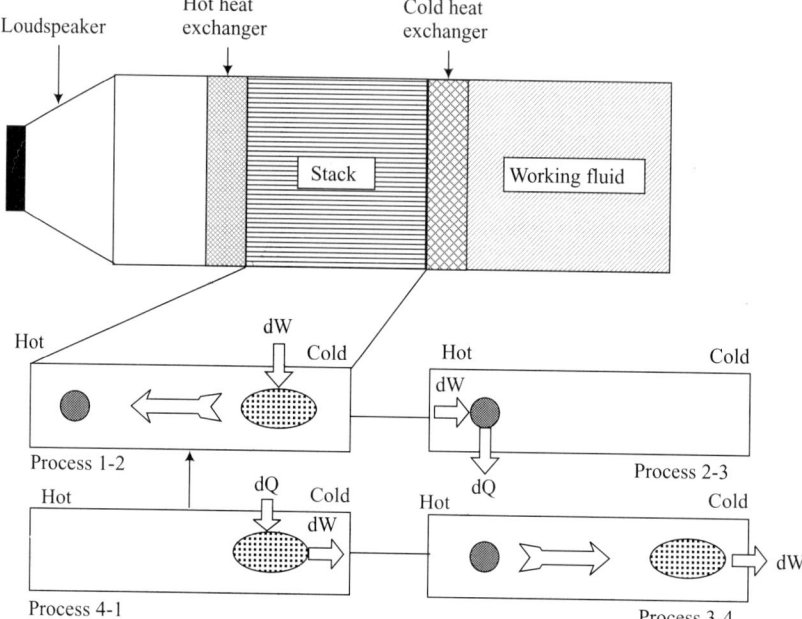

Fig. 8.3 Working Principle of a Thermoacoustic Refrigerator [17]

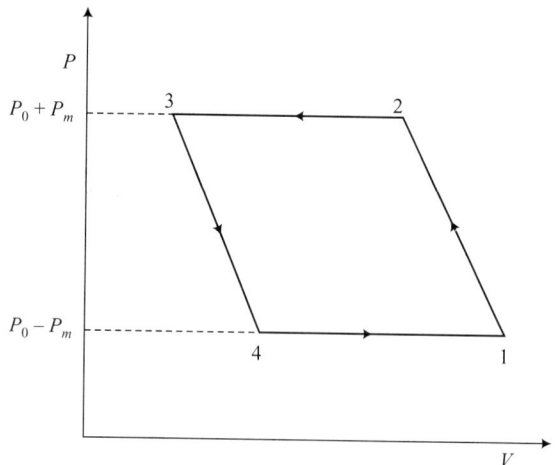

Fig. 8.4 Thermoacoustic Refrigeration Cycle

3. Process 3–4: Working fluid is expanded adiabatically while being displaced toward the pressure node.
4. Process 4–1: Working fluid is further expanded while heat is absorbed irreversibly from the stack.

8.3 TYPES OF THERMOACOUSTIC REFRIGERATORS

The thermoacoustic refrigerators can be classified, based upon the acoustic waves, into two types: standing wave stack based thermoacoustic refrigerator and travelling wave-regenerator based thermoacoustic refrigerator [26]. The difference between stack and regenerator is specified by Lautrec number (N_L). Lautrec number is a dimensionless ratio of hydraulic radius to thermal penetration depth. The porous medium is a stack for $N_L > 1$ while the porous medium is called a regeneratorfor $N_L < 1$.

8.3.1 Standing Wave Stack based Thermoacoustic Refrigerator

Standing waves are produced by the interference of incident and reflected wave confined in a closed space. It is a simple harmonic wave having frequency and wavelength of constituent incident and reflected waves but with increased amplitude as given by

$$y = -2a \sin \frac{2\pi x}{\lambda} \cos \omega t \qquad (8.1)$$

It seems to vibrate in constant position and orientation around stationary nodes and create areas of zero net displacement known as nodes. The maximum compression of the medium occurs at the halfway (between two nodes) called antinodes. This produces a large amplitude wave which has enough energy to cause visible thermoacoustic effects. The optimal resonant frequency of a wave travelling through a closed tube so as to achieve maximum heat transfer rate for thermoacoustic refrigerator is given by:

$$f = \frac{v}{4L}$$

where f is frequency, v is velocity of the wave, and L is the length of the tube.

The thermal contact between the working fluid and the stack is imperfect because of $N_L > 1$ as such, the pressure oscillations of the working fluid is perfectly isothermal at the boundary and nearly adiabatic at distances greater than thermal penetration depth away from the boundary. This imperfect contact introduces a phase shift between temperature and pressure of the working fluid. The phase shift produces the proper phasing for heat pumping along the stack. The standing wave provides that phase shift and displacement of the oscillating working fluid so as to operate stack based thermoacoustic device. Therefore, these devices are known as standing wave stack based thermoacoustic refrigerators.

Although the imperfect thermal contact provides natural mechanism to pump heat but it has a drawback of not able to achieve performance equal to ideal Carnot thermodynamic performance, even in idealized standing wave stack based thermoacoustic refrigerator, because of entropy generation during heat transfer.

8.3.2 Travelling Wave-Regenerator based Thermoacoustic Refrigerator

In regenerator ($N_L < 1$), the pressure oscillations of the working fluid are nearly isothermal within the pores by the higher heat capacity of the porous solid and the heat transfer between the working fluid and stack takes place at a very small temperature difference. Therefore, there is negligible amount of entropy generation during the heat transfer.

The cooling power absorbed by the regenerator in this type of thermoacoustic refrigerator is nearly equal to the acoustic power flow through the regenerator which is proportional to the product of velocity and phase components of the pressure of the oscillating working fluid. Because of small pores of regenerator, which increase viscous dissipation with the increase in velocity, the magnitude of pressure need to be increased while decreasing velocity in order to produce the power density required for cooling. Thus, unlike the standing wave stack based thermoacoustic refrigerator, travelling wave regenerator based thermoacoustic refrigerator is more complicated as complex passive acoustical network is required in order to provide the proper travelling wave phasing between pressure and velocity within the regenerator.

8.4 COMPONENTS OF THERMOACOUSTIC REFRIGERATOR

As we know, thermoacoustic refrigerator consists of an acoustic driver and a gas filled resonator, having a stack and two heat exchangers placed in it, placed in vacuum vessel as shown in Fig. 8.5. The detailed description of each component is given below: Tijani et al., 2002 [21].

Acoustic driver

The acoustic driver consists of a moving coil loudspeaker to generate an acoustic wave in the resonator. The acoustic driver is selected on the basis of its properties like compactness, lightweight, low losses, and high Bl-factor. It is mounted on a brass plate interfacing the bottom of the driver housing so that the heat dissipated in the driver is removed. For thermoacoustic refrigeration, following modifications are to be incorporated in the commercial loudspeaker:
1) The back plastic cover needs to be removed so that a back volume control can be mounted.
2) A rigid dome is required to generate high dynamic pressure.
3) A tapered aluminium cone is used because of lightweight and to match its dimension with the resonator.

Gas-spring system

The gas-spring system comprises a brass cylinder which is mounted on the back of the driver and is referred to as the back volume system (Fig. 8.5). This system realizes the shift between mechanical resonance frequency of the driver and the acoustical resonance frequency of the resonator by varying volume. This volume can be varied by changing the height of the piston which can be adjusted by a crank attached to the lid of the housing.

Driver housing

Driver housing consists of a cylindrical part, a bottom plate which is soldered to the cylindrical part and a cover plate (lid) which supports the gas fill port, a static pressure transducer, and the piston displacement system. It is a support to the driver, a pressure vessel for the working fluid, a feed through for the electrical leads and also provides some sound insulation.

Resonator

The resonator consists of a large diameter gas filled tube which contains the stack, a small diameter tube, a buffer volume and two heat exchangers placed in a vacuum vessel. It should be compact, lightweight and rigid. The parameters which are considered for deciding the shape and length of the

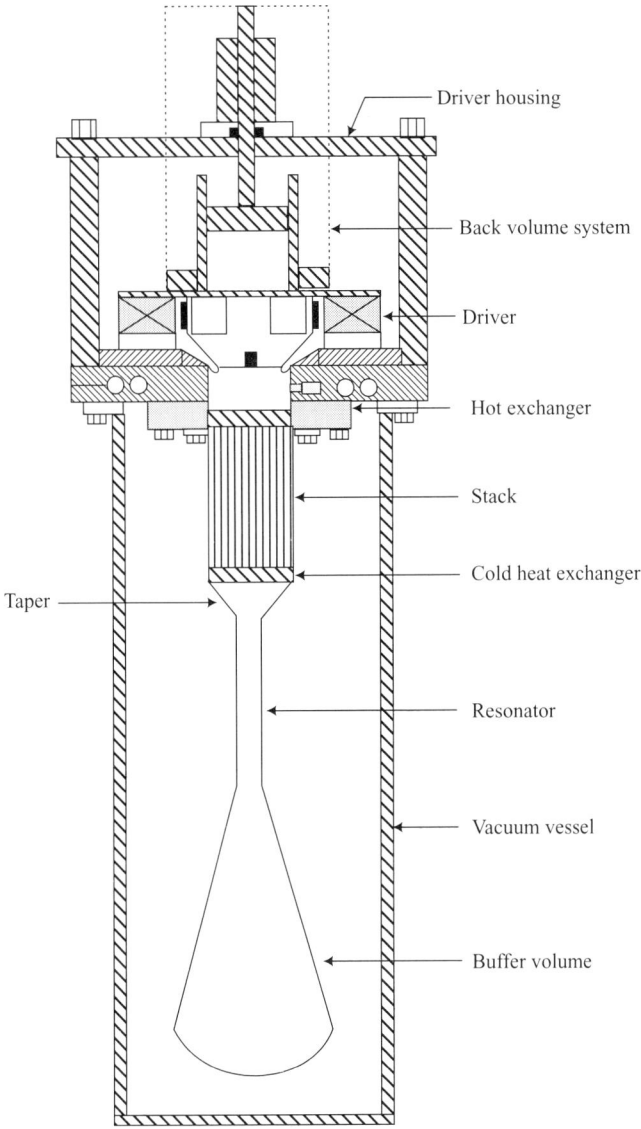

Fig. 8.5 Thermoacoustic Refrigerator Showing Different Parts (Tijani *et al.*, 2002) [21]

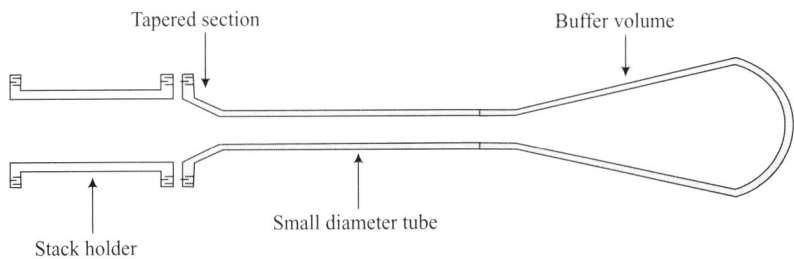

Fig. 8.6 Schematic Diagram of Resonator

334 *Alternatives in Refrigeration and Air Conditioning*

resonator are resonance frequency and minimal losses at the wall. A general shape of the resonator is shown in Fig. 8.6.

The stack holder should be rigid with low thermal conductivity (e.g., POM-Ertacetal (H) material).

A contraction of aluminium block isused to connect the large diameter stack holder to the small diameter tube known as tapered section. The small diameter tube and volume buffer is also made of aluminium.

Stack

The stack is characterized by the parameters, centre position, length and area of cross section. The stack is made of low thermal conductivity material (e.g., Mylar). There are two types of stack: spiral stack and parallel plate stack. The spiral stack consists of a long sheet winded around a rod. The parallel-plate stacks consist of parallel plates which are spaced by fishing line spacers glued between the plates. An approximate geometry of stack, i.e., spiral stack, is generally used because of the complexity of construction and fragility associated with the parallel-plate stack. The spacing between the layers is realized by fishing line spacers glued onto the surface of the sheet.

Heat exchangers

The hot heat exchanger is mounted in a copper flange which not only connects the stack holder and the driver housing but also provides a good thermal contact with the bottom of the driver housing. The cold heat exchanger is mounted in a copper ring which is then attached to the aluminum taper.

The two heat exchangers, hot and cold, differ only in length. In the heat exchangers, two copper sheets out of which one sheet is flat and the other has a sine shape are wound together to provide a sine channel structure which is chosen. The sine shape of copper sheet offers spacing and improves radial heat transfer by providing contact path.

8.5 THERMODYNAMICS OF THERMOACOUSTIC REFRIGERATION

The thermodynamic aspects provides a simplifed explanation of the temperature gradient produced by thermoacoustic effect. The overall thermodynamic performance of the refrigerator, as given by the first law of thermodynamics, is given by

$$\text{COP} = \frac{\text{Heat extracted from the cold heat exchanger}}{\text{Work done on the machine}} = \frac{Q_{load}}{E_{electric}} \quad (8.2)$$

COP is defined as the ratio of cooling load to the electric power input in the acoustic driver.

The performance of thermoacoustic refrigeration depends on electroacoustic conversion efficiency of acoustic driver, performance of stack, effectiveness of heat exchanger and acoustic power efficiency.

The electroacoustic conversion efficiency (η_{e-a}) of an acoustic driver (i.e., loudspeaker) is the ratio of acoustic power (W_{total}) to the electric power ($E_{electric}$), given by

$$\eta_{e-a} = \frac{W_{total}}{E_{electric}} = \frac{W_{stack} + W_{loss}}{E_{electric}} \quad (8.3)$$

W_{stack} is the useful work done in the stack which is responsible for the heat transfer and W_{loss} is the heat loss in heat exchanger and resonance tube.

The performance of the stack (COP_{stack}) is determined by the ratio of heat extracted from the cold heat exchanger to the work done on the stack, given by

$$COP_{stack} = \frac{\text{Heat extracted from the cold heat exchanger}}{\text{Work done on the stack}} = \frac{\dot{Q}_c}{W_{stack}} \quad (8.4)$$

The effectiveness of the cold heat exchanger (ε) is the ratio of heat transferred from the cold heat exchanger to the maximum heat that can be transferred from the cold heat exchanger, given by

$$\varepsilon = \frac{\text{Heat transferred}}{\text{Maximum heat transferable}} = \frac{Q_{load}}{Q_c} \quad (8.5)$$

The acoustic power efficiency (η_{ac}) is the ratio of acoustic power used for heat transfer to the total acoustic power, given by

$$\eta_{ac} = \frac{W_{stack}}{W_{total}} \quad (8.6)$$

Thus, the overall performance of thermoacoustic refrigerator is given by

$$COP = \frac{Q_{load}}{E_{electric}} = \eta_{e-a} \times \eta_{ac} \times \varepsilon \times COP_{stack} \quad (8.7)$$

The thermodynamic aspects of thermoacoutics were discussed by Wheatley *et al.* [13] and Swift [16]. It is assumed for simplicity of thermodynamic analysis that an average temperature gradient ∇T_m exists along the length, the working fluid is an inviscid ideal gas of vanishing Prandtl number and the pressure antinode is located at the right side of the plate and a node to the left side.

8.6 DESIGN PARAMETERS OF THERMOACOUSTIC REFRIGERATION

This section deals with different operation, working fluid and stack parameters which plays a vital role in designing a thermoacoustic refrigerator. Helium is generally used as working fluid because it has highest thermal conductivity as well as sound velocity among all inert gases. The average pressure and acoustic resonance frequency is directly proportional to the power density of the thermoacoustic refrigerator and inversely proportional to the square of thermal penetration depth. Therefore, average pressure and acoustic resonance frequency required is to be optimized considering the above two effects. The stack material should have low thermal conductivity so that minimum heat conduction through the stack plates takes place (e.g., Mylar). The thermal penetration depth (δ_k) is the thickness around the stack plates over which heat is diffused during an oscillation of gas parcel, as given by

$$\delta_k = \sqrt{\frac{K}{\rho C_p \omega}} \quad (8.8)$$

The viscous penetration depth (δ_v) is the thickness of layer of the working fluid around the stack plates which is influenced by viscous force and restrains the movement of gas parcel. The viscous penetration depth is given by

$$\delta_v = \sqrt{\frac{2\mu}{\rho\omega}} \tag{8.9}$$

where K is the thermal conductivity, μ is the viscosity, ρ is the density, C_p is the isobaric specific heat of the gas, and ω is the angular frequency of the sound wave.

The porosity of the stack (B) is a dimensionless parameter for designing the stack, also known as blockage ratio, is given by

$$B = \frac{y_0}{y_0 + l} \tag{8.10}$$

where y_0 is half the plate spacing and l is half the plate thickness.

The ratio of temperature gradient along the stack ($\Delta T_m/\Delta x$) to the critical temperature gradient (∇T_c) is a dimensionless parameter, known as normalized temperature gradient (Γ), and is given by

$$\Gamma = \frac{\Delta T_m / \Delta x}{\nabla T_c} \tag{8.11}$$

The normalized cooling power ($\dot{Q}_{cn}$) of the thermoacoustic refrigerator from cold to hot heat exchanger is given by (Tijani et al., 2002) [21].

$$\dot{Q}_{cn} = -\frac{\delta_{kn} D^2 \sin 2x_n}{8\gamma(1-\sigma)\Lambda} \times \left[\left(\frac{\Delta T_{mn} \tan x_n}{BL_{sn}(\gamma-1)} \right) \frac{1+\sqrt{\sigma}+\sigma}{1+\sqrt{\sigma}} - (1+\sqrt{\sigma} - \sqrt{\sigma}\delta_{kn}) \right] \tag{8.12}$$

The normalized acoustic power ($\dot{W}_n$) used by the stack for heat transfer is given by

$$\dot{W}_n = \frac{\delta_{kn} D^2 L_{sn}}{4\gamma} \left[B(\gamma-1)(\cos^2 x_n) \left(\frac{\Delta T_{mn} \tan x_n}{BL_{sn}(\gamma-1)(1+\sqrt{\sigma})\Lambda} - 1 \right) - \frac{\sqrt{\sigma}(\sin^2 x_n)}{B\Lambda} \right] \tag{8.13}$$

where

$$\Lambda = 1 - \sqrt{\sigma}\delta_{kn} + \frac{1}{2}\sigma\delta_{kn}^2, \tag{8.14}$$

δ_{kn} is normalized thermal penetration depth, i.e., $\delta_{kn} = \delta_k/y_0$, D is drive ratio defined as the ratio of amplitude of dynamic pressure to the average pressure (i.e., $D = p_o/p_m$), x_n is normalized stack position (i.e., $x_n = k x$), ΔT_{mn} is normalized temperature difference (i.e., $\Delta T_{mn} = \Delta T_m/T_m$), $\gamma =$ is ratio of isobaric to isochoric specific heat, ρ is Prandtl number, B is porosity of the stack and L_{sn} is normalized stack length (i.e., $L_{sn} = kL_s$).

The performance of the stack is expressed in coefficient of performance, i.e., COP, defined as the ratio of the heat transferred in the stack to the used acoustic power and is given by

$$\text{COP} = \frac{\dot{Q}_{cn}}{\dot{W}_n} \tag{8.15}$$

8.7 APPLICATIONS AND MERITS OF THERMOACOUSTIC REFRIGERATION

Thermoacoustic refrigerators have the potential to provide cooling up to cryogenic temperatures. Thus, it includes broad area of application such as to generate heating, air conditioning, cooling electronic boards. It also has potential application in low capacity domestic and commercial refrigerators, freezers and cabinets to preserve food. Thermoacoustic driven prime movers can be used to generate electricity or to drive refrigerators. Its main advantage is that it has no moving part, no tight tolerance, and is used environment-friendly (inert gases) as working fluid which makes it potentially reliable and low cost alternative to the conventional compressor based refrigerators.

Thermoacoustic technology has drawn attention of researchers worldwide towards its application of heat engines, heat pumps and refrigeration. A number of practical thermoacoustic devices are built so far mostly at LANL, Naval Postgraduate School (NPS) in Monterey (California), and at Pennsylvania State University. The Space Thermoacoustic Refrigerator (STAR) [27] on the space shuttle Discovery in 1992 to produce pump 4 W of heat and then the Shipboard Electronics Thermo-Acoustic Cooler (SETAC) on the warship USS Deyo in 1995 to provide 400 W of cooling power [28]. A large chiller called TRITON was developed at Pennsylvania State University to provide 10 kW of cooling power for navy ships [29].

A Thermo-Acoustic Driven Thermo-Acoustic Refrigerator (TADTAR) was built at both LANL and NPS. The TADTAR uses a heat driven prime mover instead of a loudspeaker to generate the sound necessary to drive the refrigerator [30]. One step further toward clean energy, besides no moving parts, a solar driven TADTAR was built at NPS which has a cooling capacity of 2.5 W [31].

LANL developed a cryogenic refrigeration technology called Thermo-Acoustically Driven Orifice Pulse Tube Refrigeration (TADOPTR) of 2 kW cooling capacity which uses a pulse-tube refrigerator driven by a natural gas powered thermoacoustic prime mover to liquefy natural gas [32]. A TADOPTR was also built at Tektronix in order to provide cooling to the electronic components in which electric heater driven prime mover provide 1 kW to the pulse-tube refrigerator [33].

A thermoacoustic refrigerator is built at CSIR (Council for Scientific and Industrial Research) in the Republic of South Africa for preserving blood and urine samples on the space shuttle.

Solar thermal and solar photovoltaic (PV) operated thermoacoustic refrigeration are renewable energy based options for meeting cooling load demand. In solar thermal based TAR system, the solar air heater is used to generate acoustic power (i.e., conversion of flowing hot gas into sound wave). In solar PV based TAR system, the acoustic driver (e.g., loudspeaker) is energized by supplying PV generated electricity. As both PV (used with fixed slope angle) and TAR doesn't have any moving part, the PV operated TAR system will emerge out as a maintenance-free and clean refrigeration technique.

The advantages of thermoacoustic refrigeration are as follows:
1) The working fluid, utilized in thermoacoustic refrigerators, is usually a mixture of xenon and helium (inert gas) which are non-toxic and environment-friendly working gases.
2) There is no moving part in TAR, so it is reliable, maintenance-free, no wear and tear losses and have long lifetime.
3) It uses simple and commercially available materials.
4) Large range of working temperature is possible.
5) Compact size and less maintenance is desirable.

8.8 CONCLUSION

A novel refrigeration system based on the combined concepts of thermodynamics and acoustics known as thermoacoustic refrigeration system has been described. The chapter discusses about the working principle of thermoacoustic cycle and classification of thermoacoustic refrigerators based on standing wave stack and travelling wave-regenerator concepts. The component level detail of thermoacoustic refrigerator have been presented and various design parameters have been discussed. The major applications and merits have also been highlighted in the chapter.

REFERENCES

1. B. Higgins, Nicholson's Journal I, 130 (1802).
2. A.A. Putnam and W.R. Dennis, Survey of Organ-Pipe Oscillations in Combustion Systems, J. Acous. Soc. Am. 28, 246 (1956).
3. P.L. Rijke, Notizübereineneue Art, die in einer an beidenendenoffenenRöhreenthalteneLuft in Schwingungenzuversetzen, Ann. Phys. 107, 339 (1859).
4. Lord Rayleigh, The theory of sound, 2nd edition, Vol. 2, Sec. 322 (Dover, New York, 1945).
5. N. Rott, Thermoacoustics., Adv. Appl. Mech. 20, 135 (1980).
6. N. Rott, Thermally driven acoustic oscillations, part III: Second-order heat flux, Z. Angew. Math. Phys. 26, 43 (1975).
7. N. Rott, Thermoacoustic heating at the closed of an oscillating gas column, J. Fluid Mech. 145, 1(1984).
8. K.T. Feldman, Jr., Review of the literature on Rijkethermoacoustic phenomena, J. Sound Vib. 7, 83 (1968a).
9. C. Sondhauss, Ueber die Schallschwingungen der Luft in erhitztenGlas-Röhren und in gedecktenPfeifen von ungleicherWeite, Ann. Phys. 79, 1 (1850).
10. K.T. Feldman, Jr., Review of the literature on Sondhaussthermoacoustic phenomena, J. Sound Vib. 7, 71 (1968).
11. W.E. Gifford and R. C. Longsworth, Surface heat pumping, Adv. Cryog. Eng. 11, 171 (1966).
12. P. Merkli and H. Thomann, Thermoacoustic effects in a resonant tube, J. Fluid Mech. 70, 161 (1975).
13. J.C. Wheatley, T. Hofler, G.W. Swift, and A. Migliori, An intrinsically irreversible thermoacoustic heat engine, J. Acoust. Soc. Am. 74, 153 (1983a)
    Experiments with an intrinsically irreversible thermoacoustic heat engine, Phys. Rev. Lett. 50, 499 (1983b).
14. J.C. Wheatley, T. Hofler, G.W. Swift, and A. Migliori, Understanding some simple phenomena in thermoacoustics with applications to acoustical heat engines, Am. J. Phys. 53, 147 (1985).
15. G.W. Swift, Thermoacoustic engines, J. Acoust. Soc. Am. 84, 1146 (1988).
16. G.W. Swift, Thermoacoustic engines and refrigerators, Encyclopedia of Applied Physics 21, 245 (1997).
17. P. Bansal, E. Vineyard, O. Abdelaziz, Status of not-in-kind refrigeration technologies for household space conditioning, water heating and food refrigeration, Int. J. of Sustainable Built Environment 1, 85–101 (2012).
18. J.J. Wollan, G.W. Swift, Development of a thermoacoustic natural gas liquefier–Update, Proc. Ann. ACM-SIAM Symp. Discrete Algorithms, 371–378 (2001).
19. S.L. Garrett, J. A. Adeff, T.J. Hofler, Thermoacoustic refrigerator for space applications, J. Thermophys. Heat Transfer 7, 595–599 (1993).
20. T.J Berhow, Construction and performance measurement of a portable thermoacoustic refrigerator demonstration apparatus, MS Thesis, Physics Department, Naval Postgraduate School, Monterey, CA (1994).

21. Tijani, M.E.H., J.C.H. Zeegers and A.T.A.M. Dewack. The optimal stack spacing for thermoacoustic refrigeration, J. Acoustic. Soc. Am., 112, 128 (2002).
    ———Design of thermoacoustic refrigerators, Cryogenics, 42, 49 (2002).
    ———Construction and performance of a thermoacoustic refrigerator, Cryogenics, 42, 59-66 (2002).
22. M.E. Poese, R.W. Smith, S.L. Garrett, R. van Gerwen, P. Gosselin, Thermoacoustic refrigeration for ice cream sales, In: Proc. 6th IIR Gustav Lorentzen Conf. (2004).
23. E.C. Nsofor, A. Ali, Experimental study on the performance of the thermoacoustic refrigeration system, Appl. Therm. Eng. 29, 2672–2679 (2009).
24. I. Paek, J.E. Braun, L. Mongeau, Evaluation of standing-wave thermoacoustic cycles for cooling applications, Int. J. Refrig. 30, 1059–1071 (2007).
25. F. Zink, J.S. Vipperman, L.A. Schaefer, Environmental motivation to switch to thermoacoustic refrigeration, Appl. Therm. Eng. 30, 119–126 (2010).
26. S.L. Garrett, Resource Letter: TA-1: Thermoacoustic engines and refrigerators, Am. J. Phys., Vol. 72, No. 1, 11-17(2004).
27. S.L. Garrett, J.A. Adeff, and T.J. Hoßer, .Thermoacoustic refrigerator for space applications., J. of Thermophysics and Heat Transfer, 7, 595 (1993).
28. S.C. Ballister, and D.J. McKelvey, Shipboard electronics thermoacoustic cooler, MS Thesis, Naval Postgraduate School, Monterey, CA, (1988).
29. R.A. Johnson, S.L. Garrett, and R.M. Keolian, Thermoacoustic Cooling for Surface Combatants, Naval Engineers Journal, 112, 335 (2000).
30. J.C. Wheatley, G. W. Swift, and A. Migliori, The natural heat engines, Los Alamos Science,14, 2 (Fall 1986).
31. J.A. Adeff and T.J. Hoßer, Design and construction of a solar, thermoacoustically driven, thermoacoustic refrigerator, J. Acoust. Soc. Am. (Acoustics Research Letters online)107, L37 (2000).
32. G.W. Swift, Thermoacoustic natural gas liquefier, Proc. DOE Natural Gas Conf. (Fed. Energy Tech. Cent., Morgantown, West Virginia, 1997).
33. K.M. Godshalk, C. Jin, Y.K. Kwong, E.L. Hershberg, G.W. Swift, and R. Radebaugh, Characterization of 350 Hz thermoacoustic driven orifice pulse tube refrigerator with measurements of the phase of the mass flow and pressure, Adv. Cryog. Eng. 41, 1411 (1996).

CHAPTER 9

Vapour Adsorption Systems: Physisorption and Chemisorption Systems

9.1 INTRODUCTION

This chapter provides basic description and status of vapour adsorption systems, including, physisorption refrigeration and Chemisorption heat pump systems. Such systems are suitable for refrigeration applications and for solar space conditioning as well.

Adsorption is a surface absorption phenomenon which can be classified into two groups: (1) physical adsorption (physisorption) and (2) chemical adsorption (chemisorption). Physisorption is a reversible process and mainly caused by dispersion-repulsion force and electrostatic forces between adsorbate (gas) molecules and atoms which compose the adsorbent (porous substance) surface. Most of the vapour adsorption processes applicable to thermal or cooling systems mainly utilize physisorption. Chemisorption systems include dry absorption, solid-gas absorption and metal-hydride refrigeration systems.

Adsorption cooling cycle is based on surface adsorption and utilises external thermal energy for the desorption process. The evaporating substance (the sorbate) uses the spaceheat as a source, while it is being absorbed by the absorbing material (the sorbent). The sorption process produces heat of sorption. The term absorption process is volume absorption used for transfer of sorbate and sorbent into a homogenous phase (e.g., liquid), while adsorption is the term used if the transfer occurs as surface absorption to a non-homogeneous phase (e.g., crystals + adsorbed water). Sorption machines can also store energy for given period of time without thermal insulation. The stored energy can then be used to produce heating or cooling power in an ecofriendly way.

The substance which can exchange a transformable component from or to a fluid mixture is called sorbent and process of removing the transformable component from the fluid mixture is known as adsorption, while the reverse process of transferring the transferable component from the fluid mixture is called desorption and the component transferred from fluid is called sorbate.

Vapour Adsorption Systems: Physisorption and Chemisorption Systems 341

9.2 DESCRIPTION OF PHYSISORPTION SYSTEM:VAPOUR ADSORPTION COOLING CYCLE

A basic adsorption cycle (Fig. 9.1(a)) consists of an evaporator, a condenser and a regenerator/adsorber. In the adsorption phase, the sorbate in the evaporator picks up space heat (Q_0) and evaporates. At the same time, the sorbent releases the heat of adsorption (Q_A). In the regeneration phase, the sorbate is removed from the adsorbent by taking up the heat (Q_H). By removing the heat (Q_C) in the condenser, the sorbate cools down and condenses, thereby flowing back to the evaporator through the receiver and valve (V).

An adsorption-desorption cycle is well represented on P-T-M (pressure-temperature-adsorbed mass) diagram as shown in Fig. 9.1(b). There are four successive processes occurring in the cycle (Luo and Feidt, 1992).

Process A-B: The regenerator rich in adsorbed refrigerant is heated at a constant mass $M(T_a)$ from an initial ambient temperature (T_a) to a temperature (T_s) and a pressure (P_c) which is determined by the condenser temperature.

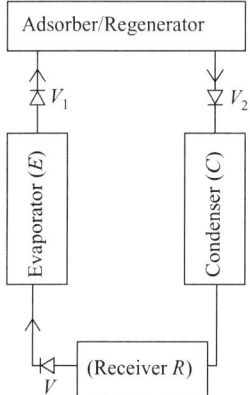

Fig. 9.1(a) Basic Adsorption Cooling Cycle

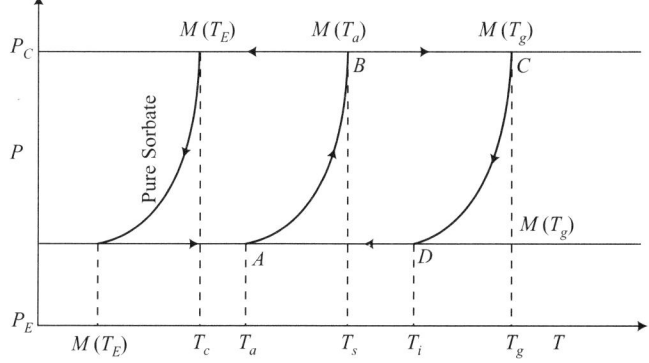

Fig. 9.1(b) P-T-M Diagram of the Basic Adsorption-Desorption Cycle

Process B-C: With external energy input to the regenerator, the refrigerant is desorbed at a constant pressure P_c while the temperature of the regenerator increases to its maximum value T_g; control valve V_2 is kept open during this process. The refrigerant (sorbate) is condensed in an air or water cooled condenser (C).

Process C-D: The regenerator/adsorber, having a weak concentration of refrigerant, cools down at a constant adsorbed mass, thus reducing the vapour pressure to P_E; the valve V_2 is closed in this process.

Process D-A: The condensed refrigerant (sorbate), collected in the receiver (R), flows towards and enters the evaporator (E). The usual evaporation process of refrigerant in the evaporator provides the refrigeration effect. During this process, throttle valve (V) and valve V_1 are open and the refrigerant (with direct flow to the regenerator) is adsorbed at a constant pressure (P_E) until state-A is reached again.

9.2.1 Thermodynamic Processes of Adsorption Cycle

The basic adsorption system works on the adsorption of a refrigerant vapour (adsorbate) into an adsorbent bed at low pressure and subsequent desorption at a high pressure by external heating of the adsorbent bed. In the simplest case, an adsorption refrigerator can be considered as two vessels connected to each other, one of which is filled with adsorbent and adsorbate as shown in Fig. 9.2(a).

At the beginning stage 1, the adsorbent bed is always at a low pressure and temperature and saturated with vapour at this point. The adsorbent bed is externally heated and the refrigerant starts to desorb from the adsorbent bed at stage 2, which raises the system pressure. Desorbed refrigerant vapour condenses in the second vessel rejecting heat of condensation. while, the hot adsorbent bed (stage 3) is cooled back to ambient temperature at stage 4 causing the refrigerant in the evaporator to vaporise and then re-adsorb on the bed. The basic adsorption refrigeration cycle consists of four thermodynamic processes which can be shown by a Clapeyron diagram (lnP vs –1/T) in Fig. 9.2(b).

The cycle begins with state A at T_{a2} where the maximum amount of refrigerant is absorbed. The adsorbent is at low temperature and at a low pressure at state T_{a2}. Along the line AB (T_{a2} to T_{g1}), the adsorbent is heated and desorbs refrigerant vapour isosterically (i.e., at constant adsorbed mass on the adsorbent), and the mass of desorbed refrigerant is small relative to the total refrigerant mass adsorbed.

Continued heating from T_{g1} to T_{g2} desorbs more refrigerant, forcing it to the condenser until state T_{g2} is attained, at which desorption ceases. This step is isobaric desorption. At the same time, the hot adsorbent is cooled isosterically causing depressurization (P_C to P_E) along the line CD, (Lambert, 2007). When the pressure drops to P_E, refrigerant in the evaporator starts to vaporise producing cooling effect and then it flows to the adsorbent bed. Cooling of the adsorbent and adsorbtion of the sorbate continues until the bed is saturated with refrigerant, hence completing the cycle. This process DA is isobaric adsorption. These processes are explained as follows. (Demir et al., 2008)

(a) Isosteric heating
On heating the adsorbent bed, the temperature is increased from T_{a2} to T_{g1} (shown in Fig. 9.2(b)) and the heat transferred given by

Vapour Adsorption Systems: Physisorption and Chemisorption Systems 343

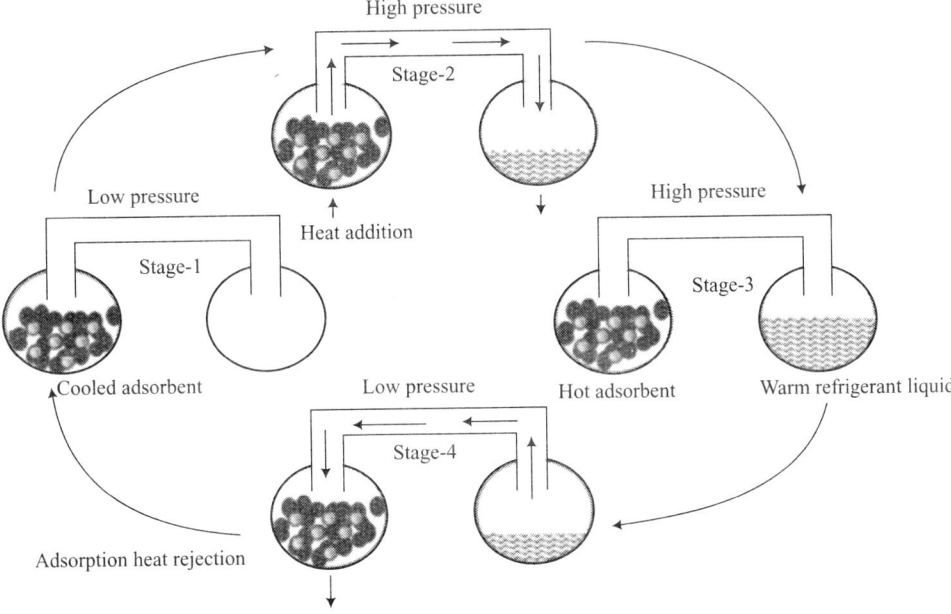

Fig. 9.2(a) Basic Adsorption Cycle Stages (Critoph and Zhong, 2005)

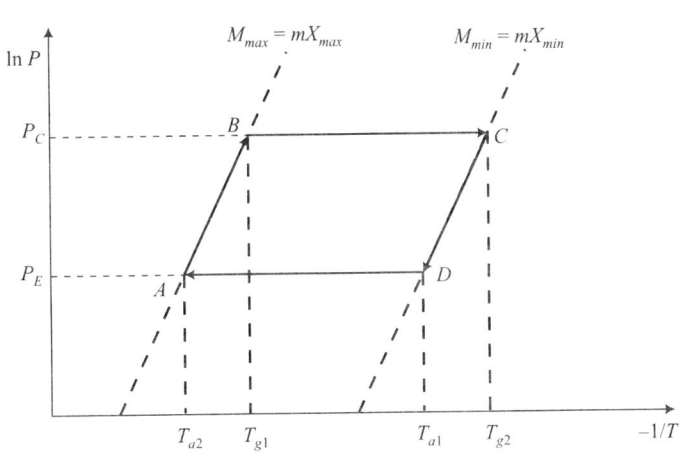

Fig. 9.2(b) P-T-M Diagrams for Basic Adsorption Cycle. (Wu *et al.*, 2011)

$$Q_{ih} = \int_{T_{a2}}^{T_{g1}} [m(C_{pad\,bent} + xC_{pad\,bate}) + m_{ad\,mat} \times C_{p\,ad\,mat}] dT \qquad (9.1)$$

(b) Isobaric desorption
After the heating process the desorption process is started at a constant pressure and the heat of desorption is given by

$$Q_{ide} = \int_{T_{g2}}^{T_{gz}} [m(C_{pad\,bent} + x_{min}C_{pad\,bate}) + m_{ad\,mat} \times C_{pad\,mat}] dT + \int_{T_{g1}}^{T_{g2}} m\Delta H\, dT \qquad (9.2)$$

where 'x' is the refrigerant to adsorbent mass ratio.

(c) Isosteric cooling
As the bed attains maximum temperature T_{g2}, the adsorbent is allowed to cool from T_{g2} to T_{a1}, causing pressure to reduce from P_C to P_E.

$$Q_{ic} = \int_{T_{gz}}^{T_{a1}} [m(C_{pad\,bent} + x_{min}C_{pad\,bate}) + m_{ad\,mat} \times C_{pad\,mat}] dT \qquad (9.3)$$

(d) Isobaric adsorption
When the adsorbent bed is cooled from T_{a1} to T_{a2} and connected to the evaporator, heat of adsorption is released, as given by

$$Q_{iad} = \int_{T_{a2}}^{T_{a2}} [m(C_{pad\,bent} + xC_{pad\,bate}) + m_{ad\,mat} \times C_{p\,ad\,mat}] dT + \int_{T_{a1}}^{T_{a2}} m\Delta H\, dT \qquad (9.4)$$

Thus, the cooling effect is calculated as the latent heat of evaporation of the adsorbate, minus the sensible heat of the adsorbate entering the evaporator at condenser pressure.

$$Q_{refri} = m(x_{max} - x_{min})(L(T_e) - \int_{T_{evp}}^{T_{con}} C_{pl}(T) dT] \qquad (9.5)$$

where C_{pl} is the specific heat of the adsorbate in the liquid phase and L is the latent heat of evaporation of adsorbate. On the basis of the above described equations, the COP for cooling operation can be calculated as the ratio of the useful refrigeration effect produced and heat input to the adsorbent bed as given by

$$COP = \frac{Q_{refri}}{[Q_{in} + Q_{ide}]} \qquad (9.6)$$

When assessing the performance of solar cooling system, it is also desirable to evaluate the ratio of cooling capacity to the costs of materials and manufacture. Therefore, another performance index (other than COP) is used as the ratio of cooling capacity to the cost of material and also known as specific cooling potential (SCP) as defined by

$$\text{Specific Cooling Potential} = \frac{Q_c}{C \times A} \qquad (9.7a)$$

where, Q_c is the cooling capacity, C is the material cost per unit area of the adsorber and A is the area of the adsorber. This parameter SCP has a better economic character and performance evaluation than COP. Another parameter of interest is expressed as specific cooling power defined by

$$\text{Specific Cooling Power} = \left[\frac{Q_c}{m_a \cdot \text{cycle time}}\right] \qquad (9.7b)$$

which is related to the mass of the adsorbent and cycle time.

9.3 DEVELOPMENTS IN SOLAR ADSORPTION COOLING

Solar energy can be directly utilized for operating adsorption cooling systems for the desorption process. Solar adsorption refrigeration is an alternative cooling option to replace the conventional vapour compression cooling system. However, the COP of the solar adsorption refrigeration system is very low, and hence it is desirable to develop these systems with higher COP. Vapour absorption systems are in general bulky and require high temperature flat plate/concentrating or evacuated tube solar collectors. Selection and performance optimization of solar collectors is also important in the adsorption refrigeration systems. This section elaborates the state of the art of closed cycle solar adsorption cooling system.

In closed cycle adsorption system, the phase change process required in the cycle is promoted through the surface energy involved in the sorption process. These systems are not fully developed as much as absorption systems probably because of the more requirements for the sorbent material and also more difficult heat exchange problem. In the beginning around 1920-30, silica gel as sorbent and sulphur dioxide as the refrigerant was used, but later adsorption systems use the combination of natural zeolite and water to create the required pressure ratio for cycling the refrigerant between the liquid and vapour phases. Zeolites have cage-like structure and high internal surface area and thus capable of adsorbing large quantities of refrigerant vapour even at room temperature. Also the heat of vaporization of water is the highest (up to 10 times) and so zeolite-water system is the most efficient system requiring only a small quantity of zeolite for system operation.

As shown in Fig. 9.3, a zeolite-water closed cycle adsorption solar cooling system is once-a-day cycle. The adsorbent is activated during the day cycle using solar heat and the water vapour is

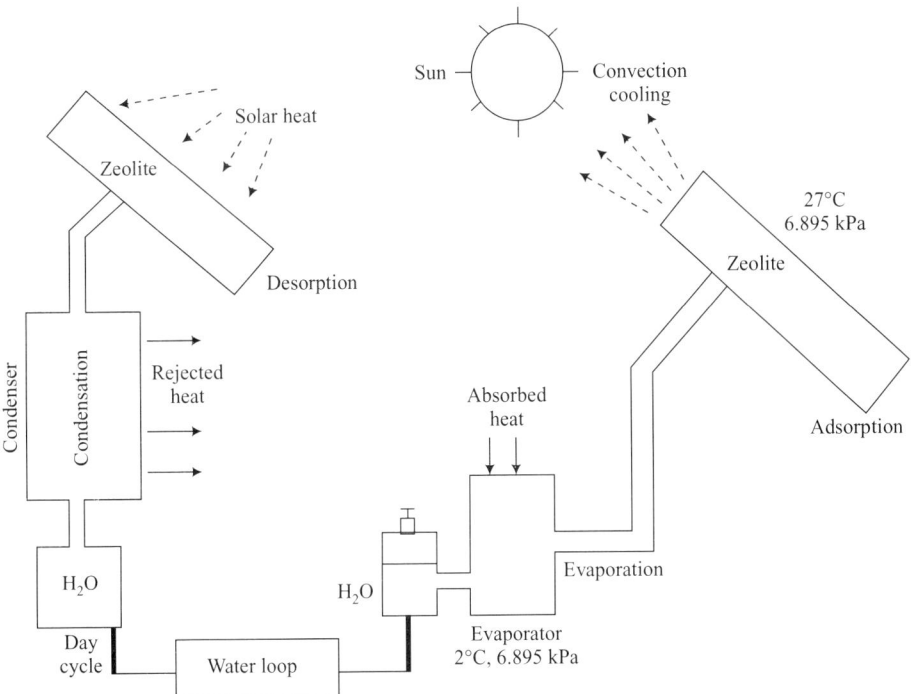

Fig. 9.3 Schematic Diagram Illustrating Day & Night Cycles for Zeolite-Water Systems (After Techernev, 1976)

condensed to liquid water by rejecting the heat of condensation to the ambient. The heat sink temperature determines the saturation pressure and hence the percentage of moisture retention. The cooling takes place in the night cycle, when the adsorbent bed (now cooler) is in equilibrium with a much lower water vapour pressure in the enclosed space. The water condensed in the day cycle can be evaporated by heat absorbed from the building space air which is passed through an evaporative heat exchanger. The heat of adsorption of the vapour in the bed must be rejected to the ambient in order to keep the bed at its minimum temperature and thereby keeping maximum moisture retention capacity.

Since the adsorption process is extremely temperature sensitive, the amount of adsorbed vapour decreases when temperature is increased above the room temperature. Techernev (1976, 1978, 1980) developed water-zeolite refrigeration system and found that chabazite type zeolite gives the best overall efficiency. Meunier *et al.* (1979a and b) have investigated zeolite systems for solar powered climatization and ice-production. Taoda *et al.* (1982) have studied a zeolite system for cooling and dehumidification. It must be mentioned that the operating pressure of zeolite collector system changes from day to night in the ratio of 10:1 and acts like a heat pump with fixed compression ratio, it can be used for both heating and cooling applications. In the conventional way, heating is produced during the day and cooling is produced in the night. Therefore, in view of demands for heating and cooling, it is necessary to provide some type of storage in the system. In a combined heating/ cooling zeolite system the condenser and evaporator are integrated in a single unit which is cooled by an external water loop. In winter operation water vapour is desorbed from the solar heated zeolite and condensed in the condensate tank during the day thereby rejecting condensation heat to the water loop. This heat can be used for domestic water heating as well as space heating. Excess of hot water may also be stored for use during night. In the cooling operation during summer, water from the storage tank is circulated into the condenser/evaporator unit during the night. Water evaporates and the vapour is adsorbed on the cool zeolite surface. The heat required for evaporation is taken from the water loop, thus cooling it for use in air conditioning. The chilled water may be stored in the same storage tank which is also used for storage of hot water during winter. Thus, change over from one season to the other is achieved by simple use of valves.

9.3.1 Solar Collector Aspects in Adsorption Cooling Systems: An Overview

The design of solar collector is a significant aspect for solar adsorption cooling applications. It depends on the temperature level required by the process. This can be achieved by collectors which are either flat plate, parabolic trough or with evacuated tubes. This section reviews various solar collector options used for adsorption refrigeration systems.

(a) Flat Plate Solar Collector
There are many parameters affecting the performance of solar collectors, such as solar radiation, wind speed, ambient temperature, number of glass covers, coating material, etc. Among these parameters, the researchers have reported the effects of single- or double-glass covers and simple black coating or selective coating materials. For the practical design of solar refrigerator, one should use double glass covers and selective coating material to improve the performance of solar refrigerator.

To enhance the heat transfer inside the adsorbent bed, Li *et al.* (2002) used number of fins in the flat plate solar collectors. The top surface of the adsorbent bed is coated with black paint to enhance absorption of solar radiation and, the adsorbent bed could reach maximum temperature of 110°C with COP range of 0.12-0.14.

Boubakari *et al.* (2000) developed a combined collector- condenser unit by using a single glazed collector and condenser connected by a flexible tube with an evaporator.

Instead of two valves necessary to fulfil the adsorption and desorption process of a basic refrigeration system, a no valve, flat plate collector based solar ice maker was also developed by Li *et al.* (2004) and the performance was tested for given solar radiation condition. The results showed that 6.0-10 kg of ice was produced with solar radiation of about 17-20 MJ/m².

Liu *et al.* (2005) also designed a simplified adsorption refrigeration system as shown in Fig. 9.4. which reduces the cost of the chiller, and made it more reliable as there were minimum moving parts, reducing air infiltration loss.

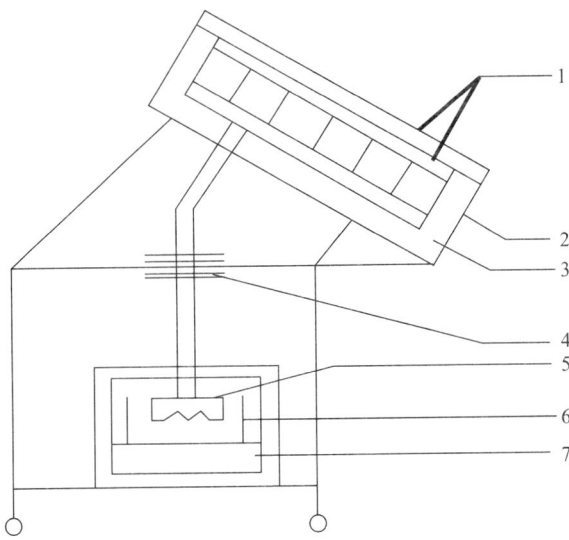

Fig. 9.4 Solar Powered Icemaker without Valves: (1) Cover Plate; (2) Adsorbent Bed; (3) Insulation; (4) Condenser; (5) Evaporator; (6) Water Tank; (7) Cold Box (Liu *et al.*, 2005)

Louajari *et al.* (2011) presented the performance results of the finned tube solar adsorption machine as shown in Fig. 9.5. The COP is found to increase from 7.5% (for tube without fins) to 11.1% (for tube with fins). The cycled mass of the refrigerant is also higher for finned tube than the tubes without fins, which shows that fins in the solar collector increases COP of the system.

(b) Compound Parabolic Collector

Solar adsorption cooling systems are usually based on the flat plate collector, whereas little attention has been paid on concentrating solar collectors. Headley *et al.* (1994) constructed CPC with an aperture area of 2.0 m². To increase desorption temperatures in excess of 120°C, but it is found to be causing conversion of methanol to dimethyl ether, which is a non-condensable gas inhibiting both condensation and adsorption.

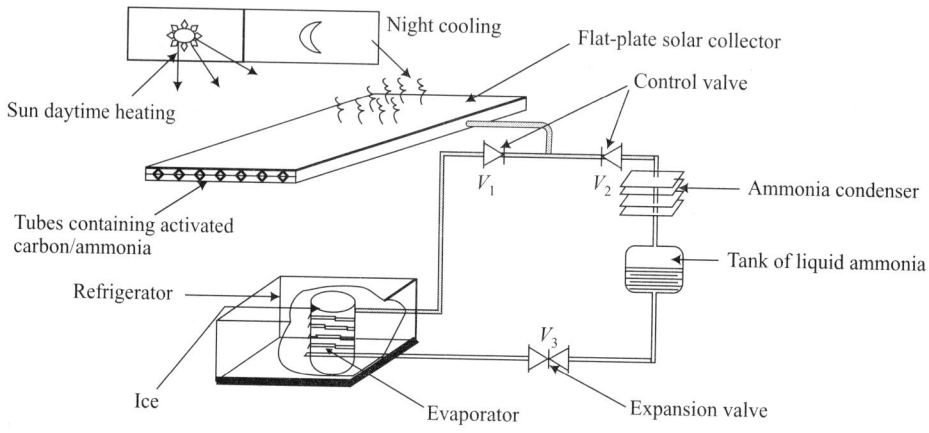

Fig. 9.5 Schematic of Solar Finned Tube System. (Louajari *et al.*, 2011)

Manuel *et al.* (2007) presented a compound parabolic collector (CPC) which focussed solar radiation on the four tubular receivers containing the sorption bed with total collection area of 0.55 m². The maximum and minimum bed temperatures were 116°C and 38°C respectively. The experimental results showed low solar COP ranging from 0.078-0.096.

Thermal design of solar powered adsorption refrigeration system was presented by Tashtoush *et al.* (2010). The system consists of an evacuated glass tube adsorber bed focussed with a parabolic solar concentrator. The adsorber is heated upto 125°C by solar energy collected by a parabolic solar concentrator.

Balghouthi *et al.* (2012) proposed an approach to heat the adsorption bed via reflectors. The testing of the module is mainly focussed on the heating of adsorption bed using four types of bed and reflector arrangements. The refrigerator could give daily ice production of 6.9 and 9.4 kg/m² and net solar COP of 0.136 and 0.159 in cold and hot climate respectively.

(c) Evacuated Tube Collector

Evacuated tube solar collectors have also been proposed for adsorption cooling systems. Niemann *et al.* (1997) designed and constructed an evacuated tube collector (ETC) coupled with an external parabolic concentrator to operate a large adsorption refrigeration system. The envisaged evacuated tube collectors of 1.6 m² area could give fluid temperature up to 170°C.

Mahesh (2010) presented a solar powered adsorption refrigerator as shown in Fig. 9.6 with an ETC solar collector area of 2 m², which can produce the COP of about 0.15-0.23. The speciality of the proposed system is that adsorption bed consisting of charcoal- methanol is separate and immersed into a water bath, which is heated directly by vacuum tube solar collector, and the adsorption bed could reach the temperature up to 110°C.

9.3.2 Selection of Adsorbent-Adsorbate Materials in Adsorption Systems

Working fluid pairs, i.e., adsorbent-refrigerant are one of the most important considerations for adsorption refrigeration system, because working conditions and environmental issues depend on

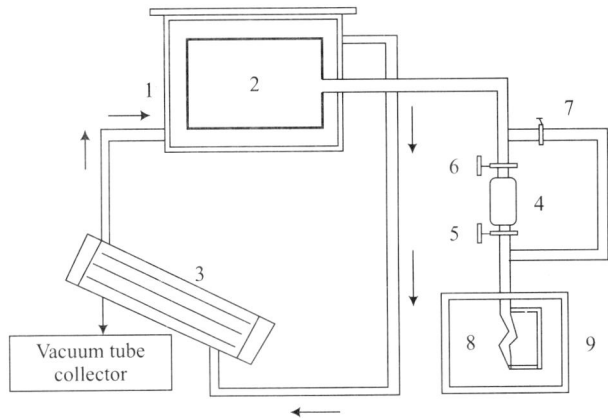

Fig. 9.6 Schematic Diagram of the Pilot Scale Solar Vacuum Tube Collector Adsorption Refrigeration Unit. (1) Outer Chamber of Adsorption Bed; (2) Inner Chamber of Adsorption Bed; (3) Flat Plate Collector; (4) Condenser; (5,6,7) Valves; (8) Chilling Chamber; (9) Evaporator. (Mahesh, 2010)

them. In adsorption refrigeration systems, the ideal refrigerant should have the following characteristics

  a. Low specific heat and high latent heat of evaporation
  b. Good thermal stability and high thermal conductivity
  c. Chemically stable in the working temperature range
  d. The molecular dimension should be small enough to allow energy adsorption
  e. Non-toxic, non-flammable and non-corrosive.

On the other hand, an ideal adsorbent should have the following characteristics:
  a. High adsorption and easy desorption capacity
  b. Good thermal conductivity and low specific heat capacity
  c. Chemically compatible with appropriate refrigerant and construction materials
  d. Reversibility of adsorption process for many cycles.

To develop adsorption cooling system with low cost, the design of adsorbent beds and selection of adsorbent-adsorbate materials is a significant process for cooling applications. Several studies have also been carried out, both experimentally and theoretically for selection of adsorbent-adsorbate materials, although the cost of adsorbent-adsorbate materials still make them non-competitive for commercialization. Therefore, recent research investigations are focussed on cost-effective materials and on increasing the efficiency of these systems. A wide variety of sorbent/sorbate materials is possible for adsorption refrigeration system. However, most common working pairs used water as the refrigerant due to high heat of vaporization as compared to other possible refrigerants. Furthermore, water is non-toxic and easy to handle. The disadvantage of water is due to its limited application above 0°C. Therefore, for ice making or refrigeration below 0°C, methanol or ammonia is used as a refrigerant with a variety of solid adsorbents. The following sections give an overview of working adsorption pair materials.

An adsorbent-refrigerant working pair for a solar refrigerator requires the following desirable characteristics:

350 *Alternatives in Refrigeration and Air Conditioning*

1. A refrigerant with a large latent heat of evaporation and low desorption temperature
2. A working pair with high thermodynamic efficiency and compact size
3. A small heat of desorption under the envisaged operating pressure and temperature conditions
4. A low thermal capacity of the working pair
5. Thermal and chemical stability of the pair

This section summarizes the most widely used working pairs, viz., zeolite-water, silica gel-water, activated carbon-methanol, and activated carbon-ammonia, and the performance comparison of adsorbent-adsorbate pair is made on the basis of given heat source temperature.

(a) Zeolite-water

The development of sorption refrigeration systems using zeolite-water is based on the pioneering work of Tchernev (1978). At the end of 1980s, Grenier et al. (1988) designed a solar adsorption air conditioning syatem using zeolite-water working pair having mass density of 22-15 kg/m^3. With of the heat source temperature of 325°C, the value of COP and specific cooling capacity (SCP) were found to be 0.26 and 95.0 W/kg, respectively.

A thermally powered prototype adsorption cooling system using natural zeolite-water was investigated by Ismail Solmus (2011). The mass and volume of working pair were 1.87 kg and 2.5 litre. The COP and SCP values of the experimental prototype were 0.25 and 6.4 W/kg respectively, for the heat source temperature of 150°C.

From the above studies, it is found out that zeolite requires larger quantities of adsorption material and higher temperatures. Table 9.1 shows some of the achieved performance of solar adsorption refrigeration systems. It is found that zeolite-water pair is not suitable with flat plate solar collector for adsorption refrigeration system. However, zeolites have other unique properties, viz., their adsorption isotherms have extremely nonlinear pressure dependence, as reported by Tchernev (1982), which is an important consideration in solar adsorption refrigeration applications.

Table 9.1 Some Performance Data of Zeolite-Water Solar Adsorption Refrigeration System

Authors	Mass of Adsorbent-adsorbate	COP	Heat source
Techrnev (1978)	50 kg/m^2	0.15	325°C
Greiner et al. (1988)	15-22 kg/m^2	0.1	93°C–103°C
Lu et al. (2004)	140 kg/m^2	0.21	220°C–250°C
Ismail Solmus (2011)	1.87 kg/m^2	0.25	150°C
Dong Wu (2011)	45-10.5 kg/m^2	0.26	320°C

(b) Silica gel-water

Silica gel-water is an another working pair, allowing cycles to be driven by much lower heat-source temperatures. Novel silica gel-water adsorption chillers with two single bed systems were built and tested in Shanghai Jiao Tong University (SJTU). Each adsorber bed contains 52 kg of silica gel. The refrigerating capacity and the COP of the chiller are found to be 8.69 kW and 0.388 for the heat source temperature of 82.5°C. In an innovative approach, a compact adsorption chiller integrated with a closed wet cooling tower was designed by Chen *et al.* (2010). The system performance was investigated for the heat source temperature of 85°C with 65 kg of silica gel and 169 kg of water. The COP and SCP were found to be 0.51 and 10.76 kW respectively.

(c) Activated carbon-methanol

Activated carbon-methanol is one of the most widely used working pair in adsorption systems, because of their large adsorption capacity, low desorption temperature and high latent heat of methanol. Table 9.2 illustrates the comparative performance results of silica gel-water and other pairs based adsorption refrigeration systems. Mahesh (2012) developed a prototype solar powered adsorption refrigeration system with an activated carbon/methanol pair and evacuated tube solar collector area of 2m². A 12 kg of activated carbon produced cooling COP of 0.23 for the maximum heat source temperature of 110°C.

Buchter *et al.* (2003) tested an adsorption ice maker consisting of 40 kg of A.C. and 7.0 litre of Methanol for the heat source temperature from 30°-100°C. The outcome of this prototype was compared with similar system of Pons and Greiner (1987) and Boubakri *et al.* (1992a and 1992b) in Morocco. The limitation of the system was mainly due to the ambient temperature, an increase in the ambient temperature beyond 23°C with reduced system performance.

Table 9.2 Adsorbent-Adsorbate Pair used in the Solar Adsorption Refrigeration System

Adsorbent-Adsorbate pair	System COP	References
Silica gel-water	0.2-0.3	Henning and Glaser (2003)
Silica gel-water	0.16	Hildbrand *et al.* (2004)
Silica gel-water	0.1-0.13	Luo *et al.* (2007)
Activated carbon-Ammonia	0.25	Critoph (1993)
Activated Carbon-Methanol	0.16	Passos *et al.* (1989)
Activated carbon-Methanol	0.1-0.12	Boukari *et al.* (1992)
Activated Carbon-Methanol	0.1-0.12	Sumathy and Zhongfu (1999)
Domestic charcoal-Methanol	0.14-0.16	Khattab (2004, 2006)
Activated carbon-Methanol	0.15-0.23	Mahesh (2010)

Wang *et al.* (2003) tested two different types of adsorber beds, one filled with solidified activated carbon and another one with granular activated carbon, and reported the performance of solidified A.C. adsorber better than that of granular A.C. since the heat transfer coefficient is much lower for the granular A.C., COP and specific cooling power of the solidified A.C. system were 0.125 and 16 W/kg and for the granular A.C. system were 0.104 and 13.1 W/kg respectively with heat source temperature of 110°C.

(d) Activated Carbon-Ammonia

The use of charcoal-ammonia is seen more visible mainly during 1990s. A small solar adsorption refrigerator was built and tested by Critoph (1993). The collector containing 17 kg of active carbon and 1.60 kg of ammonia was tested with 150°C heat source temperature and cooling produced was

4 kg of ice per day. The working pair of activated carbon-ammonia was earlier developed by Miles and Shelton in 1996 and the COP obtained varied between 0.19 and 0.42, for ambient temperatures from 20°C–35°C.

Fadar et al. (2009) proposed a novel solar adsorptive cooling system operated with a parabolic trough collector. The authors used 5–30 kg of activated carbon and its performance was tested for the heat source temperature of 70–170°C. It was found that COP increases with increase in adsorbent mass and then COP decreases. The increase in adsorbent mass induces high adsorption of ammonia initially and desorption of large amounts of ammonia in the subsequent process. This produces more cooling and results in high COP.

In order to develop a consolidated solidified adsorbent bed with better heat and mass transfer performance, Wang et al. (2011) also studied and compared solidified A.C. and granular A.C. The granular activated carbon was found better than solidified carbon because of the reduced mass transfer coefficient of ammonia at low saturated pressure. However, the practical disadvantages of activated carbon-ammonia systems are the high pressure requirement, bulkiness of the refrigerator, and the corrosive nature of the refrigerant (ammonia).

9.4 CHEMISORPTION SYSTEMS

A number of new technologies are also being developed to use chemical reactions to recover low grade energy (waste industrial exhaust gases, etc.). Among them, the monovariant chemical heat pump (CHP) can use low grade thermal energy for cooling, heating, energy upgradation and energy storage purposes. As discussed, both vapour absorption and adsorption systems are divariant physisorption systems, and require both pressure and temperature to be specified, while chemisorption systems are monovariant systems and require only one state variable (like pressure) to be specified for system operation.

The CHP utilizes the chemical reaction to store the energy in the chemical bond and release it whenever required. In principle, it consists of two different chemical reactions (more precisely, univariant processes), each running at two different temperatures.

The simplest form of a CHP consists of two vessels (or chemical reactors) connected with each other. Let first vessel be H and second vessel be L. The vessel H contains a solid or a liquid complex between a non-volatile substance N_H and a volatile substance V. The vessel L contains condensed volatile substance (or a weaker complex between another non-volatile substance N_L and volatile substance V).

When heat is supplied to the vessel H, the compound V decomposes into $V(l)$ (V is in liquid phase) and $V(g)$ (V is in the gaseous phase). Due to pressure difference $V(g)$ flows into vessel L, where it condenses to $V(l)$ and releases heat. This process is called generation/desorption process and occurs between temperatures T_H and T_M ($T_H > T_M$). After this, the vessel L is cooled and reduced in pressure. Now, heat is absorbed by vessel V from the surroundings which causes the vaporization of $V(l)$ to $V(g)$. This $V(g)$ flows to vessel H, where it combines with N_H to form compound $N_H V$ and releases heat. This is called absorption process and occurs between T_M and T_L ($T_M > T_L$). After this, again the vessel H is heated. A cycle is completed when these processes occur in series. Thus, there are three temperatures T_H, T_M and T_L.

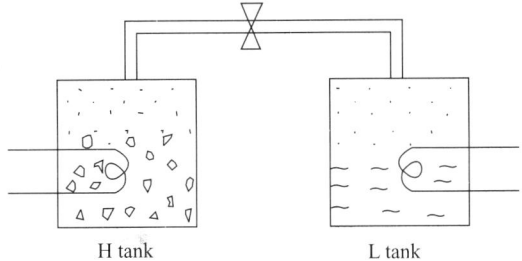

Fig. 9.7 Two Vessel Configuration of a CHP System for Thermal Energy Storage

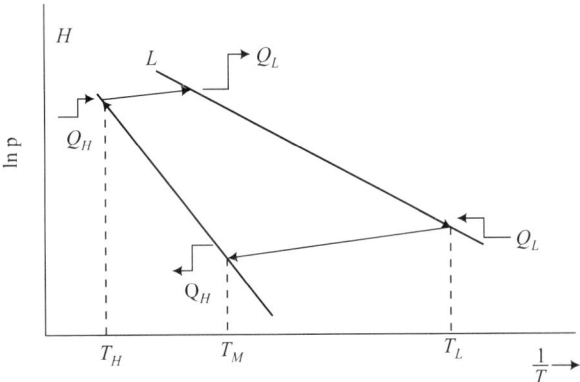

Fig. 9.8 Vapour Pressure as a Function of Temperature

Since all chemical components of the CHP working fluid are considered pure substances, the vapour pressure (P) of volatile substance P in the two vessels depends only on the temperature (T) and not on the concentration. Assuming the vapour of volatile substance V as an ideal gas, we have the following relations

$$lnp = \frac{-\Delta H}{RT} + \frac{\Delta S}{R} \tag{9.8}$$

where, ΔH and ΔS are the enthalpy and entropy changes of the reaction. Figure 9.8 shows the linear dependency of lnp on $1/T$. The slope of the line is equal to $-\Delta H/R$ and the intercept is equal to $\Delta S/R$.

A CHP can be operated in either cooling or heating mode, once the regeneration process is completed. In heating mode the ambient temperature is T_L and $T_M > T_L$. The vapour pressure of volatile substance V in vessel L at is infinitely greater than the vapour pressure of V in the vessel H at temperature. Thus, the heat will be extracted from the surroundings. This heat vaporizes the $V(l)$ to $V(g)$. This $V(g)$ reacts with N_H to form $N_H V$ complex. The heat of reaction is released from the system. The net result is that the heat is pumped from the lower temperature T_L to the higher temperature T_M, where it can be used for heating. In cooling mode, the process is same as that in the heating mode, but the ambient temperature is T_M.

The general classification of CHP is shown in Fig. 9.9. Several CHP systems have been developed depending on applications for heating, cooling, energy upgradation and storage. Some proposed

CHP systems are listed in Table 9.4. However, several other combinations of cycles are also under studies. The main problem lies in providing efficient means of heat transfer, which is prominent in solid-gas system. Hence, liquid-gas systems are preferred over solid-gas systems.

9.4.1 Thermodynamic Analysis of CHP Systems

Table 9.3 Illustration of Some CHP Cycles

S. No.	High Temperature Cycle	Low Temperature Cycle	Use
1.	$Ca(OH)_2(s) \leftrightarrow CaO(S) + H_2O(g)$	$H_2O(l) \leftrightarrow H_2O(g)$	Heating/Cooling
2.	$MgCl_2 \cdot 6NH_3(s) \leftrightarrow MgCl_2 \cdot 2NH_3(s) + 4NH_3(g)$	$CaCl_2 \cdot 8NH_3(s) \leftrightarrow CaCl_2 \cdot 4NH_3(s) + 4NH_3(g)$	Heating/Cooling
3.	$MgCl_2 \cdot 4H_2O(s) \leftrightarrow MgCl_2 \cdot 2H_2O(s) + 2H_2O(g)$	$H_2O(l)\, H_2O(g)$	Heating/Cooling

As already discussed, the overall operation of a CHP system is considered to be composed of two different subcycles as shown in Fig. 9.10. The first subcycle is the high temperature cycle involving N_H, which operates between T_H and T_M ($T_H > T_M$). It can be treated as a heat engine (HE) cycle. Second sub cycle is the low temperature cycle involving N_L (or V (g)). It operates between T_M and T_L ($T_M > T_L$). It can be treated as a heat pump (HP) or refrigerator (R) cycle. The two subcycles are coupled together and run simultaneously. For example, the part CD of HE cycle occurs simultaneously with the part EF of HP cycle. Thus, a CHP is nothing but a chemical heat pump driven by chemical heat engine or a chemical refrigerator driven by a chemical heat engine.

The maximum efficiencies for a heat engine and heat pump (or a refrigerator) are given by Carnot formulas.

It is a assumed that all the processes are reversible. It is also assumed that the heat of reactions is independent of temperature. Thus, the sensible heat to be supplied to the system (DA and FG in Fig. 9.10.) and the one to be recovered from it (BC and EH) are equal and therefore excluded from the analysis. Carnot cycle is illustrated in Fig. 9.11.

Maximum efficiency of a heat engine cycle is given by

$$\eta_{HE} = \frac{W}{Q_1} = \frac{(T_H - T_M)}{T_H} \quad (9.9)$$

For heat pump cycle

$$\eta_{HP} = \frac{Q_3}{W} = \frac{T_M}{(T_M - T_L)} \quad (9.10)$$

Table 9.4 Status of Metal Hydride Based Cooling Systems Around the World (Srinivasa Murthy, 2007)

Sl. No.	Place	Year	Alloys used	Type	Mass (kg)	Capacity (kW)	COP
1.	Southern California Gas Co. USA	1982	$LaNi_5/MmNi_{4.15}Fe_{0.85}$	R	3.6	0.6	–
2.	Solar Turbines Int., USA	1982	$LaNi_{4.7}Al_{0.3}/MmNi_{4.15}Fe_{0.85}$	R	3.6	0.6	–
3.	SeKisui Chem, Japan	1983	$LaNi_{4.7}Al_{0.3}/LaNi_{4.85}Al_{0.15}$	R	90	–	0.42
4.	Chuo Denki Kogyo, Japan	1983	$LaNi_{4.65}Al_{0.35}/MmNi_{4.15}Fe$	R	40	1.75	–
5.	JMC & Kongakuin Uni., Japan	1983	$LaNi_{4.65}Al_{0.35}/MmNi_{4.15}Fe$	R	40	1.3	0.3
6.	IIT, Technion, Japan	1984	$LaNi_{4.7}Al_{0.3}/MmNi_{4.15}Fe_{0.85}$	R	90	22.8	–
7.	Kurimoto, Japan	1985	$LaNi_5/LaNi_{4.7}Al_{0.3}$	HP	20	0.6	–
8.	IKE, Stuttgart, Germany	1985	$LaNi_{4.7}Al_{0.3}/MmNi_{4.65}Fe_{0.35}$	HP	64	–	–
9.	Sanyo Electrical, Japan	1986	MMNiMnAl/MmNiMnCo	HP	64	3.0	–
10.	JMC & Kongakuin Uni., Japan	1986	$MmNi_{4.4}Mn_{0.5}Al_{0.05}$ $MmNi_{4.7}Mn_{0.15}Lm_{0.95}Ni_5$	R	48	4.6	–
11.	Ergenics Inc. USA	1989	$LaNi_{4.5}Al_{0.5}(CFM)Ni_5$	R	2.6	–	0.33
12.	Korea Advanced Institute	1993	$Zr_{0.9}Ti_{0.1}Cr_{0.9}Fe_{1.1}/Zr_{0.9}Ti_{0.1}Cr_{0.6}Fe_{1.4}$	R	4.5	0.683	–
13.	IIT Madras, India	1996	$ZrMnFe/MmNi_{4.5}Al_{0.5}$	R	1.5	0.1	0.2-0.4
14.	Aircond & Environ. Control Lab, Korea	1996	$LaNi_{4.7}Al_{0.3}/MmNi_{4.15}Fe_{0.85}$	R	–	–	–
15.	Research Institute of SIA Lutch, Russia	1996	$LaNi_{4.6}Al_{0.4}/MmNi_{4.15}Fe_{0.85}$	HP	3.0	0.15-0.2	0.17-0.2
16.	Thermal Electric Devices, Inc.., New Mexico, USA	1997	$LaNi_5$	C	1	1.5(150 s cooling)	–
17.	Thermal electric Devices, Inc.., New Mexico, USA	1998	$Ca_{0.4}Mm_{0.6}Ni_5$	C	1	2.2(150 s cooling)	–
18.	State Research Institute of Scientific and Industrial Association, Russia	2002	$LaNi_{0.4}MmNi_{4.15}Fe_{0.85}$	R	3	0.15	–
19.	Korea Advanced Institute of Science and Technology, South Korea	2001 2002	$Zr_{0.9}ti_{0.1}Cr_{0.55}Fe_{1.45}$	C	1	0.41	1.8

356 *Alternatives in Refrigeration and Air Conditioning*

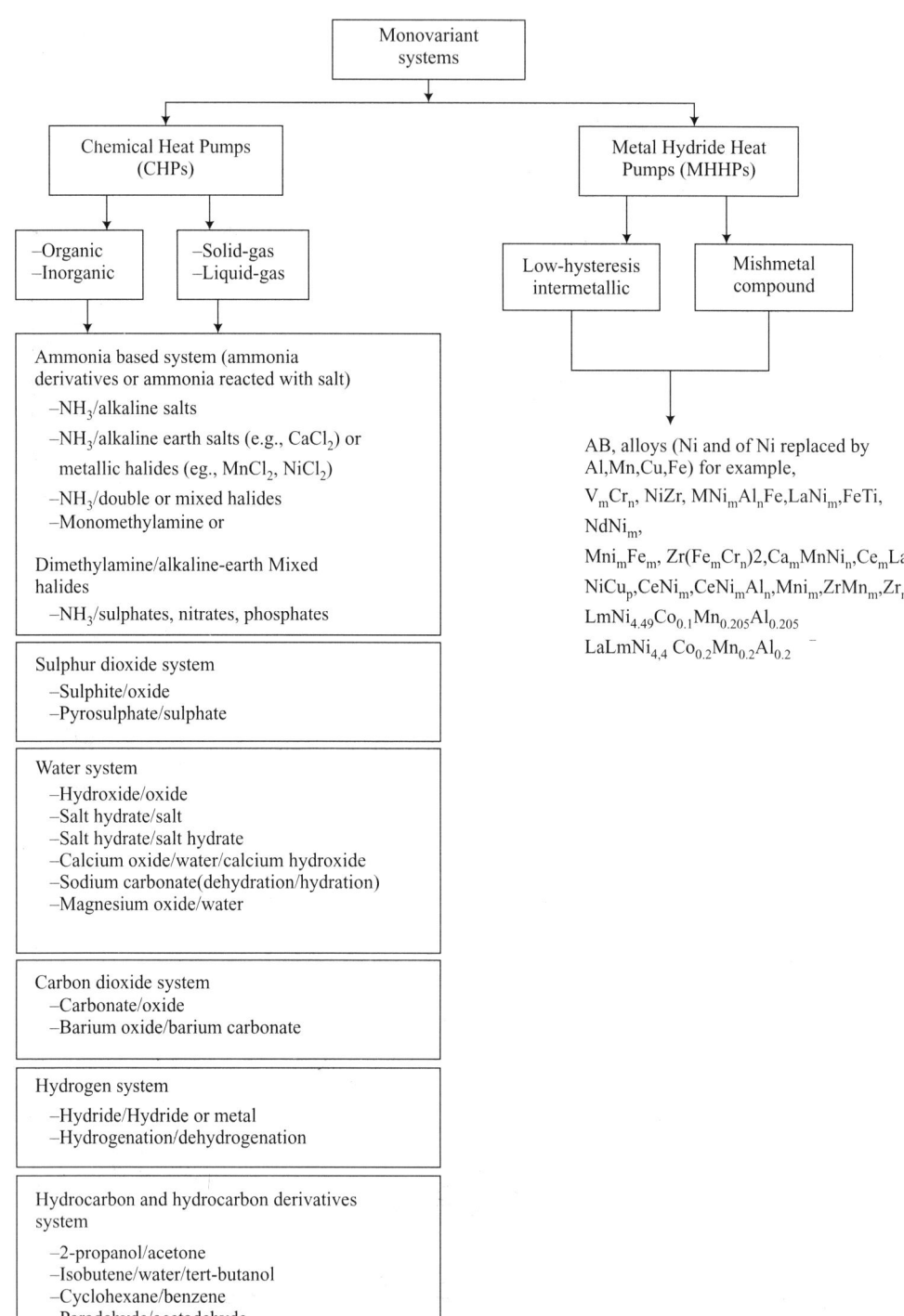

Fig. 9.9 Classification of CHP & MHHP (Wongsuwan *et al.*, 2001)

Vapour Adsorption Systems: Physisorption and Chemisorption Systems 357

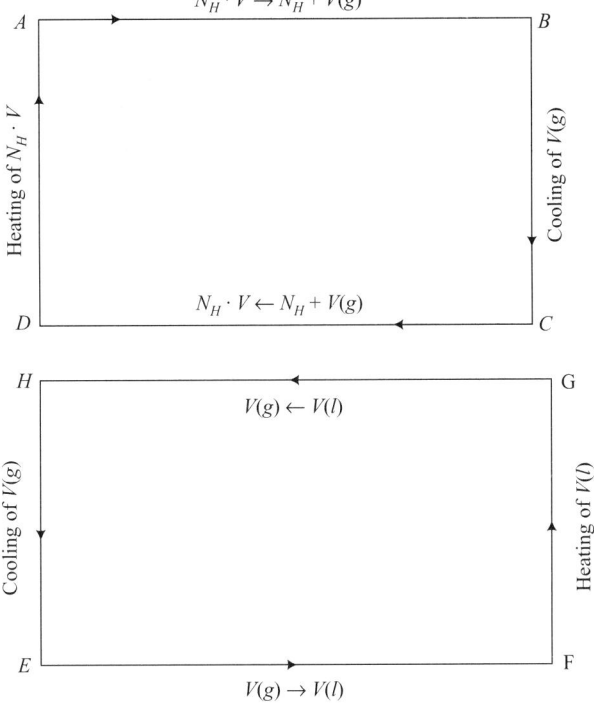

Fig. 9.10 High and Low Temperature of a CHP

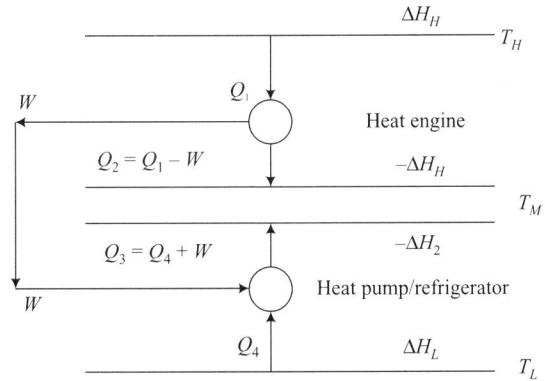

Fig. 9.11 Carnot Analysis of a CHP Cycle

For heating mode, the overall efficiency is defined as the ratio of heat obtained for heating purposes $(Q_2 + Q_3)$ to the heat supplied (Q_1) for regeneration. Therefore, maximum overall efficiency for heat pump in heating mode is given by

$$\eta_{heat,\max} = \frac{(Q_2 + Q_3)}{Q_1} = \frac{Q_2}{Q_1} + \frac{Q_3}{Q_1} = \frac{Q_2}{Q_1} + \eta_{He} \cdot \eta_{HP} \tag{9.11}$$

Putting the value of η_{HE} and η_{HP} from equations (9.9) and (9.10) in equation (9.11), we get the following:

$$\eta_{heat, max} = \frac{T_M}{T_H} \cdot \frac{(T_H - T_L)}{(T_M - T_L)} \qquad (9.12)$$

It is evident from equation (9.12), that the efficiency increases with the increase in regeneration temperature T_H and decrease with heating temperature T_M as the ambient temperature T_L is fixed. In case of cooling mode, the analysis is similar to that of the heating mode. The maximum efficiency of the refrigeration cycle is the ratio of heat withdrawn (Q_4) to the work provided (W).

$$\eta_R = \frac{Q_4}{W} = \frac{T_L}{(T_M - T_L)} \qquad (9.13)$$

The overall efficiency for cooling configuration is as follows:

$$\eta_{cool, max} = \eta_R \cdot \eta_{HE} = \frac{T_L}{T_H} \cdot \frac{(T_H - T_M)}{(T_M - T_L)} \qquad (9.14)$$

From equation (9.14), it is obvious that the efficiency increases with the increase in T_H and T_L and the ambient temperature is fixed as T_M.

The actual efficiency of CHP is determined from the Fig. 9.12, as follows:

$$\eta_{heat} = \frac{-(\Delta H_H + \Delta H_L)}{-\Delta H_H} \qquad (9.15)$$

$$\eta_{cool} = \frac{\Delta H_L}{\Delta H_H} \qquad (9.16)$$

9.5 METAL HYDRIDE SYSTEMS

Many metals, alloys, and intermetallic compounds react reversibly with hydrogen at ambient temperature and modest pressure to form compounds called hydrides. The volume density of hydrogen in these compounds is very high. Thus, this offers a primary way of hydrogen storage. While the hydrogen adsorption-desorption characteristics are used for hydrogen storage, compression and purification, the reaction enthalpy changes may be applied for thermal energy storage, heat pumps (or refrigerators) and heat transformers.

Metal hydride systems are also adsorption based chemisorption systems (or MHHP systems). In chemical adsorption, various working pairs (adsorbent + adsorbate) like metal hydride/hydrogen, metal chloride/ammonia, metal oxides/oxygen, etc. are used.

9.5.1 Basic Principle of MHHP

Figure 9.12 shows the simple metal hydride single-stage heat pump. It consists of two reactors (vessels/tanks) filled with two different substances, which can adsorb and desorb hydrogen. They are therefore, referred to as metal hydride pairs.

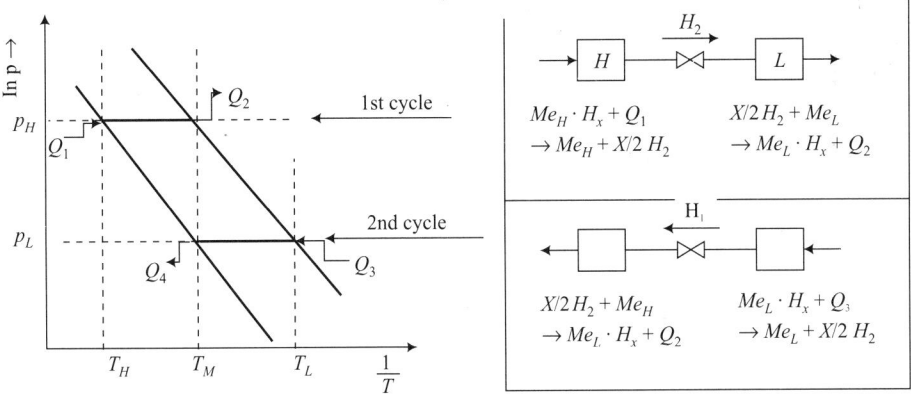

Fig. 9.12 Operating Principle of Single-stage MHHP (Kelvin *et al.*, 2003)

Let the two reactors be *H* and *L*. Reactor *H* is filled with one metal compound Me_H and reactor *L* contains another metal compound Me_L. Hydrogen is the actual working substance. The machine is operated at three different temperature levels T_H, T_M and T_L ($T_H > T_M > T_L$) and two different pressure levels p_H and p_L ($p_H > p_L$).

Initially, metal hydride $Me_H H_x$ is there in reactor *H* at temperature T_H. Heat input to reactor *H* causes desorption of hydrogen from the metal hydride. Hydrogen flows to reactor *L*, where hydrogen combines with metallic alloy Me_L to form $Me_L H_x$ at temperature (T_M) and releases adsorption enthalpy. This is the first half cycle. The reactor *L* is then cooled to temperature T_L, where second half cycle starts. In the second half cycle, heat input is given to metal hydride $Me_L H_x$ causing desorption of hydrogen from it. The free hydrogen then flows to reactor *H*, which is already cooled to temperature (T_M). At this temperature, Me_H adsorbs hydrogen to form $Me_H H_x$ and releases enthalpy of adsorption. Metal hydride Me_L is again heated to temperature (T_H). Thus, the cycle is completed. The working of the system with only two reactors on this cycle is generally intermittent because the reactors have to be sensibly cooled and heated between the two half cycles.

Since the hydrogen absorbing materials can just be intermetallic composites, a large number of compositions have been developed as the working substances. The performance of MHHP systems is measured in terms of coefficient of performance (COP) and specific cooling power (SCP), which largely depends upon the thermodynamic and thermophysical properties of the metal hydride pairs. Alloys based on lanthanum, mischmetal, titanium, vanadium, zirconium, etc. are generally preferred, as shown in Table 9.4, because of their desirable properties.

In general, the alloys must have high enthalpy of formation, low specific heat, high hydrogen adsorption capacity, high thermal conductivity and fast reaction kinetics. The other factors for due consideration are equilibrium pressure, hysteresis, activation characteristics, cost, availability and life, etc.

9.5.2 Thermodynamic Analysis of MHHP

AS already mentioned earlier, the performance of a MHHP system can be analyzed by its coefficient of performance (COP), in case it is working in cooling mode and by its coefficient of amplification (COA), when it works in heating mode.

So, in case of cooling mode, we have

$$\text{COP} = \frac{Q_3}{Q_1} \qquad (9.17)$$

where, Q_3 is the heat removed at lowest temperature T_L, from the space to be cooled and Q_1 is the heat added at highest temperature T_H.

For heating mode, we have the following relation from Fig. 9.12.

$$\text{COA} = \frac{(Q_2 + Q_4)}{Q_1} \qquad (9.18)$$

The specific cooling power (SCP) is yet another parameter to evaluate the performance of MHHP. It is defined as:

$$\text{SCP} = \frac{Q_3}{m_a} \qquad (9.19)$$

In the above equation, is the cooling power for refrigeration cycle and is the adsorbent mass in one adsorbent bed.

Adsorption and desorption of hydrogen in MHHP are exothermic and endothermic reactions. Therefore, removal and supply of heat at appropriate rate is very important. Similarly, mass transfer rates of hydrogen from one reactor to another affects its performance. Various analytical studies have been done on heat and mass transfer processes in metal hydride beds. It is found that the effects of convection are dominant while that of radiations are negligible, especially for low temperature beds.

The COP of MHHP can be enhanced by recovering heat and mass, whereas the SCP can be improved by advance adsorption technologies. The adsorbents in MHHP are solids. Therefore, it is difficult to recover internal heat in MHHP unlike in conventional wet vapour absorption systems which have solution heat exchanger to recover the heat. Internal recovery of heat is possible by two stages. One method is the direct recovery of heat while the other method involves mass recovery. Kevin et al. (2003) reported an increase of 10-15% in COP by recovering mass and heat in MHHP systems.

9.6 CONCLUSION

An extensive review of the adsorption systems related to better utilization of solar energy for production of cooling energy is presented. Among the various types of adsorption refrigeration systems, emphasis is made on the bed design options, the different adsorbent-adsorbate materials and the different solar collector systems. The literature reveals that the performance of various adsorption system varies over a wide range. Most of the works are experimental in nature with varying operating conditions. Some of them use sophisticated solar collectors, while others used the adsorbent itself, contained in a transparent tube, as the solar heat adsorbing bed, but the most efficient configuration seems to consist of metallic flat-plate solar collectors, single or double glazed, covered with a selective coated surface, filled with the adsorbent bed and evacuated tube solar collectors. The choice of solar collectors is an important factor for the adsorption refrigeration system and it depends on the level of temperature required by the process.

Another concern in the performance of the adsorption cooling system is mainly on the working pairs. A well designed system should have the characteristics of large adsorption capacity, temperature variation, and more flat desorption isotherm and refrigerant fluid whose choice depends on the desired evaporator temperature. It must possess high latent heat of vaporization and small molecular dimensions to ensure an easy adsorption. Since the refrigeration system operates at a temperature below 0°C, thus methanol seems to be a good adsorbate choice, because it can evaporate at temperature below 0°C, its enthalpy of vaporization is high, its molecule is small enough to be easily adsorbed into micropores, and its working pressure is always lower than the atmospheric one, which means a safety factor in case of leakage. As a result activated carbon-methanol is the most widely used adsorbent reported in the literature due to its extremely high surface area and micropore volume. The operating temperature is one of the limitations in the operation of activated carbon-methanol pair cooling system. At heat source temperature more than 120°C, the methanol decomposes into dimethyl, yielding low cooling performance, due to the chemical decomposition.

Silica gel-water system requires the operation under vacuum condition, which poses practical difficulties due to water not being used at ice-making temperature. Therefore, it is a better choice for air conditioning application only. Zeolite-water pair requires a regeneration temperature of above 200°C and activated carbon-ammonia pair also requires more than 150°C for its regeneration. These temperatures are not obtained by simple flat plate collectors and hence evacuated tube solar collector or concentrating collector systems are used. It is to be noted that such solar collectors increase the initial cost of the adsorption cooling system. Although activated carbon-methanol pair works at a low regeneration temperature and it is more suitable for ice production and freezing application. Comparing with activated carbon-ammonia system, activated carbon-methanol system is a vacuum based system but it is safer than using activated carbon-ammonia pair, because activated carbon-ammonia adsorption system operates in the high pressure state, and ammonia is a corrosive and flammable refrigerant and thus not so reliable if it leaks.

It is thus concluded that activated carbon-methanol is an ideal working pair for adsorption based solar cooling, because of its high COP and low regeneration temperature, low freezing point, and no corrosion problem. In addition maximum adsorption capacity of the activated carbon-methanol pair is an advantage as compared to the other adsorbent-adsorbate materials.

Another series of sorption systems using chemical reactions, viz., chemisorption systems including the chemical heat pumps and metal-hydride heat pumps have also been described and analysed in this chapter. These systems are useful for recovering low grade thermal energy, viz., solar energy and industrial waste heat, etc. The chemisorption systems are monovariant in contrast to the physisorption systems, which are divariant, requiring both pressure and temperature parameters to be specified. Such systems are also useful for storage of energy in chemical bond and releasing it whenever required. Metal-hydride refrigeration technology is emerging as a potential option for solar cooling and heating of buildings. Although these systems seem to be promising and attractive emerging options for solar cooling, but not much research and development work is reported in the literature.

REFERENCES

A. Sakoda and M. Suzuki, Simultaneous transport of heat and adsorbate in closed type adsorption cooling system utilizing solar heat, ASME J. Sol. Energy Eng. 108, 239, 1986.

B.B. Saha, A Akisawa, T Kashiwagi, Solar/waste heat driven two-stage adsorption chiller: the prototype, Renewable Energy, Volume 23, Issue 1, May 2001, pp. 93-101.

Boubakri, J. J. Guilleminot, and F. Meunier, "Adsorptive solar powered ice maker: experiments and model," Solar Energy, Vol. 69, No. 3, pp. 249-263, 2000.

Boubakri, M. Arsalane, B. Yous, L. Ali-Moussa, M. Pons, F. Meunier, J.J. Guilleminot, Experimental study of adsorptive solar-powered ice makers in Agadir (Morocco)—1. Performance in actual site, Renewable Energy, Volume 2, Issue 1, February 1992, pp. 7-13.

Boubakri, M. Arsalane, B. Yous, L. Ali-Moussa, M. Pons, F. Meunier, J.J. Guilleminot, Experimental study of adsorptive solar-powered ice makers in Agadir (Morocco)—2. Influences of meteorological parameters, Renewable Energy, Volume 2, Issue 1, February 1992, pp. 15-21.

C.J. Chen, R.Z. Wang, Z.Z. Xia, J.K. Kiplagat, Study on a silica gel-water adsorption chiller integrated with a closed wet cooling tower, International Journal of Thermal Sciences, Volume 49, Issue 3, March 2010, pp. 611-620.

C.J. Chen, R.Z. Wang, Z.Z. Xia, J.K. Kiplagat, Z.S. Lu, Study on a compact silica gel-water adsorption chiller without vacuum valves: Design and experimental study, Applied Energy, Volume 87, Issue 8, August 2010, pp. 2673-2681.

Catherine Hildbrand, Philippe Dind, Michel Pons, Florian Buchter, A new solar powered adsorption refrigerator with high performance, Solar Energy, Volume 77, Issue 3, September 2004, pp. 311-318.

Critoph, R. E., and Y. Zhong. "Review of trends in solid sorption refrigeration and heat pumping technology." Proceedings of the Institution of Mechanical Engineers, Part E: Journal of Process Mechanical Engineering 219.3, 2005: pp. 285-300.

D.I. Tchernev, Solar air conditioning and refrigeration systems utilizing zeolites, in Proceedings of Meetings of Commissions E1-E2 (International Institute of Refrigeration, Jerusalem,), pp. 209-215, 1982.

D.I. Techernev, Solar applications of natural zeolites, Proceedings, 2nd workshop on use of Solar Energy for Cooling of Buildings, California, (USA), pp. 307, 1976.

D.I. Techernev, Solar refrigeration utilizing zeolites, Proceedings inter-society, Energy Conversion conference, 2, p. 119, 1978.

Daniel J. Miles, Sam V. Shelton, Design and testing of a solid-sorption heat-pump system, Applied Thermal Engineering, Volume 16, Issue 5, May 1996, pp. 389-394.

E Willers, M Groll, Evaluation of metal hydride machines for heat pumping and cooling applications: Evaluation des machines à hydrure métallique dans les applications de pompes à chaleur et de refroidissement, International Journal of Refrigeration, Volume 22, Issue 1, January 1999, pp. 47-58.

E.E. Anyanwu, C.I. Ezekwe, Design, construction and test run of a solid adsorption solar refrigerator using activated carbon/methanol, as adsorbent/adsorbate pair, Energy Conversion and Management, Volume 44, Issue 18, November 2003, pp. 2879-2892.

E.F. Passos, J.F. Escobedo, F. Meunier, Simulation of an intermittent adsorptive solar cooling system, Solar Energy, Volume 42, Issue 2, 1989, pp. 103-111.

El Fadar, A. Mimet, M. Pérez-García, Modelling and performance study of a continuous adsorption refrigeration system driven by parabolic trough solar collector, Solar Energy, Volume 83, Issue 6, June 2009, pp. 850-861.

El Fadar, A. Mimet, M. Pérez-García, Study of an adsorption refrigeration system powered by parabolic trough collector and coupled with a heat pipe, Renewable Energy, Volume 34, Issue 10, October 2009, pp. 2271-2279.

F. Buchter, Ph. Dind, M. Pons, An experimental solar-powered adsorptive refrigerator tested in Burkina-Faso, International Journal of Refrigeration, Volume 26, Issue 1, January 2003, pp. 79-86.

H.L. Luo, R.Z. Wang, Y.J. Dai, J.Y. Wu, J.M. Shen, B.B. Zhang, An efficient solar-powered adsorption chiller and its application in low-temperature grain storage, Solar Energy, Volume 81, Issue 5, May 2007, pp. 607-613.

Hasan Demir, Moghtada Mobedi, Semra Ülkü, A review on adsorption heat pump: Problems and solutions, Renewable and Sustainable Energy Reviews, Volume 12, Issue 9, December 2008, pp. 2381-2403.

Henning, H.M., and H. Glaser. Solar assisted adsorption system for a laboratory of the University Freiburg. http://www.bine.info/pdf/infoplus/uniklaircontec.pdf

Ýsmail Solmuþ, Bilgin Kaftanoðlu, Cemil Yamalý, Derek Baker, Experimental investigation of a natural zeolite–water adsorption cooling unit, Applied Energy, Volume 88, Issue 11, November 2011, pp. 4206-4213.

Kevin Abraham, M. Prakash Maiya, S. Srinivasa Murthy, Performance analysis of a single stage four bed metal hydride cooling system, part B: influence of heat recovery, International Journal of Thermal Sciences, Volume 42, Issue 1, January 2003, pp. 79-84.

Kevin Abraham, M. Prakash Maiya, S.Srinivasa Murthy, Performance analysis of a single stage four bed metal hydride cooling system, part A: influence of mass recovery, International Journal of Thermal Sciences, Volume 42, Issue 1, January 2003, pp. 71-77.

Kevin Abraham, M. Prakash Maiya, S. Srinivasa Murthy, Performance analysis of a single stage four bed metal hydride cooling system, part C: influence of combined heat and mass recovery, International Journal of Thermal Sciences, Volume 42, Issue 11, November 2003, pp. 1021-1027.

L. Luo, M. Feidt, Thermodynamics of Adsorption Cycles: A Theoretical Study Heat Transfer Engineering Vol. 13, Issue 4, 1992.

L.W. Wang, J.Y. Wu, R.Z. Wang, Y.X. Xu, S.G. Wang, X.R. Li, Study of the performance of activated carbon–methanol adsorption systems concerning heat and mass transfer, Applied Thermal Engineering, Volume 23, Issue 13, September 2003, pp. 1605-1617.

M. Li, R.Z. Wang, Y.X. Xu, J.Y. Wu, A.O. Dieng, Experimental study on dynamic performance analysis of a flat-plate solar solid-adsorption refrigeration for ice maker, Renewable Energy, Volume 27, Issue 2, October 2002, pp. 211-221.

M. Balghouthi, M.H. Chahbani, A. Guizani, Investigation of a solar cooling installation in Tunisia, Applied Energy, Volume 98, October 2012, pp. 138-148.

M. Li, C.J. Sun, R.Z. Wang, W.D. Cai, Development of no valve solar ice maker, Applied Thermal Engineering, Volume 24, Issues 5–6, April 2004, pp. 865-872.

M. Li, R.Z. Wang, Heat and mass transfer in a flat plate solar solid adsorption refrigeration ice maker, Renewable Energy, Volume 28, Issue 4, April 2003, pp. 613-622.

M. Niemann, J. Kreuzburg, K.R. Schreitmüller, L. Leppers, Solar process heat generation using an ETC collector field with external parabolic circle concentrator (PCC) to operate an adsorption refrigeration system, Solar Energy, Volume 59, Issues 1–3, January–March 1997, pp. 67-73.

M. Pons, J.J. Guilleminot, Design of an experimental solar powered Solis-adsorption powered ice maker, J. Sol. Energy Eng. 108, pp. 332, 1986.

M. Pons, P.H. Greiner Experimental data on Solar powered ice maker using activated carbon-methanol adsorption pair, ASME. J. Sol. Energy Eng. 109, 303, 1987.

Mahesh, A., and S.C. Kaushik. "Solar adsorption cooling system: An overview." Journal of Renewable and Sustainable Energy 4, 2012: 022701.

Mahesh. A, Studies on Solar adsorption cooling system, Ph.D. Thesis, Madurai Kamaraj University, 2010.

Manuel I. González, Luis R. Rodríguez, Jesús H. Lucio, Evaluation of thermal parameters and simulation of a solar-powered, solid-sorption chiller with a CPC collector, Renewable Energy, Volume 34, Issue 3, March 2009, pp. 570-577.

Meunier. F; et al., On the use of zeolite water intermittent cycle for solar climatization of buildings, Proceedings, ISES congress, Atlanta, p. 739, 1979b.

Meunier. F; et al., Solar cooling through cycles using Microporus solid adsorbents, Proceedings, ISES congress, Atlanta, p. 696, 1979a.

Michael A. Lambert, Design of solar powered adsorption heat pump with ice storage, Applied Thermal Engineering, Volume 27, Issues 8–9, June 2007, pp. 1612-1628.

Mohamed Louajari, Abdelaziz Mimet, Ahmed Ouammi, Study of the effect of finned tube adsorber on the performance of solar driven adsorption cooling machine using activated carbon-ammonia pair, Applied Energy, Volume 88, Issue 3, March 2011, pp. 690-698.

N.M. Khattab, A novel solar-powered adsorption refrigeration module, Applied Thermal Engineering, Volume 24, Issues 17–18, December 2004, pp. 2747-2760.

O.StC. Headley, A.F. Kothdiwala, I.A. McDoom, Charcoal-methanol adsorption refrigerator powered by a compound parabolic concentrating solar collector, Solar Energy, Volume 53, Issue 2, August 1994, pp. 191-197.

P. H. Grenier, J. J. Guilleminot, F. Meunier, and M. Pons, solar powered Solid adsorption cold store, J. Sol. Energy Eng. 110, 192 (1988).

R. E. Critoph, Laboratory testing of an ammonia/carbon solar refrigerator, ISES (Solar World Congress, Budapest, Hungary, 1993), pp. 23-26.

R. E. Critoph,, and Y. Zhong. Proceedings of the Institution of Mechanical Engineers, Part E: Journal of Process Mechanical Engineering, pp. 285-300, (2005).

R.Z. Wang, Z.Z. Xia, L.W. Wang, Z.S. Lu, S.L. Li, T.X. Li, J.Y. Wu, S. He, Heat transfer design in adsorption refrigeration systems for efficient use of low-grade thermal energy, Energy, Volume 36, Issue 9, September 2011, pp. 5425-5439.

S.C.Kaushik, Solar refrigeration and Space Conditioning, Divyajyoti Prakashan publishers, 1989.

Tashtoush G., M, Jaradat Mohd., and Al-Bader S. 'Thermal Design of Parabolic Solar Concentrator Adsorption Refrigeration System' Applied Solar Energy, Vol. 46, No. 3, pp. 203-212. 2010.

Wei-Dong Wu, Hua Zhang, Chuan-lin Men, Performance of a modified zeolite 13X-water adsorptive cooling module powered by exhaust waste heat, International Journal of Thermal Sciences, Volume 50, Issue 10, October 2011, pp. 2042-2049.

Wei-Dong Wu, Hua Zhang, Chuan-lin Men, Performance of a modified zeolite 13X-water adsorptive cooling module powered by exhaust waste heat, International Journal of Thermal Sciences, Volume 50, Issue 10, October 2011, pp. 2042-2049.

X.Q. Zhai, R.Z. Wang, Experimental investigation and performance analysis on a solar adsorption cooling system with/without heat storage, Applied Energy, Volume 87, Issue 3, March 2010, pp. 824-835.

X.Q. Zhai, R.Z. Wang, J.Y. Wu, Y.J. Dai, Q. Ma, Design and performance of a solar-powered air-conditioning system in a green building, Applied Energy, Volume 85, Issue 5, May 2008, pp. 297-311.

Y.L. Liu, R.Z. Wang, Z.Z. Xia, Experimental performance of a silica gel–water adsorption chiller, Applied Thermal Engineering, Volume 25, Issues 2–3, February 2005, pp. 359-375.

Y.Z. Lu, R.Z. Wang, S. Jianzhou, Y.X. Xu, J.Y. Wu, Practical experiments on an adsorption air conditioner powered by exhausted heat from a diesel locomotive, Applied Thermal Engineering, Volume 24, Issue 7, May 2004, pp. 1051-1059.

Z Tamainot-Telto, R.E Critoph, Adsorption refrigerator using monolithic carbon-ammonia pair, International Journal of Refrigeration, Volume 20, Issue 2, March 1997, pp. 146-155.

Z.F. Li and Sumathy, A solar-powered ice-maker with the solid adsorption pair of activated carbon and methanol, Int. J. Energy Research. 23, 517 (1997).

Z.V. Baiju, Theoretical and Experimental Investigations on Solar Vapour Adsorption Refrigeration System, Ph.D. Thesis, NIT Calicut, Kerala (2013).

CHAPTER 10

Evaporative and Desiccant Based Air Conditioning Systems

10.1 INTRODUCTION

The main task of an air conditioning system is to maintain the air in a space at a certain temperature and humidity level. There are two portions (sensible and latent heat) of the cooling load. Several sources contribute to the sensible heat portion of the cooling load like the heat conduction through the building envelope, heat released by the lighting systems, the energy which is carried into the building with infiltration of air and heat released by people. There should also be controlled exchange of the air inside the space with fresh air from outside. This air has to be cooled to the room air state.

The second portion of the cooling load is called the latent heat load. It is due to moisture either carried into the building by air infiltration or released by people or vegetables. The air required for ventilation may also need to be cooled and dehumidified before entering the building.

In conventional air conditioning system the latent heat load is smaller than the sensible heat load. Nevertheless, it is the cause of the poor overall coefficient of performance in a conventional cooling system. In these systems the air to be conditioned is blown through a heat exchanger. For air conditioning, water (typically of temperature 5°C) is pumped through this heat exchanger piping. This way air is cooled down to its dew point temperature. Part of the water vapour in the air condenses. This condensation process is continued until, the air has reached the desired humidity ratio. Finally, this dehumidified cold air has to be reheated to the desired supply air temperature. The energy to reheat the air is generally available from waste heat, but the primary sub-cooling requires a lot more energy as compared to a thermodynamically optimal process.

In hot and dry climate, evaporative cooling is generally used in buildings for thermal comfort conditions. The rate of cooling is directly related to the rate of water evaporation, which is very effective for hot and dry climate. However, for hot and humid climate, there is low water evaporation and hence this kind of cooling is not very effective.

For this type of situation, air conditioning is to be based on the use of adsorbents/ desiccants for adsorbing moisture from the air and then evaporative cooling of air. For places located near sea shore where temperature and humidity are high during summer and it rains during winter, an air conditioning system based on dehumidification followed by adiabatic evaporative cooling is a viable proposition. Thus, in general, for thermal comfort, air conditioning comprises proper control of temperature, humidity, air movement and purity of air. An adsorption cooling system is suitable when a large portion of cooling load is used for latent heat removal.

The desiccant cooling process is basically similar to that associated with the absorption system. The main difference is that the refrigerant is adsorbed onto the surface of the adsorbing material rather than being absorbed into the chemical solution. The refrigerant is then driven out by heating which is the role of solar energy in this system. The liberated refrigerant vapour is then condensed and throttled to produce the cooling effect. The refrigerant is then readsorbed onto the carrier material to have a closed cycle. Solar powered adsorption cooling system is an attractive alternative to conventional air conditioning for hot and humid climates.

When the outside design conditions and the desired inside conditions are specified, the total sensible heat load and total moisture load will determine the condition and quantity of the entering air. When the dew point temperature for the entering air is low, it may be necessary to cool the cooling water even by refrigeration. After the moisture has been removed from the air, it may be necessary to cool it further, possibly using evaporative cooling which is essentially an adiabatic process, with no heat flow. As air is passed through water spray, some of the water may evaporate increasing the latent heat content of the air while lowering the sensible heat in the air. The air leaving an evaporative cooler has a lower dry bulb temperature and a higher relative humidity than when it entered. The process can be approximated by a constant wet bulb temperature line on a psychrometric chart.

10.2 AIR PSYCHROMETRICS

In a conventional cooling unit, the cooling coil is required both to remove moisture and to lower the temperature of the air to be circulated in the space to be conditioned. In some cases, the air is over-cooled to achieve proper moisture control and a reheat coil is used to achieve comfortable dry bulb temperature.

Figure 10.1 depicts the basic process of adsorption air conditioning on a psychrometric chart. In the conventional method of dehumidification (Path A) the entering air initially in state 'a' is sensibly cooled to a temperature below its dew point in order to achieve the desired degree of moisture and temperature level. This dehumidified air at lower temperature, is sensibly reheated to a desired comfort state 'd'. In desiccant type of solar dehumidification (Path B), a desiccant has to be used to lower the process of air humidity. The air being adsorbed gains some heat in the dehumidifier due to the liberation of heat of adsorption. If the adsorption process is kept adiabatic, the enthalpy in state 'b' will be larger than that in state 'a'. In case of cross-cooled dehumidifier, the enthalpy of air in state 'b' is lower than that in state 'a' and the process air in state 'b' is moderately warm and has low humidity. It is then sensibly cooled in heat exchanger to state 'c' to reduce its enthalpy. In the final state, it is adiabatically humidified in an evaporative cooler to state 'd' which is now the conditioned air.

The cooling capacity of the system is defined as the difference between enthalpy per unit mass of air leaving/entering the conditioned space. Thus, following Sharma (1984);

Cooling capacity = Enthalpy in state 'a' – Enthalpy in state 'd'

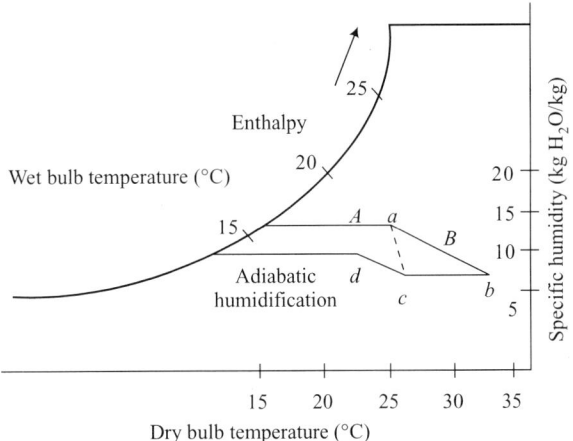

Fig. 10.1 Conventional Dehumidification and Desiccant Dehumidification Cycles on Psychrometric Chart, Sharma (1984)

$$q = (h_a - h_d) \tag{10.1}$$

It may be mentioned here that process c-d is an adiabatic humidification process in the evaporative cooler and hence it is a constant enthalpy process, i.e.

Enthalpy at state 'c', (h_c) = Enthalpy at state 'd', (h_d)

This yield,

$$\text{Cooling capacity} = \text{Enthalpy at 'a'} - \text{Enthalpy at 'c'} \tag{10.2}$$

The process a-b is the dehumidification process and b-c is the sensible cooling process. In both these processes, cooling occurs. The latent heat is removed in the dehumidifier and the sensible heat is removed in the heat exchanger. Therefore,

Specific cooling capacity = Change in enthalpy in the dehumidifier + Change in enthalpy in the heat exchanger

The total cooling capacity of the device is the product of the mass flow rate of the process air and above cooling capacity per unit mass of air.

The thermal coefficient of performance (COP) is defined as

$$\text{Thermal COP} = (\text{Total cooling capacity/Total heating supplied}) \tag{10.3}$$

As solar processes are time dependent and intermittent in operation, the temperature and energy transfer rates vary with time. Therefore, COP based on long term performance is taken as integrated value over a fixed period of time. Generally, the latent heat load is much smaller than the sensible heat load and it is the main cause of the poor COP in conventional cooling systems.

In the basic desiccant cycle a typical embodiment of air to be conditioned is adiabatically dried with a resultant increase in temperature as the latent heat of its moisture is converted to sensible heat. The airstream is then cooled nearly to ambient in its dry condition to reduce its enthalpy and passed through a humidifier which cools it by conversion of sensible to latent heat at the reduced enthalpy.

In the solar dehumidification cycle, outside air is first dehumidified and mixed with the return air to achieve the desired moisture level. The cooling coil is used to do only sensible cooling to lower the dry bulb temperature of the mixture. In the dehumidification process, there occurs cyclic adsorption and desorption modes each with a fixed period. The adsorbent/desiccant bed is usually hot at the end of desorption and is not ready for subsequent adsorption and hence it requires pre-cooling. Similarly, the desiccant bed at the end of adsorption is cool and needs preheating before the desorption mode. The process of cooling the desorbed bed and heating the adsorbed bed is known as purging process.

It is also desirable to describe the component-evaporative cooler, which is also used in desiccant cooling systems. Although the principles of evaporative cooling are well known and there are many well tested evaporative coolers commercially available as described by Watt (1963). A brief description is given here.

10.3 EVAPORATIVE COOLING

An evaporative cooling system is an air conditioning system consisting of evaporative coolers, fan(s), filters, dampers, controls, pumps and others. An evaporative cooler could be a stand-alone cooler or installed in an air system as a component. There are three types of evaporative coolers known in the literature (Wang et al., 2000):

(a) Direct evaporative coolers, (b) Indirect evaporative coolers, and (c) Direct-indirect evaporative coolers

10.3.1 Direct Evaporative Cooler

In the direct evaporative systems water evaporates directly into the supply airstream and then cooling it, but it also increases its humidity. These are suitable only in hot and dry climates. However, if the air intake is of high humidity, the supply air will be close to saturation condition and the evaporation effect may be less due to the outside moisture level, the air can take up.

In a direct evaporative cooler, the airstream to be cooled directly comes in contact with the water spray or wetted medium as shown in Fig. 10.2(a). Evaporative pads are made of wooden fibres with necessary treatment. The saturation effectiveness ε_{sat}, that assesses the performance of a direct evaporative cooler is defined by

$$\varepsilon_{sat} = (T_{ai} - T_{ae})/(T_{ai} - T_{ae}^*) \qquad (10.4)$$

where T, T^* are the dry bulb temperature and thermodynamic wet bulb temperature of airstream. Subscript 'ai' indicates the entering air and 'ae' indicates the exiting air. Normally, ε_{sat} ranges between 0.65 and 0.85 at a water-air mass flow ratio of 0.1 to 0.40.

10.3.2 Indirect Evaporative Coolers

In an indirect evaporative cooler, the process airstream to be cooled is separated from a wetted surface by a flat plate or tube wall as shown in Fig. 10.2 (b). A process airstream flows over the wetted surface so that liquid water is evaporated inside the tube and extracts heat from the process-

Evaporative and Desiccant Based Air Conditioning Systems 369

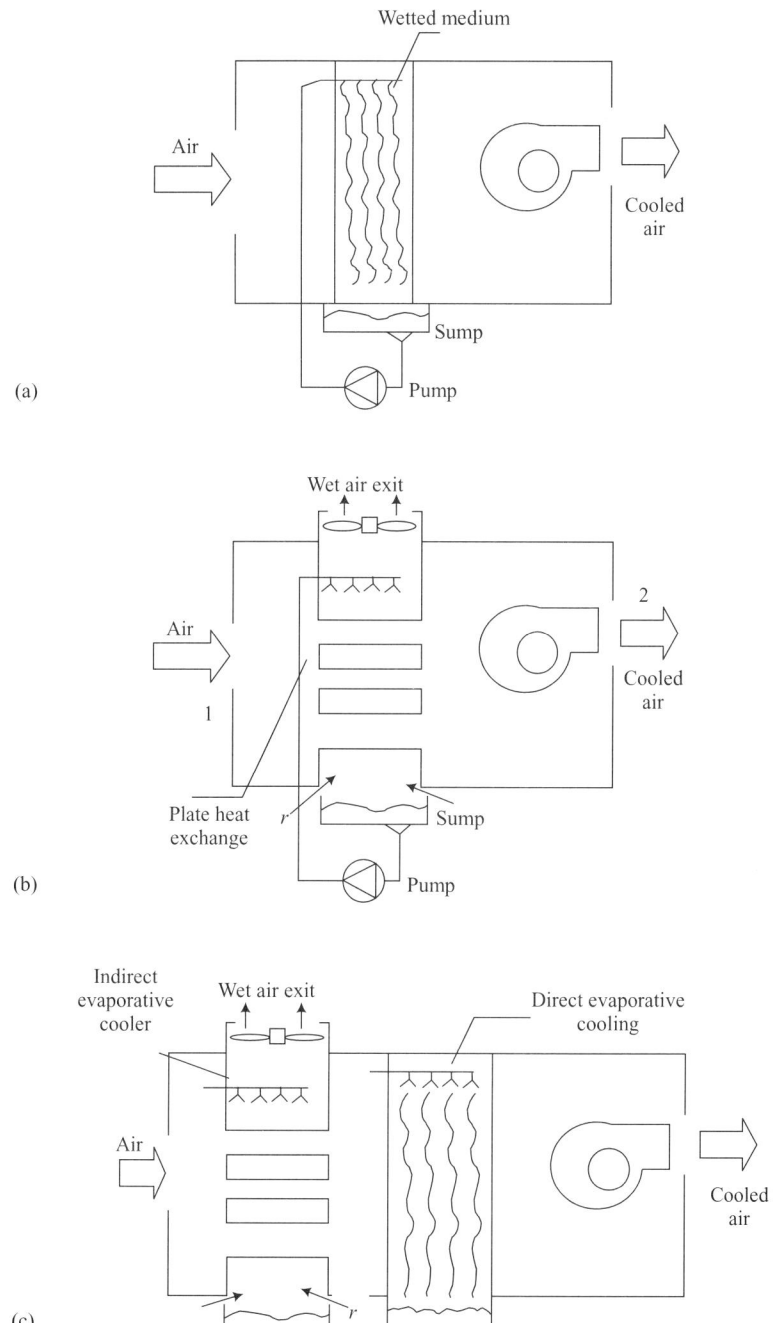

Fig. 10.2 Types of Evaporative Coolers: (a) Direct, (b) Indirect and, (c) Indirect-direct (Wang *et al.*, 2000)

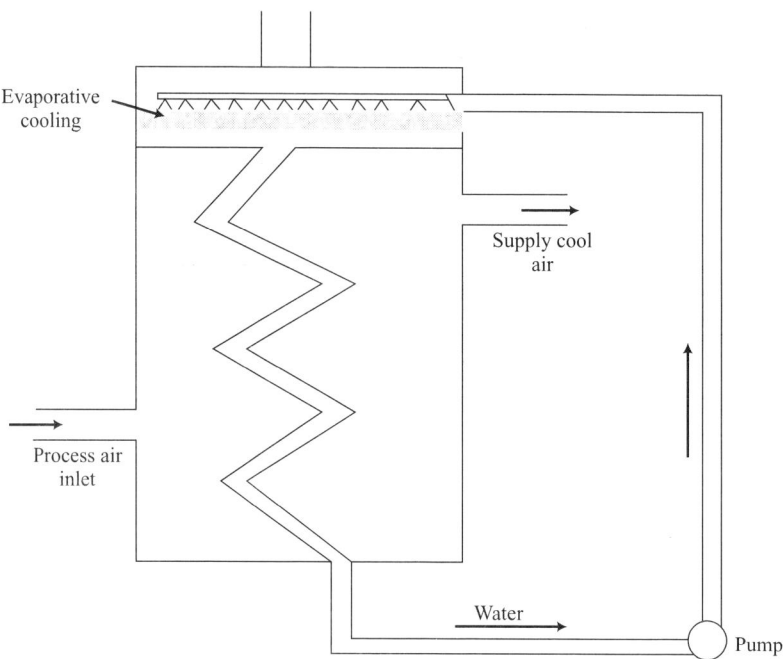

Fig. 10.2(d) Regenerative Evaporative Cooling

airstream through the flat plate or tube wall. The airstream is not in direct contact with the wetted surface. The main part of indirect evaporative coolers is a plate heat exchanger with horizontal passages for cooled air and vertical passages for wet air and water. Similar to a direct evaporative cooler, there are fan(s), water sprays, circulating pump, air intake, dampers, controls, etc., as other parts of the system (Wang *et al.*, 2000).

In an indirect evaporative cooling process air humidity ratio remains constant.

The performance of an indirect evaporative cooler can be assessed by its performance factor e_{in}, defined as:

$$e_{in} = (T_{ca.i} - T_{ca.e})/(T_{ca.i} - T_{s.a}) \tag{10.5}$$

where $T_{ca.i}$, $T_{ca.e}$ are the temperatures of cooled air entering and leaving the indirect evaporative cooler, and $T_{s.a}$ is the temperature of the saturated air film on the wet air side which is little higher than the wet bulb temperature of the entering air.

10.3.3 Direct-Indirect Evaporative Cooler

It must be mentioned that in a direct evaporative cooling, air passes over water through wet spray and is adiabatically cooled and humidified. Its dry bulb temperature approaches its wet bulb temperature. The cooled air is then available to provide comfortable conditions. On the other hand, in an indirect evaporative cooling system, the evaporatively cooled air cools the process air through the heat exchanger rather than being supplied directly. This gives the advantage that water vapour is not added to the air through air-to-air counter flow heat exchangers in which one side of the heat exchange surface is kept wet, allowing evaporative cooling to take place inside the heat exchanger itself rather than in a separate unit.

A mixed mode direct-indirect evaporative cooler is in fact indirect-direct two-stage evaporative cooler as shown in Fig. 10.2(c), in which at the first stage indirect evaporative cooler is used in series with a second-stage direct evaporative cooler for the purpose of increasing the evaporation effect.

The saturation effectiveness ε_{sat} and performance factor are both closely related to the air face velocity flowing through the air passage. For a direct evaporative cooler, face velocity is usually less than 300 cm/s to reduce drift carry over. For an indirect evaporative cooler, face velocity (v_s) is usually between 300 and 500 cm/s.

Nowadays a novel concept of regenerative cooling is also being proposed, where in air is cooled indirectly by water spray and the water is evaporatively cooled by the incoming air. This principle is used in wet surface heat exchangers for cooling and dehumidification coils, as shown in Fig. 10.2(d).

10.4 DESICCANT MATERIALS

Desiccants have been used as adsorbents to provide dry air for variety of industrial process and desiccant dehumidification for use in air conditioning systems is an extension of this application. Mostly the adsorbents used are either solid or liquid desiccants. Although the basic process of cooling is the same but the system design and operation as well as possible improvements are quite different. Desiccant cooling system has the potential to be of low cost since the sorbent materials or sorbents are less expensive. As the capacity of any sorbent is limited, the material will have to be regenerated, so that a cycle of sorption-desorption is formed. Desorption will require some form of energy input which may be done even by solar energy. Any of the following materials can be used as sorbent.

(a) Solid adsorbent: Silica gel, alumina, molecular sieves, activated bauxite, zeolite, etc.
(b) Liquid adsorbent: Aquaeous solution of $LiCl$, $CaCl_2$, $LiBr$ and their mixtures, TEG solution, etc.

Solid adsorbents retain condensed liquid molecules on their surface and do not undergo any physical or chemical change during adsorption process but these materials are poor heat conductors so the amount of refrigerant adsorbed deteriorates with time. Liquid adsorbents/desiccants are the solutions of certain substances which are diluted by the absorbed liquid during the sorption process. A liquid desiccant system uses a hygroscopic salt liquid so that any air passing through a spray of the desiccant leaves at a low relative humidity. The spray temperature must be kept high for regeneration so that the dew point of the air leaving is higher than the air entering. Liquid desiccants require lower regeneration temperature (~90°C) feasible with flat plate solar collectors. Solid desiccants like silica gel, molecular sieves, alumina or hygroscopic salts with submicroscopic pores which usually provide high degree of dehumidification and have the capacity to adsorb water vapour on account of their porous structure and large surface areas (100m^2 per gm of material). A comparative study of different solid adsorbents has been made by Lavan *et al.* (1978). Figure 10.3 (a) and (b) show the adsorption characteristics in terms of percentage of water adsorbed as a function of relative humidity and the regeneration temperature. It can be seen from these figures that:

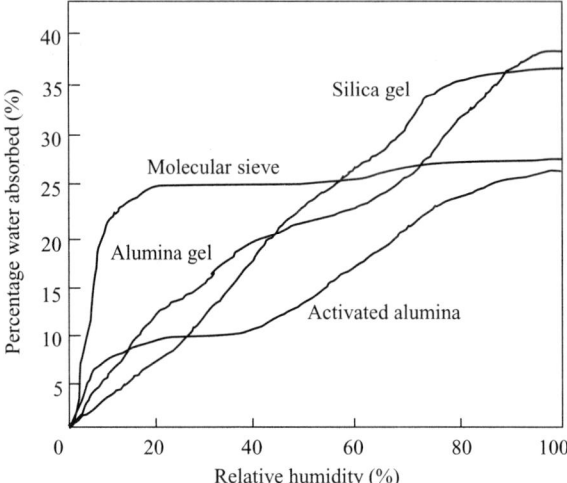

Fig. 10.3(a) Adsorption Percentage vs. Relative Humidity

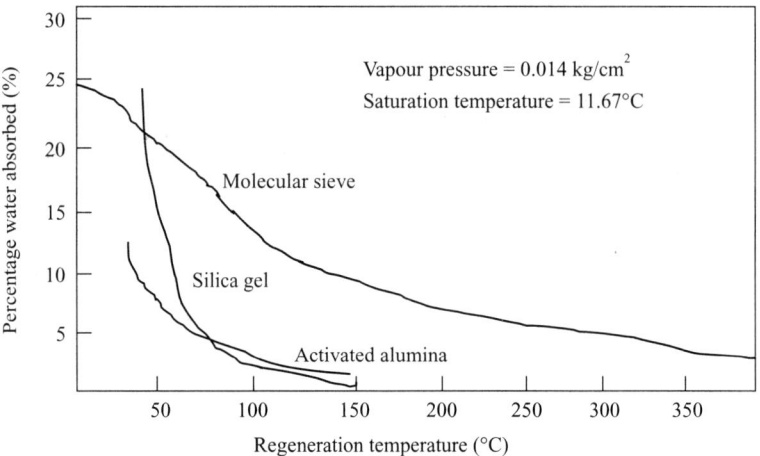

Fig. 10.3(b) Adsorption Percentage vs. Regeneration Temperature

(i) Molecular sieves have better adsorption rate at lower relative humidities and at higher regeneration temperatures.

(ii) Silica gel can be desorbed at relative lower temperature and their equilibrium water capacity is small. Furthermore, silica gel has good adsorption capacity only at lower temperatures and the adsorption capacity decreases as the bed temperature decreases. Therefore, it is necessary to keep the adsorbent bed at low temperature during dehumidification by removing the heat of adsorption.

Thus, from the point of view of solar operation, systems using molecular sieves require high performance solar collectors/concentrators while system using silica gel can be operated with simple flat plate solar collectors.

Liquid desiccants utilize solutions of LiCl, LiBr, $CaCl_2$ and various glycols with water and in general have less adsorption capability than the solid desiccants (Sharma, 1984). The relative merits are as follows:

(i) Heat and mass transfer coefficients from liquid surface are higher.
(ii) The power requirements for the blower/fan, etc., are lower since air is circulated through a loosely packed spray tower instead of thick solid granular bed.
(iii) The dew point depression is high thereby producing cooler and dried air.
(iv) The absorption-desorption process is a continuous cycle.
(v) Air sterilization is possible as desired for special applications in hospital air conditioning systems.
(vi) Energy is stored as chemical energy in the form of concentrated solution rather than thermal energy. This allows for greater energy storage option and reduces dependence on auxiliary source.
(vii) Liquid desiccants can be regenerated in open flow thin film solar collectors which are less expensive and of simple design.
(viii) Liquid desiccants can also be used for heating.

The desirable characteristics of the liquid absorbents suitable for air dehumidification should be as follows:

(a) The absorbent should be highly soluble in water.
(b) The solution should be capable of maintaining low pressure of water vapours, e.g., a 45% solution of $CaCl_2$ has an equilibrium water vapour pressure of 12 mm Hg at 35°C.
(c) The desiccant solution should not crystallize at temperatures ranging from 5°C to 8°C. (below the temperature range used in the operating cycle). LiBr, LiCl and tryethylene glycol satisfy requirement of producing sufficiently low vapour pressure without crystallization in the operation.
(d) The absorbent should be inexpensive, chemically stable, odourless, non-corrosive and non-flammable, $CaCl_2$ is quite cheap, NaOH is fairly corrosive and glycols are better absorbents than most of the salts. LiBr is a very expensive absorbent.
(e) As desired for open cycle systems, it is necessary that the partial vapour pressure of the absorbent itself should be negligible and thereby no loss of absorbent due to evaporation during desorption process. Glycols have the problems of vapour carry over but some salts are quite satisfactory absorbents from this point of view.

10.5 DEVELOPMENTS IN SOLAR DESICCANT COOLING SYSTEMS

Solar desiccant cooling is attractive due to lower regeneration temperature levels required for various adsorbents. The dehumidifier removes enough moisture to meet the latent heat portion of the cooling load. Moisture is adsorbed on a drying agent, which can be regenerated using solar heat.

The concept of solar desiccant cooling is well established. Most of the research and development work for application of the adsorption process to solar cooling application has been with the use of solid/liquid desiccants/adsorbents. A brief discussion of open and closed cycle options is given here.

10.5.1 Solid Desiccant Open Cycle Cooling Systems

A solid desiccant cooling system consists of a dehumidifier, an evaporative cooler, a number of heat exchangers and blowers. The major component in the desiccant cooling system is the desiccant dehumidifying bed, which removes moisture by adsorption from the air being processed in the system. Air from the space to be conditioned or from outside is passed through the dehumidifier, so that the emerging airstream is hot and dry. It is then cooled in a heat exchanger (recouperator) followed by evaporative cooling in an air washer. Since evaporative cooling is a constant enthalpy process, the evaporation of water takes place by the sensible cooling of the airstream. This cooled air is then sent to the space to be air conditioned. To regenerate the desiccant bed, ambient air is passed through the heat exchanger where it is heated by the dehumidified dry and hot process air. Heat energy input is also made available at the second heat exchanger, bringing the temperature to a level, at which the desiccant bed is well regenerated. The early work on solid desiccant system was undertaken by Dunkle (1965) at Commonwealth Scientific Industrial Research Organisation (CSIRO) in Australia and also by Lunde (1976) in USA. There is currently considerable interest in examining the potential applications of the solid desiccant systems. Since the process of adsorption is exothermic, the temperature of the adsorbent bed increases in an adiabatic adsorption process which reduces the adsorption capability of the bed. It is desirable to cool the adsorption bed during adsorption process. The loss of adsorption capacity due to heating of the bed (in adiabatic process) can be reduced by using material which retains its capacity at the higher temperature. Molecular sieves are such materials. Another possible solution to avoid the decrease in bed adsorption and desorption capacity due to excessive temperature swing is to use interstage cooling and interstate heat exchangers. Roy and Gidaspaw (1972, 1974) proposed cross cooling and cross flow regenerator and Worek and Lavan (1980, 1982) later developed and tested the unit of a cross flow dehumidifier as shown in Fig. 10.4. The system consists of two fixed bed cross cooled silica gel dehumidifiers (one for providing cooling and the other for regeneration) and rectangular channels in cross flow arrangement. The process stream channels are lined with silica gel sheets as shown in Fig. 10.4. In the dehumidification cycle, air from the room flows through the process channels and humidified ambient air is passed through the cooling channels. This lowers the desiccant temperature and, in turn, improves the sorption process. In the regeneration cycle, hot air from the solar collectors is passed through the process channels. The advantage of the cross-cooled concept is the ability to provide adequate cooling with relatively low regeneration temperatures. The testing results indicated that the prototype dehumidifier can provide effective cooling with a reasonable COP and low pressure drop using relatively low regeneration temperatures. Hence, the cooling unit can be operated during

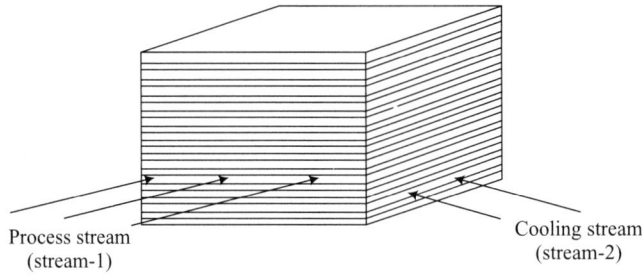

Fig. 10.4 Schematic of Cross Flow Desiccant Dehumidifier (Lavan, 1981)

a longer period of the day along with higher solar system efficiency than solar systems requiring higher activation temperatures. There is little degradation in system performance with the increase of ambient humidity and hence this dehumidifier is suitable for operation at wide range of geographic locations. This cross cooled system can be operated in a recirculating or in ventilating mode as desired.

In the former case, air from the room is passed through the process channels lined with adsorbent sheets. The cross cooling is accomplished by humidified ambient air. While in the ventilation mode, ambient air is passed through process channels and humidified room air is passed through the cross cooling channels. The process stream leaving the dehumidifier is sensibly cooled and adiabatically humidified in an evaporative cooler. This brings the process airstream to the desired level of temperature and humidity. In the desorbing cycle, the saturated desiccant bed is regenerated by hot air available from the flat plate solar collectors.

There is not much practical data information available on solid desiccant cooling systems which utilise solar energy to regenerate an adsorbent, with additional sensible cooling of the air accomplished by evaporative cooling. Solar Munters Environmental Control (MEC) system has been developed by Institute of Gas Technology, Chicago, Illinois (USA). The system is quite similar to that of Dunkle (1965) at CSIRO (Australia). Both the systems use two open process airstreams which are thermally coupled through rotary regenerators. The building airstream is dried in a desiccant bed, cooled in a regenerative heat exchanger and refrigerated by evaporative cooling. The air drier is reactivated by an outside airstream which is solar heated to supply the desorption energy. The solar MEC device requires a gas fired boost to achieve the required level of desorption. In both the cases, the desiccant bed operates in an adiabatic process. Experimental units are still under research and development stage and the only one available is that reported by Neilson *et al.* (1978).

Neilson *et al.* (1978) have stimulated seasonal solar operation of a Munters Environmental Control (MEC) device, in the Miami climate. The process is shown in Fig. 10.5(a) and is traced on the psychrometric chart of Fig. 10.5(b) where the state points are numbered corresponding to the various processes. Ambient air is dried and heated by dehumidifier, from 1 to 2, regeneratively cooled by exhaust air from 2 to 3, evaporatively cooled from 3 to 4, and introduced into the building. Exhaust air at state 5 moves in the countercurrent direction and is evaporatively cooled to

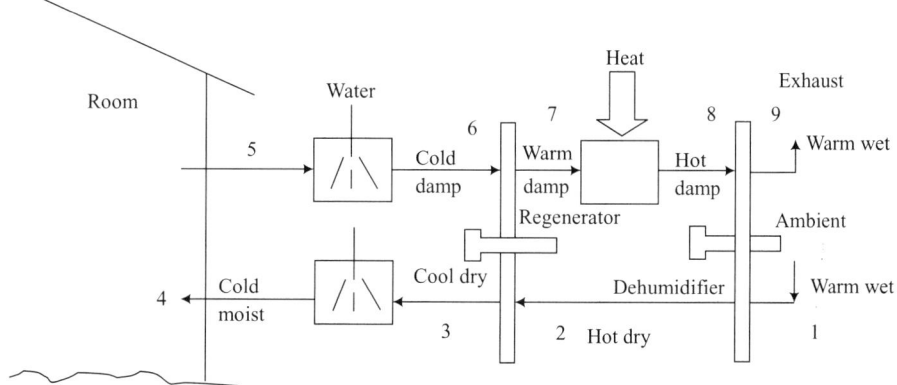

Fig. 10.5(a)   Schematic of Solar MEC System (Neilson *et al.*, 1976)

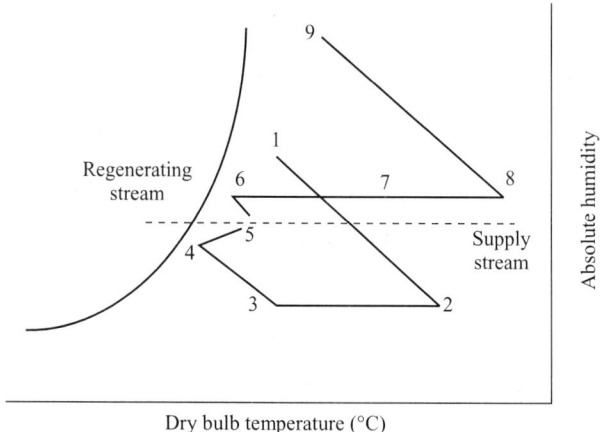

Fig. 10.5(b) Solar MEC Cycle on Psychrometric Chart (Neilson, 1976)

6, heated to 7 by the energy removed from the supply air in the regenerator, heated by solar or other source to 8 and then passed through the dehumidifier (desiccant) where it regenerates the desiccant. The temperature of the regeneration stream (from the solar heating system) was below 65°C for approximately 60% of the time and was about 80°C only for 6% of the time regardless of the collector area. This suggests that flat plate collector may be used to operate a cooling cycle of this type.

Solid desiccant cooling systems suffer from the high pressure drops and poor heat exchange between the spent and regenerated desiccant streams. Using air heating collectors, past experience indicated lower COP and hence efforts are being made to improve system COP and reduce pressure drops in current designs.

10.5.2 Liquid Desiccant Solar Cooling Systems

A conventional liquid desiccant solar cooling system is based on absorption dehumidification process and consists of an absorption tower, desiccant regenerator and evaporative cooler as major components. Dehumidification process takes place in the absorption tower. The solution which absorbs moisture from air in the absorption tower becomes diluted and it should be regenerated for recycling in the desiccant regenerator. As early as 1955, Lof (1985) developed a liquid desiccant system using triethylene glycol (TEG) solution as shown in Fig. 10.6. Lof (1974) studied solar operation of dehumidification system in which the drying agent liquid triethylene glycol is sprayed into an absorber where it picks up moisture from the building air. It is then pumped through the sensible heat exchanger to a stripping column where it is sprayed into a stream of solar heated air (in the solar operation). The high temperature air removes water from the glycol, which then returns to the heat exchanger absorber. Heat exchangers are provided to recover sensible heat, maximize the temperature in the stripper, and minimize the temperature in the absorber. Eliminators are used to remove glycol spray from the airstreams. This type of cycle, operated by hot airstream, is marketed commercially and used in hospitals and other large installations. Solar operation has also been studied and a variation using the collectors as the stripper is also described by Collier (1979).

Evaporative and Desiccant Based Air Conditioning Systems 377

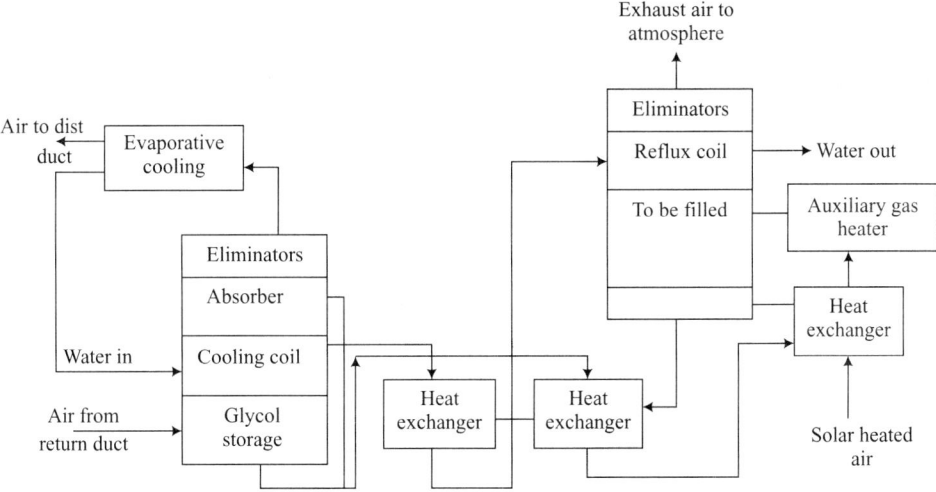

Fig. 10.6 Schematic of an Open Cycle Air Conditioning System (using TEG) After (Lof, 1985)

Sheridan (1961) proposed a refrigeration/air conditioning system for industrial cooling application (see Fig. 10.7). It consists of two spray towers—one for dehumidification of the room air and the other for regeneration of the absorbent solution. The weak lithium chloride solution from the dehumidifier is pumped through a heat exchanger from which it gains heat from strong solution returning from the regenerator.

It then proceeds to be further heated by the hot air from the regenerator before being sprayed into the regeneration tower. Here it contacts atmospheric air which has been heated by hot water from the solar collector. The solution is concentrated and then returned to the humidifier being

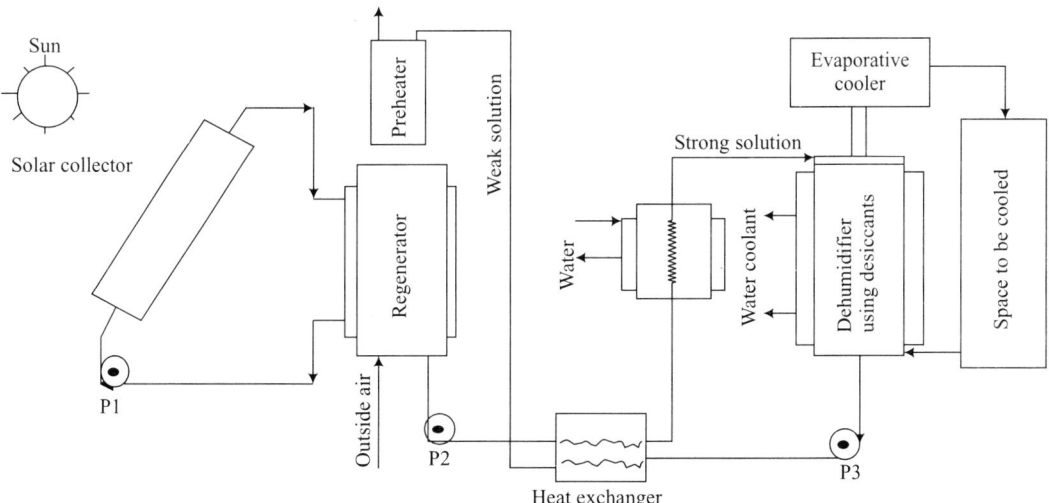

Fig. 10.7 Schematic of Absorption-dehumidification Solar Air Conditioning System (Sheridan, 1961)

cooled on the way by the weak solution and a separate water cooler, on being sprayed into the dehumidifying tower, the solution extracts moisture from the room air because of its lower vapour pressure. The room air after being dehumidified is passed to an evaporative cooler. The air in the regenerator gain moisture and then gives up heat to incoming fluid.

Lof (1984) and Peng and Hewell (1981) designed and developed advanced liquid desiccant cooling system using TEG as a desiccant solution and operated the system in different modes. The cooling water is generated internally by evaporative cooling by processing the dry any cold air over the recirculating water.

Evaporative cooler used in liquid desiccant system is of spray type consisting of small nozzles to spray water which is collected at the bottom of the tower and recirculated. Contact between the flowing airstream and the sprayed water causes the heat and mass transfer between air and water results in cooling and humidification of the air. The other components in both the flow modes of operation are identical. The absorber is an ordinary finned tube heat exchanger with a falling film distributor on the top. The desiccant solution flows down the finned tube surface as a falling film while the air flows in the counter flow direction through the flow channels between the fins. The solution regenerator involves desorption of water vapour from the weak solution due to contact with ambient air.

Since the cooling produced by a liquid desiccant system is directly proportional to the degree of regeneration of the weak solution (i.e., the amount of water evaporated from the regenerator), it is necessary to select the most efficient design of the regenerator. Most commonly used regenerators are either open surface falling film regenerator or closed circulation regenerator. In terms of performance, air-circulation type regenerator has much higher rate of regeneration and needs smaller regenerator length for the same process conditions. This may be attributed to higher mass transfer coefficient and lower heat loss coefficient in air circulation. However, this regenerator involves higher capital cost and simple design but requires high maintenance due to dirt and rains. Nevertheless, open cycle liquid desiccant cooling system are promising and needs further development work in this direction.

10.5.3 Hybrid Air Conditioning

An efficient way of achieving air conditioning is possible, if we remove the latent heat load by desiccants and the sensible load by conventional vapour compression machine. This idea has been proposed and explored by many researchers by combining the VCR/VAB systems with either solid/liquid desiccant units. A simple psychrometric chart for VCR cycle, desiccant cycle and hybrid cycle air conditioning is shown in the Fig. 10.8(a-c). Hybrid air conditioning depends neither on an evaporative cooling nor moisture condensation for latent heat removal. It uses a desiccant system to control the moisture and conventional VCR cycle to control the temperature within the comfortable range. This improves the overall performance of the air conditioning system because both the solid/liquid desiccants and VCR cycles are then operating at their respective improved performance and there is significant energy savings as compared to the conventional VCR air conditioning alone as reported by Dhar et al. (2001) and Kumar (1990).

Thus, hybrid air conditioning offers many advantages like higher COP, higher energy savings, reduced condenser area, higher dehumidifying efficiency and nominal increase in hardware and blower requirements over conventional VCR systems. In some cases, for the same supplied air, the

Evaporative and Desiccant Based Air Conditioning Systems 379

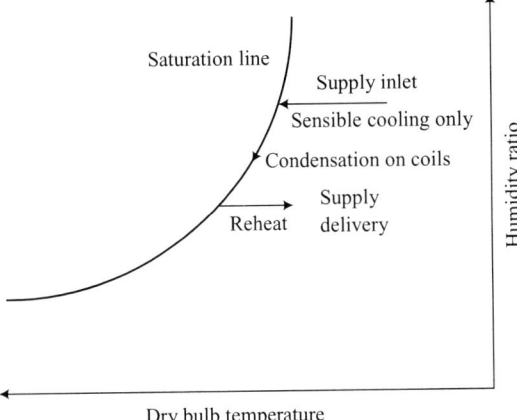

Fig. 10.8(a) Conventional V-C Air Conditioning

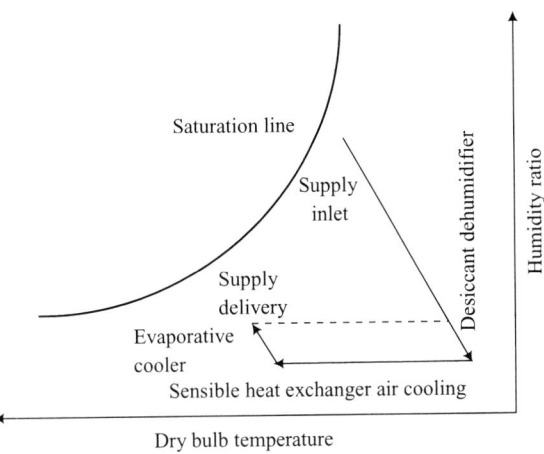

Fig. 10.8(b) Desiccant Air Conditioning

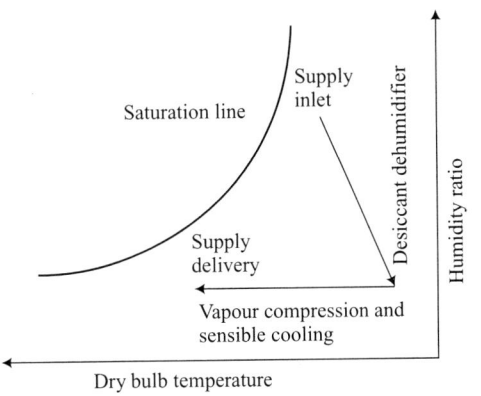

Fig. 10.8(c) Hybrid Air Conditioning

hybrid system may have little lower COP than either of the VCR cycle and desiccant cycles, but can save more energy upto 40-50%, over conventional VCR system at higher latent heat load conditions. This can also downsize the VCR system components within the hybrid system. Thus, there is great potential of retrofitting and integrating the desiccant cooling system with conventional vapour compression system which can reduce the electric power consumption significantly and can operate satisfactorily for the year round air conditioning. For industrial applications, where maintenance of moderate to low temperature with low humidities are needed, the solar desiccant dehumidification system can be coupled with compression or absorption refrigeration system. Such a combination is more efficient and less expensive.

10.6 CONCLUSION

Both evaporative and desiccant cooling systems are described and analyzed for air conditioning application. Evaporative cooling is suitable for hot and dry climates, while desiccants based air conditioning is good for hot and humid climates. Types of evaporative coolers are reviewed and their performance parameters are given. The cooling capacity and COP for air conditioning systems are defined with the help of psychrometric chart. Both solid and liquid desiccants based air conditioning and dehumidification processes are explained. Desirable characteristics and merits/demerits of solid/liquid desiccant materials are mentioned. Developments in solid and liquid desiccants cooling systems are reported as available in the literature.

A solar driven desiccant cooling system involves alternative adsorption and desorption processes and consists of a desiccant bed dehumidifier, an evaporative cooler and number of air-air or air-water heat exchangers and blowers. Both the ventilation and recirculation modes of operation are possible and effective in providing cooling of the air in a building. The major advantages of these systems are easy adaptability of solar energy input, high efficiency, low manufacturing cost, low regeneration temperature and no air leakage or corrosion and no environmental degradation. Solid desiccants cooling systems are more suitable for hot and humid climate, while liquid desiccant cooling systems have the advantage of lower regeneration temperature. However, no cost-effective design has emerged so far and more R&D work is still needed to (i) reduce the auxiliary power requirements by developing low pressure bed designs and increased air system efficiencies; (ii) reduce the cost of desiccant materials; and (iii) and develop cost effective heat exchangers, for large cooling capacity units for air conditioning. Hybrid air conditioning systems provide efficient air conditioning with higher COP and significant energy savings, and found to be suitable for all climates with low cost.

REFERENCES

Collier R.K. The analysis and simulation of an open cycle absorption refrigeration system, Los Alamos Scientific Laboratory, report No. LA-UR-78-1729 and *ibid Solar Energy*, 23, pp. 357-366, (1979).

Tchernev, D.I. Solar air conditioning and refrigeration systems utilizing zeolites, in Proceedings of Meetings of Commissions E1-E2 (International Institute of Refrigeration, Jerusalem), pp. 209-215, (1982).

Techernev, D.I. Solar applications of natural zeolites, Proceedings, 2nd workshop on use of Solar Energy for Cooling of Buildings, California, (USA), p. 307, (1976).

Techernev, D.I. Solar refrigeration utilizing zeolites, Proceedings inter-society, Energy Conversion conference, 2, p. 119, (1978).

Duffie, J. A., Lof G.O.G, and Chunk. A study of Solar Air Conditioner, *Mech. Engg.* 85, p. 31, (1963).

Dunkle, R.B, A method of solar airconditioning, Transactions, *Mechanical and Chemical Engg.* Australia, 1, p. 73-78, (1965).

http://www.itiomar.it/pubblica/dispense/MECHANICAL%20ENGINEERING%20HANDBOOK/Ch09.pdf

Lavan, Z. and Gidspaw, Development of a solar desiccant dehumidifier", Proceeding 3rd Annual Solar Heating and Cooling, R & D melting. Chicago, Illinois (U.S.A.), (1978).

Lof G.O.G., and Tybont R.A. The design of cost optimized system for residential heating and cooling by solar energy, *Solar Energy*, 16, No. 9, (1974).

Lof G.O.G., House heating and cooling with solar energy, published in solar energy notes, University of Wisconsin, pp. 33-36, (1984).

Lunde, P. Solar desiccant airconditioning system using silica gel, Proc. 2nd Workshop on use of Solar energy for cooling of buildings." California (U.S.A.), p. 280, (1976).

Meunier.F; *et al.*, On the use of zeolite water intermittent cycle for solar climatization of buildings, Proceedings, ISES congress, Atlanta, p. 739, (1979b).

Meunier, F; *et al.*, Solar cooling through cycles using microporus solid adsorbents, Proceedings, ISES Congress, Atlanta, p.696, (1979a).

Nielsen, J.S. *et al.*, Simulation of the performance of open cycle desiccant solar cooling systems, Solar Energy, 21, 4, p. 273, (1976).

Dhar, P.L. Singh, S.K. Studies on solid desiccant based hybrid air-conditioning systems, *Applied Thermal Engineering*, 21, 2, pp. 119-134, (2001).

Peng, C.P. and Howell, J.R., Analysis of open inclined surface solar regenerator for absorption cooling applications: Comparison between numerical and analytical models, *Solar Energy*, 28, No. 1, pp. 265-268, (1981).

Roy, D. and Gidaspow, D. A cross-flow regenerator, *Chem. Engg. Science*, 17, p. 779, (1972).

Roy, D. and Gidaspow, D. Non-linear coupled heat and mass exchange in a cross floow regenerator, *Chem. Engg. Science*, 29, p. 2101, (1974).

Kaushik, S.C. Solar refrigeration and Space Conditioning, Divyajyoti Prakashan Publishers, (1989).

Sharma, J.K., Solar desiccant cooling, Chapter-I in Vol. II, Reviews in Renewable Sources of Energy, Ed. By Sodha *et al.*, Wiley Eastern Publishers, (1984).

Sheridan, N.R., Industrial evaporative cooling solar air conditioning, *The Int. J. of Engg.*, Australia, pp. 47-52, (1961).

Taoda, H., *et al.*, A zeolite system for cooling and dehumidification, *Solar Energy*, 8, pp. 27-37, (1982).

Wang S.K., Frank Kreith, and Paul Norton, Lavn Zalman. Air conditioning and refrigeration engineering. CRC Press, (2000).

Watt, J.R. Evaporating air conditioning, Industrial Publication Press, New York. (1963).

Worek, W.M. and Lavan. Z. Cross cooled desiccant dehumidifier, Proc. Annual Meeting, American Section of the Solar Energy Society, p. 224, (1980).

Wouk, W.M. and Lavan Z. Performance of cross cooled disiccant dehumidities prototype. Report No. DES 81-1I.I.T. Chicago, J. Solar Energy, 104, p. 187, (1982).

Index

A

Absorption recompression refrigeration (ARR) system 155
    thermodynamic analysis of 156, 162
Acoustic streaming model 230
Activated carbon-ammonia 351
Activated carbon-methanol 351
Actual thermoelectric refrigeration system 310
Adsorption cooling systems 346
Adsorption systems
    adsorbent-adsorbate materials in 348
Air psychrometrics 366
Alternate vapour compression refrigeration cycles 23
Alternative cooling technologies, environmental considerations and 5
Alternative refrigerants 16, 18, 22, 24
    areas of application 24
Azeotropic mixtures/blends 18

B

Bitzer refrigerant report (16) 22

C

Cascade refrigeration system 67
    thermodynamic analysis of 70
Chemisorption systems 340
Chemisorption systems (CHP) 352
    thermodynamic analysis of 354
Compound parabolic collector 347
Compression absorption refrigeration (CAR) system 161
Compression-absorption cascade refrigeration (CACR) system 170
    thermodynamic analysis of 171
Counter flow vortex tube 225

D

Desiccant materials 371
Direct evaporative cooler 368
Direct-indirect evaporative cooler 370
Dual loop flow type double effect vapour absorption system 146

E

Efficiency defect 40
    in VCR system components 44, 50
Ejector integrated absorption refrigeration system 216
Ejector integrated vapour compression refrigeration 206
Ejector integrated vapour compression system 208
Ejector refrigeration system 201
    booster assisted 203
    performance improvement 202
    with heat exchangers 202
    with mechanical compressor 203
Ejector
    applications of 186, 196
    efficiencies of 199
    operation of 200
    parts of 196
    types of 196
    working principle of 196
Energy analysis 10
Energy and exergy analysis, thermodynamic basis of 7
Evacuated tube collector 348
Evaporative cooling 368
Exergetic efficiency 11
Exergy analysis 10

Exergy balance 9
    and exergetic efficiency 38
Exergy destruction ratio (EDR) 40
Exergy transfer forms 8
Experiment based models 231

F
Flat plate solar collector 346

G
Global warming potential (GWP) 20
Gouy-Stodola relation 10

H
Half effect generation water-lithium bromide vapour 134
Heat exchanger models 229
Heat pump models 230
Hilsch tube 7, 225
Hybrid air conditioning 378
Hybrid compressor ejector refrigeration system 203

I
Indirect evaporative coolers 368

K
Kelvin relationships 299
Kyoto Protocol (1997) 6

L
Life cycle climate performance (LCCP) 22
Liquid absorbents 373
Liquid desiccant solar cooling systems 376
Liquid desiccants 373
Lorenz cycle 30

M
Maxwell's demon 7, 225
Metal hydride systems 358
    thermodynamic analysis of 360
Modified double effect combined ejector-absorption refrigeration cycle 217
Momentum exchange models 227

Multi-component refrigerants 18
Multistage thermoelectric refrigeration systems 304
Multistage vapour compression refrigeration system 53

N
Near azeotropic refrigerant mixture/blends 18
Nernst and Ettinghausen effects 299
Non-dimensional exergy destruction 39
Numerical simulation models 231

O
Ozone depletion potential (ODP) 20

P
Parallel flow double effect generation VAR system 113
Peltier effect 298
Physisorption system 340
    description of 341
Pinch point 30
Pure refrigerants 18

Q
Quaternary refrigerant mixtures 18

R
R-507, R-422B, R-422C, R-422D, R-407A, R-407C, R-404A, R410A 23
Ramming effect 239
Ranque tube 7, 225
Ranque-Hilsch theory 226
Refrigerant mixtures 18
Refrigerants, evolution of 14
Refrigeration scientists, challenges for 6
Refrigerator models 230

S
Seebeck effect 297
Seebeck voltage 299
Series flow double effect generation VAR system 94, 97
Silica gel-water 350

Single component refrigerants 18
Single effect generation VAR system
    thermodynamic analysis of 95
Solar adsorption cooling 345
Solar desiccant cooling systems 373
Solar thermoelectric cooler 317
Solar thermoelectric refrigerator
    design aspects of 315
    problems of 316
Solid desiccant open cycle cooling systems 374
Solid state thermionic converters 324
Standing wave stack based thermoacoustic refrigerator 331
System, exergy of 8

T

Thermoacoustic cycle 328
Thermoacoustic refrigeration
Thermoacoustic refrigeration cycle processes of 329
Thermoacoustic refrigerators 7, 327, 331, 328
    advantages of 337
    applications and merits of 337
    components of 332
    design parameters of 335
    thermodynamics of 334
Thermodynamic system, energy analysis of 7
Thermoelectric cooling module device 297
Thermoelectric effect 297
Thermoelectric element
    improvement in 321
    parameters for 319
Thermoelectric materials 318, 319
Thermoelectric refrigeration 7, 296
    analysis of 300
Thermomagnetic effects 297
Thomson effect 298
Total equivalent warming impact (TEWI) 21
Transcritical carbon dioxide compression refrigeration cycle 30

Transcritical CO_2 compression refrigeration system with ejector-expansion device (TCCRSEJT) 214
    thermodynamic cycle of 216
Travelling wave-regenerator based thermoacoustic refrigerator 331
Triple effect generation water-lithium bromide vapour absorption cooling systems 121
Two-stage thermoelectric cooler 307
Two-stage vapour compression refrigeration (VCR) system 54
    performance of 58
Two-stage VCR cycle, thermodynamic analysis of 54
Typical constant area/Velocity mixing ejector 198
Typical thermoelectric domestic refrigerators 311

U

Uniflow vortex tube 226

V

Vapour adsorption cooling cycle 341
Vapour adsorption systems 340
Vapour compression refrigeration (VCR) system 31
    description of 35
    performance 40
    thermodynamic analysis of 37
VCR cycle, thermodynamic analysis of 31
Vortex effect
    models of 226
    theory of 226
Vortex refrigerator 7, 225
Vortex tube 7, 225
    exergy analysis of 252
    modeling of temperature separation in 236
    phenomena in 232
Vortex tube integrated refrigeration and air conditioning 259

Vortex tube integrated transcritical CO_2 compression refrigeration and air conditioning systems 260, 264

Vortex tube integrated vapour compression refrigeration system 276
    components of 278

Vortex tube refrigeration systems 225

W

Water-lithium bromide single effect generation VAR 92

Water-lithium bromide solution, properties of 99

Z

Zeolite-water 350

Zeotropic (non-azeotropic) refrigerant mixture/blends 18